AF410828

6.69x9.61_PS-1NG_M   9783036513867
2172-3172244
843515LV8B

New Pathways for Community Energy and Storage

New Pathways for Community Energy and Storage

Editors

Henny J. van der Windt
Ellen van Oost
Binod Koirala
Esther van der Waal

MDPI • Basel • Beijing • Wuhan • Barcelona • Belgrade • Manchester • Tokyo • Cluj • Tianjin

Editors

Henny J. van der Windt
Integrated Research on Energy,
Environment and Society
Faculty of Science and
Engineering
University of Groningen
Groningen
The Netherlands

Ellen van Oost
Department of Science,
Technology and Policy Studies
Faculty of Behavioural,
Management and Social Sciences
University of Twente
Enschede
The Netherlands

Binod Koirala
Energy Transition Studies
TNO
Amsterdam
The Netherlands

Esther van der Waal
Integrated Research on Energy
Environment and Society,
Faculty of Science and
Engineering
University of Groningen
Groningen
The Netherlands

Editorial Office
MDPI
St. Alban-Anlage 66
4052 Basel, Switzerland

This is a reprint of articles from the Special Issue published online in the open access journal *Energies* (ISSN 1996-1073) (available at: www.mdpi.com/journal/energies/special_issues/New_Pathways_Community_Energy_Storage).

For citation purposes, cite each article independently as indicated on the article page online and as indicated below:

LastName, A.A.; LastName, B.B.; LastName, C.C. Article Title. *Journal Name* **Year**, *Volume Number*, Page Range.

ISBN 978-3-0365-1386-7 (Hbk)
ISBN 978-3-0365-1385-0 (PDF)

Contents

About the Editors

Henny J. van der Windt

Dr. Henny van der Windt is an associate professor at the Faculty of Science and Engineering, University of Groningen (Integrated Research on Energy, Environment and Society (IREES), previously Science and Society Group). He is an expert in the relationship between sustainability and science, with specific focus on energy (responsible innovation, local energy initiatives and the energy transition) and nature conservation (participative planning, visions on nature, national parks, nature inclusive agriculture). Dr. van der Windt also worked on other themes concerning science and society interactions, such as coastal management, genomics and nanotechnology. He holds a M.Sc. in Ecology and Ph.D. in Natural Sciences.

Ellen van Oost

Dr. Ellen van Oost is an associate professor at the faculty of Behavioural, Management and Social Sciences, University of Twente. She is an expert in the history of ICT, the construction of users, gender and technology. Dr. van Oost is a member of the center for Telematics and Information Technology and Institute for Governance studies, both at the University of Twente. She is also engaged with diverse professional organizations ($s, EASST, Dutch Association for Women's Studies) and the Netherlands Graduate Research School of Science, Technology and Modern Culture. Dr. van Oost holds MSc in mathematical engineering and PhD in the field of gender and technology.

Binod Koirala

Dr. Binod Koirala is a scientist at Energy Transition Studies, TNO Energy Transition. Between 2017–2019, he was postdoctoral researcher at Department of Science, Technology and Policy studies, University of Twente. He holds an Erasmus Mundus Joint Doctorate (Ph.D.) in Electrical Engineering as well as Sustainable Energy Technologies and Strategies (2017) from the TU Delft (the Netherlands), the KTH Royal Institute of Technology (Sweden), and the Universidad Pontificia Comillas (Spain), a M.Sc. degree in Renewable Energy Management (2011) from the University of Freiburg, Germany and Bachelor of Technology in Electrical Engineering. He has in-depth knowledge of modelling, simulation and optimization of integrated energy systems and markets, distributed energy resources including energy storage as well as energy system integration. His research interest lies in technical, socio-economic and institutional aspects of energy transition.

Esther van der Waal

Dr. Esther van der Waal is a postdoctoral researcher at University of Groningen. She has obtained her Ph.D. at the group for Integrated Research on Energy, Environment and Society (IREES, previously Science and Society Group) at the University of Groningen. She studied Human geography, Spatial planning and Socio-political sciences of the environment at the Radboud University Nijmegen, with a specialization in European and Environmental Spatial Planning. Her research interests include local energy initiatives, socio-technical innovation, social impact analysis, social studies of energy sustainability, local embedding of technology, energy policy, and interactive and participative planning as well as governance.

Preface to "New Pathways for Community Energy and Storage"

Local communities are increasingly taking active roles and emerging as new actors in the energy system. Community energy and energy storage may enable effective energy system integration and get maximum benefits of local generation leading to more flexible and resilient energy supply systems, and playing an important role in achieving renewable energy and climate policy objectives. The central role of citizens is also reflected in recent EU policy. The clean energy for all package of European Union through the 2018 recast of the European Renewable Energy Directive (RED II) and the 2019 recast of the Electricity Market Directive (EMD II) which defines and promotes renewable energy communities and citizen energy communities, respectively. These developments prompted the Universities of Groningen and Twente to organize an international conference on New Pathways for Community Energy and Storage from 6-7 June 2019 in Groningen. In this book, we summarize the different topics covered in this international conference in the form of the 14 articles published in this special issue on the same topic. Both the special issue and the conference aimed at addressing important developments and challenges related to local energy transitions and the role of community energy and energy storage therein. Based on the contributions to the conference and this issue several conclusions can be drawn. We suggest experimenting with new storage technologies, such as more sustainable electric and thermal storage systems, and introducing new types of modelling which are closely connected to real-world experiments. In addition, we advise to conduct more comparative studies, to see what kinds of community energy and community energy storage work in different contexts and what can be learned from new approaches, such as regulatory sandboxes for energy initiatives or combined community ownership models for various energy technologies. We think adaption of national policy, legislation and technology-society interactions is required to enable community energy and community energy storage. One of the key questions is how to combine energy transition and stakeholders involvement in such a way that issues such as access to energy, citizens empowerment and energy poverty are addressed.

Henny J. van der Windt, Ellen van Oost, Binod Koirala, Esther van der Waal

Editors

MDPI

Editorial

New Pathways for Community Energy and Storage

Binod Prasad Koirala [1,2,*], Ellen C. J. van Oost [2], Esther C. van der Waal [3] and Henny J. van der Windt [3]

1   Energy Transition Studies, TNO Energy Transition, 1043 NT Amsterdam, The Netherlands
2   Department of Science, Technology and Policy Studies, University of Twente,
    7522 NB Enschede, The Netherlands; e.c.j.vanoost@utwente.nl
3   Integrated Research on Energy, Environment and Society, Faculty of Science and Engineering,
    University of Groningen, 9747 AG Groningen, The Netherlands; e.c.van.der.waal@rug.nl (E.C.v.d.W.);
    h.j.van.der.windt@rug.nl (H.J.v.d.W.)
*   Correspondence: binod.koirala@tno.nl

Citation: Koirala, B.P.; van Oost, E.C.J.; van der Waal, E.C.; van der Windt, H.J. New Pathways for Community Energy and Storage. *Energies* **2021**, *14*, 286. https://doi.org/10.3390/en14020286

Received: 1 December 2020
Accepted: 27 December 2020
Published: 7 January 2021

Publisher's Note: MDPI stays neutral with regard to jurisdictional claims in published maps and institutional affiliations.

1. Introduction

Worldwide, the energy landscape is changing. Energy transition has now been on the agenda of most of the governments, companies, non-governmental organizations, investors and other stakeholders around the world. It is not only focused on decarbonisation, but also on technology improvements and integration, policies, business models and citizens' engagement. Local communities are increasingly taking active roles and emerging as new actors in the energy system. In some European countries, in particular Denmark, Germany, the Netherlands and the United Kingdom, local energy communities are already considered important stakeholders in the energy system. For example, many local energy initiatives own or manage solar panels, wind turbines, micro-grids or large scale integrated systems collectively. The central role of citizens is also reflected in recent EU policy. The clean energy for all package of the European Union, through the 2018 recast of the European Renewable Energy Directive (RED II) and the 2019 recast of the Electricity Market Directive (EMD II), define and promote renewable energy communities and citizen energy communities, respectively.

At the same time, energy storage has also become one of the key building blocks of the energy transition because of the growing need to balance the supply and demand mismatch, resulting from more decentralized and variable renewable energy production, in particular by wind turbines and PV panels as well as increasing electrification of end-use sectors such as transport and heating. Therefore, both community energy and storage are related to the move from a centralized to a more decentralized and democratized energy system, in which parts of production, delivery and management take place at the local level through active citizens and local stakeholders' engagement. The demand for new technical systems that combine generation and storage at the local level will increase. Accordingly, roles and responsibilities are also shifting and there is an increasing need for new revenue models, organizational forms, decision-making processes as well as partnerships between private partners, governments, and civil society organizations. Accordingly, new and innovative socio-technological energy and storage configurations are emerging at local and regional levels.

This broadening of the transition theme prompted the Universities of Groningen and Twente to organize an international conference on New Pathways for Community Energy and Storage from 6 to 7 June 2019 in Groningen. In this editorial, we summarize the different topics covered in this international conference as well as the papers of this follow-up Special Issue on the same topic. Both the Special Issue and the conference aimed to address important developments and challenges related to local energy transitions and the role of community energy and energy storage therein.

All contributions to this issue focus on the role of energy communities, energy storage, or both. Nine contributions investigate the potential and constraints of energy cooperatives, citizens energy and community energy [1–9]. Three contributions discuss both community energy and local energy storage [10–12]. Some of them take individual households as the point of analysis [10], while others explicitly discuss national [2–5,11] or European Union (EU) governance [6–9], or the issues of energy justice and social-technological dynamics [2,3,12]. Approaches vary from modelling [13,14] to theory development [1] and empirical studies, including case studies and surveys. Two papers explore the way agent-based modelling can be used for local storage [13,14]. Broadly, the articles published in this Special Issue can be categorized into three themes: firstly, articles that focus on the understanding of the community energy dynamics; secondly, articles that study the specific new energy element in communities: energy storage; and thirdly, articles that address the institutional aspects of the interface between community energy and broader society (policy, tools, ethics, impacts, etc.).

To put the contributions to this Special Issue and the conference in perspective, the rest of this editorial article is organized based on these three broad categories, namely, community energy dynamics (Section 2); modelling, implementation and use of community energy storage (Section 3); and institutional aspects of community energy and storage (Section 4). Each section starts with a brief introduction to the relevant articles in the Special Issue. In Section 2 on community energy dynamics, we will consider developments, impact and definitions of community energy. In Section 3, we will describe developments, characteristics and perspectives of community energy storage. Section 4 delves into options for new types of coordination and the need to take into account energy justice. Eventually, we formulate some thoughts on new pathways in community energy and storage research and policy, and required configurations, in Section 5.

2. Community Energy Dynamics

Four of the articles in this Special Issue focus on community energy dynamics.

Wagemans, Scholl and Vasseur (2019) study the governance role of local renewable energy cooperatives in facilitating the energy transition [9]. The authors claim that energy cooperatives may contribute to the current decentralization movement and to a "just energy transition." Based on their empirical work in the Dutch province of Limburg, the authors identify five key governance roles that cooperatives take up in the facilitation of the energy transition: (1) mobilizing the public, (2) brokering between government and citizens, (3) providing context specific knowledge and expertise, (4) initiating accepted change and (5) proffering the integration of sustainability.

While it is possible and meaningful to conduct in-depth studies on community energy and various functions of long-standing energy cooperatives in countries such as Denmark and Germany, development has so far been less advanced in other countries. Candelise and Ruggieri (2020) focus on the development of community energy in Italy [4]. The authors review Italian initiatives and present three case studies to explore conditions for development and success of community energy initiatives. The authors found that small initiatives are largely dependent on national photovoltaics policy support. Only larger initiatives that were able to operate at a national scale, developing multiple projects and differentiating their activities, have managed to continue growing at the time of discontinuity of policy support and contraction of the national renewable energy market.

Gorroño-Albizu (2020) explores the options for new types of consumer ownership at the national level and introduces new types of smart energy systems to integrate several energy sectors [5]. She studied the ability of cross-sector consumer ownership at different locations in the power distribution system in Denmark. The results indicate that distant and local cross-sector integration will be necessary to reduce overinvestments in the grid. In addition, consumer co-ownership of wind turbines and power-to-heat units in district heating systems may provide advantages over separate common ownership regarding local acceptance and attractiveness of investments. This requires, however, improvements of the current institutional incentive system in Denmark. In particular, the current EU policy-based regulation that dictates unbundling of energy sectors/services could be an impediment to implementing cross-sector ownership solutions, such as the one presented in this study, but also in other EU countries.

One of the papers explicitly stated the need for new theoretical frames to analyze and understand the role of energy communities. Gregg et al. (2020) aimed to synthesize aspects of sustainable transition theories with social movement theory to gain insights into how what they call "collective action initiatives" mobilize to bring about niche-regime change in the context of the sustainable energy transition [1]. The authors discuss how these energy initiatives can be described within both sustainability transition theories, such as the Multi-Level Perspective, and Strategic Niche Management, and Social Movement Theory, which focusses on how social movements share interests, give shape to the identity of their organization, and mobilize resources. Making use of both traditions, the authors adapt and apply a mobilization model to gain insight into the dimensions of mobilization and upscaling of these initiatives. By doing so, they show that their expanding role is a function of their power acquisition through mobilization processes.

As can be seen in above mentioned paper, one of the characteristics of the energy transition is its democratization through the recognition of the role that citizens, citizens' groups or communities play or might play in the decision-making, design and management of energy. In our view, the involvement of communities in future energy systems is important because it leads to a more effective, democratic and inclusive energy transformation, it stimulates decentralized energy systems and it leads to more efficient generation that is closer to the points of consumption, thus leading to fewer networks, and may provide energy at lower prices than their commercial counterparts.

In addition, energy cooperatives may play a significant role in the production of renewable energy, sharing new types of socio-technological innovation, acceptance and acceleration of this transition or making it more just and democratic.

3. Modelling, Implementation and Use of Community Energy Storage

Five of the articles in this Special Issue focus on community energy storage.

Hoffmann and Mohaupt focus on the perspectives of consumers and residents in Germany on "community energy storage" [11]. They found owners of photovoltaics systems to be receptive to the idea of community energy storage because they assume this is more resource- and cost-efficient than residential storage. Owners ask for professionally managed operation and maintenance, as well as transparency of operation and management. They fear potential disadvantages such as increased coordination with neighbors, increased data security risks and fear that other participants treat common acquisitions less carefully. The authors think that abating these perceived disadvantages can help to increase the acceptance of community energy storage. The owners are also interested in monitoring, energy management and other services, as part of the storage system. It is suggested that multi-use storage systems are developed, including various ancillary services for energy networks.

It is not the perception of energy technology by residents, but the impact of this type of technology on households that is central in the contribution of Kloppenburg, Smale and Verkade (2019) [10]. They discuss how residential energy storage technologies such as home batteries can enable householders to contribute to the energy transition, but also afford new roles and energy practices for householders. The authors regard energy systems as sociotechnical configurations and use the term "mode" to understand and classify the different ways in which households use technology and give meaning to it. Their results point to five emerging storage modes in which householders can play a role: individual energy autonomy; local energy community; smart grid integration; virtual energy community; and electricity market integration. They argue that, for householders, these storage modes facilitate new energy practices such as providing grid services, trading, self-consumption, and sharing of energy. Several of the storage modes enable the formation of prosumer collectives but will change relationships with other actors in the energy system. The authors discuss how householders face new dependencies on information technologies and intermediary actors to organize the multi-directional energy flows which battery systems unleash. Because energy storage projects are currently provider-driven, they advocate giving more space to experiments with mixed modes of energy storage that both empower householders and communities in the pursuit of their own sustainability aspirations and serve the needs of emerging renewable energy-based energy systems.

Koirala, van Oost and van der Windt (2020) studied the interaction between energy technology development and societal actors, including engaged citizens [12]. They analyze the rise of two new storage technologies, the seasonal thermal storage Ecovat system and the sea salt battery of DrTen, and the way they were implemented by local energy cooperatives. The authors show how both technologies received support from governmental agencies, DSOs, universities, energy cooperatives and technology funders because of their sustainability and their ability to apply them at the local level for balancing of the grid. In practice, however, some unexpected problems arose. National regulations turned out to be financially disadvantageous for storage systems. In addition, it took a while to integrate the battery into the energy system. The possible side effects of the building of the Ecovat system caused more resistance of locals than expected. From local energy cooperatives, it asked a lot of effort to judge and take decisions on these types of new storage technologies. The authors conclude that socio-technical alignment of various actors and factors, human as well as material, national and local, is a key element in building new socio-technical configurations. During this process, new storage technologies, communities and embedded values are being developed and adapted.

Fouladvand, Mouter, Ghorbani and Herder (2020) developed an abstract agent-based model for local generation and distribution of thermal energy by community-driven initiatives [13]. These types of initiatives remain largely unaddressed in the literature, although thermal energy applications cover 75% of the total energy consumption in households and small businesses. The authors studied four factors that influence the formation and continuation of thermal energy communities: neighborhood size, minimum number of members required, satisfaction of households and number of drop-outs. Their modelling indicates correlations between this type of community formation and the percentage of households that joined, and with the satisfaction of households.

Mir Mohammadi Kooshknow, den Exter and Ruzzenenti (2020) argue that development of electricity storage systems is hindered by a lack of viable business models, as well as high levels of uncertainty in technological, economic, and institutional factors [14]. The authors discuss barriers to and uncertainties in the development of storage systems in the Netherlands, and provide a theoretical foundation for combining agent-based modeling and exploratory modeling analysis as a method to test and explore business models for electricity storage systems. The authors suggest using their agent-based model as foundation of detailed design and for testing electricity storage system business models in the Netherlands and worldwide.

As observed in these articles, there is an urgent need to find new, efficient and affordable ways to balance the supply and demand of energy. The power grid in western industrialized and urbanized countries is heavily burdened by all local and national initiatives for solar and wind energy, which increases the need for balancing. So far, however, it has turned out to be difficult to find appropriate solutions.

Despite the need to stimulate storage projects, many questions remain unanswered. What role storage may play in the balancing of the energy system, which forms of storage are promising, at what temporal and spatial scales, and how are they geared to existing systems? In addition, the way they are organized and by whom, and the distribution of costs and benefits are also open questions.

4. Institutional Aspects of Community Energy and Storage

Five of the articles in this Special Issue focus on institutional aspects.

Several authors studied the impact of legislation on citizens' energy and community energy. Horstink et al. (2020) investigated the implementation of the two new EU energy

Directives in nine EU countries [7]. The ambition of the European Union is to establish an "Energy Union" that is not just clean, but also fair and inclusive: citizens actively interact with the energy market, such as prosumers. Although prosumerism in relation to renewable energy sources has been growing for at least a decade, the two new EU Directives are intended to legitimize and facilitate its expansion. The authors identified several internal and external obstacles to the successful mainstreaming of renewable energy prosumerism, among them a mismatch of policies with the needs of different prosumer types, potential organizational weaknesses as well as slow progress in essential reforms such as decentralizing energy infrastructures.

Lowitzsch (2019) also takes the new EU legislation as a starting point [8]. He introduces consumer stock ownership plans (CSOPs) as the prototype business model for renewable energy cooperatives. Based on the analysis of 67 cases of consumer (co-) ownership, he demonstrates the importance of flexibility of business models to include heterogeneous co-investors. In Europe, this is needed, he thinks, for meeting the requirements of the new European Union energy Directives. In this paper, it is shown that CSOPs—designed to facilitate scalable investments in utilities—facilitate co-investments by municipalities, SMEs, plant engineers or energy suppliers. They may enable individuals, and also low-income households, to invest in renewable projects. Employing one bank loan instead of many micro loans, CSOPs reduce transaction costs and enable consumers to acquire productive capital, providing them with an additional source of income. The author stresses the importance of a holistic approach, including the governance and the technical side, for the acceptance of renewable energy cooperatives on the energy markets.

Because current legislation is one of the main hurdles for community energy, some countries started experiments with new legislation. van der Waal, Das and van der Schoor (2020) studied some Dutch examples of so called "regulatory sandboxes", participatory experimentation environments for exploring revision of energy law [3]. These sandboxes allow for a two-way regulatory dialogue between an experimenter and an approachable regulator to innovate regulation and enable new socio-technical arrangements. The authors looked at the way power roles and power relations changed during these experiments. They researched the Dutch executive order called "experiments decentralized, sustainable electricity production" that invites homeowners' associations and energy cooperatives to propose projects that are prohibited by extant regulation. Local experimenters can, for instance, organize peer-to-peer supply and determine their own tariffs for energy transport in order to localize, democratize, and decentralize energy provision. The authors use Ostrom's concept of polycentricity to study the dynamics between actors that are involved in and engaging with the participatory experiments. They conclude that these sandboxes are not sufficient to improve the potential of bottom-up, participatory innovation in a polycentric system. They think a better inter-actor alignment is required, providing more incentives, and expert and financial support for the bottom-up initiatives.

The issue of energy justice is covered by two papers. Kluskens, Vasseur and Benning (2019) aim to provide insight in what kinds of participation and distribution are perceived as most just and most likely to create local acceptance of wind parks, in particular in the Dutch province of Limburg [2]. Their analysis, using and operationalizing the concepts of procedural justice and distributive justice, demonstrate that different kinds of participation in different phases of the process are preferred; for instance, consultation or sharing of information. The same is true for different aspects of distribution of costs and benefits. The results indicate that the most preferred modes of participation do not necessarily cover all aspects of procedural justice. The authors also identified factors which influence perception of procedural and distributive justice. For instance, there are clear differences between the distribution of benefits between a privately developed wind park and a cooperatively developed wind park.

Hanke and Lowitzsch discuss the way vulnerable consumers may be better included in the energy transition, making use of new European legislation [6]. According to the authors, the empowerment of consumers to participate in renewable energy communities has great potential for a just energy transition; but, in practice, vulnerable consumers remain underrepresented in regional energy projects. The new European directive on energy obliges the European member states to facilitate the participation of vulnerable consumers and support their inclusion in the so-called "enabling framework" of the EU to promote and facilitate the development of renewable energy communities. However, the type and specific design of corresponding measures remains unclear so far. The authors stress the need to understand how vulnerability affects participation in renewable energy communities. They argue that both individual vulnerable consumers as well as energy communities need incentives and support to boost the capacity of these communities to include underrepresented groups.

As described in these articles, because of the rise of renewables and decentralized energy systems, and changing roles of traditional and new players, responsibilities and configurations will change as well. The rise and popularity of the term of prosumer, someone who consumes and produces energy, express the growing role of citizens. In addition, other new words such as prosumager show the changing role of citizens from passive energy consumers to more active participants in production, consumption and storage activities in the energy system.

5. Discussion and Conclusions

The aim of the conference as well as the Special Issue was to explore new pathways for community energy and storage. Several papers described new technologies and options for socio-technological configurations; for instance, the sea salt battery of DrTen as part of a local energy system, and Ecovat, a new type of thermal storage. Other papers discussed new methods of governance, new methods to take energy justice into account, and options for new legislation and technology–society interactions. Finally, some papers presented suggestions for new theoretical approaches and new types of modelling. Most striking, however, were the gaps that we found, between theory and practice, between modelling and real-world situations and between different theories and scientific approaches.

Regarding the theoretical aspects, transition theory, energy justice, energy governance, social movement theory, energy economy, commons and polycentric decision-making, approaches taking practices as starting point for analyses, and studies on user-inspired and responsible innovation, provide a strong basis in the analysis of community energy and storage. This enables further study on different ways in which various types of energy communities may contribute to the energy transition at various levels—the household level, the community level and "higher" levels—by co-governance, by co-design of technology, by introducing new values and by developing new types of ownership and economic participation.

Concerning empirical studies, we agree with several authors to continue conducting comparative studies, to see what kinds of community energy and community energy storage work in different national and regional contexts and why. The impact of "external" factors, such as different subsidy or tax schemes, may be studied further, as well as the relevance of "internal factors" relating to the manner of internal functioning, such as decision-making and inclusion of people with different gender, cultural and socio-economical backgrounds. Most important, however, is to study the way these initiatives succeed or fail to survive and the way they link internal and external strengths and opportunities. Studies on the experiences of various long existing energy communities with different types of ownership, private-community and public-community, and off grid systems in, for instance, Asia and North-America, may be inspiring for European cases. At the same time, studies on urban community owned sustainable smart grids in Europe may be useful for rural and non-rural areas in other continents.

In addition, we suggest to start modelling on community energy storage in close connection to real-world experiments, in different circumstances. As suggested by several authors, studies in which modelling of local energy systems is combined with case studies in neighborhoods, real pilots and the development of business models should continue and extend. Scenario building may be the next step, based on modelling, in-depth case studies and intensive interactive sessions with stakeholders, varying not only in terms of aims, storage technologies and governance options, but also in cost–benefit distributions. Because many storage technologies seem to be expensive and not always environmentally friendly, studies and experiments with various types of storage at different temporal and spatial scales are required, using not only proven techniques.

Pathways will differ from country to country, or even from region to region. Looking at the articles, we conclude that it is wise to study and develop community energy and storage transition pathways for each country separately. For the Netherlands, for instance, the development of integrated, citizen owned or governed energy systems started only recently. We suggest studies on different local systems, varying in type of community ownership and management, in grid connection and in storage system. Part of that should be further studied in terms of the functioning of new regulatory sandboxes to create more community-friendly legislation. In Germany, community energy and storage has been further developed. Here, studies may concern the willingness and ability of different types of house owners and tenants in different types of neighborhoods and using various type of storage, and the consequences of the changing legislation. For Denmark, which has developed even further when it comes to community energy, further study is required into new community ownership models, combining various energy technologies such as district heating, PV, wind energy and storage.

Most of the articles suggested the adaptation of national legislation to enable community energy and/or community storage. At least in the Netherlands, Denmark, Germany and Italy, the present support incentives are going to change, which makes community energy less profitable and attractive. Powerful private companies seem to be taking the lead, up to now, in successful community energy countries such as Germany and Denmark. If the new European legislation will enable citizen and community energy, and will stimulate experiments with regulatory sandboxes and new social business models, this will strongly stimulate community energy. However, is not only about policy and legislation: also energy companies, DSOs and others have to reflect on their roles. For many other European and non-European countries, one of the first research questions is how to mobilize citizens and to combine energy transition with social and justice issues, such as access to energy, citizens' empowerment and energy poverty.

In many countries, citizens are willing to participate in the energy system of the future. The growth of decentralized renewable energy, in general, has consequences for other, more traditional parties. The role of the state seems to be crucial, but has become less prominent, however, after the privatization and liberalization of energy markets in many countries. It is unclear what role traditional parties, DSOs, large energy and electricity companies may play in the new energy systems and if they are willing and able to find ways to include citizens in a proper way. Clearly, the relation between traditional and new energy actors is still in flux. The growing professionalization of community energy combined with adequate policy measures is important to secure a healthy balance and fruitful collaboration in the future.

Both community energy and community energy storage are new pathways in the energy landscape that will grow immensely in the coming decades. Furthermore, both will likely take on many different guises. We expect that the coming decade will be decisive regarding the questions of whether community energy, owned and governed collectively by citizens, will develop as a new and influential economic–social–technological configuration and whether other societal stakeholders begin a learning process which might result in a redistribution of roles and responsibilities. Yet, there are technological and social challenges of integrating energy storage in the largely centralized present energy system, which demands socio-technical innovation. Continuous and intensive attention is required to the societal dimensions of community energy and energy storage applications and the technological aspects of social innovation around community-based distributed generation and energy storage technologies. As variable renewables are also becoming big business for the traditional regime actors, this will add additional challenges for the local communities. It is more important to move political and moral topics of energy transition more to the core, both in science and society. Energy transition is not only about technological transitions, but also about transitions towards a new economy which is more fair, inclusive, democratic and sustainable as well as away from market domination and inequality. It means that future energy systems after transition should be ecological, inclusive and fair. In addition to emergence of active citizens, there are also new challenges such as populist opposition to climate change and energy transition. It is important to overcome these challenges by stimulating community energy and storage, in regulations and technology development, that fit the local scale.

Author Contributions: Conceptualization, H.J.v.d.W., E.C.J.v.O., E.C.v.d.W. and B.P.K.; methodology, B.P.K., E.C.J.v.O. and H.J.v.d.W.; writing—original draft preparation, H.J.v.d.W.; writing—review and editing, B.P.K., E.C.J.v.O., E.C.v.d.W. and H.J.v.d.W.; project administration, H.J.v.d.W.; funding acquisition, E.C.J.v.O. and H.J.v.d.W. All authors have read and agreed to the published version of the manuscript.

Funding: This research is part of the Community Responsible Innovation in Sustainable Energy (CO-RISE) project. and is funded through the socially responsible innovation program of The Netherlands Organization for Scientific Research, the Netherlands. Grant number: NWO-MVI 2016[313-99-304].

Acknowledgments: We are thankful to all the conference participants for their active contribution to the conference and this Special Issue.

Conflicts of Interest: The authors declare no conflict of interest. The funders had no role in the design of the study; in the collection, analyses, or interpretation of data; in the writing of the manuscript, or in the decision to publish the results.

References

1. Gregg, J.S.; Nyborg, S.; Hansen, M.; Schwanitz, V.J.; Wierling, A.; Zeiss, J.P.; Delvaux, S.; Saenz, V.; Polo-Alvarez, L.; Candelise, C.; et al. Collective Action and Social Innovation in the Energy Sector: A Mobilization Model Perspective. *Energies* **2020**, *13*, 651. [CrossRef]
2. Kluskens, N.; Vasseur, V.; Benning, R. Energy Justice as Part of the Acceptance of Wind Energy: An Analysis of Limburg in The Netherlands. *Energies* **2019**, *12*, 4382. [CrossRef]
3. Van der Waal, E.C.; Das, A.M.; van der Schoor, T. Participatory Experimentation with Energy Law: Digging in a 'Regulatory Sandbox' for Local Energy Initiatives in the Netherlands. *Energies* **2020**, *13*, 458. [CrossRef]
4. Candelise, C.; Ruggieri, G. Status and Evolution of the Community Energy Sector in Italy. *Energies* **2020**, *13*, 1888. [CrossRef]
5. Gorroño-Albizu, L. The Benefits of Local Cross-Sector Consumer Ownership Models for the Transition to a Renewable Smart Energy System in Denmark. An Exploratory Study. *Energies* **2020**, *13*, 1508. [CrossRef]
6. Hanke, F.; Lowitzsch, J. Empowering Vulnerable Consumers to Join Renewable Energy Communities—Towards an Inclusive Design of the Clean Energy Package. *Energies* **2020**, *13*, 1615. [CrossRef]
7. Horstink, L.; Wittmayer, J.M.; Ng, K.; Luz, G.P.; Marín-González, E.; Gährs, S.; Campos, I.; Holstenkamp, L.; Oxenaar, S.; Brown, D. Collective Renewable Energy Prosumers and the Promises of the Energy Union: Taking Stock. *Energies* **2020**, *13*, 421. [CrossRef]
8. Lowitzsch, J. Consumer Stock Ownership Plans (CSOPs)—The Prototype Business Model for Renewable Energy Communities. *Energies* **2019**, *13*, 118. [CrossRef]
9. Wagemans, D.; Scholl, C.; Vasseur, V. Facilitating the Energy Transition—The Governance Role of Local Renewable Energy Cooperatives. *Energies* **2019**, *12*, 4171. [CrossRef]
10. Kloppenburg, S.; Smale, R.; Verkade, N. Technologies of Engagement: How Battery Storage Technologies Shape Householder Participation in Energy Transitions. *Energies* **2019**, *12*, 4384. [CrossRef]
11. Hoffmann, E.; Mohaupt, F. Joint Storage: A Mixed-Method Analysis of Consumer Perspectives on Community Energy Storage in Germany. *Energies* **2020**, *13*, 3025. [CrossRef]
12. Koirala, B.P.; van Oost, E.; van der Windt, H. Innovation Dynamics of Socio-Technical Alignment in Community Energy Storage: The Cases of DrTen and Ecovat. *Energies* **2020**, *13*, 2955. [CrossRef]
13. Fouladvand, J.; Mouter, N.; Ghorbani, A.; Herder, P. Formation and Continuation of Thermal Energy Community Systems: An Explorative Agent-Based Model for the Netherlands. *Energies* **2020**, *13*, 2829. [CrossRef]
14. Mir Mohammadi Kooshknow, S.A.R.; den Exter, R.; Ruzzenenti, F. An Exploratory Agent-Based Modeling Analysis Approach to Test Business Models for Electricity Storage. *Energies* **2020**, *13*, 1617. [CrossRef]

Article

Joint Storage: A Mixed-Method Analysis of Consumer Perspectives on Community Energy Storage in Germany

Esther Hoffmann * and Franziska Mohaupt

Institute for Ecological Economy Research (IÖW), 10785 Berlin, Germany; franziska.mohaupt@ioew.de
* Correspondence: esther.hoffmann@ioew.de

Received: 30 April 2020; Accepted: 8 June 2020; Published: 11 June 2020

Abstract: In this paper, we analyze consumer attitudes toward and interest in community energy storage (CES) in Germany, based on five focus group discussions and an online survey of private owners of photovoltaic (PV) systems, as well as written surveys and workshops with the residents of two residential developments where CES has been installed. We find that owners of PV systems are generally receptive to the idea of CES but are unfamiliar with it. They assume that CES is more resource- and cost-efficient than residential storage and appreciate the idea of professionally managed operation and maintenance, but are skeptical of whether fair and transparent distribution and billing can be realized. Consumers express a need for ancillary services, such as monitoring, information or energy management, but the interest in such services, however, is strongly dependent on their perception of the costs versus potential savings.

Keywords: energy storage; community storage; battery storage; consumer; user; prosumer; acceptance; energy services

1. Introduction

One option to increase the utilization of energy from volatile renewable sources is implementation of energy storage at the household level [1]. Homeowners often install such energy storage systems to enhance energy consumption from their own photovoltaic (PV) systems (self-consumption). Currently, residential storage is the dominant technical option [2] in Germany, and as of 2018, more than 100,000 residential storage systems have been installed [3]. Nevertheless, the ongoing integration of renewable power sources calls for more flexible solutions [4,5].

In the context of energy transition, local communities in cities, towns and villages worldwide are coming together to create their own future vision of how to achieve an energy supply based on 100 percent renewable energies, as well as other climate mitigation goals [6]. Many of these communities are engaged in diverse energy-related activities and projects [7], including in part the installation of community energy storage (CES), which provides greater flexibility, higher efficiency, and lower losses as compared to residential energy storage [8]. CES has the potential to strengthen the role of local communities and may generate collective benefits such as a higher penetration of renewables and increased self-consumption [9]. Most research dealing with CES deals solely with the technical or economic aspects (e.g., [8–11]), with little analysis given to the perceptions and acceptance of potential users—for exceptions, see [12,13] for municipalities and energy companies, or [14–17] for consumer views. Existing studies on consumer views were conducted without including consumers who have practical experience with community battery storage. Moreover, they mostly rely on a single research method or a combination of two methods. In contrast to these studies, we use a mixed method approach with an online survey, focus groups and research in two neighborhoods where residents had practical experience with CES systems. We examine the advantages and disadvantages of CES from a consumer viewpoint. CES can foster an increased sense of neighborhood belonging and common identity [18,19], but it requires both organizational services and monitoring concepts. To date, only a few studies on CES and these ancillary energy services exist. We aim at filling this gap through studying consumer views on ancillary services that can be combined with CES systems. The consideration of consumer needs and viewpoints can help to improve these services.

Since future energy systems and new technologies may have characteristics that could conflict with basic human needs, such as autonomy and privacy [15], it is important to investigate their social acceptance. Given that the implementation of energy projects is generally a "community involvement issue" [20] (p. 828), an important insight has been that social acceptance is usually greater when user integration takes place at an early stage of the development process [21]. User integration is especially beneficial in the context of sustainable innovations that "often require changes in user behavior" [21] (p. 27). When key factors that can facilitate the adoption of these innovations are identified, it is possible to design new energy products "in a way that they may easily be integrated into users' habits and everyday life" [21] (p. 27). Furthermore, given the significance of small-scale investors as a necessary source of funding for a successful energy transition [22], it is important to develop business models that make private investment in CES attractive. This article deals with the acceptance of different storage solutions as well as the pros and cons of CES and related energy services. We consider the following research questions:

(1) What do PV system users expect from an energy storage system?
(2) What are the conditions under which they would be willing to invest in storage and ancillary services?
(3) Under which conditions would they favor CES over residential storage?

We focus on Germany, where the share of electrical power from renewable sources is close to the European average [23] and several CES pilot projects already exist. Electricity storage in Germany is presently attractive to consumers, as current regulations favor PV self-consumption over grid feed-in.

We set our analysis of CES acceptance within the greater conceptual framework of a discussion of technology acceptance in general, followed by a summary of two discussions in the literature that are relevant for our research: research on consumer preferences regarding storage technologies in general, and the role of energy services related to storage systems as well as respective community solutions. The following chapter summarizes the recent literature on these two issues.

1.1. Dimensions of Technology Acceptance

Whenever questions arise about the transition from one technological pathway to another, the acceptance of those affected must be considered [24]. According to Hildebrand et al. [25] and Schweizer-Ries et al. [26], technology acceptance has two dimensions: validation (positive to negative), and engagement (passive to active) (see Figure 1).

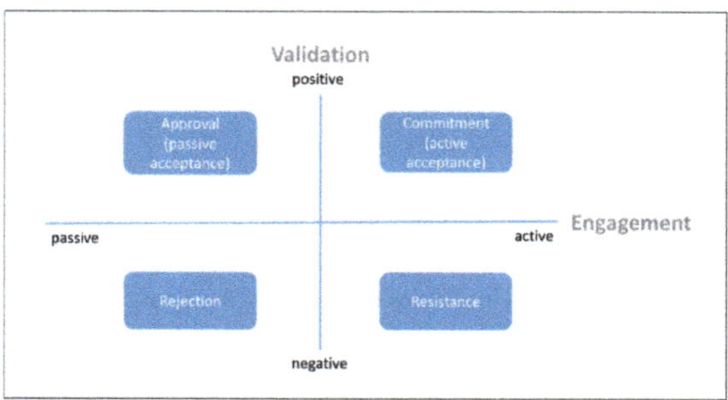

Figure 1. Dimensions of acceptance, translated and adapted from [24,26].

Among those who rate a technology positively, there is usually a group that approves of the technology but takes no further action (passive acceptance) [27]. Only when the willingness to act is transcended and individuals take observable action, such as making a personal investment in the technology [27], or its purchase and use [28], can one speak of active acceptance of a new technology [26]. In the case of energy transition, this active acceptance includes, for example, the installation of a PV system, thus, turning the consumer into a prosumer.

German consumers are being increasingly confronted with the growing number and decentralized distribution of renewable energy-based power plants [25,29]. Despite the rapid growth in numbers of such decentralized renewable energy plants within a relatively short period of time, a majority of the German population still approves of the energy transition [25,30]. To fully realize the transition, however, from a fossil fuel-based pathway to zero carbon, the mere absence of resistance, or even passive acceptance, will not suffice; active participation is often needed, i.e., commitment (upper right quadrant in Figure 1). To implement renewable and decentralized energy systems on a large scale, the need for active acceptance is inevitable. Hence, with regard to the acceptance of CES, we are mainly interested in those who generally approve of the idea but have not taken action by now (upper left quadrant in Figure 1). Therefore, we focus on how such active acceptance can be achieved.

1.2. Consumer Perceptions Regarding Energy Storage

Recent studies on the social acceptance of storage technologies in Europe and Australia have focused on the public's attitudes in general towards energy storage systems and their perception [15–17]. Specific insights into the determinants of active acceptance at the individual level, i.e., concrete action such as investment participation [24], remain scarce.

In the past few years, the topics of social acceptance and public perception of storage technologies [14,31], as well as consumer preferences [16], have received increasing attention in academia. While many researchers have focused on residential storage solutions [32–35], recently, interest in the public perception of CES has grown as well [36,37]. Some researchers have carried out comparative studies of the public perception of residential energy storage vs. community energy storage: Ambrosio-Albalà [14] discussed scenarios with residential and with community energy storage systems in focus groups and identified influencing factors for adopting residential vs. community battery storage. Kalkbrenner [16] analyzed preferences for different storage systems. MVV Energie AG et al. [38] conducted a field test with different models for storing electric energy and analyzed the perceptions of participants.

Participants in the abovementioned empirical studies (queried via focus groups [14,15,38], semi-structured interviews [33], online surveys [31,34], choice experiments [16,32] and a field-test [38]) were usually open to engaging with energy storage technologies. They felt positively about the technology, believing that they would generally benefit from its deployment, either through the reduction in individual costs [14], because the technology would ensure a secure electricity supply, or strengthen their country's national economy [31]. Their main expectations were to increase their own self-consumption of renewable energy and realize savings in electricity costs [32,33,38]. Increased independence from electricity suppliers [16,32,33,38] as well as protection against possible increases in electricity prices [34,35] were also cited. Profitability would seem to be the dominant argument [14,31], yet current research suggests that investment in an energy storage system does not yet pay off financially [11,39]. Granted, investment in the combination of PV and energy storage can be fully amortized—although the payback period would be shorter without the storage—and investment in energy storage in combination with energy-saving measures does pay off [39]. In addition to economic arguments, ideological reasons, such as making a positive contribution to the energy transition [33–35], or more generally, to climate and environmental protection [32,34], play an important role—albeit, as supporting and not as standalone arguments.

Given that personal financial gain is one of the primary drivers, respondents in the studies viewed the associated costs of an energy storage system as a primary impediment [14,18,32].

Furthermore, some were not willing to engage in energy storage until the technology matures [32]. Ambrosio-Albalá et al. [14] also found that study participants who lived on property where they did not intend to reside for a longer period of time were unwilling to install an energy storage system. Other uncertainties about the future strengthened this attitude. Jones et al. [31] found that the acceptance of storage system depends on trust in the developers.

Kloppenburg et al. [17] characterized energy storage as having an ambivalent impact regarding an active role for citizens in the energy system. The technology supports householders in achieving greater independence from suppliers, but at the same time, it increases their dependence on intermediaries and the information technologies needed to steer and manage local storage.

Comparative studies dealing with acceptance of residential versus community energy storage have shown mixed results. Kalkbrenner [16] concluded that a potential market exists for both storage concepts, but that consumers prefer ownership over use rights. Ambrosio-Albalá [14] found that participants were skeptical about CES because of a tragedy of the commons dilemma and the fear of unequal usage; moreover, participants did not like the idea of a third party operating the CES system. They thus concluded that institutional design is important to create trust in the operator and that information should be provided on how fairness can be reached [14]. Participants in the survey conducted by MVV Energie AG et al. [38] preferred CES over residential storage. In focus groups, Soland et al. [15] found that study participants were in favor of CES because of safety and environmental concerns and doubts about the cost–value ratio of residential energy storage.

Since all of the CES solutions that exist today are pilot projects [4,5,37,38], the approval ratings for community solutions must, for now, be considered primarily the evaluation of an idea. Hence, more insights are needed to better indicate and employ appropriate factors fostering community solutions.

1.3. Energy Services for CES

While recommendations have been developed for the design of services for end users in the smart grid [40], to date, no literature exists on smart energy services for CES. Kloppenburg et al. [17] mentioned that they were not able to include community driven storage solutions in their research because of the limited number of examples. Furthermore, in none of the studies evaluated were energy services for CES explicitly included in participant discussions. Nevertheless, some of the studies include statements by participants suggesting that they would only consider using such a system if certain energy services were offered along with it, i.e., monitoring and operation.

To establish trust, Schnabel and Kreidel [41] advocate that such services be offered by a professional third party. Ambrosio-Albalá et al. [14] found that participants would only want CES "if it could be guaranteed that the use would be equal and fair" (p. 145). There was a fear that some community members might exceed their fair share, and thus, the perceived need for an individual assessment of the energy performance of each community member [14]. In accordance with Schnabel and Kreidel [41], the systems that monitor energy flows and usage patterns must ensure the desired level of fairness and transparency and provide users with their specific energy use data as a basic service; furthermore, an overview of individual energy use is a prerequisite for forecasting and other energy services. With the information from monitoring, users have the opportunity to actively engage with their power consumption data and entertain suggestions for efficiency improvements [41]. Finally, such "information systems help to deliver customers the feeling of consuming self-produced electricity also at times when their PV module is not generating electricity" [5] (p. 498).

Several authors [5,9,16,37,42] have dealt with the general structure of possible business models. Regarding potential operators of a community storage system, the studies mention utility companies, aggregators, energy service companies and housing companies—all of which would manage the entire local energy system including energy storage. Parra et al. state that "in this context, the development of new policies and business models including different services provided by CES and creating win-win situations for customers (who generate their own energy locally) and other stakeholders should be pursued" [42] (p. 744).

Summarizing the existing literature on the implementation of CES, it is clear that it is technically feasible to operate such systems. Although the findings are based exclusively on the results of pilot projects, it is also clear that an essential component of a successful operation is the monitoring and communication of energy flow data to the system users. Most studies consider ancillary services that allow for transparent monitoring of storage usage to be important.

While the focus so far has been on technical feasibility, there is a lack of assessment of consumer willingness to pay, motivation, and potential disincentives. Both the research and the practical implementation of necessary ancillary services related to energy and storage management are still in their infancy. Thus, we began our investigation with a close look at consumer attitudes toward and interests in CES and ancillary energy services.

In the following, we present a mixed-method analysis consisting of focus group discussions, an online survey, and research in two residential neighborhoods (survey, workshops) (Chapter 2). We present our results in Chapter 3: We first analyze general views on storage solutions (Section 3.1), and then, assess users' perceptions of pros and cons of residential and community energy storage solutions (Section 3.2). We then look at consumer interest in the energy services necessary for realizing community solutions (Section 3.3). Finally, we discuss the results (Chapter 4) and draw some conclusions (Chapter 5).

2. Materials and Methods

We used a mixed-method approach that combined field research from two neighborhoods with five focus group discussions and an online survey of owners of a PV system. All research was conducted in Germany.

2.1. Neighborhoods

We studied two exemplary residential neighborhood developments, chosen because of their contrasts in terms of PV ownership and CES participation. The first, a recent development located in Groß-Umstadt in Hessen, Germany, comprises 82 properties for 90–100 residential units (mostly single- and two-family houses and a few apartment buildings). The total area of the neighborhood is 43,000 m^2. The neighborhood development plan stipulates that each house must have a PV system with a minimum installed capacity of 5 kWp. Furthermore, each household must have the capability of storing energy. Residents can choose to connect to a CES system installed by ENTEGA or invest in a private residential storage solution. Entega is a municipally owned utility offering electricity, gas, water, district heating, and energy services with about 570,000 energy customers [43,44]. The CES system has a gross capacity of 115 kWh and a charging capacity of 250 kW. All residential units in Groß-Umstadt, as well as the CES system, are connected to the public utility grid. In 2019, 25 households were connected to the CES system.

The second neighborhood is a condominium development in Cologne, Germany. It comprises nine apartment buildings, with approx. 170 inhabitants living in 74 residential units. The total area of the neighborhood is 11,236 m^2. The development is organized as a homeowners' association, with members owning their own apartments and co-owning common property. The association documents provide for a community energy concept; association members automatically become members of the energy community as well. The homeowners' association has a shared PV system with a nominal capacity of 225 kWp, as well as a CES system with a gross capacity of 96 kWh and a charging capacity of 18 kW, and heat pumps. The CES system and the residential units in the development constitute an internal microgrid, i.e., the CES system is not connected to the public utility grid.

2.2. Research Methods

We conducted empirical research in the two neighborhoods as well as on a general level, allowing us to compare attitudes and perceptions between users already familiar with CES and those without prior experience. The following formats were used to research consumer attitudes toward CES.

In the two sample neighborhoods, we conducted a written survey of the residents in early 2018 to investigate their attitudes regarding the energy transition, as well as their specific requirements from energy storage services. In October 2018, possible energy services for community energy storage were discussed with inhabitants of the two sample neighborhoods during two user innovation workshops. In these workshops, energy storage users discussed their expectations with respect to CES, their preferences for residential versus community energy storage (Groß-Umstadt only), their views on taking over the energy storage when the funding ends (Cologne) and their perceptions and ideas regarding ancillary services. The workshops were documented, and we relied on the recorded minutes for our analyses.

At roughly the same time, in September 2018, we conducted an online survey of owners and users of PV systems in Germany. The survey aimed at identifying attitudes towards CES, conditions for its acceptance and interest in ancillary services. The invitation to participate in the survey was distributed by co2online to its newsletter subscribers and to contacts in its consumer database. Co2online is as a non-profit consulting company that supports private households in conserving energy and reducing CO_2 emissions. Since all participant contacts were through co2online, we assume their interest in renewable energies to be above average. A total of 474 individuals participated in the survey, of which 94.5 percent were PV system owners. The online survey included 33 questions concerning: (1) the use of energy storage and corresponding motives, (2) preferred forms of use of energy storage, (3) attitudes toward and requirements for CES, (4) energy storage services and preferred operators of CES, and (5) sociodemographic data. We conducted descriptive statistical analyses (e.g., T-tests). For analyzing ranking questions, we calculated the *average rank* for each answer choice following Equation (1).

$$Average\ rank\ =\ \frac{x_1 w_1 + x_2 w_2 + \cdots + x_n w_n}{Total\ response\ count} \tag{1}$$

w = *weight of ranked position.*
x = *response count for answer choice.*

Weights are assigned in reverse i.e., the respondents' most preferred choice (ranked No. 1) receives the highest *weight*, and their least preferred choice (ranked in the last position) receives a *weight* of 1. This means that for a question with eight answers, the most preferred choice receives a *weight* of 8.

In November 2018, we held two focus group discussions with PV system owners, one in Berlin and one in Dusseldorf, in order to gain insights into their attitudes towards CES. The participants, likewise recruited via co2online, discussed: (1) their general attitudes towards community energy storage vs. residential storage, (2) appropriate operating models, and (3) possible ancillary services.

In addition, we conducted a second round of three focus group discussions, in November 2019, to discuss: (1) pros and cons of energy storage in general, (2) information relied on when making investment decisions regarding energy storage, (3) attitudes towards CES vs. residential storage, and (4) possible energy services with a focus on flexible shares of CES. The focus groups took place in Hamburg, Berlin, and Munich, with participants recruited by a market research institute. The focus group discussions were recorded and transcribed. We developed a coding system and conducted a content analysis using MaxQDA.

Table 1 gives an overview of the various sample characteristics.

Table 1. Sample characteristics.

Method	Attribute	Characteristics
Online survey	Number of participants	N = 474
	Gender	91% male, 6% female, 3% not stated
	Experience with PV and energy storage	95% PV system owners; 25% rely on or have access to an energy storage system
	Living situation	98% homeowners (about two-thirds single-family dwellings, one-fifth two-family, and a small share of apartments); 1% tenants
Focus groups	Number of participants	2018: Berlin, N = 9; Dusseldorf, N = 7; 2019: Berlin, N = 8; Hamburg, N = 12; Munich N = 9
	Gender	2018: Berlin, 6 male, 3 female; Dusseldorf, 7 male, 0 female; 2019: Berlin, 5 male, 3 female; Hamburg, 7 male, 5 female; Munich, 7 male, 2 female
	Experience with PV and energy storage	All participants were PV system owners; some were also storage users: 2018: Berlin: 3; Dusseldorf: 1; 2019: Berlin: 5; Hamburg: 9; Munich: 7; in Munich and Hamburg some were joint owners of a storage system together with others in their house (Hamburg 2019: 7; Munich 2019: 3).
	Living situation	Homeowners (house or apartment)
Questionnaires (neighborhoods)	Number of participants	Groß-Umstadt: N = 18, Cologne: N = 35
	Gender	Groß-Umstadt: 11 male, 5 female, 2 not stated Cologne: not collected
	Experience with PV and energy storage	Groß-Umstadt: PV system owners; CES users (N = 11) Cologne: users of community energy system including CES
	Living situation	Homeowners (Groß-Umstadt: house, Cologne: apartment)
User innovation workshops (neighborhoods)	Number of participants	Groß-Umstadt: N = 20 (from 14 households), Cologne: N = 15
	Gender	Groß-Umstadt: 14 male, 6 female, Cologne: 8 male, 7 female
	Experience with PV and energy storage	Groß-Umstadt: PV system owners, CES users Cologne: users of community energy system including CES
	Living situation	Owners (Groß-Umstadt: house, Cologne: apartments)

As shown in the table, the samples show specific characteristics. In almost all methods, the share of male respondents was much higher. In the focus groups and the innovation workshops, we strove to attain an equal share of female respondents, but without success. Since the use of a PV system was a selection criterion, it is not surprising that almost all of the respondents—regardless of survey method—were home (or apartment) owners.

3. Results—What Do Users Expect from Energy Storage?

3.1. Interest in Energy Storage and Motivation for Investing in Them

PV system users also expressed a substantial interest in energy storage systems: In the online survey, 87.4 percent of the respondents without their own energy storage system expressed an interest in investing in such a system. The online survey included a ranking question on the motive for having bought (respondents with storage) or being interested in an energy storage system (respondents without storage) (see Table 2). In both groups, the most important motive clearly is increased self-consumption. For those respondents already using an energy storage system, this was followed by contributing to climate protection (ranked second) and independence from the electricity supplier (third); for those respondents not using energy storage, the ranking of these two motives was reversed. However, in both cases, the values of the *average rank* for both motives was close to each other. For those respondents already using an energy storage system, this was followed by the desire to reduce the burden of rising energy prices (fourth), and increasing the use of renewable energies (fifth); for those respondents not using energy storage, the ranking of these two reasons was reversed. Finally, the desire to use an innovative product or to increase property value as motives for investment were ranked sixth and seventh respectively by both groups and have been notably ranked lower.

Table 2. Ranking of motives for storage acquisition.

Motive	Respondents without Storage System (N = 276)		Respondents with Storage System (N = 111)	
	Average Rank	*Position*	*Average Rank*	*Position*
I can increase self-consumption of renewable energy.	5.802	1	6.275	1
I contribute to climate protection.	4.243	2	4.091	3
I want to become more independent from the electricity supplier.	4.234	3	4.149	2
I can reduce the burden of rising energy prices.	3.901	4	3.268	5
I increase the use of renewable energies.	3.856	5	3.975	4
I use an innovative product	2.243	6	1.663	6
The storage system increases my property value.	1.613	7	1.496	7
Other	0.306	8	0.417	8

Most respondents in the online survey were concerned that such investments will not pay off. They consider the financial risks of an energy storage system to be serious (64 percent indicated the risk to be very significant or significant). This was followed by ecological risks (use of critical/scarce resources, 61%), limited recycling (55%), and environmental stress related to production (46%). Respondents currently without access to an energy storage system ranked all risks significantly higher ($p < 0.001$, T-test).

In the focus groups and the workshops in the neighborhoods, participants were interested in energy storage and some were willing to make the investment. During discussions about motives for investing in an energy storage system, one important aspect was cost savings. In this context, respondents considered increased self-consumption to be a means to reduce costs and—to a lesser extent—to become less dependent on energy suppliers. Although respondents mentioned different motives during discussion, the rankings from both the focus groups and the workshops show that, in the end, a profitable investment is more important than other aspects, e.g.:

"As I said, it depends. As soon as energy storage becomes cheaper, I'll be able to increase my own [self-] consumption and reduce expenses. But currently such storage systems are simply too expensive."

(Munich 2019—M2, I-105).

In light of the fact that private energy storage systems to date have not been financially viable [39], the question arises as to why some people nevertheless invest in energy storage. The results of the 2019 focus groups showed that participants simply did not consider all relevant parameters influencing the cost of storage operation—e.g., cycle life, efficiency, and depth-of-charge characteristics—when purchasing their storage system. Only a few participants knew the cost of their storage system. Additionally, no one was able estimate the specific costs per kWh of their own energy storage system. The 2019 focus groups also revealed that those participants who had invested in an energy storage system did not monitor cost reductions after installation.

In the 2019 focus groups, we additionally inquired as to how participants with energy storage had arrived at their decision to buy and the information they had relied upon for their decision. Most drew upon a combination of information sources, such as information from the Internet, an energy utility, installers, or even neighbors and friends.

"In our street some already had [a storage system], so we started talking about it […], and they named some companies. I did some Internet research […] and then I contacted several companies."

(Hamburg 2019—HH9, I-82).

Most had requested bids from one or more installation firms or an energy utility and made their decision after considering factors such as cost, their impression of reliability of the offer, and their own trust in the seller's competence. Moreover, the perceived quality of the seller's guidance was important.

"The [companies] consulted us and then we decided. […] We bought [from the firm] where the consultation was better and [the assumptions being made] seemed more plausible."

(Hamburg 2019—HH9, I-82).

However, several participants were not satisfied with the advice they received. Due to a lack of knowledge, they felt overly dependent on the providers.

"Well, there were some surprises. It was not very nice because they simply hadn't thought it through and we of course didn't have the necessary knowledge."

(Berlin 2019—B7, I-97).

Some participants perceived a need for better information regarding energy storage.

"Well, it would be nice if someone provided real information and advice instead of only providing a sales quote because I don't really have a clue."

(Munich 2019—M3, I-108).

3.2. Comparison of Storage Systems and Requirements Concerning Community Storage

Respondents in the online survey showed a clear preference for residential energy storage (see Figure 2): about two-thirds (64%) preferred to own and six percent to rent a residential energy storage system. About half of those who preferred their own residential storage system suggested that with a residential storage system, they would be more independent and have better control over the storage system. Moreover, about 15% feared that a CES causes bureaucracy and coordination with neighbors, and about 13 percent stressed that they were the only residents with a PV system in their neighborhood.

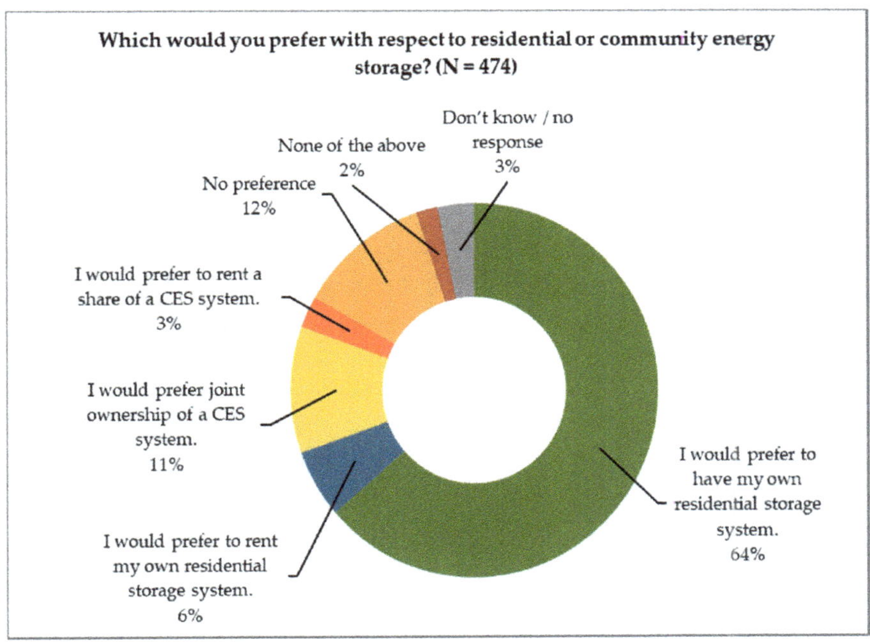

Figure 2. Preferred type of storage usage.

Only 14 percent preferred CES, with 11 percent preferring joint ownership with other residents and three percent rental participation. Those respondents who were in favor of community solutions mainly argued that these are more cost-efficient and require less or no personal investment and achieve better capacity utilization. A few also noted that one large storage system rather than several smaller ones is more resource-efficient. Those who prefer renting stressed the lack of a need for personal investment and professionally managed operation and maintenance.

We asked whether the respondents could imagine participating in CES if it was offered to them. There was a substantial receptiveness to CES, we found, with 64 percent agreeing with such a proposition and only 22 percent disagreeing (see Figure 3). This result shows a general openness towards CES. However, it is not possible to deduce from the stated willingness to act whether the respondents would actually take part. In the following, we will go into more detail about what respondents think about advantages and disadvantages of CES and factors influencing the decision for CES.

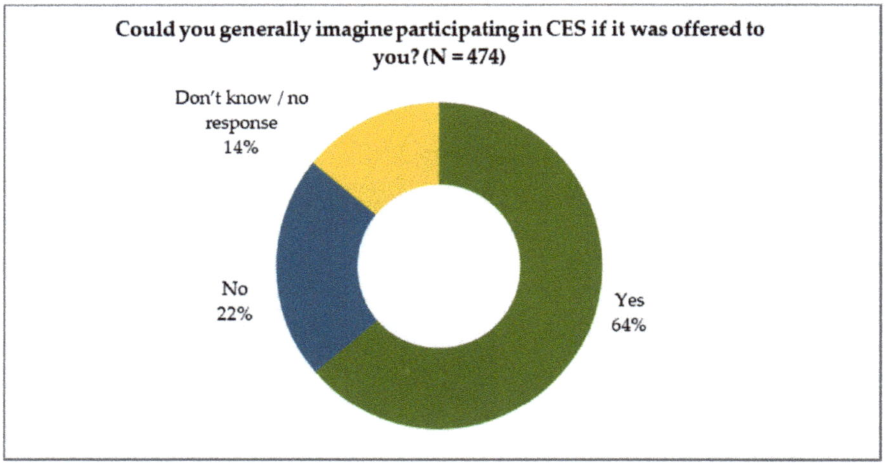

Figure 3. Assessment of willingness to participate in a CES project.

Our other investigations also showed significant receptivity to and interest in CES. In the focus groups, we asked participants whether they would prefer a residential or a community storage system, provided that community storage was not more expensive. The majority of participants chose CES, and most preferred share rental of a CES over owning.

The surveys in the two neighborhoods we studied showed differences: The development in Cologne has a community energy concept that integrates a PV system, energy storage and heat pumps; thus, it is not possible for residents to operate an individual energy storage system. Here, all participants were accordingly in favor of community storage. In Groß-Umstadt, where residents not currently participating in CES also responded, the interest in residential energy storage was almost as high as that for community storage. However, only seven households not participating in the CES project participated in the survey and the high approval rate cannot be transferred to the majority who did not participate. In the innovation workshop in Groß-Umstadt, where only participants of the CES project took part, almost all were in favor of CES.

In all of the surveys, we analyzed respondents' perceptions of the advantages and disadvantages of CES versus residential storage. Although most participants in the online survey preferred residential energy storage, many did perceive certain advantages of CES (see Figure 4). They assumed that a CES can better balance power fluctuations in the grid, and thereby, increase net stability; there was also the assumption that CES is more cost-effective than residential storage. Still, respondents also perceived disadvantages. Many were concerned that CES requires too much coordination with neighbors, and that such common property acquisitions are generally treated less carefully by users. Slightly less than a third assumed that with CES there would be an increased data privacy risk. Other arguments for CES were less relevant, e.g., more respondents disagreed than agreed with the potential for increased fire risk with residential storage, or the idea that acquisition and operation should be left to a third party.

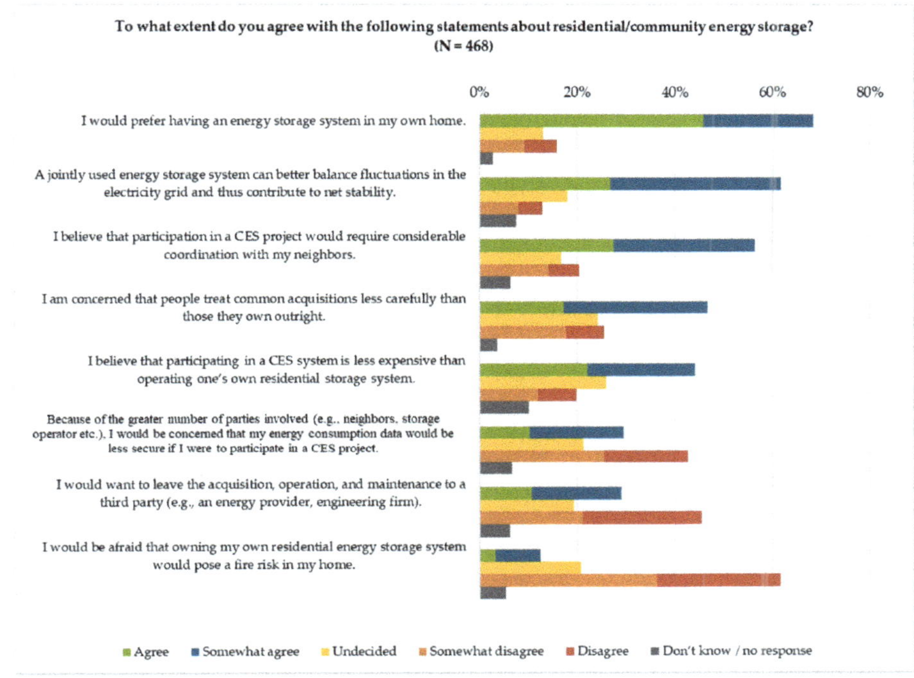

Figure 4. Assessment of (community) energy storage.

A comparison of those with a preference for community or residential energy storage shows that those respondents who prefer CES agree significantly more on the advantages of community storage and significantly less on the disadvantages (T-test, $p < 0.001$). However, differences are not significant for the fear of fire risk. A comparison between respondents who already own an energy storage system and those who do not shows almost no significant differences. Respondents owning an energy storage system, nevertheless, showed significantly less agreement on the fear of fire risk (T-test, $p < 0.001$) and data security (T-test, $p < 0.05$).

In the online survey, we asked under which conditions the respondents would be interested in participating in a CES system (ranking question, see Table 3). By far, the most important condition was that the rental fee for such storage be less than the reduction in electricity costs attained using the storage system. The next important condition was that respondents could freely choose their electricity provider, the third that the storage operator would ensure a higher proportion of renewable energies in the neighborhood, and the fourth that the storage operator uses the storage systems for grid relief. Ranked last were the stipulations that the storage is offered in combination with energy consulting and that the storage contract could be cancelled upon relatively short notice. This shows that once again costs are an important factor; however, respondents are open to other arguments.

Table 3. Conditions for being interested in participating in a CES system (N=332).

Conditions	Average Rank	Position
Rental fee for CES should be less than the reduction in electricity cost attained using the CES system.	5.777	1
Users can freely choose their electricity provider.	3.452	2
Storage operator should ensure higher proportion or renewable energies in the neighborhood.	3.130	3
Storage operator should use the storage system for grid relief.	2.729	4
The storage is offered in combination with energy consulting.	2.277	5
The storage contract could be cancelled upon short notice.	1.967	6
Other	0.497	7

In the surveys in the two neighborhoods, the view of the advantages of CES was quite similar to the online survey; however, an even larger share of respondents assumed that CES is more cost-efficient. The residents were somewhat less in agreement about the disadvantages. Participants in the focus groups, in particular, expected economic advantages and assumed that larger solutions would have lower specific costs and can be offered at a better price.

"I think community energy storage is a great thing and if there was a system in my district I would participate. My requirement is that it has to be cheaper. The price is the kicker."

(Berlin 2019—B5, III-61).

Focus group participants moreover assumed that CES is more efficient and more reliable and can better be used to provide grid relief. Their arguments for renting were that professional staff would take care of operation and maintenance, and that they would bear less of the investment cost and financial risk:

"With community energy storage I would in effect be safe from failure. When the storage system breaks down […], it's not my problem."

(Dusseldorf 2018—D3, I-105).

"It's simply more reliable […] and also better maintained."

(Dusseldorf 2018—D1, I-144).

Moreover, some participants appreciated that less knowledge and effort on their part is necessary. They assumed that CES is a good option for those who have less knowledge about or interest in technical details, but who still want to participate in the energy transition.

"If it's not more expensive I would choose community energy storage because then I don't have any work with it."

(Munich 2019—M6, III-77).

"[…] Whenever we purchase something that we don't understand, we either have to acquire knowledge, or we have to find someone who explains it to us and that can be very time-consuming. Therefore, in terms of the required expertise community energy storage would be an advantage for me."

(Berlin 2018—B2, I-62).

A few participants in the focus groups mentioned that CES may increase the sense of community in the neighborhood, and additionally, is a visible sign that a district is engaged in supporting the energy system transition:

"Community energy storage also brings a kind of community spirit: together we are doing something for the environment. Those are the factors that are important to develop acceptance [of the energy transition]."

(Dusseldorf 2018—D5, I-219).

Many focus group participants, however, doubted that a community storage system could be realized in their neighborhood; they assumed that there are not enough PV systems. Their main concern regarding CES was whether equal and fair billing could be realized. They expressed a concern that others would use the renewable electricity they produced while they were left to buy "expensive electricity" from the energy utility:

"The most important thing would be that there is an equitable distribution. If I constantly feed in [electricity] and others only withdraw, this has to be recorded and compensated in terms of costs."

(Hamburg 2019—HH1, II-21).

Some participants moreover stressed that a CES system would reduce their PV-based independence from external suppliers:

"With such a large [storage] system, one again becomes dependent on some kind of operator."

(Munich 2019—M2, II-76).

In the 2018 focus groups, in which many participants had a strong technical interest in renewable energies, some participants were more in favor of residential storage, which, they argued, offered better control and greater freedom in making decisions. One participant was moreover critical because he feared that CES users would feel less responsible with respect to energy consumption and environmental protection.

"Regarding community energy storage, I would be afraid that we all would once again say: 'Electricity comes from the electrical outlet.' And this educational effect—of being the master of my own technology and power consumption—would be lost."

(Dusseldorf 2018—D5, I-161).

In the user innovation workshop in Groß-Umstadt, participants were clearly in favor of CES and system rental. Their arguments resemble those of the focus groups: They would be less engaged with the storage system (e.g., technical aspects, current developments); professional staff would take care of maintenance and repair and catch problems sooner rather than later. They assume that CES is more cost-effective than residential storage. Moreover, many participants do not have the physical space needed to install their own residential storage system, and some see an advantage in not having the additional fire load in their home. Participants in the innovation workshops did not mention the fear of unequal billing and use of the CES system, which was very prominent in the focus groups. In the user innovation workshop in Cologne participants did not discuss the question, whether they prefer a community or residential storage system as the residents in the Cologne neighborhood fall under a community energy concept.

The question of who is operating the CES system also influences its acceptance (see Figure 5). Respondents to the online survey clearly trust in local and regional energy utilities (more than 80 percent showed great or medium trust). More than two-thirds of the respondents also trust in so-called energy communities and municipalities. The trust in national energy utilities and housing companies is notably lower, with almost half of the respondents indicating little or no trust in housing companies. In terms of trust, we can observe group differences between storage system owners and non-owners, as well as between people who prefer residential storage and those who prefer community storage. Those who prefer CES, in particular, have more confidence in energy cooperatives ($p < 0.001$). Those who prefer residential storage are generally more skeptical ($p < 0.01$), except for network operators and storage producers, for whom no significant differences can be found. Storage owners show higher trust ($p < 0.01$) in storage producers than non-owners and lower trust in local and regional energy utilities ($p < 0.05$).

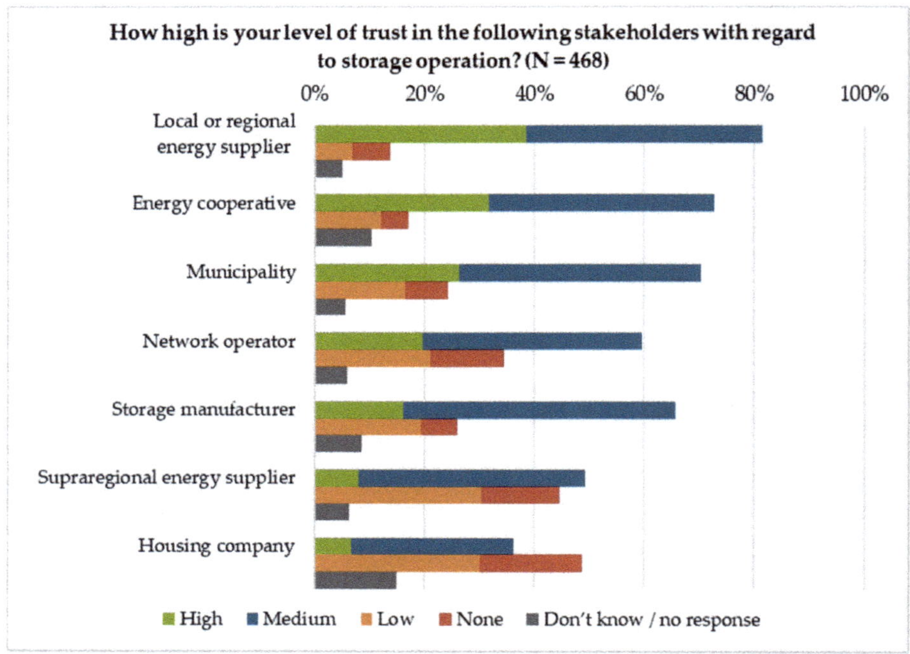

Figure 5. Trust in potential operators of an energy storage system.

Focus group participants also show strong trust in local and regional energy suppliers and energy cooperatives. They mistrust large energy suppliers and housing companies, which they assume are more likely to pursue their own economic interests and exclude customers from participation in savings.

"Housing companies would be [a possibility], but I wouldn't really trust them because, sure, they make plans and say: 'Okay, I need a 15–20 percent return on equity, everything else isn't of interest to me.'"

(Berlin 2018—B7, II-66).

"Municipal utilities that are close by and already have a relationship of trust with the customer, […] I believe they could do it."

(Dusseldorf 2018—D5, II-29).

Moreover, the participants in the 2019 focus groups argued for a short notice of cancellation (mainly between three and six months) and choice of electricity provider.

3.3. Energy Storage Services

Shared energy storage at the community level needs to be supported by ancillary services, such as renewable energy prioritization, energy flow management, and operation and maintenance. How do users perceive these services? Would they be willing to pay for them? The online survey shows a generally high level of interest in the energy services described in Figure 6, provided that no additional costs are incurred overall (agreement between 40 and 73%).

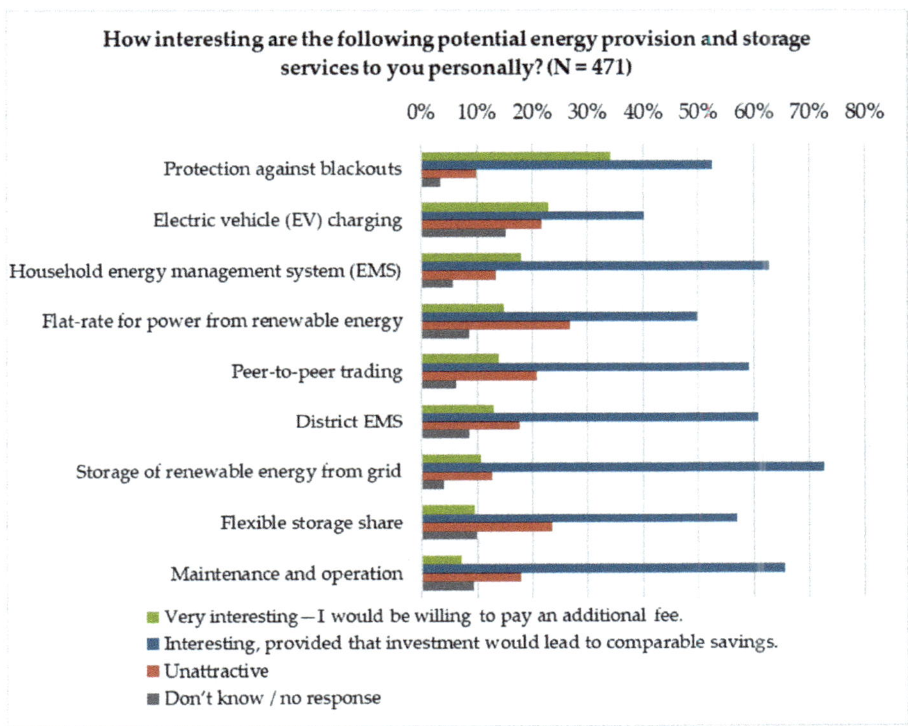

Figure 6. Assessment of various energy services.

At first glance, protection against power losses appears to be the most attractive service, despite the very low probability of blackout occurrence. In Germany, the average annual outage duration for customers served was a mere 14 min in 2018 [45]. In the online survey, it received the highest approval ratings in terms of willingness to pay an additional charge, with support from 34 percent of respondents, followed by the possibility of charging an electric car via the energy storage system (23%). The willingness to pay for the charging of electric cars and energy management for the household or the neighborhood is greater among those with their own energy storage system.

In the focus groups, some participants confirmed a strong interest in protection against blackouts, while others acknowledged that the power supply in Germany is very reliable and the occurrence of blackouts is seldom.

"For me, a very important topic is protection against blackouts. […] That is something important and would be great."

(Dusseldorf 2018—D7, III-98).

Charging of electric vehicles was positively discussed in the workshops and focus groups, although only a few participants already drove an electric car. Participants, however, were skeptical about the future possibility of bidirectional charging because they perceive this as infringing on their own options:

> "I would not like it […] if somebody could tap the electricity from my car because I can
> never know exactly whether I might have to drive somewhere."

(Berlin 2019—B5, II-55)

The focus groups in 2019 showed that most participants were unfamiliar with ancillary energy services. Almost no one had dealt with such services or even load management, nor had much thought been given to the cost of the stored electricity. This was even true of those already participating in shared energy storage (HH 2019). On the other hand, within the 2018 focus groups, some participants did have experience with ancillary services such as monitoring and energy management systems. Several participants in the 2018 focus groups used smartphone apps to monitor their power generation and consumption of self-produced electricity.

Many focus group participants estimated that electricity from the grid is more expensive than from storage and were concerned about disadvantages due to unequal use of the CES system. To avoid unfair billing, the majority highlighted the need for transparent energy management as a necessary prerequisite. Many focus group participants, however, found it difficult to imagine the management of numerous households connected to a CES system. "The allocation of stored electricity to the person who wants to use it later" (Berlin 2018—B1, I-57) was estimated to be too difficult. At the same time, all respondents were interested in the services of monitoring, energy management, and storage operation.

> "I really see added value in the monitoring service. Someone who professionally operates [an
> energy storage system] will have a great deal more experience and knowledge than someone
> who does this on his own."

(Dusseldorf 2018—D1, I-208).

In the two neighborhoods we studied, the residents described monitoring, management, and operation as necessary services, but they were hardly willing to pay for them. The primary goal in the Cologne neighborhood was to achieve possible cost savings by using CES to increase the rate of self-consumption. In this context, participants also desired greater information and indications on energy saving and times when the consumption of self-generated electricity is possible (e.g., by means of a visual display such as a "traffic light") as well as notification when unusual consumption rates or failures occur.

A particular advantage of CES is that the individual household can rent a flexible share of the storage according to its actual needs. Shared use of the CES system provides such flexibility. The CES operator can then apportion unused storage to other services, e.g., the control energy market. Focus group participants found the idea of variable storage shares very complex and in need of further explanation. Only after a longer discussion in the focus groups did a majority of participants agree that variable storage could bring advantages such as gains in flexibility through an only-as-needed capacity rental model—for example, easier adaptation to changes in household energy requirements—or the ability to plan costs. The participants linked the topic of flexible storage shares with a discussion of monitoring and maintenance and the rental model versus ownership. For many, it was not clear why the optimal storage size would differ during the year and why it was necessary to define an upper limit to the amount of electric power that could be stored at any one time. They also expressed the desire to store electricity (at least virtually) for a longer period of time, so that their summertime solar power could be used in winter. There was a greater receptivity in the Groß-Umstadt district, where the energy utility, ENTEGA, is planning to offer flexible storage shares as a service. The service has already been addressed in various customer events that ENTEGA has held in Groß-Umstadt, which has led to a greater understanding of the service there. Here, as in the Cologne housing development, some workshop participants relied on the use of analogies, such as comparing storage management to that of a bank account or shared network access with Netflix, in order to better understand the benefits.

4. Discussion

In all of the surveys, respondents described the economic arguments as being decisive for making investment decisions with respect to energy storage. Their goal is to increase self-consumption, which is perceived as a means to lower costs and increase independence. Although investment in energy storage is still economically infeasible [11], more than 100,000 consumers in Germany have already acquired such energy storage systems [3]. Given the complexity of the investment and cost-return issues, we find that some consumers made decisions based on insufficient information or relied on idealistic payback periods provided by storage providers. Furthermore, once such an energy storage system has been acquired, many consumers do not control or monitor the cost or actual savings in the energy bill. This suggests that, although economic arguments are often presented as the most important issue, many decisions are in fact only ostensibly based on economic reasoning.

The differences regarding consumer preferences for residential versus community energy storage arising among the survey methods suggest that consumers who are better informed about CES are more inclined to this sharing model. This was especially obvious in the two housing developments we surveyed, where residents had acquired real world experience with CES and were supported and supplied with information by the local storage operator.

Soland et al. [15] identified safety and ecological concerns as well as economic reasons as arguments for preferring CES. We can confirm these results regarding economic motives and partly regarding safety concerns. In our surveys, the ecological advantages of CES, however, were scarcely mentioned. With respect to economic arguments, consumers assumed that CES is more efficient and cost-effective due to economies of scale. We moreover find that consumers prefer CES because it is operated by professional staff, and hence, more reliable, which minimizes the level of knowledge and effort required from the consumer. Moreover, consumers appreciate the role of CES in providing grid support functions.

Consumers who are skeptical about CES fear that the distribution of cost will not be fair among participants—this was also found by Ambrosio-Albalá et al. [14]. Moreover, they perceive disadvantages such as increased coordination with neighbors, increased data security risks or fear that other participants treat common acquisitions less carefully. Abating these perceived disadvantages can help to increase the acceptance of CES.

In addition, there is a group of technically skilled and interested consumers who prefer to have their own storage systems in order to keep their technical devices and their energy flows under control. CES operators will have difficulties to reach this group and will only be able to convince them if they offer technically innovative solutions and attractive ancillary services such as monitoring and detailed evaluation of electricity production and consumption data.

The results of our online survey confirm Kalkbrenner's [16] findings that consumers prefer ownership over use-rights, however the focus groups and our research in Groß-Umstadt and Cologne paint a different picture. We must assume that consumers who are more familiar with and more knowledgeable about CES and the various ownership models are also open to renting. Here again, the advantages in handing off investment costs, responsibility, and risk are the dominant arguments.

CES requires ancillary energy services in order to function. In contrast to a residential storage system, the participants' energy flows into and out of the community storage must be monitored in order to provide the transparency required for billing and to ensure optimal storage operation. This topic was new for most of the participants in our inquiries and needed further explanation during discussions. Accordingly, the results of the online survey must be seen primarily as an expression of interest in such services, provided that they are not economically disadvantageous. However, it is hardly possible to deduce which of these services participants might be interested in and willing to pay for. The expressed greater willingness to pay for emergency power and e-mobility is for specific services whose direct benefits are more easily recognizable. Overall, the concern about power failures seems to be a strong incentive in the trade-off between CES and residential energy storage, even though such blackouts occur very seldom in Germany. Feedback from the two neighborhoods underlines that residents can better discuss and make decisions about storage services after they have dealt with the subject in depth. Such energy services must be explained in some detail, as the benefits of CES will only become readily apparent to consumers when they are sufficiently well informed. Comparisons of results, however, suggest that an informed discussion of ancillary energy services and energy storage is difficult to achieve if the services are not a part of an overall energy concept. People can better discuss these services when they are familiar with and understand the context in which they are to be applied. Even though it was possible to briefly outline such services in the focus groups, in-depth discussion was only possible in the two neighborhoods we surveyed.

The participants in our surveys were all owners and/or users of a PV system. These results can, thus, not be transferred to consumers not using PV. Given, however, that PV system owners are the relevant target group for CES, our results can still be considered relevant. The majority of respondents were male, and many were technically skilled or interested in the technical aspects. Further research into female consumers and those with more limited technical knowledge is necessary to gain broader insights. Our research analyzed interest in ancillary energy services for CES based on hypothetical offerings. A better insight into consumer acceptance, interest and willingness to pay would be possible if real world proposals were studied. Real world user experience with CES is still limited; here, further research is needed to compare user views and perceptions in various neighborhoods or pilot projects. The neighborhoods we studied were recent housing developments and most participants in our research were single-family or condominium owners. The questions of how to introduce CES into already developed areas and how to involve residents—owners as well as tenants—have to be studied in more detail. This study focused on Germany. Since CES is currently still in the pilot project phase, our results may be relevant for other countries as well. However, the question of who should operate CES has to be answered considering the national context, i.e., the historical development of energy supply and grid management. A transnational comparison of users' views on CES, however, would offer deeper insights into influential factors, such as laws and regulations at the national level, cultural identity, or the like.

5. Conclusions

Among users of PV systems, there exists a significant interest in energy storage systems and they perceive many advantages related to these. Although about six percent of the 1.7 million PV owners in Germany have already installed an energy storage system [3], acceptance among the remainder is rather more passive than active. This is due to barriers such as limited economic feasibility, the complexity of the topic, and lack of trust in information sources.

The provision of information and advice on energy storage makes for better decision-making. It is safe to assume that better consulting and advisory services will lead to a greater degree of active acceptance and to investments in storage facilities being made on the basis of actual costs. It is possible that with better information, less PV users will invest in storage under current conditions. However, since economic arguments are not the only motive, it can be assumed that with better information, PV users or those interested in PV systems will still be drawn to energy storage.

Those consumers who have already acquired a residential energy storage system mainly did so to increase their self-consumption; others were motivated by an interest in the technology. However, in order to motivate PV users with less technical interest or know-how to invest or to participate in CES, clear and economically feasible offerings are important. Currently, it is still difficult to compare the energy costs of PV systems combined with energy storage and many consumers are overwhelmed in the search for reliable information.

PV owners show limited interest in CES, but their interest increases when they are better informed. They need information on the comparison between residential and community energy storage regarding cost, efficiency and environmental aspects. Moreover, they need information on the operator of the CES system and on how the fair participation of different households is reached. Information on CES moreover needs to reach more PV users. There is also a need for more pilot and demonstration projects so that storage providers and potential customers can gain practical experience and learn about the advantages, disadvantages, and possible applications. Therefore, best-practice examples for community solutions must still be promoted and communicated. It is very unlikely that PV users would start a CES initiative on their own. Professional providers are needed to increase the use of CES. As mentioned by other authors [5,12,13,46], changes in the German regulations are needed to make the implementation of CES financially feasible. To increase economic feasibility, CES should be developed as multi-use storage including various ancillary energy services, and providers should additionally aim at markets other than the consumer market [41]. In setting pricing, CES providers should keep in mind that PV users tend to compare the cost of CES to that of residential storage systems.

CES providers should moreover consider the perceived disadvantages we identified: It is important to communicate how a fair and transparent billing will be reached, and how data security is guaranteed. Moreover, they should clarify that professional staff takes care of the CES system and that there is no need to coordinate with neighbors and no risk that other users damage the CES system through careless treatment.

In order to convince the group of technology enthusiasts, a professional storage operator should offer more in-depth analyses, e.g., the provision of monitoring data for the respective household, a forecast of storage performance for the next week, month or year, or make suggestions for increasing self-consumption.

One question that we have only dealt with in passing is the organizational structure for CES operation. Who should initiate CES projects? The advantages of CES can best be exploited if central monitoring and energy management systems are established. The Cologne example shows that the integration of CES into a comprehensive energy concept brings further advantages due to the combination of technologies, i.e., heat pumps and thermal storage. However, the preparation of such a centralized energy management protocol requires advanced energy planning and management qualifications. In particular, it requires specialization in the field of energy data analysis and management, which can best be realized by a professional operator. Currently, CES operators are still acquiring experience with the viable operation of CES and related energy services in pilot projects.

Local municipalities can create supportive conditions for participation in CES: in the sample neighborhood in Groß-Umstadt, the neighborhood development plan stipulated the installation of PV systems and the use of storage systems. This prescription increased the residents' interest in CES.

Even if environmental protection and support for the energy transition are not the first priorities of consumers with an interest in energy storage, the perception of doing something good for the environment or helping with climate protection can increase the willingness to invest in energy storage. Here, CES has the advantage due to lower resource needs and more options for flexibility, and these advantages should be emphasized more strongly to increase interest in and acceptance of CES. CES can provide grid services or store electricity from the grid and can, thus, contribute to net stabilization. These are benefits that are also appreciated by PV owners and will help to increase acceptance.

Author Contributions: Conceptualization, F.M. and E.H.; methodology, E.H. and F.M.; formal analysis of online survey data, E.H.; analysis of focus groups and data from the neighborhoods, E.H. and F.M.; writing—original draft preparation, F.M. and E.H.; writing—review and editing, E.H. and F.M.; visualization, E.H. and F.M.; supervision F.M. and E.H.; project administration, F.M. and E.H.; funding acquisition, F.M. and E.H. All authors have read and agreed to the published version of the manuscript.

Funding: This research was funded by the German Federal Ministry of Education and Research (Bundesministerium für Bildung und Forschung, 02K15A020). Further project results can be found at www.esquire-projekt.de.

Acknowledgments: We are grateful to our colleagues Swantje Gährs (as leader of the ESQUIRE project) and Jan Knoefel for discussing our research results. We moreover thank our project partners ENTEGA and evohaus for the opportunity to conduct empirical research in the two sample neighborhoods. We thank co2online for the realization of the online survey and for support in data analysis. We are moreover grateful to our student workers Paula Wörteler, Meike Ortmanns, and Ruth Berkowitz, who supported the project through literature research, data analysis, drafting of text passages, and drafting of visualizations.

Conflicts of Interest: The authors declare no conflict of interest. The funders had no role in the design of the study; in the collection, analyses, or interpretation of data; in the writing of the manuscript; or in the decision to publish the results.

References

1. Agnew, S.; Dargusch, P. Effect of residential solar and storage on centralized electricity supply systems. *Nat. Clim. Chang.* **2015**, *5*. [CrossRef]
2. Kairies, K.-P.; Magnor, D.; Sauer, D.U. Scientific Measuring and Evaluation Program for Photovoltaic Battery Systems (WMEP PV-Speicher). *Energy Procedia* **2015**, *73*, 200–207. [CrossRef]
3. Wirth, H.; Schneider, K. *Aktuelle Fakten zur Photovoltaik in Deutschland*; Fraunhofer ISE: Freiburg, Germany, 2020.
4. Koirala, B.P.; van Oost, E.; van der Windt, H. Community energy storage: A responsible innovation towards a sustainable energy system? *Appl. Energy* **2018**, *231*, 570–585. [CrossRef]
5. Müller, S.C.; Welpe, I.M. Sharing electricity storage at the community level: An empirical analysis of potential business models and barriers. *Energy Policy* **2018**, *118*, 492–503. [CrossRef]
6. van der Schoor, T.; Scholtens, B. Power to the people: Local community initiatives and the transition to sustainable energy. *Renew. Sustain. Energy Rev.* **2015**, *43*, 666–675. [CrossRef]
7. Brummer, V. Community energy—Benefits and barriers: A comparative literature review of community energy in the UK, Germany and the USA, the benefits it provides for society and the barriers it faces. *Renew. Sustain. Energy Rev.* **2018**, *94*, 187–196. [CrossRef]
8. Marczinkowski, H.M.; Østergaard, P.A. Residential versus communal combination of photovoltaic and battery in smart energy systems. *Energy* **2018**, *152*, 466–475. [CrossRef]
9. Koirala, B.P.; Hakvoort, R.A.; van Oost, E.C.; van der Windt, H. Community energy storage: Governance and business models. In *Consumer, Prosumer, Prosumager: How Service Innovations Will Disrupt the Utility Business Model*; Sioshansi, F., Ed.; Academic Press, an imprint of Elsevier: London, UK, 2019; pp. 209–234. ISBN 978-0-12-816836-3.
10. Lombardi, P.; Schwabe, F. Sharing economy as a new business model for energy storage systems. *Appl. Energy* **2017**, *188*, 485–496. [CrossRef]
11. van der Stelt, S.; AlSkaif, T.; van Sark, W. Techno-economic analysis of household and community energy storage for residential prosumers with smart appliances. *Appl. Energy* **2018**, *209*, 266–276. [CrossRef]
12. Gährs, S.; Knoefel, J. Anforderungen verschiedener Stakeholder an Dienstleistungen mit Quartierspeichern. Ergebnisse einer Analyse von Stakeholderinterviews. 2018. Available online: https://www.ioew.de/fileadmin/user_upload/BILDER_und_Downloaddateien/Publikationen/2018/G%C3%A4hrs-Knoefel_Anforderungen_verschiedener_Stakeholder_ESQUIRE_2018-09-17.pdf (accessed on 21 April 2020).
13. Gährs, S.; Knoefel, J. Stakeholder demands and regulatory framework for community energy storage with a focus on Germany. *Energy Policy* **2020**. accepted.
14. Ambrosio-Albalá, P.; Upham, P.; Bale, C.S.E. Purely ornamental? Public perceptions of distributed energy storage in the United Kingdom. *Energy Res. Soc. Sci.* **2019**, *48*, 139–150. [CrossRef]
15. Soland, M.; Loosli, S.; Koch, J.; Christ, O. Acceptance among residential electricity consumers regarding scenarios of a transformed energy system in Switzerland—A focus group study. *Energy Effic.* **2018**, *11*, 1673–1688. [CrossRef]
16. Kalkbrenner, B.J. Residential vs. community battery storage systems—Consumer preferences in Germany. *Energy Policy* **2019**, *129*, 1355–1363. [CrossRef]
17. Kloppenburg, S.; Smale, R.; Verkade, N. Technologies of Engagement: How Battery Storage Technologies Shape Householder Participation in Energy Transitions. *Energies* **2019**, *12*, 4384. [CrossRef]
18. Gährs, S.; Hoffmann, E. *Dienstleistungen mit Quartierspeichern*; e|m|w: Essen, Germany, 2018; pp. 68–70.
19. Hoffmann, E.; Mohaupt, F.; Ortmanns, M. Akzeptanz von Speicherdienstleistungen und Weiteren Energiedienstleistungen. Stand der Forschung aus Sozialwissenschaftlicher Perspektive. 2018. Available online: https://www.esquire-projekt.de/data/esquire/Datein/Arbeitspapier_Akzeptanz_von_Speicherdienstleistungen_und_weiteren_Energiedienstleistungen.pdf (accessed on 22 April 2020).
20. Wolsink, M. The research agenda on social acceptance of distributed generation in smart grids: Renewable as common pool resources. *Renew. Sustain. Energy Rev.* **2012**, *16*, 822–835. [CrossRef]

21. Hoffmann, E. *User Integration in Sustainable Product Development: Organisational Learning through Boundary-Spanning Processes*; Routledge: Abingdon, UK; New York, NY, USA, 2017; ISBN 978-1-351-27792-1.
22. Broughel, A.E.; Hampl, N. Community financing of renewable energy projects in Austria and Switzerland: Profiles of potential investors. *Energy Policy* **2018**, *123*, 722–736. [CrossRef]
23. European Commission. *EU Reference Scenario 2016: Energy, Transport and GHG Emissions—Trends to 2050. Main Results*; European Union: Luxembourg, 2016.
24. Schäfer, M.; Keppler, D. *Modelle der Technikorientierten Akzeptanzforschung. Überblick und Reflexion am Beispiel eines Forschungsprojekts zur Implementierung Innovativer Technischer Energieeffizienz-Maßnahmen*; Discussion Paper; Technische Universität Berlin: Berlin, Germany, 2013.
25. Hildebrand, J.; Rau, I.; Schweizer Ries, P. Akzeptanz und Beteiligung—Ein ungleiches Paar. In *Handbuch Energiewende und Partizipation*; Holstenkamp, L., Radtke, J., Eds.; Springer VS: Wiesbaden, Germany, 2018; pp. 195–209.
26. Schweizer-Ries, P.; Rau, I.; Zoellner, J.; Nolting, K.; Rupp, J.; Keppler, D. *Aktivität und Teilhabe—Akzeptanz Erneuerbarer Energie durch Beteiligung Steigern*; Forschungsgruppe Umweltpsychologie: Magdeburg/Berlin, Germany, 2010.
27. von Wirth, T.; Gislason, L.; Seidl, R. Distributed energy systems on a neighborhood scale: Reviewing drivers of and barriers to social acceptance. *Renew. Sustain. Energy Rev.* **2018**, *82*, 2618–2628. [CrossRef]
28. Huijts, N.M.A.; Molin, E.J.E.; Steg, L. Psychological factors influencing sustainable energy technology acceptance: A review-based comprehensive framework. *Renew. Sustain. Energy Rev.* **2012**, *16*, 525–531. [CrossRef]
29. Wunderlich, C. Akzeptanz und Bürgerbeteiligung für Erneuerbare Energien. Erkenntnisse aus Akzeptanz-und Partizipationsforschung. Available online: https://www.energiewende-sta.de/wp-content/uploads/2011/09/AEE_Akzeptanz-und-B%c3%bcrgerbeteiligung-EE.pdf (accessed on 22 April 2020).
30. Mast, C.; Stehle, H. *Energieprojekte im Öffentlichen Diskurs: Erwartungen und Themeninteressen der Bevölkerung*; 1. Auflage 2016; Springer VS: Wiesbaden, Germany, 2016; ISBN 978-3-658-12710-7.
31. Jones, C.R.; Gaede, J.; Ganowski, S.; Rowlands, I.H. Understanding lay-public perceptions of energy storage technologies: Results of a questionnaire conducted in the UK. *Energy Procedia* **2018**, *151*, 135–143. [CrossRef]
32. Agnew, S.; Dargusch, P. Consumer preferences for household-level battery energy storage. *Renew. Sustain. Energy Rev.* **2017**, *75*, 609–617. [CrossRef]
33. Gährs, S.; Mehler, K.; Bost, M.; Hirschl, B. Acceptance of ancillary services and willingness to invest in PV-storage-systems. *Energy Procedia* **2015**, *73*, 29–36. [CrossRef]
34. Graebig, M.; Erdmann, G.; Röder, S. *Assessment of Residential Battery Systems (RBS): Profitability, Perceived Value Proposition, and Potential Business Models*; Technische Universität Berlin: Berlin, Germany, 2014.
35. Figgener, J.; Haberschusz, D.; Kairies, K.-P.; Wessels, O.; Tepe, B.; Sauer, D.U. *Wissenschaftliches Mess- und Evaluierungsprogramm Solarstromspeicher 2.0*; Annual Report; RWTH Aachen: Aachen, Germany, 2018.
36. Smale, H.; Rowlands, I.H.; Gaede, J. *A Gap Analysis. Community Acceptance of Energy Storage Projects*; University of Waterloo: Waterloo, Canada, 2017.
37. Wawer, T.; Griese, K.-M.; Halstrup, D.; Ortmann, M. Stromspeicher im Quartier: Aktuelle Herausforderungen und Geschäftsmodelle in Deutschland. *Z. Für Energiewirtschaft* **2018**, *42*, 225–234. [CrossRef]
38. MVV Energie AG; Universität Stuttgart; Netrion; ADS-TEC. *Strombank—Innovatives Betreibermodell für Quartierspeicher*. 2016. Available online: http://fachdokumente.lubw.baden-wuerttemberg.de/servlet/is/120150/bwe13017_13020.pdf?command=downloadContent&filename=bwe13017_13020.pdf&FIS=203 (accessed on 22 April 2020).
39. Graulich, K.; Hilbert, I.; Heinemann, C. *Einsatz und Wirtschaftlichkeit von Photovoltaik-Batteriespeichern in Kombination mit Stromsparen. Kurzinformation für Verbraucherinnen und Verbraucher*; Öko-Institut e.V.: Freiburg, Germany, 2018.
40. Geelen, D.; Reinders, A.; Keyson, D. Empowering the end-user in smart grids: Recommendations for the design of products and services. *Energy Policy* **2013**, *61*, 151–161. [CrossRef]
41. Schnabel, F.; Kreidel, K. *Dienstleistungen für Gemeinschaftlich Genutzte Quartierspeicher*; Fraunhofer-Institut für Arbeitswirtschaft und Organisation IAO: Stuttgart, Germany, 2019.
42. Parra, D.; Swierczynski, M.; Stroe, D.I.; Norman, S.A.; Abdon, A.; Worlitschek, J.; O'Doherty, T.; Rodrigues, L.; Gillott, M.; Zhang, X.; et al. An interdisciplinary review of energy storage for communities: Challenges and perspectives. *Renew. Sustain. Energy Rev.* **2017**, *79*, 730–749. [CrossRef]

43. ENTEGA. *Besser Wachsen. Geschäftsbericht 2018*; ENTEGA: Darmstadt, Germany, 2018.
44. ENTEGA. *Besser Handeln. Nachhaltigkeitsbericht 2018*; ENTEGA: Darmstadt, Germany, 2018.
45. Statista. *Länge der Versorgungsunterbrechung je Stromverbraucher in Deutschland in den Jahren 2006 bis 2018*; Statista GmbH: Hamburg, Germany, 2019.
46. Gährs, S.; Knoefel, J.; Cremer, N. Politische Zielsetzungen und Rechtlicher Rahmen für Quartierspeicher—Bestandsaufnahme der Aktuellen Rahmenbedingungen und Diskurse. 2018. Available online: https://www.esquire-projekt.de/data/esquire/Datein/Arbeitspapier_Politische_Zielsetzungen_und_rechtlicher_Rahmen_fuer_Quartierspeicher-aktualisiert.pdf (accessed on 25 March 2020).

 *energies*

Article

Innovation Dynamics of Socio-Technical Alignment in Community Energy Storage: The Cases of DrTen and Ecovat

Binod Prasad Koirala [1,2,*], **Ellen van Oost** [2] **and Henny van der Windt** [3] 

1   Energy Transition Studies, TNO Energy Transition, 1043 NT Amsterdam, The Netherlands
2   Department of Science, Technology and Policy Studies, University of Twente,
    7522 NB Enschede, The Netherlands; e.c.j.vanoost@utwente.nl
3   Science and Society Group, Faculty of Science and Engineering, University of Groningen,
    9747 AG Groningen, The Netherlands; h.j.van.der.windt@rug.nl
*   Correspondence: binod.koirala@tno.nl

Received: 29 April 2020; Accepted: 5 June 2020; Published: 9 June 2020

Abstract: With energy transition gaining momentum, energy storage technologies are increasingly spotlighted as they can effectively handle mismatches in supply and demand. The decreasing cost of distributed energy generation technologies and energy storage technologies as well as increasing demand for local flexibility is opening up new possibilities for the deployment of energy storage technologies in local energy communities. In this context, community energy storage has potential to better integrate energy supply and demand at the local level and can contribute towards accommodating the needs and expectations of citizens and local communities as well as future ecological needs. However, there are techno-economical and socio-institutional challenges of integrating energy storage technologies in the largely centralized present energy system, which demand socio-technical innovation. To gain insight into these challenges, this article studies the technical, demand and political articulations of new innovative local energy storage technologies based on an embedded case study approach. The innovation dynamics of two local energy storage innovations, the seasalt battery of DrTen® and the seasonal thermal storage Ecovat®, are analysed. We adopt a co-shaping perspective for understanding innovation dynamics as a result of the socio-institutional dynamics of alignment of various actors, their articulations and the evolving network interactions. Community energy storage necessitates thus not only technical innovation but, simultaneously, social innovation for its successful adoption. We will assess these dynamics also from the responsible innovation framework that articulates various forms of social, environmental and public values. The socio-technical alignment of various actors, human as well as material, is central in building new socio-technical configurations in which the new storage technology, the community and embedded values are being developed.

Keywords: energy transition; community energy storage; responsible innovation; energy system integration; socio-technical innovation

---

1. Introduction

Currently, the energy system is at a crossroads and is going through rapid techno-economical and socio-institutional changes both at the central and the local level [1–5]. New distributed energy resources such as solar photovoltaics, wind and energy storage technologies are emerging in the energy landscape [6,7]. These changes demand the increased engagement of citizens and communities in the energy system [8–13]. Accordingly, there are new regulatory and governance changes such as new European clean energy for all packages, as well as new societal developments in the form of local energy initiatives [3,5,9,14–17]. The concept of Renewable Energy Communities (REC) and Citizen Energy Communities (CEC) were introduced in the European legislation by the 2018 recast of the European Renewable Energy Directive (RED II) and the 2019 recast of the Electricity Market Directive (EMD II), respectively [18,19]. These institutional transformations caused resulting techno-economic changes in the energy system which imply not only political and socio-economic issues in the energy system transformation, but also fundamental shifts in the way the energy system is organized and operated [2,20–22]. As innovation becomes more rapid and complex, uncertainty increases regarding the effectiveness of existing policies and regulations, as well as the permissibility of the innovations [23]. Moreover, there are serious challenges concerning their embedding in existing technological and societal frames and systems.

The transforming energy system has to be more adaptive, diverse and flexible to accomodate increasing temporal fluctuations in both supply and demand [5]. Supply fluctuations are growing with the increasing penetrations of intermittent solar and wind. Both energy demand as well as its fluctuation will rise due to the increasing electrification of different end-use sectors such as heating and transport. With higher intermittent generation through solar and wind as well as changing consumption patterns, the mismatch between supply and demand will only increase in future. Energy storage is seen as crucial for solving this mismatch, and thus is expected to gain an important place in a future sustainable energy system [21,24,25].

Storage technologies of the future have many different shapes, scale, functions and politics. As trends and developments in energy storage technologies are fast-moving, no dominant community energy storage technology has cristallized to date. Neither is it clear how these innovations will possibly affect the energy system and society as a whole. Furthermore, advances in information technology and digitalization generate a wide variety of new applications and services for energy storage. The new opportunities and challenges created by these innovations are unclear. Currently, at least three approaches can be identified: storage close to production sites, for instance configurations of wind parks and hydrogen storage facilities, storage close to consumers, such as home and neighborhood battery systems, and in-between approaches, such as configurations of electricy and thermal storage by water or gas [5,24].

In this study, we focus on analyzing local storage innovations close to (a community of) consumers, as we are interested in how energy innovations can empower local communities. Local communities simultaneously can be a breeding place for social and technological energy innovations [5,26–30]. New technologies, co-operations, markets and energy attitudes can develop, stimulating social, cultural and economic activities of the local communities. Various factors have been identified for these successes: cultural backgrounds, timely cooperation between local initiatives, technology developers and firms as well as support by the governments [31]. The innovation of windmills in Denmark and solar collectors in Austria is explained by the design of the technology, orchestrated learning processes between owner–user groups and firms, specific cultural traditions and governmental policy [32,33]. The skills and attitudes of the people involved in the initiatives and cooperation on different societal levels have also been noted as main factors [34,35].

Local energy initiatives can be seen as a specific innovative sector, characterized by its own social dynamics, values, technological preferences and learning processes. According to Seyfang et al. (2014), innovations by these types of initiatives differ from market-based innovations in several ways [36]. For these innovations, social and/or environmental needs are driving forces, which means that collective values such as locality, solidarity and sustainability outweigh efficiency and profit. The input of volunteers, grant funding and reciprocity are as least as important as business loans or commercial norms, and output in terms of greening society is at least as important as material economic results. In addition, cooperatives, and voluntary organisations are dominant organisational forms, and firms are rare. However, for these initiatives, connections to other energy actors, through intermediaries or networks, is crucial [37,38]. Thus, local communities are an interesting and relevant place to study energy innovation dynamics and the processes of socio-technical alignment, meaning giving and social learning which may constitute new innovative socio-technical configurations that may be one of the building blocks for the future energy transition.

Local communities, thus, are an interesting place to study the dynamics of energy storage innovations as the involved collective values often go beyond market values and include other social values like environmental, justice, fairness and privacy. We aim to study local energy storage innovations that allow for new roles and responsibilities for citizens, e.g., as energy prosumers or even prosumagers (combining production, storage and comsumption). This type of socio-technical innovation could grant local energy collectives more agency to realize their sustainability goals. The energy storage innovations themselves are not neutral and also embody values, have politics and exercise agency [39,40]. This paper will analyze two emerging local energy storage innovations that explicitly embody environmental and social values in their basic design, the seasalt battery of DrTen® and the thermal storage Ecovat®, which, respectively, avoid toxic or scarce elements and minimize visual impact on the landscape. Both local storage innovations are on the verge of market introduction and still have to be implemented in and adapted to use situations. As such, these two innovations allows for gaining insights into the co-shaping dynamics during the early implementation processes that may lead to new innovative and socio-technical local energy configurations, that could potentially form an important element of the future energy system transition.

Our research is embedded in the recently developed innovation policy framework of Responsible Research and Innovation (RRI) for guiding technological innovations towards strengthening social and ecological welfare next to economic goals (For example, The Netherlands Research Organization (NWO) has developed the Responsible Innovation research programme (NWO-MVI). The programme identifies the ethical and societal aspects of technological innovations at an early stage so that these can be taken into account in the design process) [5,41]. This RRI perspective is described as "taking care of the future through collective stewardship of science and innovation in the present" [42]. RRI is not only a policy framework for innovation, but also a growing field of research. To date, energy innovations are underrepresented in the RRI literature [43]. Although no reasons for this are given, one could argue that many energy innovations today strive towards a more sustainable energy system and thus already have an environmental embedded normativity. Yet, energy innovations may raise new societal tensions (e.g., large windmills on land) or future unwanted impacts like shortages of rare materials (lithium), waste problems (old windmill wings) or new social problems like energy poverty, or new forms of social inequality as not everyone can afford energy innovations and benefit from them. Our study aims to contribute to insight into possible pathways and pitfalls for a responsible energy innovation dynamics through an empirical study of the development and implementation of two sustainable storage innovations in local energy communities.

We study the innovative potential of local energy initiatives in terms of energy storage technology adoption, social embeddedness and normativity through various forms of alignment with the innovative potential of emerging energy storage technologies, including their normative social, politcal and environmental dimensions. In our research, we will address the question of how to orchestrate socio-technical alignment issues in the implementation of innovative community energy storage technologies. We aim to gain insight into the local contextualized co-shaping dynamics of local energy storage innovation and the local network of involved actor groups.

The article is organized as follows. First of all, in Section 2, a conceptual research framework is provided. In Section 3, research design and methods are outlined. Section 4 presents the case studies of emerging responsible community energy storage innovations, DrTen® and Ecovat®. Finally, Section 5 provides a conclusion and discussion on socio-technical alignment dynamics in the implementation of responsible innovations in community energy storage.

2. Conceptual Research Framework

In this section, we will elaborate the conceptual research framework. First, we will elaborate on how we use and define the concept of community energy storage. Second, we elaborate on our social constructive and co-evoluationary perspective on technological development as recently developed in science and technology studies (STS).

We refer to community energy storage as a subset of the overarching concept of "community energy" [3,8,16,20,44–62]. The exisiting energy communities may provide fruitful ground for the adoption, development and implemention of community energy storage [5]. In essence, the difference is mainly technological but there may also be minor socio-institutional differences. The storage technology enables local communities to have higher control of their energy systems. At the same time, the interactions with institutional actors as well as business models are slightly different to community energy.

Several definitions of community energy storage are available [5,21,28,30,63,64]. Robert and Sandberg (2011) see community energy storage as an intermediate solution between residential and utility-scale energy storage, whereas Parra et al. (2017) suggest that community energy storage brings benefits both consumers and system operators [21,65]. Koirala et. al. (2018a) define community energy storage as "an energy storage system with community ownership and governance for generating collective socio-economic benefits such as higher penetration and self-consumption of renewables, reduced dependence on fossil fuels, reduced energy bills, revenue generation through multiple energy services as well as higher social cohesion and local economy" [5]. To the knowledge of the authors, the research on community energy storage systems to date has a main focus on techno-economic aspects and limited attention towards societal, institutional and environmental aspects (notable exemptions are [5,13,21]) This article analyzes community energy storage from a socio-technical perspective. This approach allows to investigate interactions and dynamics between different actors and components of community energy storage. The focus is on the socio-technical alignment of community energy storage systems as well as their transformative capacity.

Pragmatic theories such as domestication theory, social practice theory and actor–network theory offer research tools to study socio-technical innovation dynamics [38,66–69]. An innovation is seen as an evolving socio-technical actor network with various material and societal actors and relations [66]. The actor network is a product of successful alignments of material as well as social and regulatory actors [66,69]. Jalas et. al. (2017) as well as van der Waal et al. (2020) highlight that experimentation opens up possibilities for participation for a wide range of actors [10,70]. Ryghaug and Toftaker (2014) combined social practice theory and a theory of domestication to study different dimensions of electric vehicle introduction in Norway [71]. From the energy transition perspective, frameworks such as technological innovation systems, multi-level perspectives as well as strategic niche management are relevant [72–75].

This research builds on these exisiting theories and frameworks and goes beyond them as it aims to stimulate social learning to improve the alignment and coordination of social and technological innovations and offers a unique opportunity to engage in and learn from reflexive social learning in aligning technical, demand and cultural articulation as a form of responsible innovation in the sustainable local energy storage technologies. We positioned our research as contributing to the innovation policy framework of Responsible Innovation. (RRI). RRI is not the conceptual framework but it does help in structuring the normative goals underlying this research.

This approach aims to stimulate "research and innovation outcomes aimed at the "grand challenges" of our time, for which they share responsibility. Research and innovation processes need to become more responsive and adaptive to these grand challenges. This implies, among others, the introduction of broader foresight and impact assessments for new technologies, beyond their anticipated market-benefits and risks" [76]. The RRI approach distinguishes four dimensions to guide the innovation process: anticipation, reflection, deliberation and responsivity [41]. Anticipation aims to gain insight in possible future societal impacts in an early phase of the innovation development. Reflection highligths and discusses social, ethical and environmental aspects of the anticipated impacts. Deliberation refers to involving relevant actor groups in the innovation process by highlighting their perspective in the challenges and uses of the new technology. Last, but not least, responsivity aims to feed back the insights and analyse of the three other dimensions into the ongoing development, implementation and societal embedding of the innovation. RRI thus broadens the technology design by including social, ethical, and environmental aspects and involving a variety of stakeholder groups [77]. However, there are also some critiques of RRI, one of the main ones being the limited availabily of indicators to measure the effects of RRI and important innovation barriers to including RRI values [78].

In each of these four dimensions, different forms of socio-technical linkages are created. When technologies are designed, assumptions are made regarding users, regulations, available infrastructures and responsibilities between various relevant stakeholders [39,77]. The notion of scripts links technological design choices (technological articulation) to expectations about users (demand articulation) and other stakeholders and regulations (political articulation). The technology developers, users, governments and other actors have their own set of assumptions and expectations. Moreover, it is important to allow the early and regular confrontation and exchange of these assumptions and expectations [77,79]. Devine-Wright et.al. (2017) studied the social acceptance of energy storage, combining market, socio-political, community and environmental aspects [69]. Energy storage is accepted or rejected in different ways in different geographical and societal contexts. Thus, it is important to consider the roles of different actors, their values, needs, expectations and interactions, as well as the materialization of technologies and their societal embedding in different contexts.

Figure 1 illustrates the way technological and societal elements are interwoven in complex socio-technical systems such as community energy storage [45,80–82]. These elements develop in co-evolutionary dynamical processes. Various societal stakeholders develop new routines and institutions embedding the new technologies by anticipation, reflexivity, deliberation and learning [41,83].

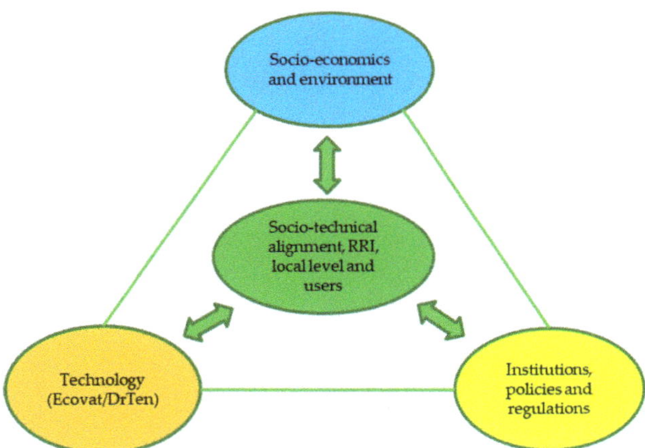

Figure 1. Research framework.

Socio-technical alignment may be seen as a process in which responsible innovation is achieved through identifying imperatives and anticipating incompatibilities in social and technical innovation and taking measures to counteract unwanted effects. Community energy storage technology also needs to overcome path dependency and the socio-technical lock-in of existing energy systems and should be related to various dimensions of society and its demands such as regulatory frames, already esisting technologies, organizations, environmental requirements and psychological issues of acceptibility [81,84]. Socio-technical alignment is central to overcoming these lock-ins and problems of acceptability.

In the next section, we will address how we studied the socio-technical alignment processes in the dynamics of community energy storage innovations by describing the development of two new technologies and by analyzing how they tried to include various values and needs, in particular citizens' involvement and environmental consideration, and to align both new storage technologies to existing technological and institutional configurations.

3. Research Design and Methods

The explorative, process-oriented research question of our study fits a qualitative approach based on a in-depth case studies [85]. The qualitative approach enables a detailed analysis of all actors and their positions and roles and the alignment dynamics. The selection of the two emerging Dutch storage innovation cases, Ecovat and DrTen, is based on sustainable, responsible design features, the high societal expectations and the developmental phase—a working prototype phase, with the intention to grow in coming years towards market introduction. The innovation was still open to accomocodate technological, social and regulatory articulations (responsiveness). The two cases are complementary in the sense that DrTen batteries focus on day/night electricity storage whereas Ecovat allows for thermal seasonal storage. These two cases allow for qualitative insight into a broad spectrum of involved actors and socio-technical alignment processes.

As community energy storage is at the early stage in the development process, conceptual tools from technology dynamics such as social actor analysis, dynamics of technological promises and expectations, script analysis and niche dynamics are applied to analyze socio-technical alignment processes of the community energy storage [86]. As an empirical source, we apply embedded case studies, based on interviews, participatory observation and document analysis. In particular, the development and initial adoption of two emerging innovative energy storage technologies in The Netherlands, Ecovat® and DrTen® has been followed [87,88]. For each storage innovation, we followed for a longer period two use-cases in two Dutch villages, Heeten (DrTen) and Wageningen Benedenbuurt (Ecovat), where the innovator-entrepreneurs collaborate with local communities, citizens, energy system actors and local government. These local use-cases offer a context in which different actors, both incumbent energy system actors as well as new energy actors, work together for a sustainable and decentralized energy future including an innovative community energy storage technology [89,90].

The data collection was carried out between November 2016 and April 2020. The embedded-case of Benedenbuurt was followed from November 2016 to April 2020, whereas the embedded-case of Gridflex Heeten was followed from March 2017 to April 2020. In the case of Benedenbuurt, we observed and participated in all project meetings in the period 2017–2018 and in three local initiative meetings. In addition, we interviewed key actors of the cooperative, the project team and the municipality. In the case of Gridflex Heeten, we attended various project meetings as well as the three information meetings for the participating residents. For the Benedenbuurt case, we collected several documents, which include the minutes of the local initiatives, feasibility studies, webpages and news articles, whereas for the gridflex Heeten we collected several documents including project proposal, flyers and webpages. For data on the technological and organizational development, we held interviews with technical sales manager of ECOVAT, marketing director of DrTen and initiators of both initiatives. We also collected and studied academic publications and other documents: two journal publications [91,92], three conference papers [22,93,94], several expert reports, two patents [95,96], several presentations, two master-theses [97,98] and one PhD thesis [99] related to ECOVAT and, for Dr.Ten batteries, three journal publications [5,100,101], three conference papers [22,102,103], several presentations and one PhD thesis [104].

The analysis of the collected data was not processed digitally, nor coded, but used to heuristically construct a qualitative understanding of the innovation dynamics, by focusing on identifying the relevant actor groups, their changing definitions, articulated meanings, agenda and process roles as well as crucial successful and failed socio-material alignments and re-alignments.

4. Emerging Energy Storage Technologies: The Cases of DrTen® and Ecovat®

Current energy storage have several issues such as high costs, limited capacity and life time, use of rare earth or polluting materials, geographical dependency (e.g., pumped hydro and compressed air) and safety issues [105–107]. Sustainable, cheap and reliable energy storage is still a challenge [107]. In this context, two promising community energy storage innovations are emerging: DrTen® for short-term electrical energy storage and Ecovat® for seasonal thermal energy storage [87,88]. The DrTen® seasalt battery promises a sustainable, clean, and relatively cheap storage of electricity and can be applied at the level of households and communities. Seasonal themal storage Ecovat® stores heat in the summer, and this can be retrieved in winter. This storage system functions at the level of neighborhoods.

In this section, we will describe and analyse the socio-technical alignment dynamics of these two community energy storage innovations. Section 4.1 provides a more elaborate description of the case technologies and the evolving companies. In Section 4.2, we give an overview of the innovation dynamics summarized in a timeline overview. Section 4.3 describes the relevant actors and stakeholders and the way they contributed to the storage innovation as well as their mutual relations. Section 4.4 provides a more detailed analysis of a use case, a pilot that intended to implement the storage innovation. In Section 4.5, socio-technical alignment dynamics and strategies are elaborated.

4.1. Key Characteristics

Ecovat® is developed as reliable and affordable solution for solving the seasonal energy gap in (solar) renewables. As illustrated in Figure 2, Ecovat® is a large seasonal thermal storage with a smart control software. The physcial system of Ecovat® consists of large subteraranean buffer tank, heat exchangers, energy management systems, district heating networks and communication networks. Based on weather forcecast, actual electricity market prices and anticipated heat and electricity demand, the Ecovat® software can optimally operate the system. Thermal energy is stored as hot water in a large subterranean buffer tank. Test results show energy losses of less than 10% over six months [88]. The heat sources can be renewables (solar thermal) and geo-thermal as well as waste heat and electricity. The electricity should preferably come from renewable sources like solar or wind. This could also increase the rate of self-consumption of locally generated renewable electricity in single- or multi-apartment buildings as well as neighbourhoods [108]. It also has the potential to provide a local balance between supply and demand and provide 100% renewable heating and cooling. The subterranean buffer tank does not impact landscapes and is almost maintenance-free, as the system has no moving parts. The expected life expetancy for an Ecovat® system is 50 years. Through smart integrated infrastrucutes for heat and electricity, it has potential economic value propositions such as peak shaving of heating networks, congestion management, balancing of the electricity network, increased self-consumption of local generation, avoided grid-reinforcement costs due to the electrification of the residential heating sector, better utilization of waste heat, reduced energy prices, maximum use of renewables and minimum environmental impacts.

Figure 2. Ecovat® as large subterranean buffer tank.

DrTen® provides safe, clean and affordable energy storage solutions. DrTen® seasalt batteries have seasalt as the main salt in the electrolyte and carbon electrodes. Currently, it has reached an energy density of about 35 Wh/kg, comparable to about 20 Wh/kg for a lead acid battery with largest market share worldwide [87]. As materials used in seasalt batteries are green and low-costs, the prices are expected to be lower than existing batteries upon mass production. The battery is now in pilot production. It can be deep-discharged and charging and discharing cycles of more than 64000 have been recorded. To make it affordable, the batteries were originally housed in simple plastic cups, rainwater pipes followed by more professional boxes with pouch cells, also becoming more professional with some inspiration from green food packages andli-ion batteries. With materials still coming from China, Israel, Germany, the Netherlands and the US, cell-manufacturing has been relocated to Israel while the batteries are still being assembled manually to systems in the Netherlands. A future production lines are foreseen to be automated, with one cell per 5-20 seconds, leading to about 7 MWh per year. Currently, DrTen® batteries are being tested in several local energy pilots, such as gridflex Heeten [89] and Israel, scaling up materials and production with more massive implementation.

4.2. Key Processes in Innovation Dynamics

At around 2013, both technologies started in small firms, both specialized in sustainable technology, but had a larger portfolio. Ecovat® was founded by Aris de Groot, a successful architect and designer in sustainable buildings. DrTen® was owned by chemical technologist Marnix ten Kortenaar, who had worked at Delft University of Technology and a large chemical company. The initial idea for DrTen® batteries originated during his visit to Africa, based on fundamental research he has done in 1994 at Delft University of Technology, followed by various lab scale prototypes development between 2008 – 2013 and first simple prototype in 2014. Both took initiatives to develop their technologies, together with the technical universities of Delft and Twente and universities of applied sciences Avans and Hanze, involving Master and PhD students. In addition, they started pre-engineering or pilot projects together with municipalities, governmental science and technology funders, local stakeholders and grid operators. Because many of these parties were looking for innovative, sustainable and environmental friendly technologies which could be applied at local levels, both companies were attractive to cooperate with.

Below, we will describe the main activites of DrTen® and Ecovat® between 2013–2019. During 2013–2019, both innovations participated in and won several innovation prize contests. DrTen® received the prestigious Terlouw innovation prize in 2013, two Blauw tulp accenture innovation award in 2014, was seen as the most sustainable and innovative SME in the Netherlands (Squarewise) and belonged to the top 100 innovations in the Netherlands (RTL Z) in 2015. In addition, DrTen became successful in joining events and projects, such as pitches during Yes Delft (2015), a Turkish innovation week (2015) and the so-called Kenniskring smart energy of Innovation Centre Green Economy Noord-Veluwe (IGEV) (2016). The German magazine Bizz energy selected ECOVAT as innovation of the month (2015), it received the innovation award for sustainable energy from DSO Stedin and leading Dutch environmental organization Natuur and Milieu (2016), and it won the FLEXCON energy startup challenge (2017) and Enpuls flex energy gap challenge (2018). In 2019, Ecovat was included in the list of mission innovations, a global initiative to accelerate clean energy innovation [109].

In both cases, the focus has been not only on technology development but also on aligning the innovation with societal needs and requirements. DrTen started five pilot projects in the Dutch provinces of Zeeland, Gelderland, Groningen and Overijssel, while Ecovat started seven pre-engineering feasibility studies or pilots in the Dutch provinces of Zuid-Holland, Limburg, Noord-Brabant, Groningen and Gelderland and one in Germany. Most of these projects were co-financed by public funders, such as the EU and national funds for innovation, regional development or energy transition. In 2016, Dr. Ten joined a relatively large, publicly funded research consortium, concerning a pilot project on smart microgrids and local energy markets, a collaboration of academia, one of the largest DSOs (Enexis), a local energy cooperation, an ICT company and Buurkracht, an organization which specialized in citizen engagement in local energy projects, called Gridflex Heeten (this pilot project will serve a embedded use case, see Section 4.4). In various other subsidized energy projects, DrTen was welcomed as a storage technology partner, such as an INTERREG program, funded by the EU, Dutch ministry of economics and climate, North Rhein Westphalia ministry of economics, innovation, digitalization and energy and several Dutch provinces, the Northern Climate Summit in the province of Groningen to develop pilots, and COOBRAA/AVANS projects in the cities of Breda and Tilburg, for the development of sustainable concepts such as the 'neighborhood battery' and an "autarkic (tiny) house" [110]. Although a marketable version of the battery still remained to be created, the sea salt battery of DrTen kept invoking the genuine interest of various governmental actors and other societal parties. For instance, in 2019 the province of Groningen asked DrTen to start pilots in this province, and in the province of Gelderland, DrTen won the Veluwe innovation prize for its battery, as a promising concept for sustainable, self-supporting urban smart grids.

The development of the ECOVAT® seasonal warm and cold storage system showed various similarities. In 2014, during the construction of the demo, Dutch innovation programs TKI and REAP subsidized the company, followed by subsidies from the Dutch innovation-funding organization RVO in 2015 for the energy system integration of the ECOVAT network balancing system. Meanwhile, it received support from BOM business development in the province of Noord-Brabant, worked together with various higher education organizations and joined the Dutch storage platform Energy Storage NL. Ecovat developed the system further, both technically and economically, resulting in the production of pre-fab elements in 2015, wall components in 2016 and several pre-engineering projects, for instance in Wageningen Benedenbuurt and Arnhem Ons Dorp. The finalization of the demo plant in Uden took place in 2017. In 2016, the company started participating in new platforms and projects, such as the Frisse Dingen, Dutch platform for sustainable innovation, the flexible heat and power H2020 project consortium and the talent for energy transition project. After the patent for the wall part of Ecovat in 2016, the system as a whole was patented in 2018 and certificated in 2018 (ISO 9001 and VCA certificates). In early 2020, Ecovat was also certified as a B corporation, making the startup company one of the worldwide frontrunners of for-profit companies with a high social and environmental performance [111].

A next step in the development of Ecovat was the improvement of the software, funded by the European regional development fund and regional funds and a Berenschot study on the system consequences and saving potential of ECOVAT and robotization of the production in 2018.

In 2019, ECOVAT successfully launched an issue of shares in NPEX (€1.26 million). Despite an intensive preparatory trajectory successfully aligning all relevant actors and aspects (local government, politics, safety, cost efficiency proof etc.) of the first large-scale Ecovat Ons dorp project in Arnhem (claimed to become the most energy innovative neighborhood of NL), it was cancelled unexpectedly in June 2019 by the commissioner (SIZA) as the number of houses reduced from 550 to 175, making Ecovat no longer cost-effective.

In conclusion, during the step-by-step development of these technologies, both companies succeeded in organizing networks, cooperation with all kinds of societal parties, and receiving financial and other support. The two technologies and companies have some striking similarities. Both are seen as promising technologies for environmental friendly energy provision at local levels. Various societal actors such as universities, grid providers and governments expressed their support by funding and cooperation, and both technologies have won various awards. Both innovative storage technologies have similar organizational characteristics as well. They are both developed in-house, led by their 'inventor-entrepreneur'—in both cases a creative, socially engaged individual (one an engineer and the other an architect). Both technologies now have reached the stage of working prototypes and both startup companies developed their own commercial production processes. However, the companies and technologies differ as well. There are two main differences. First, Ecovat offers an integral solution. The company offers the main technology, options for financing, smart control as well as operation services and contracts. Although Ecovat is flexible in functionalities and size, a minimum size and low temperature district heating networks are required to make it financially more attractive. DrTen does not offer an integrated solution. The company is confined to the technical functioning of the battery and does not regard smart control (ICT) as part of its business. Second, there is a difference regarding the financial position of both companies. Ecovat is financially more robust (in 2018: turnover EUR 4 million and profit of EUR 1 million) [112]. Their pilot projects generate larger amounts of cash-flow (e.g., TKI EUR 4 million), and in April 2019 the company successfully issued shares (EUR 1.26 million) in the Dutch NPEX stock exchange. DrTen (turnover about EUR 1 million) finances the development and production of testbatteries primarily through participation in publicly funded energy pilot projects and with consultancy and demo-battery assigments.

4.3. Actor Analysis

For both technologies and companies, the similar type of actors are relevant or even crucial. We already mentioned the importance of municipalities, other governments, grid operators and universities for the funding, cooperation and further development of the technologies. In this context, actors refer to people or citizen's organisations, industries, or other private parties and governmental institutions that can affect or are affected by these technologies. Directly affected actors are households and communities, as well as energy system actors which are related to the installation of the technology, such as municipalities and grid operators where it will be installed as well as material suppliers and distributors. Public authorities, such as regulatory agencies, such as ACM, authorities for consumers and market, and DNV-GL, responsible for standard setting, as well as ministries and municipalities, may directly affect technological development through the introduction of new rules or subsidies or by creating enabling or inhibiting an environment for technology implementation. Both technologies have been certified by DNV-GL. Knowledge institutions, technology or material providers and transport companies may also be influential. In particular, in the case of Ecovat, the alignment of new and already existing technologies for the construction and functioning of the storage system was an important elment of technology development. Both DrTen® and Ecovat® joined the Dutch industry association EnergystorageNL, which is actively lobbying for a better regulation for energy storage in the Netherlands and support or aligned storage options. Figures 3 and 4 illustrate an overview of the various actors that play an important role in the socio-technical alignment of DrTen® and Ecovat®, respectively.

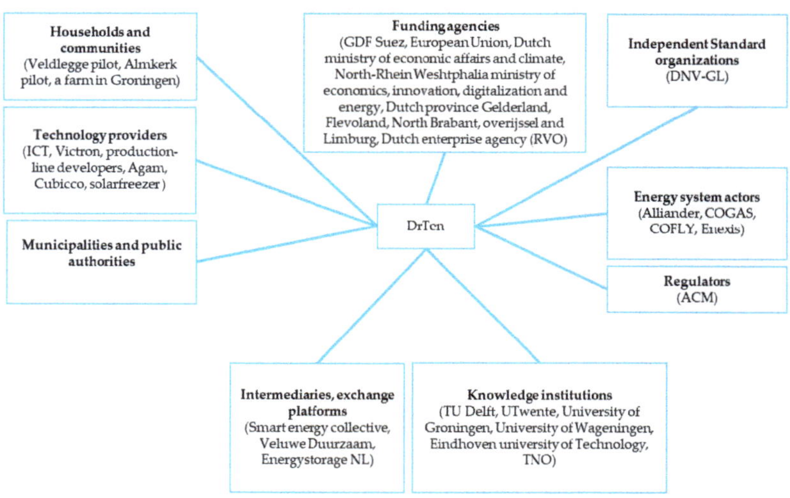

Figure 3. Actor mapping of DrTen®.

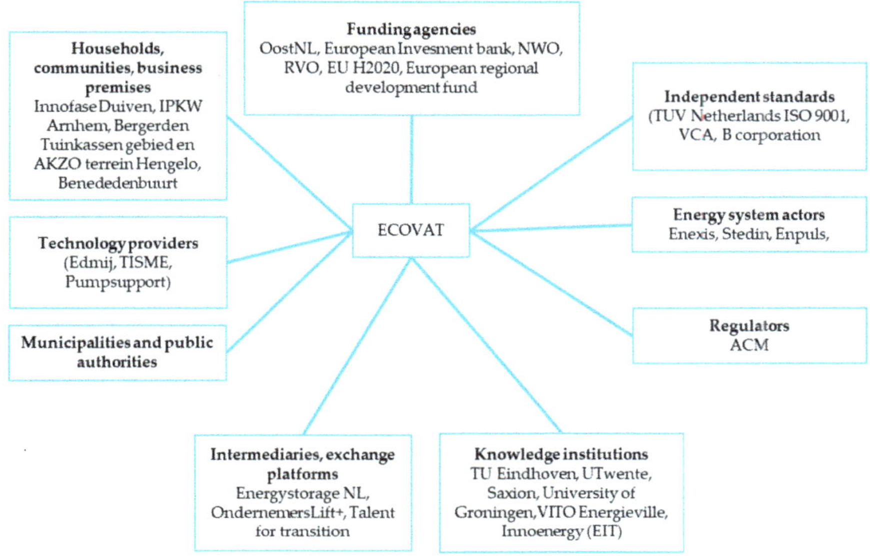

Figure 4. Actor mapping of Ecovat®.

The relationship with energy system actors like TSOs and DSOs is more complex and ambigious. Ecovat can fulfill a variety of functions in the future energy system. Although ecovat® itself is a stand-alone technology, a preference for a total system concept by the technology developer is observed due to techno-economic complexities such as high upfront costs as well as operational requirements. As technical functionalities are starting to become clear, different application areas are being envisoned, namely short-term storage, seasonal storage, storage of waste heat and electricity and the production and transport of heat. Based on these funcationalities, power to heat as well as heat to power application are foreseen, although the latter will be feasible only under a very high share of renewable energy in the energy mix for efficiency reasons. Ecovat® also has the potential to take active participation in balancing markets, mainly to avoid the curtailement of renewables or peak demand due to electrication of the heating sector. For example, a recent study shows the cost-saving potential of Ecovat® due to avoided grid reinforcement and peak power plants [113]. Recently, potential application in agri-food sector has also been explored.

The focus of DrTen® to date has been on the technology development of sea-salt batteries for home owners and neighbourhoods. Increasing attention has been paid to the balance of the components such as charge controllers, energy management systems, battery management systems and inverters, and in some pilotprojects, Dr. Ten batteries were seen as a tool for balancing and peakshaving in smart local microgrids (Heeten and Interreg project). DrTen® has to work together with several technology developers in order to make it interoperable with the existing balance of components in the grid market.

4.4. Use Cases in Local Communities

The adoption process of Ecovat® and DrTen® community energy storage by local communities in the Netherlands was followed in the form of embedded case studies. At this level of envisioned use of the new technology in real life situations, both use cases showed a high level of alignment dynamics. A potential technology adoption of Ecovat® was discussed in a series of meetings in the local community of Benedenbuurt in the city of Wageningen. DrTen® seasalt batteries were foreseen to be installed in one household first followed by 7 and then 24 out of the 47 households in the neigbourhood of Veldegge, in the village of Heeten, part of a pilot of the so-called Gridflex project [89]. As delays were faced in ICT/technology integration, 3 houses were set so far but the rest are expected to follow in the near future with improved quality. In Table 1, observations from both of these use cases are presented.

Table 1. Observations from community use cases Benedenbuurt (Ecovat®) and Heeten (DrTen®).

	Ecovat® (Benedenbuurt)	DrTen® (Heeten)
Initiators	Charismatic, creative idealist (resident) (*heteregoneous citizens*)	Combination of smart strategist (Escozon) and charismatic resident (Endona) (*heterogeneous citizens*)
Stakeholders/organisation	Only local stakeholders, little involvement of traditional energy regime actors	Involvement of combination of local stakeholders and energy regime actors
Organisation/problem definitions	Fuzziness regarding who is responsible, inequalities of paid and volunteer workers, ad hoc incidental yet successful financing/problem definitions shift easily	Different roles are clear, All projects participants and materials are financed. Much alignment in problem definitions.
Involvement of users/residents	Users become more strongly organised during the process/now growing group of residents on drivers seat, discussions on ownership	Users/residents: very high participation but mainly in limited user role. More active contribution can grow in next phase of project
Tensions/conflicts	Between energy cooperative and interested commercial actors	DSO's interest initially in conflict with vision of 'net-zero behind the transformer'
Material redefinitions	Reframing Ecovat® as logistic problem (number of trucks)	Battery safety was foregrounded/slow development of the linkage of seasalt battery to theICT system caused tension in the project but was improved later.

In the following paragraph, we will present three elements of the alignment dynamics. First, the orgnization and empowerment of the citizen group, second, the role of the companies, and third, the involvment of other actors at the local level. The wider societal context, the flexibility of the technology and the technologcal infrastructure will be discussed in next paragraphs.

In both neighbourhoods, Heeten and Benedenbuurt, active citizens took the initiative for a more sustainable local energy system. They had other things in common. Both were interested in new technologies, which resulted in the plan to install sea salt batteries (Heeten) or Ecovat® (Benedenbuurt). Both shared an entrepeneurial attitude and are active networkers. Despite these similarities, development in both villages followed different pathways, resulting in different roles for the storage technologies. Benedenbuurt is a typical bottom-up citizens' initiative. Engaged citizens found each other during a sustainability street challenge in 2015. When the sewage pipes in the neighborhood had to be replaced, one creative citizen developed the initial idea to install a district heating network with an Ecovat® and associated system as heat source. He contacted the Wageningen municipality, who were very supportiveof this idea, as it fitted well in their ambitous sustainabilty policy. Because the Housing association owned a substantial number of houses in the neighbourhood (a total of about 450 households), they were asked to join. Soon, a working group was created with representatives of the Wageningen municipality, the housing corporation, the citizens and a representative from a neighbouring energy collective. Simultaneously, in the neighbourhood itself the initiating resident together with a small group of involved residents took the initiative to create an energy cooperative in the neigbourhood. The co-operative Warmtenet Oost Wageningen (WoW) was founded in 2018 and has, in early 2020, about 150 members (one third of the households in Benedenbuurt).

After ample discussion, the working group agreed on the replacement of the gas grid by a collective district heating system, in which Ecovat® was supposed to play a crucial role. Subsequently, various gatherings were organized. The working group visited Ecovat® company several times to see the pilot in Uden and to discuss the option to install a Ecovat® in the neigbourghood. Ecovat® was asked to make a design for an Ecovat® and to present it in the town hall of Wageningen. Ecovat® presented various options for citizens to participate, as a co-owner or shareholder, for instance, and suggested options for the topology of the heating grid. In general, the citizens and local policy makers welcomed the innovative idea of an Ecovat® seasonal storage that used the summer heat to warm dwellings in winter. However, the question of Ecovat®'s suitability in this 1950s neighborhood with the wide spread of low-rise houses and a low degree of heat insulation of the houses was also on the table from the beginning. High-temperature heating and a wide heat infrastructure is not optimal for Ecovat®. A concentration of well insulated houses makes a Ecovat®-based heating system more efficient. For that reason, the working group asked a consultancy to make several scenarios for heating the neighborhood, one of which was Ecovat®. This scenario study made clear that it is very difficult to make scenarios because of many uncertainties. For that reason, it is hard to say which scenario is most risky in terms of finances, but individual heat pumps turned out to be the most expensive. The consultancy found that in terms of sustanaibility Ecovat® is the most attractive option. However, it was concluded that, for this neighborhood, with its widespread, poorly insulated houses, Ecovat® was not the first-choice option. In addition, some inhabitants feared the many heavy trucks needed for transporting the soil out for the construction of the large Ecovat® in the middle of the neigbourhood. After consulting the key members of the working group, the housing corporation, the municipality and the citizens, it was decided to focus on other options, in particular a high-temperature system, heat-cold storage and a central heat pump. This proposal was also presented and accepted in one of the residents' meetings. At that time, it was proposed to start negotiations with various commercial companies, such as Engie, but not all citizens agreed on that, because they were afraid to lose control of the project. To be able to work on the project in a professional way, the Wageningen municipality decided to pay some of the initiators of the project.

Therefore, citizens appeared to be able to organize around a technology, i.e., Ecovat, and to gather relevant knowledge, inspired by some entrepeneurial technology and the interested and passionate citizens. They were able to grow and to involve several other local actors, firstly the housing association and the municipality and to discuss with the Ecovat company, supported by a consultancy. Despite close cooperation between citizens, the municipality and the housing association, the division of responsibilities was not always clear. It took time to found an energy cooperative, to develop a team of paid professionals and to define the different roles of volunteers compared to these professionals. In addition, tensions arose about the inclusion and roles of interested commercial actors, such as traditional energy regime actors. However, close collaboration combined with strong citizen involvement led, in autumn 2018, to a successful application as a 'voorbeeldwijk' (for example, a natural-gas free neighbourhood) for the Dutch policy to stop natural gas heating completely by 2050, involving millions of euros in subsidy [114]. By that time, Ecovat® did not figure anymore in the plans.

In Heeten, a small group of highly involved citizens realized, via the local energy cooperative (Endona) and energy service company (Escozon), various 'big' sustainability projects. One such project is the installation and exploitation (local self-consumption) of a solar PV field in Heeten. Another is an exemption to the Dutch electricity law (e.g., allowing experiments such as local grids and local energy markets). A third project is the Gridflex project that aimed to experiment with local flexibility and a local energy market in a Heeten neigbourhood of Veldegge with 47 relatively new houses. For the Gridflex project, a consortium with the Endona, Escozon, a DSO (Enexis), an ICT company, the University of Twente, Buurkracht (an organization specialized at activating groups of residents for sustainable energy) and DrTen®. The aim was to explore options for grid flexibility options at local level, by using storage with batteries and tge coordination of the use and production of renewable energy among residents. The residents of the neigbourhood were asked to participate in this project by changing their behaviour and allow technical adjustments. In the period 2016–2020, the consortium partners came together regularly to discuss the project progress. One of the main problems was the functioning of the sea-salt battery in real-life conditions. Despite good test results in lab conditions, it took a long time to make the battery function in real-life conditions. In particular, the safety of the battery asked for some discussions, as well as the rate of charging and discharging as grid operators, users, and technology suppliers needed time to set agreements on steering and user protocols. It led to a different version of the battery, the powdered battery (higher energy density) and non-powdered coated version with higher (dis)charging rates. Moreover, linking the DrTen batteries to regular Battery Management System and inverters developed for li-ion batteries was a real issue and took time. This failed linkage caused some tension in the consortium, as the cause of the malfunctioning system was difficult to find. In the end, with half a year prolongation and help of some external experts, the roadmap to integration was found. As a temporary solution, virtual batteries were simulated and, in the last half year, a few Li-ion batteries were also installed in addition to DrTen batteries. DrTen batteries could act on local level soon but more professional integration took more than two years. First systems are running with success since the end of 2019.

Other goals of the Gridflex project, such as gaining insight into the dynamics of local energy markets and the optimization of flexibility in co-ordination with the users faded into the background. The active involvement of the Veldegge residents in performing energy flexibility and experiment with local energy markets was hardly realized. This was especially disappointing for the GridFlex consortium, as Endona and Buurkracht put a lot of effort into succesfully realisering a staggering 100% participation in the Veldegge neighbourhood at the start of the project. The residents had a positive attitude towards the DrTen battery, although other types of batteries were also welcomed. However, the residents had no possibility of controlling the batteries; they only got information on the battery status through the app.

In the end, the project could conclude that peakshaving combined with more self-consumption of self-produced solar leads to 10–20% less cost at the transformator (depending on the reimbursment by the DSO under the experimentation conditions). The 'earned money' was given to the residents and they decided to allocate this money to a community goal, the purchase of a AED for their neighborhood. For DrTen® this pilot in the end was fruitful as they learned a lot about aligning the new battery to existing and available control technology. The ongoing discussions on 'false' expectations and the discongruent definitions of what a 'working battery' entails, further highlighted the importance of socio-technical alignment.

Besides this choice for a particular technology, the problems with technical alignment and the relatively small role of citizens, we observed that, just as in the Benedenbuurt case, tensions arose in the Heeten case about the inclusion and roles of interested commercial actors and traditional energy regime actors, because of conflicting visions (e.g., net-zero behind the transformer, see Table 1). In contrast to the Benedenbuurt case, the local government was not included, which may reduce its moral, political and financial involvement and responsibility.

4.5. Socio-Technical Alignment Dynamics

In both energy storage technologies, socio-technical alignment dynamics have been observed. The innovator of DrTen® was looking for a cheap and environmentally friendly way to store energy to provide affordable energy access in Africa. Accordingly, certain values such as environmental friendliness, safety and afforability are already embedded in the design of DrTen® batteries. At the same time, DrTen® batteries are engaged in several pilots and research projects such as gridflex Heeten as well as the Germany–Netherland cross-border project (INTERREG) [89,115]. In the gridflex project, DrTen® had to align interoperability issues with other energy management systems, battery management systems and inverter technologies developers. Moreover, to improve the charging and discharging rate, the cell configuration of the batteries had to be changed and powdered DrTen® had to be developed. In this process, the safety standard for residential use through DNV-GL was also obtained. DrTen® also obtained membership of Dutch industry association, EnergystorageNL, which is currently lobbying for better regulation for energy storage in the Netherlands [116].

Environmental values such as sustainable and reliable heating and cooling are embedded in the design of Ecovat®. In several engagements of Ecovat® in pilot projects (e.g., Ons Dorp), feasibility studies (e.g., Benedenbuurt) and reports are observed. A working prototype of small Ecovat® has been successsfully tested in Uden. Technical innovation based on this demonstration includes improved construction methods, roofs as well as a hybrid system with a traditional exchanger for peak demand. For logistical reasons, manufacturing has been moved from Uden to Oss, close to the waterways. Given the very high upfront costs, Ecovat® developed a total systems concept including financing, energy management systems (ecovat software 2.0), and operation and distribution through subsidaries. To avoid the very long time needed for planning approvial (approximately 40 weeks), Ecovat® strategically managed to be included in the new urban master planning of Hague.

Despite their potential, both technologies have to overcome some problems to improve their alignment with societal actors, structures and processes. Ecovat® was welcomed enthusiastically in Benedenbuurt. However, Ecovat® could not meet the requirements of the local stakeholders here, who asked for a high-temperature heating system. In addition, several citizens feared the construction activities and some labeled Ecovat as a "logistic nightmare". In Arnhem, where Ecovat® succesfully negotiated with a large care institution, the municipality and many other relevant stakeholders, and received permission to place a Ecovat at the care institution, the court decided to forbid the chosen transport route of trucks through their neighborhood. A group of local residents feared burdening and unsafe transport of the approximately 5000 m^3 of soil by heavy trucks [117]. An alternative transport route was available, but not put into practice, as the commissioner stopped the whole project.

Ecovat® is well aligned to existing technological systems. The Ecovat® system can be connected to all kinds of heat, information and electricity systems. The only problem is the localisation and design of buildings, in particular in old neighborhoods. Old and widespread small buildings are difficult to align with Ecovat®, compared to the green field of compact, concentrated new buildings. It is, however, possible to use Ecovat® in a high-temperature district heating network (thus avoiding investments in the renovations/insulation of households) but this will be expensive and less sustainable. This is also related to the design of the Ecovat® configuration as a whole: the interwovenness of size, efficiency and logistics. The ecovat® system wins a great deal of cost-effiency by increasing the size. There is a tendency to increase the minimum size too (in 2016: 15 m diameter and 15 m deep) In 2020: 30 m diameter and 30 m deep). This too implies a huge increase in transport and logistic needs during the construction period, which clearly can raise strong objections from local residents. Although Ecovat® tries to limit the nuisance for the neighborhood, e.g., by prefabrication of elements, this aspect is likely to remain a sensitive element in aliging dynamics as it can easily invoke resistance, especially in residential areas. A continuous sharp eye and ear is crucial in the alignment strategy.

DrTen® was welcomed by local citizens. Here, too, citizens are, in general, positive about its environmental friendliness. Ownership is not a problem; batteries can be owned both by individuals as well as communities. The local energy community in Heeten highly values and stimulates local ownership of the local energy infrastrucutres. DrTen batteries still face interoperability challenges with existing technological systems, but progess is being made lately. The specific battery characteristics cause difficulties in the integration of existing smart steering technology of the batteries at first.

Because of the abandoning of Dutch natural gas, heat production will increasingly be electrified. The rising share of green electricity production implies rising seasonal gaps, which probably will make Ecovat® more profitable. It is not certain how markets and the regulation of markets will develop, however [118]. Crucial in Ecovat's business model is the long-term availability of cheap surplus wind and solar energy, which will eventually outweigh the high investment costs. DrTen® can also be more profitable in future. Now, electricity storage is financially not interesting due to FIT regulation, but this regulation is being gradually phased out in the period 2021–2030.

Both Ecovat® and DrTen® also claim avoided future network costs for DSOs and TSOs, as less grid enforcement is needed in the case of the widespread implementation of (seasonal) local storage. However, the storage costs are made by the local actors and citizen communities, and there is no clear regulation how distributed (future) profits and costs will be aligned to different stakeholders.

5. Conclusions and Discussion

There are techno-economic as well as socio-institutional challenges for implementing innovative community energy storage technologies in the energy system. In community energy storage, both technical and social innovation go hand in hand. The dynamics of interaction between the actors and technological innovation processes in community energy storage makes its implementation complex.

In the process of the adoption and use of energy storage technologies in local energy communities, new user-inspired innovations are possible. Such innovation can be in the governance and operation of the energy storage system or on further technological improvement based on the feedback of users and other stakeholders. A careful alignment of technical–technical, socio-technical, and social–social articulation is required for the successful integration of community energy storage in the energy system. Socio-technical alignment is critical, as technology shapes the society, and society in turn influences the technology development. Enabling regulatory, policy and market environment are also important for the socio-technical alignment of innovative energy storage technologies. A level playing field can be created for energy storage, for example, by removing the double taxes, as storage is still considered as a load, as well as by a fair costs and benefits allocation of avoided network costs due to energy storage among consumers/prosumers and network system operators.

Our cases showed that, in different contexts, and regarding different technologies, problems with both technological and social alignment arise. Both technologies are promising in terms of sustainability and locality. However, both technologies faced resistance or problems, sometimes unexpectedly. An RRI analysis of DrTen® and Ecovat® demonstrated several technological-economic, regulatory and social challenges and requirements at various levels. For all these aspects and levels, most actors are in the first stage of a learning process.

Ecovat® is a well developed technology and fits well in existing technological systems. However, it has been difficult to implement it to date, because of high investment costs, some unclarity or uncertainty on participation options and storage policy in the long-run and the duration and thoroughness of the construction work. Modification of the technology, more options for participation and early negotiations regarding the means of construction could improve alignment. As we have seen, however, stable social and governmenental configuration at a local level are required to enable these types of technology and to organize learning processes.

DrTen® sea salt batteries gave good results in first peak-shaving experiments (both consumption and solar-PV generation peaks) but integrationin existing technological systems took longer time. This was one of the main reasons that implementation of only few systems was possible though improvements are expected.Yet, the pilot project in Heeten learned DrTen and other project actors a lot about any technical misalignment. The technology is flexible and easy to implement at household and neighbourhood levels. Making the battery part of a local flexible energy use-production-storage system requires new governance models and learning processes at the local level. Regulation and certainity on prices are crucial contextual factors for further development.

In our view of community energy storage cases as responsible inclusive innovations, we made some interesting observations (tensions) in socio-technical alignment. First, radical innovations are more likely from a non-regime actor, new actors in the energy system. In the case of Heteen, these new actors were the energy service company Escozon, energy co-operative Endona, technology proivders DrTen® and ICT. In the case of Benedenbuurt, local energy co-operative and Ecovat® were the new actors. Second, radical innovation is also about empowering and engaging citizens and user communities in the process, thereby creating new socio-technical configurations. For example, in the case of the Heteen battery, control was not allowed, but it gave the neighborhood full decision on financial benefits allocation, which was a result of network costs reductions due to peak shaving in the neighbourhood microgrid. For Benedenbuurt, participaton options for an Ecovat® were given, but the Benedenbuurt energy co-operative/residents were, at the beginning, too inexperienced to handle such big upfront investment costs (EUR 3.5 million excluding the costs of a district heating network). However, such a capability was quickly developed with the help of government subsidy for a natural gas-free neighbourhood pilot. Both cases show that, with the availability and support of energy service companies (third-party experts), the local energy initiatives can grow to ownership and exploitation.

The relative newcomers Ecovat and DrTen introduced interesting and promising sustainable technologies, which may help to solve energy storage problems at a local scale. Both companies have been able to build networks around their technology, including energy cooperatives, large companies, municipalities and DSOs. For that reason, both have been able to further improve their technology. The involved energy cooperatives have been heavily involved in the pilots around these technologies. This resulted in empowerment because they were seen as interesting partners, which could help to test and develop new technologies, and because they developed as experts in local energy systems. For that reason, energy cooperatives may be a stimulating factor for social-technological innovations and in the energy transition (see [38]), but this requires quite a long period of involvement, patience and capacity-building to be able to co-create technologies which go along with relatively high investments in infrastructure or interoperability with existing technologies.

Author Contributions: Conceptualization, B.P.K., E.v.O. and H.v.d.W.; methodology, B.P.K., E.v.O. and H.v.d.W.; writing—original draft preparation, B.P.K.; writing—review and editing, B.P.K., E.v.O. and H.v.d.W.; visualization, B.P.K.; supervision, E.v.O. and H.v.d.W.; project administration, H.v.d.W.; funding acquisition, E.v.O. and H.v.d.W. All authors have read and agreed to the published version of the manuscript.

Funding: This research is part of the Community Responsible Innovation in Sustainable Energy (CO- RISE) project [29] and is funded through the social responsible innovation program of The Netherlands Organization for Scientific Research, the Netherlands. Grant number: NWO-MVI 2016[313-99-304].

Acknowledgments: We are thankful to the Gridflex andBenedenbuurt project consortium for the possibilities to take part in all the meetings, their openness to our questions and access to the project documentation. In addition, we are grateful to the studied SME's Ecovat and Dr Ten for providing open and transparent project information.

Conflicts of Interest: The authors declare no conflict of interest. The funders had no role in the design of the study; in the collection, analyses, or interpretation of data; in the writing of the manuscript, or in the decision to publish the results.

References

1. Kelly, S.; Pollitt, M.G. The local dimension of energy. In *The Future of Electricity Demand: Customers, Citizens and Loads*; Pollitt, M.G., Ed.; Cambridge University Press: Cambridge, UK, 2011; pp. 249–275.

2. Vargas, M.; Davis, G. *World Energy Scenarios 2016*; World Energy Council: London, UK, 2016.

3. Koirala, B.P.; Koliou, E.; Friege, J.; Hakvoort, R.A.; Herder, P.M. Energetic communities for community energy: A review of key issues and trends shaping integrated community energy systems. *Renew. Sustain. Energy Rev.* **2016**, *56*, 722–744. [CrossRef]

4. Devine-Wright, P. Community versus local energy in a context of climate emergency. *Nat. Energy* **2019**, *4*, 894–896. [CrossRef]

5. Koirala, B.P.; van Oost, E.; van der Windt, H. Community energy storage: A responsible innovation towards a sustainable energy system? *Appl. Energy* **2018**, *231*, 570–585. [CrossRef]

6. Kyriakopoulos, G.L.; Arabatzis, G. Electrical energy storage systems in electricity generation: Energy policies, innovative technologies, and regulatory regimes. *Renew. Sustain. Energy Rev.* **2016**, *56*, 1044–1067. [CrossRef]

7. Manfren, M.; Caputo, P.; Costa, G. Paradigm shift in urban energy systems through distributed generation: Methods and models. *Appl. Energy* **2011**, *88*, 1032–1048. [CrossRef]

8. Koirala, B.P.; Araghi, Y.; Kroesen, M.; Ghorbani, A.; Hakvoort, R.A.; Herder, P.M. Trust, awareness, and independence: Insights from a socio-psychological factor analysis of citizen knowledge and participation in community energy systems. *Energy Res. Soc. Sci.* **2018**, *38*, 33–40. [CrossRef]

9. Kalkbrenner, B.J.; Roosen, J. Citizens' willingness to participate in local renewable energy projects: The role of community and trust in Germany. *Energy Res. Soc. Sci.* **2016**, *13*, 60–70. [CrossRef]

10. Van der Waal, E.C.; Das, A.M.; van der Schoor, T. Participatory Experimentation with Energy Law: Digging in a 'Regulatory Sandbox' for Local Energy Initiatives in the Netherlands. *Energies* **2020**, *13*, 458. [CrossRef]

11. Gregg, J.S.; Nyborg, S.; Hansen, M.; Schwanitz, V.J.; Wierling, A.; Zeiss, J.P.; Delvaux, S.; Saenz, V.; Polo-Alvarez, L.; Candelise, C.; et al. Collective Action and Social Innovation in the Energy Sector: A Mobilization Model Perspective. *Energies* **2020**, *13*, 651. [CrossRef]

12. Horstink, L.; Wittmayer, J.M.; Ng, K.; Luz, G.P.; Marín-González, E.; Gährs, S.; Campos, I.; Holstenkamp, L.; Oxenaar, S.; Brown, D. Collective Renewable Energy Prosumers and the Promises of the Energy Union: Taking Stock. *Energies* **2020**, *13*, 421. [CrossRef]

13. Kloppenburg, S.; Smale, R.; Verkade, N. Technologies of Engagement: How Battery Storage Technologies Shape Householder Participation in Energy Transitions. *Energies* **2019**, *12*, 4384. [CrossRef]

14. Dilger, M.G.; Konter, M.; Voigt, K.-I. Introducing a co-operative-specific business model: The poles of profit and community and their impact on organizational models of energy co-operatives. *J. CoOper. Organ. Manag.* **2017**, *5*, 28–38. [CrossRef]

15. EU Clean Energy for All Europeans—Energy—European Commission. Available online: /energy/en/topics/energy-strategy-and-energy-union/clean-energy-all-europeans (accessed on 24 July 2019).

16. Van der Schoor, T.; Scholtens, B. Power to the people: Local community initiatives and the transition to sustainable energy. *Renew. Sustain. Energy Rev.* **2015**, *43*, 666–675. [CrossRef]

17. Lowitzsch, J. Consumer Stock Ownership Plans (CSOPs)—The Prototype Business Model for Renewable Energy Communities. *Energies* **2019**, *13*, 118. [CrossRef]

18. *RED II Directive (EU) 2018/2001 of the European Parliament and of the Council of 11 December 2018 on the Promotion of the Use of Energy from Renewable Sources*; Publications office of the European Union: Luxembourg, 2018.

19. *EMD II Directive (EU) 2019/944 of the European Parliament and of the Council of 5 June 2019 on Common Rules for the Internal Market for Electricity and Amending Directive 2012/27/EU*; Publications office of the European Union: Luxembourg, 2019.

20. Koirala, B.P. Integrated Community Energy Systems. Ph.D. Thesis, Delft University of Technology, Delft, The Netherlands, 2017.

21. Parra, D.; Swierczynski, M.; Stroe, D.I.; Norman, S.A.; Abdon, A.; Worlitschek, J.; O'Doherty, T.; Rodrigues, L.; Gillott, M.; Zhang, X.; et al. An interdisciplinary review of energy storage for communities: Challenges and perspectives. *Renew. Sustain. Energy Rev.* **2017**, *79*, 730–749. [CrossRef]

22. Koirala, B.P.; van Oost, E.; van der Windt, H. *Socio-Technical Innovation Dynamics in Community Energy Storage*; CESUN Conference: Tokyo, Japan, 2018; pp. 1–8.

23. Maclaine Pont, P.; van Est, R.; Deuten, J. *Shaping Socio-Technical Innovation through Policy*; Rathenau Institute: Den Hague, The Netherlands, 2016.

24. *EnergystorageNL National Actionplan on Energy Storage and Conversion*; ESNL: Zoetermeer, The Netherlands, 2019.

25. *IRENA Renewables and Electricity Storage: A Technology Roadmap for REMAP 2030*; International Renewable Energy Agency: Abu Dhabi, UAE, 2015.

26. De Torres, P.M.A. An overview on strategic design for socio-technical innovation. *Strateg. Des. Res. J.* **2018**, *11*, 186–192. [CrossRef]

27. Konrad, K.; Böhle, K. Socio-technical futures and the governance of innovation processes—An introduction to the special issue. *Futures* **2019**, *109*, 101–107. [CrossRef]

28. Dong, S.; Kremers, E.; Brucoli, M.; Rothman, R.; Brown, S. Techno-enviro-economic assessment of household and community energy storage in the UK. *Energy Convers. Manag.* **2020**, *205*, 112330. [CrossRef]

29. Koirala, B.P.; van Oost, E. Local Energy Communities: Responsible Innovation Towards Sustainable Energy (CO-RISE). In *Proceedings of the CTIT Symposium 2017: IoT is Ready. What about Us*; University of Twente: Enschede, The Netherlands, 2017.

30. Koirala, B.P.; Hakvoort, R.A.; van Oost, E.C.; van der Windt, H.J. Community Energy Storage: Governance and Business Models. In *Consumer, Prosumer, Prosumager*; Academic Press: Cambridge, MA, USA, 2019; pp. 209–234. ISBN 978-0-12-816835-6.

31. Garud, R.; Karnøe, P. Bricolage versus breakthrough: Distributed and embedded agency in technology entrepreneurship. *Res. Policy* **2003**, *32*, 277–300. [CrossRef]

32. Ornetzeder, M.; Rohracher, H. User-led innovations and participation processes: Lessons from sustainable energy technologies. *Energy Policy* **2006**, *34*, 138–150. [CrossRef]

33. Ornetzeder, M.; Rohracher, H. Of solar collectors, wind power, and car sharing: Comparing and understanding successful cases of grassroots innovations. *Glob. Environ. Chang.* **2013**, *23*, 856–867. [CrossRef]

34. Seyfang, G.; Park, J.J.; Smith, A. A thousand flowers blooming? An examination of community energy in the UK. *Energy Policy* **2013**, *61*, 977–989. [CrossRef]

35. Bomberg, E.; McEwen, N. Mobilizing community energy. *Energy Policy* **2012**, *51*, 435–444. [CrossRef]

36. Seyfang, G.; Hielscher, S.; Hargreaves, T.; Martiskainen, M.; Smith, A. A grassroots sustainable energy niche? Reflections on community energy in the UK. *Environ. Innov. Soc. Transit.* **2014**, *13*, 21–44. [CrossRef]

37. Hargreaves, T.; Hielscher, S.; Seyfang, G.; Smith, A. Grassroots innovations in community energy: The role of intermediaries in niche development. *Glob. Environ. Chang.* **2013**, *23*, 868–880. [CrossRef]

38. Van der Waal, E.; van der Windt, H.; van Oost, E. How Local Energy Initiatives Develop Technological Innovations: Growing an Actor Network. *Sustainability* **2018**, *10*, 4577. [CrossRef]
39. Verbeek, P.-P. Materializing Morality: Design Ethics and Technological Mediation. *Sci. Technol. Hum. Values* **2006**, *31*, 361–380. [CrossRef]
40. Winner, L. Do artefacts have politics? *Daedalus* **1980**, *109*, 121–136.
41. Owen, R.; Stilgoe, J.; Macnaghten, P.; Gorman, M.; Fisher, E.; Guston, D. A Framework for Responsible Innovation. In *Responsible Innovation*; Owen, R., Bessant, J., Heintz, M., Eds.; John Wiley & Sons, Ltd.: Chichester, UK, 2013; pp. 27–50. ISBN 978-1-118-55142-4.
42. Stilgoe, J.; Owen, R.; Macnaghten, P. Developing a framework for responsible innovation. *Res. Policy* **2013**, *42*, 1568–1580. [CrossRef]
43. Timmermans, J. Mapping the RRI Landscape: An Overview of Organisations, Projects, Persons, Areas and Topics. In *Responsible Innovation 3*; Asveld, L., van Dam-Mieras, R., Swierstra, T., Lavrijssen, S., Linse, K., van den Hoven, J., Eds.; Springer International Publishing: Cham, Switzerland, 2017; pp. 21–47. ISBN 978-3-319-64833-0.
44. Rogers, J.C.; Simmons, E.A.; Convery, I.; Weatherall, A. Public perceptions of opportunities for community-based renewable energy projects. *Energy Policy* **2008**, *36*, 4217–4226. [CrossRef]
45. Walker, G.; Devine-Wright, P.; Barnett, P.; Burningham, K.; Cass, N.; Devine-Wright, H.; Speller, G.; Barton, J.; Evans, B.; Heath, Y. Symmetries, expectations, dynamics and contexts: A framework for understanding public engagement with renewable energy projects. In *Renewable Energy and the Public: From Nimby to Participation*; Earthscan; Routledge; Taylor & Francis: London, UK, 2010; pp. 2–14. ISBN 1844078639.
46. Walker, G. What are the barriers and incentives for community-owned means of energy production and use? *Energy Policy* **2008**, *36*, 4401–4405. [CrossRef]
47. Walker, G.; Devine-Wright, P. Community renewable energy: What should it mean? *Energy Policy* **2008**, *36*, 497–500. [CrossRef]
48. Walker, G.; Simcock, N. Community Energy Systems. *Int. Encycl. Hous. HomeElsevier* **2012**, *1*, 194–198.
49. Wirth, S. Communites matter: Institutional preconditions for community renewable energy. *Energy Policy* **2014**, 1–11.
50. Koirala, B.; Chaves Ávila, J.; Gómez, T.; Hakvoort, R.; Herder, P. Local Alternative for Energy Supply: Performance Assessment of Integrated Community Energy Systems. *Energies* **2016**, *9*, 981. [CrossRef]
51. Koirala, B.; Hakvoort, R. Integrated Community-Based Energy Systems: Aligning Technology, Incentives, and Regulations. In *Innovation and Disruption at the Grid's Edge*; Academic Press: Cambridge, MA, USA, 2017; pp. 363–387. ISBN 978-0-12-811758-3.
52. Koirala, B.P.; Hakvoort, R.A.; Ávila, J.P.C.; Gómez, T. Assessment of Integrated Community Energy Systems. In Proceedings of the 2016 13th International Conference on the European Energy Market (EEM), Porto, Portugal, 6–9 June 2016; pp. 1–6.
53. Acosta, C.; Ortega, M.; Bunsen, T.; Koirala, B.; Ghorbani, A. Facilitating Energy Transition through Energy Commons: An Application of Socio-Ecological Systems Framework for Integrated Community Energy Systems. *Sustainability* **2018**, *10*, 366. [CrossRef]
54. Caramizaru, A.; Uihlein, A.; European Commission; Joint Research Centre. *Energy Communities: An Overview of Energy and Social Innovation*; Publications Office of the European Union: Luxembourg, 2020; ISBN 978-92-76-10713-2.
55. Van der Schoor, T.; Scholtens, B. *Scientific Approaches of Community Energy: A Literature Review*; Centre for Energy Economics Research (CEER): Groningen, The Netherlands, 2019; ISBN 978-94-034-1657-1.
56. Van der Schoor, T.; van Lente, H.; Scholtens, B.; Peine, A. Challenging obduracy: How local communities transform the energy system. *Energy Res. Soc. Sci.* **2016**, *13*, 94–105. [CrossRef]
57. Mendes, G.; Ioakimidis, C.; Ferrão, P. On the planning and analysis of Integrated Community Energy Systems: A review and survey of available tools. *Renew. Sustain. Energy Rev.* **2011**, *15*, 4836–4854. [CrossRef]
58. Romero-Rubio, C.; de Andrés Díaz, J.R. Sustainable energy communities: A study contrasting Spain and Germany. *Energy Policy* **2015**, *85*, 397–409. [CrossRef]
59. De Vries, G.W.; Boon, W.P.C.; Peine, A. User-led innovation in civic energy communities. *Environ. Innov. Soc. Transit.* **2016**, *19*, 51–65. [CrossRef]

60. Moroni, S.; Alberti, V.; Antoniucci, V.; Bisello, A. Energy Communities in a Distributed-Energy Scenario: Four Different Kinds of Community Arrangements. In *Smart and Sustainable Planning for Cities and Regions*; Bisello, A., Vettorato, D., Laconte, P., Costa, S., Eds.; Springer International Publishing: Cham, Switzerland, 2018; pp. 429–437. ISBN 978-3-319-75773-5.

61. Gui, E.M.; MacGill, I. Typology of future clean energy communities: An exploratory structure, opportunities, and challenges. *Energy Res. Soc. Sci.* **2018**, *35*, 94–107. [CrossRef]

62. Espe, E.; Potdar, V.; Chang, E. Prosumer Communities and Relationships in Smart Grids: A Literature Review, Evolution and Future Directions. *Energies* **2018**, *11*, 2528. [CrossRef]

63. Barbour, E.; Parra, D.; Awwad, Z.; González, M.C. Community energy storage: A smart choice for the smart grid? *Appl. Energy* **2018**, *212*, 489–497. [CrossRef]

64. Van der Stelt, S.; AlSkaif, T.; van Sark, W. Techno-economic analysis of household and community energy storage for residential prosumers with smart appliances. *Appl. Energy* **2018**, *209*, 266–276. [CrossRef]

65. Roberts, B.P.; Sandberg, C. The Role of Energy Storage in Development of Smart Grids. *Proc. IEEE* **2011**, *99*, 1139–1144. [CrossRef]

66. Jolivet, E.; Heiskanen, E. Blowing against the wind—An exploratory application of actor network theory to the analysis of local controversies and participation processes in wind energy. *Energy Policy* **2010**, *38*, 6746–6754. [CrossRef]

67. Hargreaves, T. Practice-ing behaviour change: Applying social practice theory to pro-environmental behaviour change. *J. Consum. Cult.* **2011**, *11*, 79–99. [CrossRef]

68. Aune, M.; Godbolt, Å.L.; Sørensen, K.H.; Ryghaug, M.; Karlstrøm, H.; Næss, R. Concerned consumption. Global warming changing household domestication of energy. *Energy Policy* **2016**, *98*, 290–297. [CrossRef]

69. Devine-Wright, P.; Batel, S.; Aas, O.; Sovacool, B.; Labelle, M.C.; Ruud, A. A conceptual framework for understanding the social acceptance of energy infrastructure: Insights from energy storage. *Energy Policy* **2017**, *107*, 27–31. [CrossRef]

70. Jalas, M.; Hyysalo, S.; Heiskanen, E.; Lovio, R.; Nissinen, A.; Mattinen, M.; Rinkinen, J.; Juntunen, J.K.; Tainio, P.; Nissilä, H. Everyday experimentation in energy transition: A practice-theoretical view. *J. Clean. Prod.* **2017**, *169*, 77–84. [CrossRef]

71. Ryghaug, M.; Toftaker, M. A Transformative Practice? Meaning, Competence, and Material Aspects of Driving Electric Cars in Norway. *Nat. Cult.* **2014**, *9*, 146–163. [CrossRef]

72. Geels, F.W.; Schot, J. Typology of sociotechnical transition pathways. *Res. Policy* **2007**, *36*, 399–417. [CrossRef]

73. Geels, F.W. Ontologies, socio-technical transitions (to sustainability), and the multi-level perspective. *Res. Policy* **2010**, *39*, 495–510. [CrossRef]

74. Markard, J.; Hekkert, M.; Jacobsson, S. The technological innovation systems framework: Response to six criticisms. *Environ. Innov. Soc. Transit.* **2015**, *16*, 76–86. [CrossRef]

75. Schot, J.; Geels, F.W. Strategic niche management and sustainable innovation journeys: Theory, findings, research agenda, and policy. *Technol. Anal. Strateg. Manag.* **2008**, *20*, 537–554. [CrossRef]

76. Von Schomberg, R. A Vision of Responsible Research and Innovation. In *Responsible Innovation*; Owen, R., Bessant, J., Heintz, M., Eds.; John Wiley & Sons, Ltd.: Chichester, UK, 2013; pp. 51–74. ISBN 978-1-118-55142-4.

77. Schot, J. Constructive Technology Assessment. Available online: https://www.encyclopedia.com/science/encyclopedias-almanacs-transcripts-and-maps/constructive-technology-assessment (accessed on 28 February 2019).

78. De Hoop, E.; Pols, A.; Romijn, H. Limits to responsible innovation. *J. Responsible Innov.* **2016**, *3*, 110–134. [CrossRef]

79. Akrich, M. The de-scription of technical objects. In *Shaping Technology/Building Society: Studies in Sociotechnical Change*; Bijker, W.E., Law, J., Eds.; The MIT Press: Cambridge, MA, USA; London, UK, 1992; pp. 205–224.

80. Crettenand, N.; Finger, M. The Alignment between Institutions and Technology in Network Industries. *Compet. Regul. Netw. Ind.* **2013**, *14*, 106–129. [CrossRef]

81. Bauwens, T. Socio-Technical Lock-in and the Alignment Framework: The Case of Distributed Generation Technologies. *Compet. Regul. Netw. Ind.* **2015**, *16*, 155–181. [CrossRef]

82. Bowman, D.M.; Stokes, E.; Rip, A. (Eds.) *Embedding New Technologies into Society: A Regulatory, Ethical and Societal Perspective*; Pan Stanford Publishing: Singapore, 2017; ISBN 978-981-4745-74-1.

83. Schot, J.; Rip, A. The past and future of constructive technology assessment. *Technol. Forecast. Soc. Chang.* **1997**, *54*, 251–268. [CrossRef]

84. Unruh, G.C. Escaping carbon lock-in. *Energy Policy* **2002**, *30*, 317–325. [CrossRef]
85. Yin, R.K. *Case Study Research: Design and Methods*, 5th ed.; SAGE: Los Angeles, CA, USA, 2014; ISBN 978-1-4522-4256-9.
86. CTA-Toolbox CTA Toolbox. Available online: https://cta-toolbox.nl/ (accessed on 2 December 2019).
87. DrTen Seasalt Battery. Available online: http://www.drten.nl/zeezout-batterij/?larg=en (accessed on 26 July 2017).
88. Ecovat Ecovat®. Available online: http://www.ecovat.eu/?lang=en (accessed on 4 July 2017).
89. Gridflex GridFlex. 2018. Available online: http://gridflex.nl/ (accessed on 8 June 2020).
90. WOW District Heating Co-Operative in East Wageningen. Available online: https://cooperatiewow.nl/ (accessed on 2 December 2019).
91. De Goeijen, G.; Smit, G.; Hurink, J. An Integer Linear Programming Model for an Ecovat Buffer. *Energies* **2016**, *9*, 592. [CrossRef]
92. Goeijen, G.; Smit, G.; Hurink, J. Improving an Integer Linear Programming Model of an Ecovat Buffer by Adding Long-Term Planning. *Energies* **2017**, *10*, 2039. [CrossRef]
93. De Goeijen, G.J.H.; Hurink, J.L.; Smit, G.J.M. A Heuristic Approach to Control the Ecovat System. In Proceedings of the 2018 IEEE PES Innovative Smart Grid Technologies Conference Europe (ISGT-Europe), Sarajevo, Bosnia-Herzegovina, 21–25 October 2018; pp. 1–6.
94. De Goeijen, G.J.H.; Hoogsteen, G.; Hurink, J.L.; Smit, G.J.M. Using the Ecovat system to supply the heat demand of a neighbourhood. In Proceedings of the 2019 IEEE Milan PowerTech, Milan, Italy, 23–27 June 2019; pp. 1–6.
95. De Groot, A.W. Wall Part, Heat Buffer and Energy Exchange System. U.S. Patent No. 10,024,549, 17 July 2018.
96. De Groot, A.W. Underground Thermal Energy Storage. U.S. Patent Application No. 15/829,854, 29 March 2018.
97. Ephrati, S.; Jonker, H. A Numerical Study of the Ecovat with IFISS. Bachelor's Thesis, University of Twente, Enschede, The Netherlands, 2017.
98. Van den Bosch, R. Modelling the Effects of Different Renovation Scenarios of Apartments on the Configuration of the Ecovat Energy Storage System Finding the Optimal Economic Combination of the Ecovat System and Renovation. Master Thesis, Eindhoven University of Technology, Eindhoven, The Netherlands, 2015.
99. De Goeijen, G. Developing a Method for the Operational Control of an Ecovat System. Ph.D. Thesis, University of Twente, Enschede, The Netherlands, 2019.
100. Quintero Pulido, D.; Hoogsteen, G.; ten Kortenaar, M.; Hurink, J.; Hebner, R.; Smit, G. Characterization of Storage Sizing for an Off-Grid House in the US and the Netherlands. *Energies* **2018**, *11*, 265. [CrossRef]
101. Quintero Pulido, D.F.; Ten Kortenaar, M.V.; Hurink, J.L.; Smit, G.J.M. The Role of Off-Grid Houses in the Energy Transition with a Case Study in the Netherlands. *Energies* **2019**, *12*, 2033. [CrossRef]
102. Homan, B.; Smit, G.J.M.; van Leeuwen, R.P.; ten Kortenaar, M.V.; Ten, B.V. A comprehensive model for battery State of Charge prediction. In Proceedings of the 2017 IEEE Manchester PowerTech, Manchester, UK, 18–22 June 2017; pp. 1–6.
103. Reijnders, V.; Gerard, M.; Smit, G.; Hurink, J. Testing Grid-Based Electricity Prices and Batteries in a Field Test. In *CIRED 2018 Ljubljana Workshop on Microgrids and Local Energy Communities*; [0500] (CIRED Workshop Series); CIRED: Ljubljana, Slovenia, 2018.
104. Quintero Pulido, D.F. Energy Storage Technologies for Off-grid Houses. Ph.D. Thesis, University of Twente, Enschede, The Netherlands, 2019.
105. Whittingham, M.S. Materials Challenges Facing Electrical Energy Storage. *Mrs Bull.* **2008**, *33*, 411–419. [CrossRef]
106. Liu, J. Addressing the Grand Challenges in Energy Storage. *Adv. Funct. Mater.* **2013**, *23*, 924–928. [CrossRef]
107. Koohi-Fayegh, S.; Rosen, M.A. A review of energy storage types, applications and recent developments. *J. Energy Storage* **2020**, *27*, 101047. [CrossRef]
108. Jager-Waldau, A.; Lindahl, J.; Heilscher, G.; Kraiczy, M.; Masson, G.; Mather, B.; Mayr, C.; Moneta, D.; Mugnier, D.; Nikoletatos, J.; et al. Electricity produced from photovoltaic systems in apartment buildings and self-consumption: Comparison of the situation in various IEA PVPS countries. In Proceedings of the 2019 IEEE 46th Photovoltaic Specialists Conference (PVSC), Chicago, IL, USA, 16–21 June 2019; pp. 1701–1710.
109. MI Mission Innovation. Available online: http://mission-innovation.net/ (accessed on 13 February 2020).

110. DeCOOBRAA De COOBRAA: Sustainable Smart Grids Solutions. Available online: http://decoobraa.com/ (accessed on 25 April 2020).

111. B-Corp Certified B Corporation. Available online: https://bcorporation.net/about-b-corps (accessed on 27 May 2020).

112. Van Gastel, E. Ecovat Geeft voor 2,3 Miljoen euro Certificaten van Aandelen uit via NPEX-Effectenbeurs, Smart Storage Magazine. Available online: https://smartstoragemagazine.nl/nieuws/i18306/ecovat-geeft-voor-2-3-miljoen-euro-certificaten-van-aandelen-uit-via-npex-effectenbeurs (accessed on 25 April 2020).

113. Warnaars, J.; Kooiman, A.; den Ouden, B. *System Consequences of Ecovat: Quantification of Avoided Costs for Grid Reinforcement and Peak Power Plants*; Berenschot: Utrecht, The Netherlands, 2018; p. 15.

114. Rijksoverheid 120 Miljoen Euro Voor 'Proeftuinen' Aardgasvrije Wijken in 27 Gemeenten. Available online: https://www.rijksoverheid.nl/actueel/nieuws/2018/10/01/120-miljoen-euro-voor-\T1\textquoteleftproeftuinen\T1\textquoteright-aardgasvrije-wijken-in-27-gemeenten (accessed on 25 April 2020).

115. INTERREG INTERREG Germany Netherlands. Available online: https://www.deutschland-nederland.eu/en/ (accessed on 12 February 2020).

116. EnergystorageNL. Available online: https://www.energystoragenl.nl/ (accessed on 13 February 2020).

117. Van der Ploeg, H. *Arnhemse Wijk Heijenoord in Actie Tegen Vrachtverkeer Voor Aanleg 'Ecovat'*; de Gelderlander: Arnehm, The Netherlands, 2019.

118. Meulenkamp, E.; van Bree, T.; Geurts, A. *Verkenning Batterijen 2: Positie NL in de Waardeketen*; TNO: The Hague, The Netherlands, 2019.

energies

Article

Formation and Continuation of Thermal Energy Community Systems: An Explorative Agent-Based Model for the Netherlands

Javanshir Fouladvand [1,*], Niek Mouter [2], Amineh Ghorbani [1] and Paulien Herder [3] 

[1] Energy and Industry Section, Engineering Systems and Services Department, Technology, Policy and Management Faculty, Delft University of Technology (TU Delft), 2628 BX Delft, The Netherlands; A.Ghorbani@tudelft.nl

[2] Transports and Logistics, Engineering Systems and Services Department, Technology, Policy and Management Faculty, Delft University of Technology (TU Delft), 2628 BX Delft, The Netherlands; N.Mouter@tudelft.nl

[3] Process and Energy Department, Mechanical, Maritime and Materials Engineering Faculty, Delft University of Technology (TU Delft), 2628 CB Delft, The Netherlands; P.M.Herder@tudelft.nl

* Correspondence: j.fouladvand@tudelft.nl; Tel.: +31-(0)6-84-08-52-68

Received: 17 March 2020; Accepted: 26 May 2020; Published: 2 June 2020

Abstract: Energy communities are key elements in the energy transition at the local level as they aim to generate and distribute energy based on renewable energy technologies locally. The literature on community energy systems is dominated by the study of electricity systems. Yet, thermal energy applications cover 75% of the total energy consumption in households and small businesses. Community-driven initiatives for local generation and distribution of thermal energy, however, remain largely unaddressed in the literature. Since thermal energy communities are relatively new in the energy transition discussions, it is important to have a better understanding of thermal energy community systems and how these systems function. The starting point of this understanding is to study factors that influence the formation and continuation of thermal energy communities. To work towards this aim, an abstract agent-based model has been developed that explores four seemingly trivial factors, namely: neighborhood size, minimum member requirement, satisfaction factor and drop-out factor. Our preliminary modelling results indicate correlations between thermal community formation and the 'formation capability' (the percentage of households that joined) and with the satisfaction of households. No relation was found with the size of the community (in terms of number of households) or with the 'drop-out factor' (individual households that quit after the contract time).

Keywords: energy community; thermal energy systems; agent-based modelling and simulation; formation and continuation; critical factors

---

1. Introduction

All around the world, energy systems are going through a transition [1–3]. The energy transition, mainly fuelled by climate change, requires a concerted change in technological developments and institutional settings, while not hampering economic growth. This transition is being discussed and executed at different scales: international, national, regional and local [4,5]. Energy communities are considered key by many scholars and politicians for realizing the energy transition at the local level as they allow the generation and distribution of renewable energy at the local level [6,7].

There are different definitions of energy communities. Energy cooperatives, as more formal energy communities, enable citizens who participate to collectively own and manage renewable energy projects at the local level [8]. Based on this organizational model, participants generate, and in some cases, consume renewable energies. One of the most important aspects of energy cooperatives is that they are commercial organizations that operate in a market [8,9]. Energy communities, however, are projects/organizations that, in addition to financial benefits, also consider other aspects, such as environmental concerns [10]. In other words, apart from possible financial benefits, environmental concerns, norms and values also play important roles in energy communities [11,12]. While an energy cooperative's goal is mainly to generate financial benefits (by generating, participating in the market and consuming in some cases), in energy communities, generation and distribution of renewable energy are for the participants' consumption to address goals, including environmental concerns and financial benefits. Therefore, leadership, membership and interactions between the energy community participants are important [13].

In a broad sense, [14] defined an energy community as "a group of consumers and/or prosumers, that together share energy generation units and electricity storage". Energy communities are also presented as initiatives that focus on renewable energy generation, distribution and consumption (including considering energy-saving measures) for all involved stakeholders [15,16]. In this study as [7] defines, we consider an energy community as the combination of a technical energy system (mainly renewable energy technology) on the local level (e.g., an urban neighbourhood), its associated group of stakeholders that share common interest(s) and problem(s), and institutions (formal and informal rules) that govern these systems. Participants and stakeholders of an energy community share resources and collaborate on energy generation, distribution and conservation processes [9,12]. Typical energy community characteristics are: operation at the local scale, community engagement, participatory decision-making, involvement of local actors and distribution of financial resources [17]. Different stakeholders (including households) who decide to participate in an energy community, would have different roles, such as leader [13] or investor/shareholder [18].

Recent literature on the establishment and management of community energy systems predominantly focuses on electricity systems (e.g., [4,7,19,20]). However, thermal energy plays an important role in the urban context, as it is used for the purposes of heating, cooling, bathing, showering and cooking, covering approximately 75% of the non-transport related energy consumption among households [19,21,22]. Although heating energy cooperatives (such as district heating cooperatives) are discussed in the literature (e.g., [23–25]), thermal energy communities are relatively understudied. It is meaningful to study whether thermal energy communities are sustained over time and, if so, which factors influence their formation and continuity or decline. Such study will increase our understanding of thermal energy community (TEC) initiatives, how TECs would function and what factors are more important to consider to facilitate their formation and continuity. In this paper, we present the basis of an agent-based simulation model that provides insights into factors influencing the formation and continuation of TEC initiatives.

The structure of the paper is as follows: The next section presents the methods that were used in this research. Section 3 presents the data collection procedure. The structure of the abstract model is presented in Section 4. Section 5 discusses the model results. The model's limitations are presented in Section 6. Finally, Section 7 provides a discussion and conclusions.

2. Research Methods

A literature review and several interviews were conducted to first deepen our understanding of factors that influence the formation and continuation of thermal energy communities (TECs). The literature review was based on peer-reviewed material collected from scholarly databases, www.scopus.com and www.sciencedirect.com, using keywords including: "energy community/ies", "thermal energy community/ies", "heat energy community/ies", "thermal community energy systems", "factors of thermal energy community/ies", "formation of thermal energy community/ies" and "agent-based modelling AND thermal energy community". As the existing literature on TEC (including both thermal/heat energy systems and community energy systems) was relatively small, articles that focus on community energy systems, in general, were also included. The focus of this literature review was to provide an understanding of TECs and the factors which influence their formation and continuation. Therefore, in this step, a snowballing method was used, focusing on the most cited articles. In the next round, backward snowballing was applied, reviewing the articles that were cited in the articles found in the first round of snowballing. Furthermore, since the peer-reviewed literature related to TECs is relatively small, non-peer-reviewed documents cited in the reviewed articles were also considered, which led to a better understanding of the factors that have influence on the formation and continuation of TECs.

To delineate and focus on the important and unexplored factors, nine semi-structured interviews with main stakeholders in the Netherlands (policy makers, municipalities, community's presenters, energy companies and researchers) were conducted. These stakeholders were closely involved in projects related to local thermal energy transition in the Netherlands and were already working on TECs projects. The focus of the interviews was on TECs and on discovering the main factors and narrowing them down to a selected number of factors that influence their formation and continuation. Interviewees were explicitly asked to discuss the main factors which influence the formation and continuation of TECs. The interviews were transcribed, and the mentioned factors were extracted.

To deepen the understanding of the influence of these factors on the formation and continuation of TECs, there is a need for a set of experiments. In such experiments, measures related to the formation and continuation of TECs can be studied. However, performing these experiments in the real world would be time-consuming and costly and would have an actual, not necessarily beneficial, impact on individuals' lives [26,27]. Therefore, given the complexity of TECs and lack of possibility to perform a wide and varied set of experiments in the real world, a simulation model can provide benefits of experiments more quickly and less costly in a virtual setting. Simulation models that present a simpler version of the real world would help to demarcate certain design options or variables. [28,29].

In our research, we used agent-based modelling and simulation (ABMS) to study TEC initiatives. ABMS is an approach where a system is modelled as a collection of autonomous decision-making entities called agents who interact with each other and the environment [30–32]. In ABMS, Agent-based models consist of a collection of agents and their states, the rules governing the interactions of the agents and the environment within which they live [33,34]. ABMS was selected for our research due to the importance of actors, their decision-making process and interactions within thermal energy community systems, which aligns with the specific strengths of agent-based modelling [35,36]. Due to the complexity of the real world, an agent-based model cannot represent all of the details of a real-world decision-making process. However, ABMS could facilitate decision-makers by equipping them with insights about crucial variables affecting the decision-making process, thereby allowing decision-making in a less time-consuming and costly way. A sensitivity analysis [37] was conducted for various model parameters to explore various experimental configurations. The results of the model were evaluated through expert interviews.

3. Data Gathering

3.1. Literature Review

The literature on energy communities is mainly dominated by electricity systems (e.g., [4,19,20]). However, since thermal energy applications, such as heating, cooling, bathing, showering and cooking, cover 75% of energy consumption among the households, it is vital also to discuss thermal energy systems and communities (TECs) and their related challenges.

Based on the literature and studies, such as [13–16], we defined TECs based on three main components: a renewable energy technology (for thermal applications), involved stakeholders and related institutions. The technology component includes generation, distribution and consumption of thermal energy [38,39]. Involved actors and their roles [13,40] are related to stakeholders component. Finally, the institutional component covers both formal and informal institutions that govern an energy community [6,12,41].

TECs have technical, social and governance challenges. These challenges can be translated into factors that influence the formation and continuation of TEC initiatives. System challenges, such as system design, system efficiency and intermittency in generation and use, have been discussed in the literature (e.g., [42–44]).

Technical challenges and factors related to infrastructure and thermal technologies are discussed extensively in studies in which authors explore various technologies and integration and deployment of infrastructure in local energy systems (e.g., [20,38,45–47]). Furthermore, there are different studies related to demand-side management and its application for energy communities (e.g., [48–52]). In relation to these technical challenges, reported factors that influence the formation and continuation of (thermal) energy communities are (i) the availability of technology, such as solar thermal technology, geothermal wells or heat pumps (e.g., [53–55]), (ii) available resources for energy generation (e.g., [9,56]) and (iii) the number of households (e.g., [57,58]). Finally, (iv) the influence of the initial community size is also discussed in [57,59].

TEC initiatives also have challenges related to social, governance and economic arrangements. For instance, the involvement and analysis of stakeholders in energy communities is the focus of studies such as [5,60,61]. These studies focus on the important role of municipalities and households in energy communities. In this group, important factors for the formation and continuation of (thermal) energy communities that are discussed include trust [59,62] characteristics of participants, such as willingness to participate [63,64] or satisfaction [56,63–68].

Studies such as [5,45–47,57,67,68] are focused on the challenges and factors related to regulation and governance in energy communities. Financial aspects, such as investment, payback time and subsidies, are the focus of [17,31,56,57,63,69,70]. The size of the community and investment (e.g., [4,6,56,71]) are examples of factors in this group that influences the formation and continuation of (thermal) energy communities. Furthermore, other important factors related to interactions within the community, such as satisfaction and quitting the community (drop-out rates), are also discussed [7,17,31,71–73]. Table 1 presents the most cited studies in recent years, which explicitly focus on different factors and challenges related to energy communities.

Table 1. Studies with a focus on factors and challenges related to energy communities.

Study	Year of Publication	Focus of Study [1]	Domain of Study	Main Focused Factor
[57]	2007	Heat/ thermal energy communities	Economic, technological	Available technologies
[64]	2008	Energy communities	Social	Acceptance
[72]	2008	Energy communities	Social, institutional	Ownership
[12]	2010	Energy communities	Social	Trust
[47]	2011	Energy communities	Technological	Integration of infrastructure
[74]	2013	Energy communities	Technological, social, economic	Reliable
[22]	2013	Heat/thermal energy communities	Technological	Emission
[42]	2014	Heat/thermal energy communities	Technological	Integration of RETs
[6]	2014	Energy communities	Institutional	Incentivizing policies
[66]	2016	Energy communities	Social	Willingness to participate

Table 1. *Cont.*

Study	Year of Publication	Focus of Study [1]	Domain of Study	Main Focused Factor
[4]	2016	Energy communities	Technological, socio-economic, environmental, institutional	Intermittency in generation and demand
[9]	2016	Energy communities	Social, institutional	Incentivizing policies
[62]	2016	Energy communities	Social, institutional	Trust and justice
[53]	2017	Energy communities	Technological	Available technologies
[75]	2017	Energy communities	Social	Acceptance
[76]	2017	Energy communities	Social, institutional	Governance
[59]	2018	Heat/thermal energy communities	Economic, technological	Available technologies
[11]	2018	Energy communities	Social	Trust
[63]	2019	Energy communities	Social	Willingness to participate
[17]	2019	Energy communities	Socio-economic	Size of investments
[70]	2019	Energy communities	Social, economic	Acceptance

[1]: In the current literature, studies usually discuss energy communities as a general term for both electricity and heating systems. But the studies which are mentioned as heat/thermal energy communities, specifically focus on heating systems.

As Table 1 shows, a limited number of studies [22,42,57,59] (gray rows in the table) specifically discuss the challenges and factors of TEC initiatives in depth. The available studies mainly focus on technical challenges. In the scarce literature on influencing factors related to energy communities, factors such as the size of the community, financial aspects (e.g., cost and investment) or satisfaction of participants (with relation to financial and social aspects) are studied through empirical studies, such as [17,62,72]. However, the computer modelling of these factors is rarely explored. According to the literature, besides technical challenges, trust, governance, willingness to participate and size of the community are important factors that are discussed through joining, satisfaction and dropping out of the community participants.

3.2. Interviews

After the literature review, nine semi-structured interviews were conducted to gain a deeper understanding of TEC initiatives and to narrow down the number of factors that were found in the literature (main focused factors in Table 1) to a limited set of factors. The interviewees were stakeholders involved in the Dutch thermal energy transition, mainly at the local level. They included policymakers (municipalities of the Hague and Amsterdam), representatives of communities (from the cities of The Hague and Rotterdam), researchers and energy companies (one energy company, one network company, one consultancy firm and one energy branch organization). All these stakeholders were actively involved in the development of Dutch local heat transition. Interviewees discussed the factors for the formation and continuation of TEC initiatives (with a focus on the factors which are presented in Section 3.1.). Although interviewees elaborated on some of their ideas on a specific case study, the focus of the interviews was on an overall view related to TECs. The main topics for the interview were:

- Definition of a (thermal) energy community and its main components;
- Differences between thermal energy communities and electrical energy communities;
- Importance of thermal energy communities in the energy transition at the local level;

- Availability and suitability of renewable thermal energy technologies which can be used at the community level;
- Challenges and factors which influence the formation and continuation of thermal energy communities;
- Main social and governance challenges and factors;
- Main interactions between stakeholders;
- Challenges and factors that could influence the formation and continuation of thermal energy communities;
- Current agenda and planning for deployment of renewable thermal energy systems at the community level.

Three components of energy community definition, technologies (e.g., geothermal and solar) stakeholders (e.g., households and municipality) and institutions (e.g., energy policies and incentives), were discussed in detail in the context of TECs. From this, we extracted the main empirical challenges and factors for the formation and continuation of TECs.

Policymakers at the municipalities and researchers mainly mentioned the financial aspects (e.g., investment and payback time) and size of the neighbourhood (the number of households) as an important factor for the formation and continuation of TEC initiatives. Willingness to participate and the trust among participants, and the influence of these challenges and factors on current and future status of TECs, were also mentioned in these interviews.

Energy companies and representatives of communities also referred to the importance of drop-out processes of unsatisfied households. Although financial aspects were also mentioned, the importance of quitting the energy community when the participants were not satisfied was emphasized. Furthermore, energy companies and also policymakers discussed their ambitions for investments in local energy systems (e.g., district heating) for energy communities. As the Dutch government and municipalities have targets for natural gas free cities, stakeholders, such as municipalities and energy companies, are willing to invest in local energy systems. Energy companies and policymakers extensively elaborated on different renewable thermal energy technologies that are available for this purpose. Geothermal wells, heat pumps, bioenergy and waste-heat, were the main thermal sources in our interviews.

Among the factors which surfaced during the interviews, the size of the neighbourhood, the minimum member requirement and member interactions (such as the satisfaction of members and dropping out) were mentioned most often. Knowledge about these factors was limited, and interviewees raised questions about the influence of these factors on the formation and continuation of TEC initiatives in their current ongoing projects in the Netherlands. There are few studies about these factors, and most of them are empirical studies. The size of the neighbourhood and the minimum member requirement are discussed in empirical studies, such as [17,20,77]. Satisfaction and dropping out are discussed mainly in studies related to the characteristics of households and neighbourhoods (e.g., environmental concerns and financial status) [11].

Furthermore, interviewees reflected on the factors which were found in the literature and elaborated on them according to their own ideas. The interviews led to four factors that have an influence on the formation and continuation of TEC initiatives. These will be further explored in our modelling efforts:

- Size of the community;
- Minimum member requirement (formation capability);
- Satisfaction of members;
- Drop-out factor.

Further elaboration about these four factors will be presented in the next section.

4. Model Conceptualization

The purpose of the abstract model is to explore the relation between the four unexplored factors and the formation and continuation of TECs. In this section, first, the main components of the model are presented. Then the structure of the model is introduced. Finally, the experimental setup of our simulations and the model's outputs are discussed.

4.1. Model Components

The model consists of agents that represent households. The model also contains various technological options, and various energy plans that households can choose from. The agents join a community initiative based on their personal characteristics (financial benefits, environmental stance, willingness to participate) and their interactions with their peers in the network. Further elaboration on each of these model components, agents, different options (financial options, technological options and energy plans), network and interaction is presented next.

4.1.1. Agents

Before joining a TEC initiative, each household evaluates various options and makes a decision based on this evaluation. The options fall into the following categories:

- Financial options;
- Technological options;
- Energy plans.

Financial Options

As mentioned in the literature, financial factors play an important role in the decision making of the agents. For the model, we define a financial package as a combination of three elements:

- Investment: The agent is assigned a random number for how much it is willing to initially invest to join the TEC. Based on existing costs of thermal technologies [78–80], we defined five options for investment: 2500, 5000, 7500, 10,000 and 12,500 euro.
- Monthly payment: Besides the initial investment, the agent also needs to pay a monthly fee. Since the average monthly payment for heating purposes in the Netherlands is 110 euro/month [80,81], there are five options for monthly payments: 50, 75, 100, 125 and 150 euro. The model randomly assigns a value to an agent.
- Payback time: The households also take the payback time of their investment into account. Usually, the expectation for the payback time is between 7 and 20 years [79,81,82]. Therefore, the model randomly assigns a value from five options for payback time: 5, 10, 15, 20 and 25 years.

While deciding to join a community, each household calculates a financial package (parameter: idea-about-budget) that is based on the three financial parameters explained above (investment, monthly payment and payback time) (Equation (1)).

$$\text{Financial package} = \text{investment} + (\text{payback time} \times 12 \times \text{monthly payment}) \tag{1}$$

Technology Options

The three technological energy generation options that are implemented in the model are:

- Geothermal wells;
- Heat pumps;
- Solar thermal technology.

Solar thermal is the smallest sized technology which is used only for one building (maximum five households). Heat pump technology is the medium-sized technology which is used for up-to five buildings (maximum 20 households). Geothermal wells are the biggest sized technologies which are used for more than twenty buildings (maximum 100 households). Although there are other renewable energy technologies (e.g., bioenergy, waste heat), these three are chosen for the following reasons: (1) These three technologies represent different possible sizes for a community. (2) There are existing Dutch thermal energy communities that are working with these three options, which makes them the most viable options in this country. (3) The focus of the model is not on technological feasibility; therefore, a comprehensive set of technologies is not required.

In the model, the assumption is that the initial investment of the households is only spent on thermal energy generation. For the distribution system, the model assumption is that the infrastructure (i.e., district heating) is available for the whole neighbourhood. This assumption is endorsed in the literature [3,83,84] and in interviews with policymakers in the Netherlands. Given that the Dutch government and municipalities, such as Amsterdam and Utrecht, want to meet the targets for natural gas free cities in the coming years, they are willing to provide such infrastructure. In addition, energy companies who are already providing thermal energy for households in the conventional way (e.g., natural gas and electricity), want to be still involved in renewable thermal energy systems and would therefore support the system by providing the distribution infrastructure [6,21,66,71,73,75,84,85]. Households' monthly payment is spent on the maintenance of the energy system.

Energy Plans

The agent follows a certain energy plan. Using the results of our interviews, in a TEC initiative, there could be financial income (when more energy is generated than needed), which would need to be distributed among the members of the community. Three energy plans were implemented in our model, based on the agents' environmental-economic trade-offs [57,86,87]. The options were:

- Energy plan for maximizing renewable energy generation: In this plan, agents and the community only focus on maximising the generation of renewable thermal energy and, therefore, contribution to environmental benefit. Therefore, all the available money from the households that join afterwards (investments and monthly payments) will be used to generate as much renewable energy as possible, thus increasing the share of renewable energy in their energy mix consumption
- Energy plan for maximizing the individuals' profit: This plan only focuses on maximizing the economic benefits of the joined households. Therefore, the income is used to give financial benefits to individual households. This plan is the least environmentally friendly.
- Mixed energy plan: This plan is an option between the two other options. In this plan, fifty percent of the available money would be invested in increasing generation of renewable thermal energy, and the rest would be used for economic benefit of the joined households.

4.1.2. Decision Making

According to the literature (e.g., [11]), households have incentives, such as environmental concerns, independency, perception of belonging to a community and financial benefits, to make a decision to join energy communities. In this study, the households will make decisions mainly based on financial benefits, environmental concerns and perception of belonging to a community.

At the start of the simulation, all households calculate their own financial package (Equation (1)) and preferences based on the assigned variables. The most popular technology, energy plan and budget/financial package among households would be considered as the final plan for the TEC. After this, households select one of three main choices:

1. Decision to join the community initiative: Each individual household decides whether to join the community based on the comparison of its own idea about the budget with the selected budget of the neighbourhood, and its own energy plan and selected energy plan of the neighbourhood.
2. Decision to join an existing community: There is a possibility for the households to join a community after its formation. In this case, the household would make the decision based on two

comparisons: first, the comparison of its own idea about the budget with the budget required to join the community project, and second, its own energy plan and current energy plan of the project.

3. Decision to drop out: After the payback time, each individual household inside the community can make a decision to drop out of TEC initiatives. This decision is based on the self-satisfaction of an individual and the satisfaction of its network (Equation (2)). If the individual is unsatisfied and its network is minimally satisfied, after the payback time of a household is passed (otherwise they would make a financial loss), the household will drop out of TEC initiatives.

4.1.3. Network

Agents' interaction is determined by a social network model, which in this study is a "small-world" network [88,89]. This means households' interactions with their connections (other households in their neighbourhood) depend on the small-world social network structure, which is discussed in [31,90]. In the model, each household has up to 10 other households in its social network [31]. These 10 other households are in the same neighbourhood and are chosen randomly for each agent. According to network's assigned variables and satisfaction, the network of the agent influences its decisions regarding dropping out and joining TEC initiatives after its formation. For instance, if the network of an agent is satisfied, it would have positive influence on the decision of the agent on joining the TEC initiative.

4.2. Model Structure

In each round in the simulation, each agent makes a decision to join a TEC initiative or not. In the first step, all households in the neighbourhood are randomly assigned an available investment package and an energy plan. In the next stage, the package and the plan that are most popular among the households will be the selected options for the whole neighbourhood. Then, each individual household will decide about its participation according to the energy plan and the budgets. If the number of households who are participating in the community energy initiative is equal or higher than the required participants for the chosen technology, the community is formed.

The formation of a TEC leads to the start of the generation of thermal energy. After the formation, the important criterion to be calculated is the satisfaction of the households. The satisfaction is based on the comparison between their monthly payments and their previous energy bills and their budget, as shown in Equation (2), which means the satisfaction of a household mainly depends on the financial benefits. In other words, households are satisfied when

$$
\begin{aligned}
&((\text{Selected actual budget}) < ((\text{satisfaction level}) \times \text{idea about budget})) \text{ AND} \\
&((\text{monthly payment}) < ((\text{satisfaction level}) \times (\text{previous energy bills})))
\end{aligned}
\tag{2}
$$

If they drop out, they return to natural gas consumption. Dropping out is based on the self-satisfaction and the satisfaction of the households' network. As presented in Equation (3), if the individual agent is not satisfied and if unsatisfied households in the agent's network are more than the specific percentage (a parameter that is varied for the experiment), drop out factor * number of join households, after the contract time (payback time) of each household is passed, the household will drop out of the community (Equation (3)).

$$
\begin{aligned}
&(\text{satisfaction of an individual household is false}) \text{ AND } (\text{number of satisfied households in the} \\
&\text{household's network}) < ((\text{drop out factor}) \times (\text{households who participate in the community}))
\end{aligned}
\tag{3}
$$

There is always an opportunity to rejoin the community. The households can join an existing TEC throughout the simulation regardless of having joined before or not. Joining an existing TEC initiative is mainly based on two factors: (1) satisfaction of the household's network and (2) comparison of the required budget to join the project and the agent's preference about the budget (Equation (4)). Therefore, each individual household will join an existing TEC initiative when:

$$
\begin{aligned}
&\text{(number of satisfied households who already joined the community)} > ((\text{satisfaction join}\\
&\text{threshold}) \times (\text{number of satisfied households in the household's network}))\\
&\text{AND ((selected actual budget)} > ((\text{afterwards join factor})\\
&\times (\text{idea about budget}))
\end{aligned}
\tag{4}
$$

Figure 1 presents the oviewview of model structure.

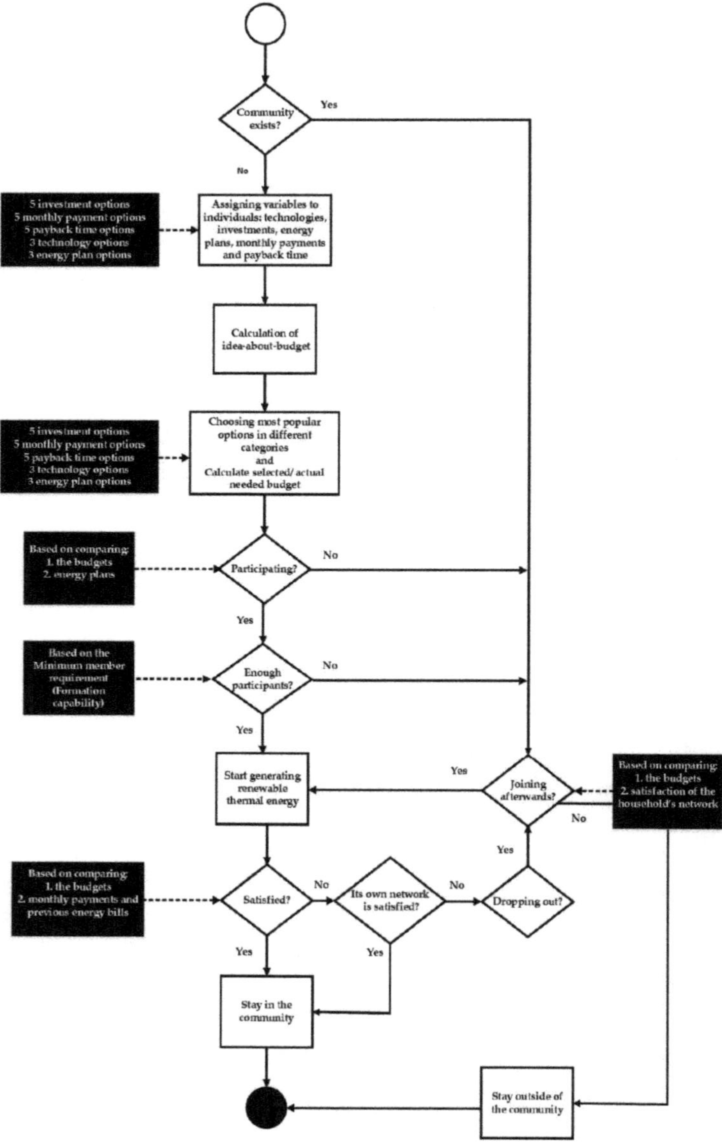

Figure 1. Overview model structure of thermal energy community (TEC) initiatives.

4.3. Experimental Setup of Simulation and Factors

As discussed, the formation and continuation of TEC initiatives can be influenced by four factors which are simulated in the agent-based model as follows:

- Number of households in the neighbourhood:

This input parameter concerns the size of the neighbourhood within which TEC initiatives may be formed. The size of the neighbourhood is equal to the number of households in that neighbourhood. For this model, the number of households has three values: 200, 500 and 700 households, representing three typical sizes of small scale neighbourhoods in the Netherlands.

- Minimum member requirement or formation capability:

This input parameter refers to the minimum percentage of households in the neighbourhood that needs to join TEC initiative at the start, to initiate a community energy system. For this model, minimum member requirement or formation capability has three values: 0.2, 0.5 and 0.8. These represent the percentage (20%, 50% and 80%) of households in the neighbourhood that should join TEC initiatives at the beginning. These values randomly selected to cover the whole range of possible values.

- Satisfaction factor:

This parameter represents the satisfaction of each individual household who has joined a TEC initiative. It is calculated based on the comparison of the initial idea about the budget and the actual invested budget, and the money they earn in terms of energy saving. If the satisfaction factor is set to a smaller number at the beginning, it means the individuals would be satisfied more easily. For this model, the satisfaction factor has three values: 0.5, 1.5 and 2.5, which will be multiplied to the other aspects of the model, such as the budget. Equation (2) illustrates how this parameter is used in the model.

- Drop-out factor:

This input parameter influences individual households that have joined TEC initiatives but drop out after the contract time. If the drop-out factor is set to a smaller number at the start, it means that the individuals would drop out more easily. For this model, the drop-out factor has three values: 0.2, 0.5 and 0.8, representing the percentage of households in agents network compared to all unsatisfied joined households (see Equation (3)).

Since the goal of TECs and also this model is to generate and distribute thermal energy based on renewable energy sources, if the agents do not participate in TECs or drop out from TECs, the conventional form, national natural gas grid, will be the source of thermal energy supply. There are four factors, and each has three options; therefore, we have $3^4 = 81$ scenarios to study. We repeated each run 100 times (to have enough experiments to decrease the influence of the parameters that agents choose randomly (e.g., number of the links with other agents). Therefore, there were 8100 runs in total. The model will run for 50 years, which is the age of an energy infrastructure and technology that is deployed, using time steps of one year.

4.4. Model Outputs

To explore the influence of these four factors on the formation and continuation of TEC initiatives, three output variables will be analysed:

- Percentage of joined households:

Percentage of joined households is an indicator of the formation of TEC initiatives. Since the experiments are in different neighbourhood sizes, this output is in percentage (Equation (5)).

$$\text{Percentage of joined households} = 100 \times ((\text{number of households who joined the community})/ \\ (\text{number of households in the neighbourhood})) \tag{5}$$

- Percentage of households who joined afterwards:

This variable captures how many of the households in the neighbourhood have joined the TEC initiatives after it has been initiated. This provides information about the process of continuation of thermal energy systems (Equation (6)).

$$\text{Percentage of households who joined afterwards} = 100 \times ((\text{number of households who joined the community afterwards})/(\text{number of households in the neighbourhood})) \tag{6}$$

- Satisfaction of the households who joined the community:

This variable reflects the satisfaction and continuation of the TEC initiatives (Equation (7)):

$$\text{Satisfaction of the households who joined the community} = 100 \times ((\text{number of joined households who are satisfied})/(\text{number of households who joined})) \tag{7}$$

5. Model Results and Discussion

In this section, we present the results of our simulation analysis. First, we give an overview of how many TEC initiatives were actually initiated in all 8100 runs. The analysis of four factors (number of households, formation capability, satisfaction factor and drop-out factor) through three outputs (percentage of joined households, percentage of households who joined afterwards and satisfaction of the households who joined the community) are shown in the next Tables. To provide a better understanding and overview, the results are first presented separately for each of the three output variables.

As Table 2 presents, the results show that in 26% of the model runs, the percentage of joined households was less than 20% of the whole neighbourhood. According to the interviews, less than 20% of joined households means the TEC is not initiated. In fact, of this 26%, in 7.5% of model runs, no household joined a TEC initiative, which shows that there was no community formation at all. In the other 18.5% of the model runs, the number of the joined households was less than 20% of the whole neighbourhood. According to the interviews, around 80% of the neighbourhood need to join to consider the TEC as established, which only happened in 5.7% of all model runs.

Table 2. Percentage of joined households in each run in 8100 runs.

Percentage of Joined Households in Each Run	What Does It Mean?	Number of Runs with This Output Out of 8100 Runs	Percentage of Runs with This Output
0–20%	No community formed or survived	2102	26%
20–80%	Some communities were formed and sustained	5530	68.3%
80–100%	Most of the neighbourhood joined a community	468	5.7%

The percentage of households who joined after the initial community was formed, is presented in Table 3.

Table 3. Percentage of households who joined afterwards in 8100 runs.

Percentage of Joined Households after Energy Community Was Formed	What Does It Mean?	Number of Runs with This Output Out of 8100 Runs	Percentage of Runs with This Output
0%	No household joined the community after it was formed	7083	87.4%
0–20%	Less than 20% of the neighbourhood joined the community after it was formed	844	10.4%
20–50%	Less than 50% of the neighbourhood joined the community after it was formed	171	2.2%

Table 3 reveals that in the majority of model runs, households did not join TEC initiatives after their formation. Out of 8100 runs, in 7083 runs, there was no household that joined TEC initiatives after their formation. In 12.6% (10.4% + 2.2%) of the model runs, fewer than 50% of households joined TEC initiatives after initiation. There was no run in which more than 50% of households join the TEC initiatives after initiation.

As presented in Table 4, the satisfaction of households who joined the community was divided mainly between no satisfaction among households or the majority of the joined households were satisfied. In 4958 runs (out of 8100 runs), there was no satisfaction among joined households. In contrast, in 2714 runs, most of the joined households were highly satisfied. These results present a polarized satisfaction, which needs further exploration to find the possible root causes.

Table 4. Percentage of satisfied households in each run in 8100 runs.

Percentage of Satisfied Households	What Does it Mean?	Number of Runs with This Output out of 8100 Runs	Percentage of Runs with This Output
0%	There is no satisfaction among households	4958	61.2%
0–80%	Some of the households are satisfied	428	5.3%
80–100%	Majority of households are satisfied	2714	33.5%

Two of the factors, i.e., the number of households and the formation capability, were analysed further to understand their influence on the model's outputs. This is shown in Figures 2 and 3.

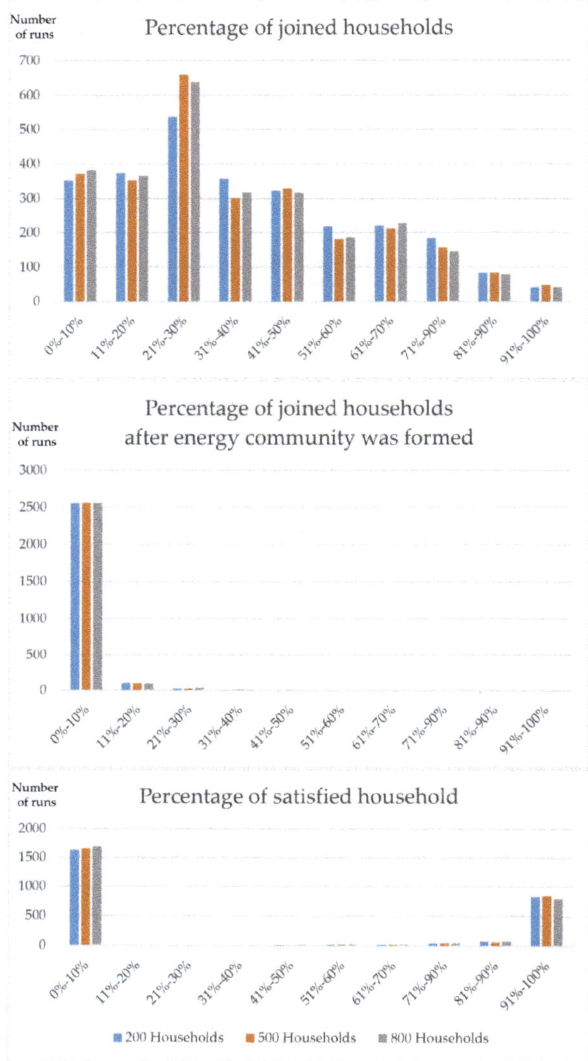

Figure 2. Influence of the number of households on the outputs.

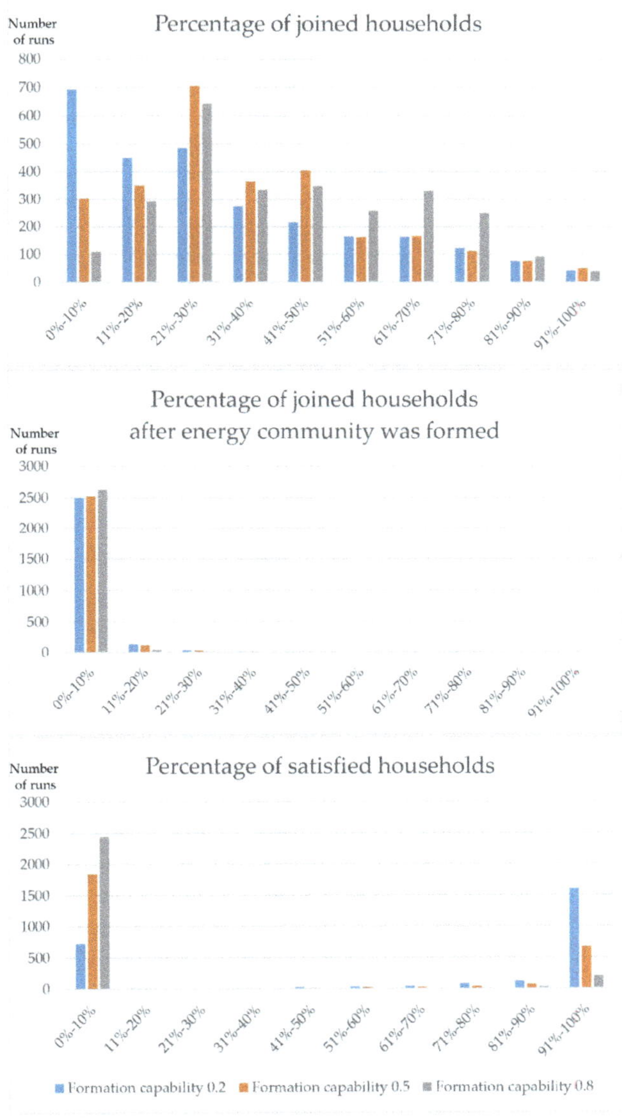

Figure 3. Influence of formation capability on the outputs.

Figure 2 shows that there was no significant impact on the outputs when the number of households was changed. Based on the model's structure and assumptions, the number of households (size of the community) did not have considerable influence on the results.

Figure 3 shows the results for the formation capability (minimum member requirement). As the figure shows, the formation capability had considerable influence on the behaviour of two outputs: the percentage of joined households and the percentage of satisfied households. When the formation capability was changed, the outputs change.

This comparison between the correlation of two factors, number of households and formation capability, on the model's outputs, shows that the influence of factors on the outputs was different.

To have better insights from the results, the correlation between each factor and each output is presented in Table 5.

Table 5. Correlations between factors and model outputs.

Factors to Explore		Percentage of Joined Households	Percentage of Joined Households Afterwards	Percentage of Satisfied Joined Households
Number of households	Pearson Correlation	−0.024	−0.001	−0.15
	Sig. (2-tailed)	0.032	0.936	0.177
Formation capability	Pearson Correlation	0.221	−0.118	−0.524
	Sig. (2-tailed)	0.000	0.000	0.000
Satisfaction factor	Pearson Correlation	0.388	0.49	0.218
	Sig. (2-tailed)	0.000	0.000	0.000
Drop out factor	Pearson Correlation	−0.002	−0.005	0.001
	Sig. (2-tailed)	0.840	0.647	0.918

The correlation of formation capability and the satisfaction factor with the model outputs was strong, highlighting the role of satisfaction in the formation of TEC. The satisfaction factor had a positive correlation with all of the model outputs, which means that the satisfaction of households would boost the formation and continuation of TEC initiatives. While the correlation between formation capability and percentage of joined households was positive, the Pearson Correlation was negative between formation capability and the other two model outputs (percentage of joined households afterwards and percentage of satisfied joined households). This means that it is important to incentivise households to join the community at the beginning of its formation because making people join later and increasing satisfaction are hard to achieve.

In contrast, the number of households and the drop-out factor did not show a strong correlation with the model outputs, especially the drop out factor. However, due to the model limitations, this needs further studies.

6. Model Limitations

Although this study brought interesting and important insights into light about the formation and continuation of TEC initiatives, it can be developed further to have more in-depth results. All four factors can be structured in the model with more details and complexity, especially the satisfaction factor and drop-out factor. Other related factors, such as available technology and economies of scale, were not captured in this version of the model. Although, size of the neighbourhood and the percentage of participants have an influence on the initial investment of the whole neighbourhood, the assumption in the abstract model is that the chosen technology would not face financial problems in the model (households will successfully provide the needed finances). To provide more insights about technical options, the techno-economic feasibility study of different heating technologies is necessary.

Factors which were already implemented in the abstract model, but are not the focus of the study, such as social aspects (e.g., trust) and financial aspects (e.g., payback time and investment), can be made data-driven to gain more insights about their impact. The technical aspects can be modelled in more detail to understand their role on the formation and continuation of TECs. This would help to have a more comprehensive overview of TECs and the related decision-making processes.

The model is abstract in the sense that the data used to build it were either qualitative (based on interviews) or general statistics (from National websites). This limits the model in exploring the influence of actual demographics and characteristics of a given neighbourhood on the formation and continuation of TEC initiatives. Furthermore, since the model did not include detailed financial specifications, the relation between financial packages and the model's results were not explored. This implies that the financing options, such as bank loans, energy company lease, governmental subsidies, and their influence were not studied

In the current version of the model, each neighbourhood had only one energy community. Theoretically, each neighbourhood can have several energy communities. Apart from the values of the household, other aspects, such as technical feasibility, play a role in choosing one of the communities in the neighbourhood for joining. Providing the opportunity for households to choose between different TECs in a neighbourhood, would provide more insights into the households' decision-making process.

To address these limitations, using other qualitative and quantitative approaches would be beneficial. Some examples are:

- Detailed interviews with the main stakeholders: This approach would help to have a better understanding of the responsibilities and strategies of different TEC's stakeholders and the dynamics between them.
- Conducting surveys: This approach would increase the awareness about the social perception and understanding about TECs. Surveys could focus on different topics, such as social acceptance and willingness to participate. Models drawn from such surveys would have a case-specific nature rather than the generalized version we presented in this paper.
- Optimization of the thermal energy system design: This approach would help to have a better technological design for TECs. Different technological designs have differences in their social and governance aspects.

7. Conclusions and Further Study

Our research aimed to increase our understanding of the formation and continuation of TEC initiatives. In this paper, we presented the basis of an agent-based model that allowed us to explore four main factors: number of households, formation capability (minimum member requirement), satisfaction factor and the drop-out factor. Their correlation with our model outputs (percentage of joined households, percentage of joined households afterwards and percentage of satisfied households) was investigated, as they are prime indicators for TEC initiative formation and continuation. This model can be deployed for studying certain factors that affect the formation and continuation of TECs. The model provides a simplified version of the real world to provide insights into the potential importance of the factors.

Our preliminary results show that while the formation capability and the satisfaction factor have a strong positive correlation with the percentage of joined households, the number of households and the drop-out factor have relatively weak correlations. Furthermore, both formation capability and the satisfaction factor show a stronger correlation with the percentage of households who joined afterwards and the satisfaction of joined households.

The satisfaction factor has a considerable positive correlation with the percentage of households who joined afterwards. Hence, the model showed that the satisfied households would influence their network to make them join the community or not to drop out of the community. Furthermore, the satisfaction factor has a positive correlation with the percentage of satisfied joined households. In contrast, the number of households and the drop-out factor have weaker correlations with the model's outputs. The negative correlation of the number of households with all the model's output needs further study. Although this model did not investigate the causality between the factors, one of the possibilities for this might be the negative impact of the size of the neighbourhood on the formation and continuation of the thermal energy communities.

Based on preliminary results, the following suggestions to policymakers and households can be made:

- The size of the neighbourhood (Number of the households) may not be the most important factor to be considered in policies related to the development of TEC initiatives. In other words, to develop policies and strategies to facilitate deployment and establishment of TEC initiatives, there are more important factors to be considered than the size of the neighbourhood.

- It appears to be important that a large enough fraction of households join a community in the beginning. The percentage of the households who join at the beginning of TEC formation seems influential for the continuation of TEC and the satisfaction of participants. Therefore, the focus may need to be on incentivizing households at the beginning to join and participate in the TEC initiatives.

- It seems relatively more important to focus on the satisfaction of the households who joined the community rather than focusing on dropping out of the households.

- If the households do not join at the beginning, it seems relatively hard to join a TEC afterwards. Therefore, it is important to try to incentivize the households in the neighbourhood as much as possible at the beginning. However, there should still be a possibility for households to join a TEC after its establishment.

The results and recommendations would provide new insights for stakeholders to focus on the important factors to further developments of TECs, which leads to the establishment of thermal energy communities. The model presented in this paper is only the start of the modelling effort required to study thermal community energy systems. We are expanding the model further to include more details that make it more representative of actual communities. For that, a more comprehensive data collection will also be pursued. For example, the literature suggests that institutional configurations of such communities are decisive factors for the success of these communities along with individual characteristics, such as willingness to contribute. These factors, among others, are included in the next version of this model. These will provide more concrete recommendations in our future work.

Author Contributions: J.F. built the agent-based model and took the lead in writing the paper. The rest of the authors (N.M., A.G., P.H.) provided feedback on the conceptualization model, verification and validation. They also provided feedback on the paper. All authors have read and agreed to the published version of the manuscript.

Funding: This research was funded by Netherlands Organization for Scientific Research, for their financial support (NWO Responsible Innovation grant—313-99-324).

Acknowledgments: The authors wish to thank the Netherlands Organization for Scientific Research for their financial support.

Conflicts of Interest: The authors declare no conflict of interest.

References

1. Păceşilă, M.; Burcea, S.G.; Colesca, S.E. Analysis of renewable energies in European Union. *Renew. Sustain. Energy Rev.* **2016**, *56*, 156–170. [CrossRef]

2. Verbong, G. *Governing the Energy Transition: Reality, Illusion or Necessity*; Routledge; Taylor & Francis Group: Abingdon, UK, 2012.

3. Kern, F.; Smith, A. Restructuring energy systems for sustainability? Energy transition policy in the Netherlands. *Energy Policy* **2008**, *36*, 4093–4103. [CrossRef]

4. Koirala, B.P.; Koliou, E.; Friege, J.; Hakvoort, R.; Herder, P. Energetic communities for community energy: A review of key issues and trends shaping integrated community energy systems. *Renew. Sustain. Energy Rev.* **2016**, *56*, 722–744. [CrossRef]

5. Van Der Schoor, T.; Scholtens, B. Power to the people: Local community initiatives and the transition to sustainable energy. *Renew. Sustain. Energy Rev.* **2015**, *43*, 666–675. [CrossRef]

6. Oteman, M.; Wiering, M.; Helderman, J.-K. The institutional space of community initiatives for renewable energy: A comparative case study of the Netherlands, Germany and Denmark. *Energy Sustain. Soc.* **2014**, *4*, 11. [CrossRef]

7. *Guide to Developing a Community Renewable Energy Project*; Commission for Environmental Cooperation: Montreal, QC, Canada, 2010.
8. Bauwens, T. Explaining the diversity of motivations behind community renewable energy. *Energy Policy* **2016**, *93*, 278–290. [CrossRef]
9. Bauwens, T.; Gotchev, B.; Holstenkamp, L. What drives the development of community energy in Europe? the case of wind power cooperatives. *Energy Res. Soc. Sci.* **2016**, *13*, 136–147. [CrossRef]
10. Walker, G.; Devine-Wright, P. Community renewable energy: What should it mean? *Energy Policy* **2008**, *36*, 497–500. [CrossRef]
11. Koirala, B.P.; Araghi, Y.; Kroesen, M.; Ghorbani, A.; Hakvoort, R.A.; Herder, P.M. Trust, awareness, and independence: Insights from a socio-psychological factor analysis of citizen knowledge and participation in community energy systems. *Energy Res. Soc. Sci.* **2018**, *38*, 33–40. [CrossRef]
12. Walker, G.P.; Devine-Wright, P.; Hunter, S.; High, H.; Evans, B. Trust and community: Exploring the meanings, contexts and dynamics of community renewable energy. *Energy Policy* **2010**, *38*, 2655–2663. [CrossRef]
13. Martiskainen, M. The role of community leadership in the development of grassroots innovations. *Environ. Innov. Soc. Transit.* **2017**, *22*, 78–89. [CrossRef]
14. Schram, W.; Louwen, A.; Lampropoulos, I.; Van Sark, W. Comparison of the greenhouse gas emission reduction potential of energy communities. *Energies* **2019**, *12*, 4440. [CrossRef]
15. Visa, I.; Duta, A.; Moldovan, M.; Burduhos, B.; Neagoe, M. Sustainable Communities. In *Solar Energy Conversion Systems in the Built Environment*; Green Energy and Technology; Springer: Cham, Switzerland, 2020.
16. Magnusson, D.; Palm, J. Come together-the development of Swedish energy communities. *Sustainability* **2019**, *11*, 1056. [CrossRef]
17. Bauwens, T. Analyzing the determinants of the size of investments by community renewable energy members: Findings and policy implications from Flanders. *Energy Policy* **2019**, *129*, 841–852. [CrossRef]
18. Heldeweg, M.; Saintier, S. Renewable energy communities as 'socio-legal institutions': A normative frame for energy decentralization? *Renew. Sustain. Energy Rev.* **2020**, *119*, 109518. [CrossRef]
19. Persson, U.; Möller, B.; Werner, S. Heat Roadmap Europe: Identifying strategic heat synergy regions. *Energy Policy* **2014**, *74*, 663–681. [CrossRef]
20. St. Denis, G.; Parker, P. Community energy planning in Canada: The role of renewable energy. *Renew. Sustain. Energy Rev.* **2009**, *13*, 2088–2095. [CrossRef]
21. van Leeuwen, R.P.; de Wit, J.B.; Smit, G.J.M. Review of urban energy transition in the Netherlands and the role of smart energy management. *Energy Convers. Manag.* **2017**, *150*, 941–948. [CrossRef]
22. Finney, K.N.; Zhou, J.; Chen, Q.; Zhang, X.; Chan, C.; Finney, K.N.; Swithenbank, J.; Nolan, A.; White, S.; Ogden, S.; et al. Modelling and mapping sustainable heating for cities. *Appl. Therm. Eng. Appl. Therm. Eng.* **2013**, *53*, 246–255. [CrossRef]
23. Werner, S. International review of district heating and cooling. *Energy* **2017**, *137*, 617–631. [CrossRef]
24. Moshkin, I.; Sauhats, A. Solving district heating optimization problems in the market conditions. In Proceedings of the 57th International Scientific Conference on Power and Electrical Engineering of Riga Technical University (RTUCON), Riga & Cesis, Latvia, 13–14 October 2016; pp. 1–6.
25. Perez-Mora, N.; Bava, F.; Andersen, M.; Bales, C.; Lennermo, G.; Nielsen, C.; Furbo, S.; Martínez-Moll, V. Solar district heating and cooling: A review. *Int. J. Energy Res.* **2018**, *42*, 1419–1441. [CrossRef]
26. Ghorbani, A.; Bravo, G. Managing the commons: A simple model of the emergence of institutions through collective action. *Int. J. Commons* **2016**, *10*, 200–219.
27. Minai, A.A.; Bar-Yam, Y. *Unifying Themes in Complex Systems, Vol. IIIB: New Research*; Springer Publishing Company: New York City, NY, USA, 2007.
28. Vespignani, A. Modelling dynamical processes in complex socio-technical systems. *Nat. Phys.* **2012**, *8*, 32–39. [CrossRef]
29. Righi, A.W.; Saurin, T.A. Complex socio-technical systems: Characterization and management guidelines. *Appl. Ergon.* **2015**, *50*, 19–30. [CrossRef]
30. Bonabeau, E. Agent-based modeling: Methods and techniques for simulating human systems. *Proc. Natl. Acad. Sci. USA* **2002**, *99* (Suppl. 3), 7280. [CrossRef] [PubMed]
31. Rai, V.; Robinson, S.A. Agent-based modeling of energy technology adoption: Empirical integration of social, behavioral, economic, and environmental factors. *Environ. Model. Softw.* **2015**, *70*, 163–177. [CrossRef]

32. Holtz, G.; Alkemade, F.; De Haan, F.; Köhler, J.; Trutnevyte, E.; Luthe, T.; Halbe, J.; Papachristos, G.; Chappin, É.; Kwakkel, J.H.; et al. Prospects of modelling societal transitions: Position paper of an emerging community. *Environ. Innov. Soc. Transit.* **2015**, *17*, 41–58. [CrossRef]

33. Holm, S.; Hilty, L.; Lemm, R.; Thees, O. Empirical validation of an agent-based model of wood markets in Switzerland. *PLoS ONE* **2018**, *13*, e0190605. [CrossRef]

34. Moglia, M.; Cook, S.; McGregor, J. A review of Agent-Based Modelling of technology diffusion with special reference to residential energy efficiency. *Sustain. Cities Soc.* **2017**, *31*, 173–182. [CrossRef]

35. Couclelis, H. The Certainty of Uncertainty: GIS and the Limits of Geographic Knowledge. *Trans. GIS* **2003**, *7*, 165–175. [CrossRef]

36. Macal, C.M.; North, M.J.; Macal, C.M.; North, M.J. Tutorial on Agent-based Modeling and Simulation Agent-based Modeling and Simulation Initiative at Argonne National Laboratory View project AGENT-BASED MODELING AND SIMULATION. In Proceedings of the Winter Simulation Conference, Orlando, FL, USA, 4 December 2005; p. 14.

37. Thiele, J.C.; Kurth, W.; Grimm, V. Facilitating Parameter Estimation and Sensitivity Analysis of Agent-Based Models: A Cookbook Using NetLogo and R. *J. Artif. Soc. Soc. Simul.* **2015**, *17*, 11. [CrossRef]

38. Jaccard, M.; Failing, L.; Berry, T. From equipment to infrastructure: Community energy management and greenhouse gas emission reduction. *Energy Policy* **1997**, *25*, 1065–1074. [CrossRef]

39. Mavromatidis, G.; Orehounig, K.; Carmeliet, J. A review of uncertainty characterisation approaches for the optimal design of distributed energy systems. *Renew. Sustain. Energy Rev.* **2018**, *88*, 258–277. [CrossRef]

40. Lapniewska, Z. Cooperatives governing energy infrastructure: A case study of Berlin's grid. *J. Co-op. Organ. Manag.* **2019**, *7*, 100094. [CrossRef]

41. Mahzouni, A. The role of institutional entrepreneurship in emerging energy communities: The town of St. Peter in Germany. *Renew. Sustain. Energy Rev.* **2019**, *107*, 297–308. [CrossRef]

42. Huang, Z.; Yu, H. Approach for integrated optimization of community heating system at urban detailed planning stage. *Energy Build.* **2014**, *77*, 103–111. [CrossRef]

43. De Boeck, L.; Verbeke, S.; Audenaert, A.; De Mesmaeker, L. Improving the energy performance of residential buildings: A literature review. *Renew. Sustain. Energy Rev.* **2015**, *52*, 960–975. [CrossRef]

44. Murphy, R.; Jaccard, M. Energy efficiency and the cost of GHG abatement: A comparison of bottom-up and hybrid models for the US. *Energy Policy* **2011**, *39*, 7146–7155. [CrossRef]

45. Brower, M.; Sawin, J.L.; Sverrisson, F.; Chawla, K.; Lins, C.; McMrone, A.; Musolino, A.; Riahi, L.; Sims, R.; Skeen, J.; et al. *Renewable 2014 Global Status Report REN21*; Reneable Energy Policy Network for the 21st Centry: Paris, France, 2014.

46. Lyden, A.; Pepper, R.; Tuohy, P.G. A modelling tool selection process for planning of community scale energy systems including storage and demand side management. *Sustain. Cities Soc.* **2018**, *39*, 674–688. [CrossRef]

47. Rees, M.T.; Wu, J.; Awad, B.; Ekanayake, J.; Jenkins, N. A total energy approach to integrated community infrastructure design. In Proceedings of the 2011 IEEE Power and Energy Society General Meeting, Detroit, MI, USA, 24–28 July 2011; pp. 1–8.

48. Arteconi, A.; Hewitt, N.; Polonara, F. Domestic demand-side management (DSM): Role of heat pumps and thermal energy storage (TES) systems. *Appl. Therm. Eng.* **2013**, *51*, 155–165. [CrossRef]

49. Wolisz, H.; Punkenburg, C.; Streblow, R.; Müller, D. Feasibility and potential of thermal demand side management in residential buildings considering different developments in the German energy market. *Energy Convers. Manag.* **2016**, *107*, 86–95. [CrossRef]

50. Tronchin, L.; Manfren, M.; Nastasi, B. Energy efficiency, demand side management and energy storage technologies—A critical analysis of possible paths of integration in the built environment. *Renew. Sustain. Energy Rev.* **2018**, *95*, 341–353. [CrossRef]

51. Arteconi, A.; Hewitt, N.; Polonara, F. State of the art of thermal storage for demand-side management. *Appl. Energy* **2012**, *93*, 371–389. [CrossRef]

52. Negro, S.O.; Alkemade, F.; Hekkert, M.P. Why does renewable energy diffuse so slowly? A review of innovation system problems. *Renew. Sustain. Energy Rev.* **2012**, *16*, 3836–3846. [CrossRef]

53. Li, Y.; Jin, M.; Li, Y. Community energy system planning: A case study on technology selection and operation optimization. *Procedia Eng.* **2017**, *205*, 2076–2083. [CrossRef]

54. Boon, F.P.; Dieperink, C. Local civil society based renewable energy organisations in the Netherlands: Exploring the factors that stimulate their emergence and development. *Energy Policy* **2014**, *69*, 297–307. [CrossRef]

55. *Energy Policies of IEA Countries—The European Union 2014 Review*; European Union: Brussels, Belgium, 2014.

56. Kumar, A.; Sah, B.; Singh, A.R.; Deng, Y.; He, X.; Kumar, P.; Bansal, R. A review of multi criteria decision making (MCDM) towards sustainable renewable energy development. *Renew. Sustain. Energy Rev.* **2017**, *69*, 596–609. [CrossRef]

57. McKim, A.E.; Randall, W.L. From Psychology to Poetics: Aging as a Literary Process. *J. Aging Humanit. Arts* **2007**, *1*, 147–158. [CrossRef]

58. Von Wirth, T.; Gislason, L.; Seidl, R. Distributed energy systems on a neighborhood scale: Reviewing drivers of and barriers to social acceptance. *Renew. Sustain. Energy Rev.* **2017**, *82*, 2618–2628. [CrossRef]

59. Schmidt, D. Low Temperature District Heating for Future Energy Systems. *Energy Procedia* **2018**, *149*, 595–604. [CrossRef]

60. Dvarionienė, J.; Gurauskiene, I.; Gecevicius, G.; Trummer, D.R.; Selada, C.; Marques, I.; Cosmi, C. Stakeholders involvement for energy conscious communities: The Energy Labs experience in 10 European communities. *Renew. Energy* **2015**, *75*, 512–518. [CrossRef]

61. Brandoni, C.; Polonara, F. The role of municipal energy planning in the regional energy-planning process. *Energy* **2012**, *48*, 323–338. [CrossRef]

62. Goedkoop, F.; Devine-Wright, P. Partnership or placation? The role of trust and justice in the shared ownership of renewable energy projects. *Energy Res. Soc. Sci.* **2016**, *17*, 135–146. [CrossRef]

63. Woo, J.; Chung, S.; Lee, C.-Y.; Huh, S.-Y. Willingness to participate in community-based renewable energy projects: A contingent valuation study in South Korea. *Renew. Sustain. Energy Rev.* **2019**, *112*, 643–652. [CrossRef]

64. Schweizer-Ries, P. Energy sustainable communities: Environmental psychological investigations. *Energy Policy* **2008**, *36*, 4126–4135. [CrossRef]

65. Perlaviciute, G.; Steg, L. The influence of values on evaluations of energy alternatives. *Renew. Energy* **2015**, *77*, 259–267. [CrossRef]

66. Kalkbrenner, B.J.; Roosen, J. Citizens' willingness to participate in local renewable energy projects: The role of community and trust in Germany. *Energy Res. Soc. Sci.* **2016**, *13*, 60–70. [CrossRef]

67. Parag, Y.; Hamilton, J.; White, V.; Hogan, B. Network approach for local and community governance of energy: The case of Oxfordshire. *Energy Policy* **2013**, *62*, 1064–1077. [CrossRef]

68. Armstrong, A.; Bulkeley, H. Micro-hydro politics: Producing and contesting community energy in the North of England. *Geoforum* **2014**, *56*, 66–76. [CrossRef]

69. Nair, G.; Gustavsson, L.; Mahapatra, K. Factors influencing energy efficiency investments in existing Swedish residential buildings. *Energy Policy* **2010**, *38*, 2956–2963. [CrossRef]

70. Vuichard, P.; Stauch, A.; Dällenbach, N. Individual or collective? Community investment, local taxes, and the social acceptance of wind energy in Switzerland. *Energy Res. Soc. Sci.* **2019**, *58*, 101275. [CrossRef]

71. Mittal, A.; Krejci, C.; Dorneich, M.C.; Fickes, D. An agent-based approach to modeling zero energy communities. *Sol. Energy* **2019**, *191*, 193–204. [CrossRef]

72. Walker, G.P. What are the barriers and incentives for community-owned means of energy production and use? *Energy Policy* **2008**, *36*, 4401–4405. [CrossRef]

73. Cipriano, X.; Vellido, A.; Cipriano, J.; Martí-Herrero, J.; Danov, S. Influencing factors in energy use of housing blocks: A new methodology, based on clustering and energy simulations, for decision making in energy refurbishment projects. *Energy Effic.* **2017**, *10*, 359–382. [CrossRef]

74. Fay, G.; Udovyk, N. Factors influencing success of wind-diesel hybrid systems in remote Alaska communities: Results of an informal survey. *Renew. Energy* **2013**, *57*, 554–557. [CrossRef]

75. Walker, C.; Baxter, J. Procedural justice in Canadian wind energy development: A comparison of community-based and technocratic siting processes. *Energy Res. Soc. Sci.* **2017**, *29*, 160–169. [CrossRef]

76. Young, J.; Brans, M. Analysis of factors affecting a shift in a local energy system towards 100% renewable energy community. *J. Clean. Prod.* **2017**, *169*, 117–124. [CrossRef]

77. Franceschinis, C.; Thiene, M.; Scarpa, R.; Rose, J.; Moretto, M.; Cavalli, R. Adoption of renewable heating systems: An empirical test of the diffusion of innovation theory. *Energy* **2017**, *125*, 313–326. [CrossRef]

78. Van Heekeren, V.; Bakema, G. The Netherlands Country Update on Geothermal Energy. In Proceedings of the World Geothermal Congress 2015, Melbourne, Austrilia, 19–25 April 2015; Volume 2013, pp. 2–7.

79. Gudmundsson, O.; Thorsen, J.E.; Zhang, L. Cost analysis of district heating compared to its competing technologies. *WIT Trans. Ecol. Environ.* **2013**, *176*, 107–118.

80. Sandvall, A.F.; Ahlgren, E.O.; Ekvall, T. Cost-efficiency of urban heating strategies—Modelling scale effects of low-energy building heat supply. *Energy Strateg. Rev.* **2017**, *18*, 212–223. [CrossRef]

81. Studies, I.C. Prices and Costs of Annex 3 Household Case Studies. Available online: https://gaap.ifpri.info/files/2012/02/Complete_Case_Studies.pdf (accessed on 1 June 2020).

82. Doelgroep, A. Investeringsreglement VvE Energiebespaarlening Nationaal Energiebespaarfonds B. Procedure. Available online: https://www.energiebespaarlening.nl/wp-content/uploads/2019/03/Investeringsreglement-VvE-Nationaal-Energiebespaarfonds.pdf (accessed on 1 June 2020).

83. Tambach, M.; Hasselaar, E.; Itard, L. Assessment of current Dutch energy transition policy instruments for the existing housing stock. *Energy Policy* **2010**, *38*, 981–996. [CrossRef]

84. Treffers, D.J.; Faaij, A.P.C.; Spakman, J.; Seebregts, A. Exploring the possibilities for setting up sustainable energy systems for the long term: Two visions for the Dutch energy system in 2050. *Energy Policy* **2005**, *33*, 1723–1743. [CrossRef]

85. Singh, H.; Muetze, A.; Eames, P. Factors influencing the uptake of heat pump technology by the UK domestic sector. *Renew. Energy* **2010**, *35*, 873–878. [CrossRef]

86. Kurahashi, S.; Jager, W. An Electricity Market Game using Agent-based Gaming Technique for Understanding Energy Transition. In *Proceedings of the 9th International Conference on Agents and Artificial Intelligence*; SciTePress: Setubal, Portugal, 2017; Volume 1, pp. 314–321.

87. Sousa, T.; Soares, T.; Pinson, P.; Moret, F.; Baroche, T.; Sorin, E. Peer-to-peer and community-based markets: A comprehensive review. *Renew. Sustain. Energy Rev.* **2019**, *104*, 367–378. [CrossRef]

88. Watts, D.J.; Strogatz, S.H. Collective dynamics of 'small-world' networks. *Nature* **1998**, *393*, 440–442. [CrossRef]

89. Amaral, L.; Scala, A.; Barthélémy, M.; Stanley, H.E. Classes of small-world networks. *Proc. Natl. Acad. Sci. USA* **2000**, *97*, 11149–11152. [CrossRef]

90. Latora, V.; Marchiori, M. Efficient behavior of small-world networks. *Phys. Rev. Lett.* **2001**, *87*, 198701-1–198701-4. [CrossRef]

Article

Status and Evolution of the Community Energy Sector in Italy

Chiara Candelise [1,2,*] and Gianluca Ruggieri [3] 

[1]   GREEN, the Centre for Research in Geography, Resources, Environment, Energy and Networks, Bocconi University, 20136 Milan, Italy
[2]   Imperial College Centre for Energy Policy and Technology (ICEPT), Imperial College London, London SW7 1NE, UK
[3]   Dipartimento di Scienze Teoriche e Applicate, Università degli Studi dell'Insubria, 21100 Varese, Italy; gianluca.ruggieri@uninsubria.it
*   Correspondence: chiara.candelise@unibocconi.it

Received: 15 February 2020; Accepted: 31 March 2020; Published: 13 April 2020

Abstract: Community energy (CE) initiatives have been progressively spreading across Europe and are increasingly proposed as innovative and alternative approaches to guarantee higher citizen participation in the transition toward cleaner energy systems. This paper focuses the attention on Italy, a Southern European country characterized by relatively low CE sector development. It fills a gap in the literature by eliciting and presenting novel and comprehensive evidence on recent Italian CE sector developments. Through a stepwise approach it systematically maps and reviews Italian CE initiatives, to then focus the attention on three specific case studies to further explore conditions for development as well as of success within the Italian energy system. The analysis presents an Italian CE sector still at its niche level, characterized by small initiatives largely dependent on national photovoltaics (PV) policy support. It also points out how only larger initiatives, able to operate at national scale, developing multiple projects and differentiating their activities have managed to continue growing at the time of discontinuity of policy support and contraction of the national renewable energy market. Recent EU and national legislative development might support revived development of CE initiatives in Italy.

Keywords: community energy; renewable energy; citizen participation; energy cooperatives

---

1. Introduction

Commitments and efforts in reducing greenhouse gas emissions as well as increasing concerns over energy security have triggered the transitioning of the European Union (EU) energy system toward a higher proportion of clean energy generation and reduction of energy use through the implementation of energy efficiency measures [1–3]. In most of the EU much of the transition to decarbonized energy systems has to date been led by major investors and large companies [4,5], but smaller players as well as citizens and local communities are increasingly playing an active role in delivering clean energy investments. Transition toward decentralized energy systems, progressive liberalization of energy markets, and technological innovation have left space for an active role of energy users, which are turning into "prosumers" or co-providers of energy services [6,7]. While consumers' participation to energy transition is increasingly concerning the policy makers [8], community energy (CE) and shared ownership approaches for investments in the energy sector have been developing worldwide [9–11]. They enable citizens to collectively develop and manage energy projects or services, presenting a different model of development and ownership than traditional business organizations [12,13].

The first CE initiatives date back to early 20th century, when rural electrification cooperatives existed in Europe in countries such as Germany, Italy, or Spain [14–16]. They have been later associated with renewable energy production with the rise of wind cooperatives in Denmark in the late 1970s and with new waves of citizens' initiatives after Chernobyl disaster in 1986 (in particular in Germany and Belgium). It is from the 2000s that they began emerging as new paradigms of people engagement in the energy transition toward renewable energy production, facilitated and driven by the last decade's energy system liberalization and transition toward more decentralized energy systems [12].

However, the degree of recognition of the potential contribution of citizens to the energy transition and the level of deployment of CE initiatives still varies considerably across Europe. CE initiatives are more common in Northern Europe, particularly in Denmark, Germany, and the United Kingdom, and far less developed in Southern Europe. Germany hosts more than 800 energy cooperatives, accounting for about 34% of the citizenship [17] whereas in countries like Spain or Greece less than 10 initiatives have been reported [16,18]. Indeed, most of the academic literature researching dynamics, drivers, and conditions for implementation of CE initiatives mainly focus on Northern European countries [19–23]. This suggests the need of deeper analysis on the status of the CE sector in Southern Europe.

The intention of this paper is to contribute to this debate by providing new evidence on the Italian CE sector, which has been to date overlooked by scholars. Magnani and Osti [24] have looked into the role of Italian civil society in energy transition, and few other contributions have studied some specific Italian CE initiatives [25,26]. However, no academic contribution has to date provided a comprehensive review of the Italian CE sector.

We use a qualitative and descriptive approach to search, analyze, and present evidence of CE initiatives that emerged in the country in the last decade. We firstly characterize the sector through a systematic review of the Italian CE initiatives which, as experienced in other northern European countries [14,21], are very heterogeneous. They can take multiple forms depending on the type and scope of their activity, the approach taken for their development as well as the level of citizens' financial involvement, ownership, and co-determination implied by their legal structure and governance. The objective of the review is providing novel data and evidence as well as a clearer characterization of CE initiatives in Italy. We then focus the attention on three specific case studies representing those larger initiatives still operating to date with the objective of further analyzing and understanding characteristics and conditions for deployment and success of CE within the Italian energy sector.

The paper is structured as follows: Section 2 defines the boundaries of the analysis and introduces the methodology adopted. Section 3 presents the results of the systematic review of the Italian CE sector and the case studies. Section 4 discusses the results of the systematic review and the comparative case studies and in Section 5 we present the conclusions, including possible future developments.

2. Materials and Methods

Civil society engagement in energy markets can take several forms [9,27] and the concept of CE is subject to different interpretations within the academic literature. Some define it in a broad sense: any sustainable energy initiative led by non-profit organizations, not commercially driven or government led [4,28], others have stressed the grassroots innovation nature of CE, as driven by civil society activists and by social and/or environmental needs, rather than rent seeking [29]. Overall, citizens' participation is commonly identified as a major defining characteristic of CE, but it can encompass a wide range of initiatives: green associations, collective purchasing of energy services, community or local authority led schemes for renewable energy implementation, community programme for energy poverty alleviation [17,30,31]. Such variety would in turn imply different levels and forms of participation and co-determination of citizens in energy services provisions. Similarly to other relevant contributions in the literature [13,29,32,33], this paper takes a specific perspective in interpreting citizens' participation in energy service provision by focusing on CE initiatives:

1.   which imply a form of citizen ownership or financing of an energy project, as well as control over the initiatives;

2. where citizens directly benefit from the outcomes of the initiative.

This study will not focus on other forms of civic engagement in the energy service provision, such as green associations, collective purchasing of energy services, and ethical consumerism, although present and active in the Italian energy ecosystem and in some instances involved in emerging CE initiatives studied in this paper [24]. The historical hydroelectric cooperatives established in Italian alpine regions at the beginning of the 20th century are also not included in the analysis. They are very specific and currently not replicable cases, functioning as a group of special legal status which in particular allow them to own and manage the local distribution network. Instead, this paper specifically looks at paradigms of citizens' financial and ownership involvement in energy initiatives which began appearing in Italy and the rest of Europe since the late 2000s [12,15]. They are mostly initiatives focused on development of renewable energy production facilities and, most of all, differentiate themselves from Italian historical cooperatives as they do not benefit from their special legal status and cannot own local distribution networks. We took a stepwise approach to investigate the Italian CE sector (Figure 1).

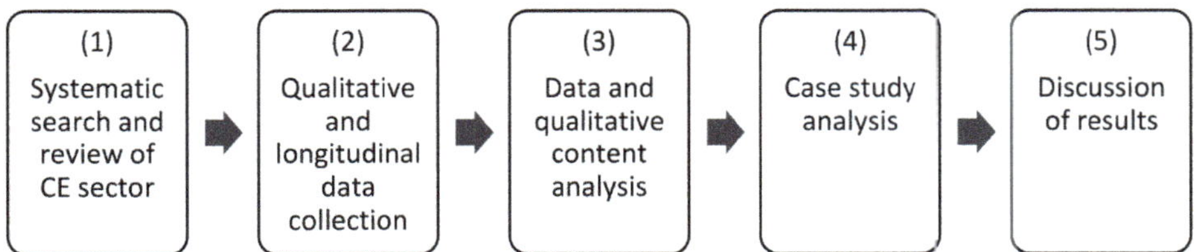

Figure 1. A stepwise approach to investigate Italian community energy (CE) sector.

The first step was a systematic search and review of CE initiatives in Italy (step 1 in Figure 1). A starting point in the search was the REScoop energy cooperatives inventory [18] which has been integrated through web-based searches as well as interviews with relevant Italian organizations and stakeholders. These included regional and national green organizations (such as Energoclub, Gas Energia), the Italian ethical bank which has financed several CE initiatives (Banca Etica) and researchers active in the field [24]. Although the majority of the population has certainly been targeted, it is realistic to assume that some initiatives have slipped through the searching net. This could in particular apply to early stage and civil society led projects not connected to relevant networks and without web presence. The systematic review allowed identification of 17 CE projects in Italy providing a level of financial and/or ownership involvement of citizens.

We then collected qualitative and longitudinal data on the identified initiatives (step 2 in Figure 1) through semi-structured interviews with one to two representatives for each of them. In some instances, further communication exchange with the representative (both in person and through emailing) allowed us to fine tune and better understand information and data gathered. We gathered data and evidence along the following dimensions:

- Dynamics of creation, including information on the timing, the proponent, and the approach adopted for the development of the initiative. We define bottom up approaches as those characterized by strong involvement and initiatives of citizens or other types of grassroots organizations in the initiation and development of the project. Top down approaches are instead those where it is an institution (i.e., a local authority or a private company) leading the process, defining structural features of the project and facilitating citizens' involvement.

- Type of activity and economics, including information on their primary activity (whether energy production, energy consumption, energy services, or a mix of those), characteristics of the projects implemented (e.g., technology type, plant size), and geographical scope of the initiatives (in particular whether citizens involved are geographically close to the project (local) or spread over the national territory (national)).

- Organizational structure, including legal form adopted (e.g., cooperative, limited company, or other forms), financing structure (i.e., self-funded, bank loan, coop funds, or a combination of those), finance instrument offered to the citizens (i.e., equity or debt) and ownership structure and level of citizens' involvement.
- Outcomes of the initiatives in terms of benefits offered to members/users, including monetary benefits (returns on investment offered, potential savings on electricity bills) and any other services and benefits accruing from the project (e.g., other energy or community services provided).

We then organized and analyzed data collected together with interviews transcripts and notes (step 3 in Figure 1). The objective of this evidence gathering was to provide a comprehensive picture of the heterogeneity of the Italian CE sector, to analyze their dynamics of creation, organizational dynamics and level and forms of citizens' engagement, their type of activity and timing, as well as their outcomes delivered.

Following on we undertook an in-depth comparative case study analysis (step 4 in Figure 1) of three specific CE initiatives in order to provide a further understanding of CE initiatives conditions for development as well as of success within the Italian energy system (step 5 in Figure 1).

3. Results of Systematic Review of Italian CE Sector

We used the evidence gathered through the systematic review of the Italian CE initiative to explore their characteristics, dynamics of development, and the forms and level of citizens' involvement. Although rather complete, the sample is relatively small (17 experiences), but nonetheless provides a snapshot of the Italian CE sector to date and highlights some trends in their characteristics and in the conditions for their development. Data and evidence gathered are presented in Appendix A (Table A1, Table A2, Table A3) and discussed in what follows.

3.1. Dynamics of Creation and Organizational Structures

In Figure 2 we present the distribution of the initiatives between top down and bottom up approaches, i.e., showing to which extent the initiatives identified have been proposed and developed by citizens or other types of grassroot organizations (bottom up) or instead by an institution that defines the project and the form of citizens' involvement. The majority of the initiatives have been proposed through a top down approach; of those, five have been proposed by a municipality and seven by a commercial actor (either a company or a municipal utility). Only five initiatives have been initiated with a bottom up approach by either a group of citizens or green associations (Figure 2).

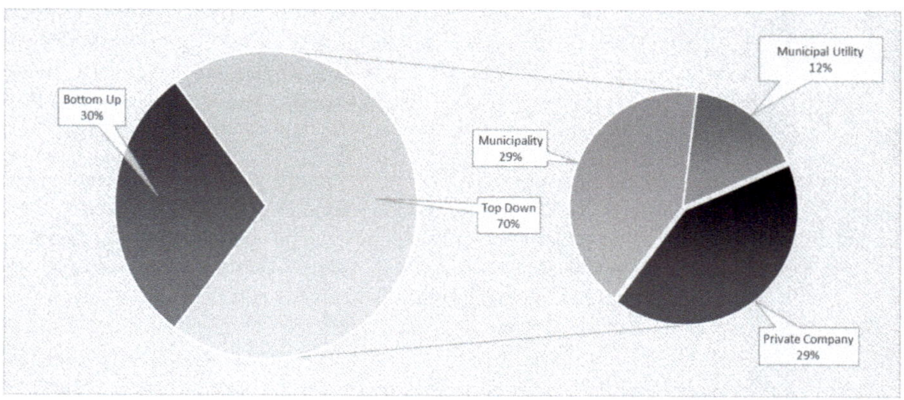

Figure 2. Dynamics of creation: top down versus bottom up approach and proponents.

The role of local authorities as facilitators of several projects also emerges, by providing the assets to develop the initiative, such as public building rooftops, or by creating the local regulatory and financing framework conditions to allow it. This reinforces literature views on their potential key position in facilitating energy transitions and influencing local energy system change [34–36].

As also experienced in other countries [12,17] the legal structure adopted varies, including limited companies, non-profit associations, and cooperatives, which account for about 60% of the sample (Table A1). Cooperatives are the legal form mostly used in the European CE sector [12,14,37,38] and are generally deemed to provide the best institutional framework for locally owned and participatory approaches to renewable energy projects. They encompass both the social and economic dimension in their scope and are characterized by a 'one head one vote' decision making process, with the aim to provide higher levels of co-determination [9,37,39,40]. However, generally speaking, the level of participation and co-determination of citizens is not determined only by the legal form adopted and the relative internal governance as defined by national laws and regulations. For example, in the case of cooperatives the 'one head one vote' may be applied only in the annual general assembly, resulting in a formal rather than a substantial approach to participation. In order to facilitate co-determination, a wider involvement and influence on the project development and management must be experienced by members of the initiative on a permanent basis and not only sporadically.

For example, Dosso Energia and Kennedy Energia are limited companies, but fully owned, financed, and managed by citizens located close to the renewable generation plant [41,42] (Table A1). Similarly, the Comunità Energetica San Lazzaro has been totally financed and managed by citizens (which also enjoy the relative economic returns and participate in the company governance) although the municipality has retained the formal ownership and the legal form adopted is an association [43]. Vice versa, evidence shows that some cooperatives may be included among initiatives reaching lower levels of participation and co-determination. They are those developed by companies and/or with a strong top down approach, e.g., Energyland, Masseria del Sole and Comunità Solare. The first two have been promoted by a company, which have firstly fully developed the renewable energy project to offer participation to citizens in a second phase. However, they reached lower levels of citizen ownership than initially planned and through longer processes than other initiatives (several months versus e.g., less than a month for Kennedy Energia [44,45]). Comunità Solare shows a similar experience, where ownership has been offered to citizens once PV systems had been already developed by local Energy Service Companies (ESCOs) resulting in very low citizens' involvement (less than 1% citizens' ownership [46]).

Overall, initiatives proposed by companies and with a strong top down approach have been developed with lower involvement of citizens and their organizational structure implies lower citizens' co-determination. This also emerges from the financing structure adopted: both the three cooperatives proposed by a company and the project proposed by a municipal utility have been initially financed through some form of project financing and then opened to citizens' financing in a second phase. Instead, initiatives promoted by communities and municipalities have been founded through direct financial contribution of citizens.

3.2. Type of Activity and Timing

CE projects have been deployed since the second half of the 2000s (Table A2), particularly since 2010 onwards. This timing coincides with the rapid increase in distributed renewable energy capacity installation in Italy as a result of the implementation of renewable energy support measures, in particular feed in tariffs (FiT) schemes for photovoltaic (PV) systems [47] (Figure 3).

Between 2008 and 2013 PV technologies have been benefiting from generous and uncapped FiT schemes [47] which have guaranteed fixed long-term tariffs and net-metering to PV system owners. Such strong policy support, combined with remarkable reductions in PV modules and installation costs since 2010 [53,54], has made PV investments quite profitable and relatively low risk in the wider context of the Italian energy sector. These favorable conditions have been a major driver for the development of Italian CE sector, opening a window of opportunity for the development of PV systems by proponents generally not equipped to deal with large, complex, and high-risk project development in the energy sector. Apart from one initiative providing electricity supply (È Nostra) and one dedicated to wind, electricity production from PV systems is in fact the primary activity across the whole sample (Table A2).

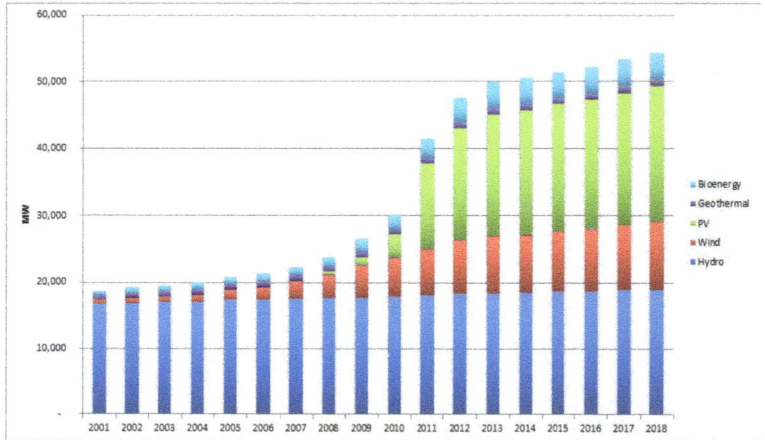

Figure 3. Renewable cumulative installed capacity in Italy (MW), 2001–2018 (data collected from reports by the Gestore dei Servizi Energetici (GSE)) [48–52].

With the reduction of FiT support in 2013 the Italian PV market has contracted (moving from 3.5 GW/year of installed PV between 2008 and 2013 to 385 MW/year in the period between 2013 and 2018, as shown in Figure 3) and the Italian CE sector with it. CE sector dependence from PV FiT incentives is clearly shown in Figure 4, which highlights how the majority of renewable energy plants have been developed between 2008, date of implementation of first FiT scheme in Italy, and 2013, date of discontinuity of FiT support to PV.

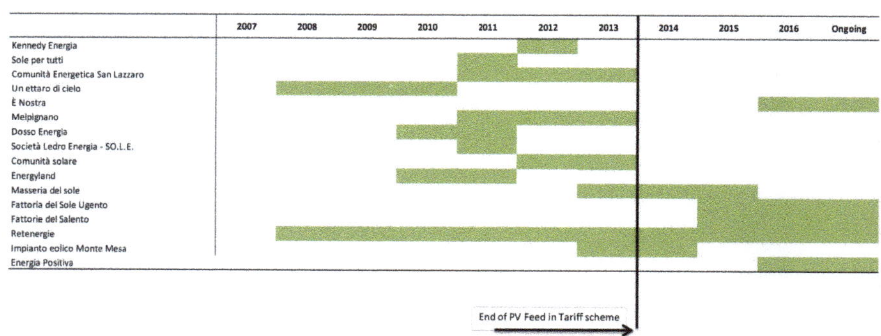

Figure 4. Timing of renewable energy plants development across CE initiatives.

Moreover, up to 2013, the Italian CE sector has been mainly characterized by the development of rather small, 'ad hoc' initiatives with a strong local focus. While PV systems installed vary in size and application, the majority are small/medium size projects, more easily developed and financed by actors with lower experience in the energy sector (see Table A2). The focus on smaller, roof mounted PV plants has also been reported by some representatives interviewed as a consequence of a deliberate choice of community or municipality led projects to focus activities on investments perceived more sustainable and with lower impact on the local environment than large ground mounted plants [41,42,55]. The largest projects (ground mounted PV systems in the megawatt range and a wind farm) have been developed by the initiatives led by commercial actors, either company or municipal utility (see also Table A1). They developed larger projects thanks to their higher internal technical knowledge and expertise which made the founding and development process easier; they were also more connected with economic networks which allow them to get access to capital more easily, making them able to develop more complex projects and bear higher risks (e.g., the risk of not raising enough capital among citizens to finance the investment).

Figure 4 shows how only a few CE initiatives have been developing renewable energy plants after the cancellation of the FiT in 2013, the larger ones and with a national scope in their activities or promoted by commercial actors: Retenergie (which has then merged with È nostra), Masseria del sole, Fattoria del Sole e Fattorie del Salento (the latter three developed by the same company, ForGreen), and Energia Positiva. Moreover, those still operating after 2013 have rarely developed new renewable energy plants and mostly focused their activity on acquiring operating PV plants on the secondary market, which are still benefiting from the FiT support. We will further analyze these initiatives in Section 3.4.

3.3. Outcomes of CE Initiatives

All the CE initiatives surveyed involve a form of financing or ownership from members against which a monetary return is offered. The returns on investment offered to citizens can vary quite substantially, from 8% to about 1% (Table A3). Such variation is particularly striking considering that most initiatives have been investing in the same energy technology, PV systems (see Table A2). This can be partly explained by the size and typology of the PV system: larger ground mounted plants allow higher economies of scale in the investment (both in terms of initial capital costs and transaction costs) and therefore higher returns than smaller roof mounted systems. However, what makes a stronger impact on the monetary returns offered to citizens is the typology of the initiative. Indeed, two distinctive typologies of initiatives emerge (Table A3):

Initiatives whose primary activity is the production of electricity from a renewable energy plant (in most cases PV) and having as their main objective the distribution among their members of the revenues accruing from the operation of a renewable generation project. The revenues are generally distributed in monetary terms or in electricity bill savings or a combination of both. These kinds of initiatives have generally developed a single renewable generation project, as unique primary activity. Higher financial returns, on average around 6%–8%, are offered by these initiatives. Sole per tutti is the only exception generating lower returns due to the inclusion of roof insulation in the initial total investment cost

Other initiatives which are set up not just to develop renewable energy plants and aggregate citizens around the relative financing and ownership, but also to offer other energy and social services to benefit both cooperative members and wider local communities. These initiatives generally offer on average lower financial returns on the investment as they tend to have more complex financing and organizational structures and, mostly, redistribute revenues from investments in renewable generation projects across a wider set of activities including those that do not generate monetary benefits for their members. An example is Retenergie which offers returns around 0%–3%, but besides fostering deployment of renewable generation plants offers to their members energy and community services, including domestic energy efficiency audits and consultancy, collective purchasing of energy services (for PV systems, storage, electric bikes, and cars as well as wider services such as discounted insurance, banking, internet provision) as well as wider community development schemes (such as information campaign or activities with schools) [46,55].

3.4. Case Studies

In what follows we focus the attention on three specific case studies: Retenergie/È nostra, WeForGreen, and Energia Positiva. They are the only CE initiatives that managed to continue activities after 2013. The following paragraphs describe and analyze the initiatives in greater detail and explore the reasons behind their success.

3.4.1. Retenergie and E'nostra

Retenergie was founded by 12 citizens in 2008 with a strong bottom up approach. Its aim was to "contribute to a new economy based on the principles of environmental sustainability, sobriety and solidarity" by promoting renewable production and supply as well as energy efficiency services [55]. By 2017 Retenergie had developed 13 projects, seven of which newly built PV rooftop plants, developed under FiT support (Table 1). Since the discontinuity of FiT support to PV in 2013 Retenergie has acquired four PV plants on the secondary market (hence plants initially developed under FiT support) and managed to develop a small wind power project (60 kW turbine located in Sardinia) and an energy efficiency project (the energy retrofit of a building in Vicenza acting as an ESCo) [56].

Table 1. Projects developed by Retenergie.

Plant Location	Secondary Market	Plant Operating Year	Operating Year by Retenergie	Total Investment Cost (k€)	Technology	Plant Size (kWp)
Piemonte, Cuneo	No	2011	2011	171	PV	50,63
Piemonte, Isola Bene vagienna	No	2011	2011	108	PV	30,38
Emilia-Romagna, Savigno	No	2011	2011	59	PV	15,51
Piemonte, Fossano	No	2011	2011	131	PV	44,65
Lombardia, San Giuliano Milanese	No	2011	2011	111	PV	29,44
Piemonte, Boves	No	2012	2012	655	PV	255,36
Piemonte, Lagnasco	No	2012	2012	44,5	PV	19,85
Sicily, Capizzi	Yes	2013	2015	499	PV	92,23
Sicily, Capizzi 2	Yes	.	.	.	PV	.
Veneto, Vicenza	NA	NA	2016	50	Energy Saving	NA
Sardegna, Nulvi	No	na	2016	330	WIND	59,99
Umbria, Bevagna	Yes	2011	2017	na	PV	47,25
Umbria, Bevagna	Yes	2011	2017	na	PV	198,65

Note: Plant operating year is the year in which the plant was installed and started operating; Operating year by Retenergie is the year in which Retenergie began operating it.

The cooperative has been growing steadily in members (Table 2) which have been progressively involved in the initiative through public meetings and campaigning in collaboration with social and environmental associations, collective purchasing groups, and other actors active in the Italian solidarity economy. It was later organized as a national initiative across territorial nodes in order to facilitate the development of local initiatives.

Table 2. Retenergie, summary of activities.

Summary of Activities	2009	2010	2011	2012	2013	2014	2015
Cumulative number of PV plants	0	0	5	7	7	7	9
Cumulative capacity installed (kWp)	0	0	171	446	446	446	630
Cumulative investment by citizens (k€)	0	0	628	1278	1278	1278	1575
Cumulative number of members	147	230	368	541	694	814	911
Return on capital	0	0	0	0	0	0	0
Return on social lending	3.5%	2.5%	2.5%	3%	3%	2%–3%	1.5%–3%

Renewable plant development has been mainly financed through members/citizens contributions (about 70% of the total investment, with the remaining 30% covered by debt) which could take two forms: (1) citizens can buy equity of the cooperative (minimum quote of 500 €) or, (2) they can finance the cooperative through social lending. In the first case returns for citizens depended on the annual profits of the cooperative and on the assembly decision on whether to redistribute them or keep them as reserve capital (to date the assembly has never earmarked any return on the capital invested, Table 2). Social lending returns were instead from 1.5% to 3% for two years to six years bonds.

Retenergie also offered a series of other services, which were granted against a membership of 50 € for those that had not already invested in the cooperative. They included discounts on different services and products (insurance, internet providers, bank services, magazines, and books) and collective purchasing groups for PV, storage systems, and electric vehicles. Retenergie had also established a network of energy advisors that offered discounted domestic energy audits to the members of the cooperative.

In 2014, Retenergie was one of the founding members of È nostra, the first electricity supply cooperative in Italy. È nostra activities started in 2015, with a membership campaign and in 2016 began to supply green electricity to its members, i.e., domestic and commercial consumers and not for profit organizations (the latter benefiting of a special tariffs). Table 3 presents the increase in members, contracts, and sales volume of È nostra between 2015 and 2018.

Table 3. Members, contracts, and sales volume of È nostra between 2015 and 2018.

Members, Contracts, and Sales Volume	2015	2016	2017	2018
Members	324	819	1662	4372*
Supply contracts	-	890	1963	3271
Energy sold (MWh)	-	1271	4270	8642

* Number of members after merging with Retenergie.

Since the beginning of the operations Retenergie and È nostra activities were closely linked: È nostra purchased from Retenergie the electricity produced by renewable energy plants and Retenergie offered to È nostra members the services provided by its network of energy advisors.

In 2018, Retenergie merged into È nostra, thus creating a cooperative able to provide both production and supply of renewable electricity and to serve a national community of prosumers, with the objective of enabling them to access sustainable electricity provision and energy services at better conditions than the traditional market.

This new EC stands on three pillars: production, supply, and energy services. The renewable electricity produced by the plants owned by the cooperative currently covers about 15% of the members' consumption and the remaining is covered with certified renewable electricity purchased on the national electricity market. Similarly to Retenergie, the new È nostra also provides energy services to its members, besides renewable electricity production and supply. The cooperative provides assistance to its members, both domestic and commercial, in designing energy efficiency measures, including energy audit, thermal plants renewal, insulation, and PV installation.

3.4.2. WeForGreen

ForGreen is a limited company born as a spinoff of an Italian multi-utility in 2010 [57] with the aim of developing PV systems and energy efficiency services. The first project, Energyland, was a 1 MWp ground mounted PV plant in Verona province. The project was initially fully financed by a local finance company (Finval) and opened to the participation of citizens afterward. It was intended mainly as a local project, addressed to people living in the Verona province. Citizens could invest in quotas of the plant, each meant to finance 1 kW of the PV plant at a cost of 3600 €, of which 1000 € was contribution to cooperative capital and 2600 € social lending. Citizens would get annually: (1) return on the capital invested (as determined by the annual assembly), here assumed to vary between 0% and 4%; (2) one twentieth of the social lending contribution, i.e., 130 € per year per quota; (3) the value of electricity bill savings over a consumption of 1000 kWh per year, per quota (for a varying electricity price, here assumed between 0.17 € and 0.20 € /kWh). Accounting for the variability of return on capital (0%–4%) and of the electricity price (0.17 € to 0.20 € /kWh), this sums up roughly to a return of 6.5% to 8.8% on the total investment (Table 4). The value of the electricity bill savings accounts for the higher share of returns offered to citizens (≈500–600 € per year). The initial aim was to involve around 333 people each contributing for 3 kW [57,58], in order to cover the full investment cost of 3.6 M€ [45]. In the end about 123 households have joined the cooperative, for a total of approximately 1 M€ (≈28% of the total investment) [45].

Table 4. Summary of Energyland offer and financial scheme.

Quota	3 kW
Initial investment	10,800 €
Capital	3000 €
Lending	7800 €
Annual return on capital (variable)	0 € to 120 € per year (0%–4%)
Annual return on lending	390 € per year (7800 €/20 years)
Annual electricity free of charge	3000 kWh per year
Value of electricity bill saving	510 € to 600 € per year (0.17–0.20 €/kWh)
Total return	6.5%–8.8%

The group of people that initiated the Energyland project decided to replicate the scheme on a national scale. In 2011 ForGreen developed a new 1 MWp PV plant in Apulia region, which was financed by the company through bank loan. In 2014 a new cooperative (Masseria del sole) was set up to give people the chance to invest in this PV plant. The financial scheme was very similar to Energyland with calculated expected returns for citizens investing of 8% (over 15 years). As in the case of Energyland, participation has been lower than initially planned, with 187 households joining the cooperative out of the about 300 initially planned [45].

Each project developed by ForGreen focuses on the development of a single plant and with the aim of supplying green electricity to its members through an electricity bill saving scheme, which represents a relevant component of the guaranteed return. The electricity produced by the PV plants is sold to an electricity supplier and each member of the cooperative gets an annual amount of kilowatt-hours free of charge for each kilowatt purchased. The change of supplier for each member is associated with the purchase of cooperatives shares, thus the size of the three cooperatives allowed ForGreen to have bargaining power on the electricity supply market. This in addition to its commercial background and other activities in the electricity sector.

In 2015 a new cooperative, WeForGreen Sharing, was founded. The cooperative now works as an umbrella for all projects. WeForGreen, besides managing the previous two projects, has developed three new projects, applying a similar structure to the previous ones: Fattoria del sole di Ugento and the two Fattorie del Salento (Table 5). These three additional PV plants are not new built, but they have been acquired by the cooperative in the secondary market of PV. They were built in 2011, thus still benefiting from FiT support. A 112 kW hydroelectric plant (named Lucense 1923) is also currently under development in Montorio, Veneto region, with expected annual production around 700 MWh. Similarly, to Retenergie/È nostra, WeForGreen has also integrated its activities with the supply of green electricity to its members. It is now possible to become member of WeForGreen in two different ways: Socio Autoproduttore (Self-Producing Member), by investing capital in the acquisition of quotas of existing generation plants, or Socio Consumatore (Consumer Member), by simply switching to ForGreeen 100% renewable electricity supply.

3.4.3. Energia Positiva

Energia Positiva was founded and promoted by one individual with the aim of "bringing to the market a participative initiative, which could bring benefits not just to the environment but to the whole collectivity". Energia Positiva started its operation in 2016, developing a new wind turbine project in Basilicata (Southern Italy). The Muro Lucano wind turbine (19.98 kWp for an expected annual production of approximately 64 MWh) required an investment of 126 k€ and was the first of a series of projects. By the end of 2019 Energia Positiva had developed 15 projects, 10 of which are PV plants benefiting of FiT support acquired on the secondary market (for a total of 1.5 MWp approximately, see Table 6). The cooperative has also acquired one additional 20 kW wind turbine and developed four energy saving projects. At January 2020, Energia Positiva reports a total investment almost 5 M€ by 415 members (average investment of about 12,000 € per member) [59].

Table 5. Projects developed by WeForGreen.

Project Name	Plant Location	Secondary Market	Plant Operating Year	Operating Year by WeForGreen	Total Investment Cost (k€)	Technology	Plant Size (kWp)
Energyland	Veneto, Cerro Veronese	No	2011	2011	3.6	PV	997,81
Masseria del sole	Puglia, Lizzanello	No	2011	2013	1	PV	997,92
Fattoria del sole di Ugento	Puglia, Ugento	Yes	2011	2015	1 000	PV	998,40
Fattorie del Salento 1	Puglia, Racale	Yes	2011	2017	NA	PV	999,60
Fattorie del Salento 2	Puglia, Ugento	Yes	2010	2017	NA	PV	997,92

Note: Plant operating year is the year in which the plant was installed and started operating; Operating year by WeForGreen is the year in which WeForGreen began operating it.

Table 6. Projects developed by Energia Positiva.

Plant Location	Secondary Market	Plant Operating Year	Operating Year by Energia Positiva	Total Investment Cost (k€)	Technology	Plant Size (kWp)
Puglia, Ortelle	Yes	2012	2019	147	PV	94,08
Puglia, Surano	Yes	2012	2019	306	PV	185,22
Piemonte, Torino	Yes	2018	2019	31	Energy Saving	115
Lombardia, Arcore	Yes	2019	2019	114	Energy Saving	343,3
Piemonte, Druento	Yes	2009	2018	72	PV	19,32
Abruzzo, Giulianova	Yes	2010	2018	71	PV	19,74
Puglia, Surbo	Yes	2012	2019	276	PV	197,76
Lombardia, Varedo	Yes	2011	2018	735	PV	573,5
Lombardia, Trivolzio	Yes	2012	2017	582	PV	187,2
Puglia, Sant'agata di Puglia	Yes	2009	2017	58	WIND	20
Piemonte, Anzola	Yes	2017	2017	16,5	Energy Saving	5,5
Piemonte, Dusino S. Michele	Yes	2011	2016	185	PV	66
Piemonte, Valfenera	Yes	2011	2016	255	PV	99,88
Piemonte, Villanova d'Asti	Yes	2013	2016	282	PV	88,5
Basilicata, Muro Lucano	No	2016	2016	126	WIND	19,98

Note: Plant operating year is the year in which the plant was installed and started operating; Operating year by Energia Positiva is the year in which Energia Positiva began operating it.

To become members of Energia Positiva individuals invest in quotas of the cooperative, which are linked to specific projects in order to become owners of a "virtual renewable energy plant". The return on the investment is guaranteed with a direct discount on the electricity bill. Energia Positiva in fact manages the electricity bill of its members, in partnership with Dolomiti Energia a national supplier of green electricity with more than 400,000 customers.

The member benefits from a discount on the electricity bill equal to 5% of the investment and can buy a maximum number of quotas equivalent to its annual electricity consumption. Furthermore, Energia Positiva has qualified as an innovative start-up, which, under the current Italian regulation, implies a tax rebate. If the member stays in the cooperative for at least three years, he or she can obtain a tax rebate equal to the 30% of the capital invested. Assuming an average customer with an annual consumption of 2700 kWh, in Table 7 we calculate possible total investment and benefits. Considering an investment of 10,500 € for a duration of 10 years, the internal rate of return is approximately equal to ≈9%.

Table 7. Energia Positiva possible total investment and benefits for an annual consumption of 2700 kWh (electricity bill 525 €/year).

Coverage of the annual electricity bill	30%	60%	100%
Number of quotas to be subscribed (each quota 500 €)	6	13	21
Total investment	3000 €	6500 €	10,500 €
Annual energy bill	525 €	525 €	525 €
Annual discount (5% of the investment)	150 €	325 €	525 €
Residual electricity bill	375 €	200 €	0 €
Tax rebate (30% of the investment)	900 €	1950 €	3150 €
Internal rate of return	≈9%	≈9%	≈9%

Energia Positiva offers membership only to domestic customers. In order to expand the activities, the promoters very recently set up a parallel cooperative (EpCo), which offers the same participation model (investment in virtual renewable energy plant to benefit from electricity bill savings) to commercial customers. To further support their activity of development and acquisition of renewable energy plants they also ran in 2019 an equity crowdfunding campaign which raised about 650,000 €.

4. Discussion

After decades of inaction the CE sector in Italy has experienced a new growth between 2008 and 2013 with the development of initiatives aimed at people engagement in the energy transition. The majority were local energy community projects, mostly developing PV plants generally of a size below 100 kW, and only very few were initiatives with wider territorial scope and able to develop megawatt size plants or different projects summing up to several hundred of kilowatts.

Despite the prevalence of the local dimension, only a few initiatives (the 24%) have been developed with a bottom up approach, hence characterized by strong involvement of citizens or other types of grassroots organizations in the initiation and development of the project. The majority have been developed with a top down approach, i.e., with an institution (i.e., a local authority or a private company) leading the process, defining structural features of the project and facilitating citizens' involvement. Among those, the role of municipalities and municipal utilities is nonetheless remarkable, which have often acted as promoters or as facilitators of the initiatives.

As also experienced within the CE sector in other European countries, the cooperative emerges as the most utilized legal form. However, evidence presented shows that, although it implies 'a one head one vote' rule, it does not necessarily bring high levels of citizens' participation in the development and in the decision process. The level of participation rather depends on the practices adopted. Overall, initiatives proposed by companies and with a strong top down approach have been developed with lower involvement of citizens and their organizational structure implies lower citizens' co-determination.

A major driver for this new wave of CE initiatives in Italy has been the implementation of the FiT scheme support, which has made PV investments quite profitable and relatively low risk, thus suitable for shared ownership projects and accessible to small scale, local projects. All but two (Energia Positiva and È nostra) of the CE initiatives mapped were established between 2008, date of implementation of first FiT scheme in Italy, and 2013, which has marked a watershed for the Italian CE sector. Since the progressive discontinuity of risk reducing support mechanisms such as FiT and the reintroduction of market-based support (such as capacity and auction-based mechanisms) the scaling up of the sector, either by developing large plants or replicating smaller projects, has proven in most cases to be challenging. The small-scale model became not any more profitable and sustainable, and new approaches were needed.

The three case studies presented managed to start (in the case of Energia Positiva) or continue their activities because they all embraced a different avenue from the small, local scale approach. Firstly, they all increased their activities by focusing on larger size projects and/or developing multiple projects. As a consequence of this evolution they enlarged the territorial scale of their activities, both by developing projects in different locations across the country and by involving members at a national scale. They thus managed to achieve economies of scale in their activities, which allowed them to involve and hire professionals as permanent staff and progressively enhance the services provided.

In a context of a contracted Italian renewable energy market (as also discussed in Section 3.2, Figure 3) they managed to develop new projects mainly focusing on the secondary market of PV, thus investing in PV plants with higher profitability and lower risk because they were still benefiting from the FiT support. More recently they have started differentiating projects activities by also developing energy efficiency projects and a wind project (Energia Positiva and È nostra).

As a result, although on one hand they lost the local dimension in project development, on the other they could integrate the proposition offered to their members with a model which combines participative renewable energy production with provision of green electricity. They managed to do so in different ways. È nostra is a proper electricity supplier and directly provides electricity to its members (which then also participate in the investments in renewable energy production) and to non-members. Energia Positiva and WeForGreen instead provide the service through an agreement with other green electricity supply companies, and link the electricity supply directly (and proportionally) to the investment of their members into the renewable production plants. Nonetheless, they all have developed a sort of 'virtual' prosumer model, which allows citizens across the entire Italian territory to support and participate in their renewable energy production projects while also directly consuming green electricity.

In summary the three case studies have managed to continue their activities after 2013 because they have grown to a national scale, have developed multiple projects, and have expanded their member base over time, also thanks to a progressive diversification of their proposition, i.e., offering a combination of production with consumption of green electricity.

A closer look at the three initiatives also highlights different approaches in their development and growth over time. Evidence presented on the Italian CE sector has highlighted two typologies of initiatives: those whose primary activity is the production of electricity from a renewable energy plant (having as their main objective the distribution among their members of the revenues accruing from its operation), and other initiatives which are set up not just to develop renewable energy plants and aggregate citizens around the relative financing and ownership, but also to offer other energy and social services to benefit both cooperative members and wider local communities. Energia Positiva and WeForGreen belong to the first typology. Indeed, Energia Positiva's growth seems to be rooted in the successful replication of a model in which the investments in single renewable energy projects are shared among members through a sort of quota system in exchange of participation in the revenues accruing from them (despite in the form of electricity bill savings). This modular approach has allowed a constant grow over time of the initiative, which has been steadily developing projects and has recently raised more finance through a crowdfunding campaign to support further expansion. A very similar approach has been followed by WeForGreen which has developed less projects than Energia Positiva, but of larger size, probably thanks to the fact that they are supported by an energy company. These typologies of initiatives are typically able to offer higher financial returns and, among all those mapped within the Italian CE sector, Energia Positiva and WeForGreen offer the highest, around 8%–9%.

Retenergie and È nostra belong to the other typology of initiatives. Retenergie was a bottom up initiative, initially constituted with the aim of promoting renewable energy production, supply, and energy services. Over the years, Retenergie activities have in fact been focusing on the development of collectively owned renewable energy plants, but also on offering energy and community services to its members. The structure of the initiative was more complex than Energia Positiva and WeForGreen, both in term of its activities (as it combined renewable electricity production projects with energy and community services) and in its financial structure and citizens' engagement process. Revenues generated by investments in renewable generation projects have been redistributed across a wider set of activities (including energy and community services), which did not generate direct monetary benefits for their members. This has resulted in lower returns offered to their members (ranging from

0% to 3%). Compared to Energia Positiva and WeForGreen, such more complex structure would make quick replication and upscale of the model less viable. Nonetheless, the cooperative managed to continue growing, probably thanks to its longer history (practically one of the first to be founded in Italy), its national scope, a large member base, and the development of an internal structure of permanent staff at the time of the contraction of the renewable energy market in Italy. In addition, the merging with È nostra has been crucial, which has allowed expansion of its member base and support of additional activities. The new È nostra that emerged from the merging has further diversified the initial proposition of Retenergie, by providing to its members not only collectively owned renewable energy production and energy services, but also electricity supply.

In conclusion the evidence presented on the three case studies highlight different scaling up strategies which are affected by the choices on the initiatives' activities and organizational structures made since the founding stage of the initiatives [60]. Energia Positiva and WeForGreen follow a growth path more focused on serving mutual interest (i.e., serving the interest of their members) while Retenergie/E Nostra scale up has been more informed by general/public interest (i.e., serving the broader interest of society) [26].

5. Conclusions

This paper elicits and presents novel evidence on CE initiatives that emerged in Italy in the 2000s, filling a gap in the literature to date. The findings of this study contribute to better understand the different phases in the development of the Italian CE sector and to explore the conditions that made some initiatives more successful than others.

The evidence presented in the systematic review depicts an Italian CE sector still at its niche level. It has been initially mainly characterized by the development of rather small, 'ad hoc' initiatives, for the majority dedicated to PV system deployment and with a strong local focus. Its development has been largely dependent on generous PV FiT schemes and since its discontinuity in 2013, only three larger initiatives have been able to keep growing and diversifying their activities (i.e., Retenergie/È nostra, WeForGreen, and Energia Positiva). This has been possible thanks to a progressive change in the business and implementation model. They have moved from a paradigm of small, local CE initiatives to a large and national scale, expanding their member base, developing multiple projects, and integrating the proposition offered to their members with other activities, including green electricity supply. This has allowed them to benefit from economies of scale, to hire permanent staff, and become more professional in their service provision.

Recently, community energy has attracted the attention of the legislator both at EU and national level, with a progressive recognition of its potential role within the EU as well as the Italian energy system. In Table 8 we summarize the most relevant legislative milestones for the Italian CE sector.

Energy communities were first mentioned within the Italian legislation and regulatory framework by the Italian Energy Strategy in 2017 and, subsequently, by the National Energy and Climate Plan in 2018. However, they were both legislative framework documents which did not imply any concrete measure to support the implementation of community energy initiatives in the country. In 2018, the Piedmont region implemented a law on energy communities, which has mainly been a declaration of intent, although politically relevant, being the first legislative initiative explicitly dedicated to the Italian CE sector. A recent call for proposal launched by RSE (a public company devoted to research on the energy system) is also acting as showcase and test of pilot projects of energy communities, here intended as local, collective self-consumption initiatives. The conclusions of these pilot experiences are likely to provide the supporting evidence for the design of new incentive schemes currently under discussion.

Table 8. Summary of recent legislative and regulation developments having an impact on the Italian energy community sector.

Date	Level	Legislative and Regulation Developments
November 2017	National	The Italian Energy Strategy is the first national document explicitly mentioning energy communities
August 2018	Regional	A new regional law promoting energy communities was approved in Piedmont
December 2018	National	The National Energy and Climate Plan wants to promote self-consumption (prosumer) and energy communities but it is not clear how (the only explicit measure highlighted is the simplification of authorization procedures)
December 2018	EU	Directive (EU) 2018/2001 on the promotion of the use of energy from renewable sources introduces and promotes renewable energy communities
June 2019	EU	Directive (EU) 2019/944 on common rules for the internal market for electricity introduces and promotes citizen energy communities
July 2019	National	New decree that re-introduces subsidies for renewable electricity (except PV)
January 2020	National	Energy communities pilot projects will be developed following a consultation paper promoted by the Energy Authority and two call for proposal by RSE (Ricerca sul Sistema Energetico, a public company devoted to research on the energy system)
February 2020	National	A provision of Law 8/2020 allows small-scale collective self-consumption of renewable energy plants below 200 kW for customers linked to the same low voltage distribution sub-grid

The process of national implementation of the two EU directives (December 2018 and June 2019) supporting two different models of energy communities (renewable energy communities and citizen energy communities) is creating the momentum for the possible design of a national legislative framework in support of the development of the CE sector. In particular, EU Directive 2018/2001 defines the framework for the implementation of place-based renewable energy communities, with the objective of fostering local self-consumption as well as collective self-consumption. The focus is on experiences that link production and consumption on a proximity base. As an initial step toward the national implementation of the EU Directive, a provision has been included in the recent Italian Law 8/2020 to allow small-scale, collective self-consumption of renewable energy plants of size below 200 kW, for customers linked to the same low voltage distribution sub-grid. A typical case is the block of flats, where the electricity produced by a collective PV plant can now be directly supplied to the customers living in the flats.

This regulatory framework goes in the direction of reducing the distance between production and consumption (with positive impacts on grid management), thus increasing the opportunities for citizens and consumers to become prosumers. The three larger Italian EC initiatives presented in the case studies have already made a step in this direction, by integrating their electricity production activities with green electricity supply. They have done so by developing different types of 'virtual' prosumer models, allowing citizens across the entire Italian territory to participate to their renewable energy production projects while also directly consuming green electricity. However, these models work on a national scale, while the evolution of the Italian regulatory framework is likely to foster the development of new small scale, local initiatives across the country.

Thus, in terms of business models, the regulation could lead to a renewed development of local, place-based energy communities. These energy communities could well be deployed by small, local initiatives which might not require a complex organizational structure, including permanent and professional staff. On the other hand, national energy communities (such as those presented in the case studies) may also be well placed to deliver new energy community initiatives, as they might benefit from economies of scale, from a deeper understanding of the energy market and regulation as well as of an internal organization supported by professional permanent staff. An open question remains regarding how they will be able to reconcile the national, larger size of their business models with the dynamics of community engagement at the local level, including the possibility of guaranteeing a high level of participation of their members in the decision processes.

In conclusion, the national evolution of the regulatory framework for energy communities joint with the renewed national support to renewable energy, implemented in July 2019, will progressively shape the CE sector in Italy, which might be on the verge of a profound evolution. As of February 2020, only a first step has been taken by the national legislators (Law 8/2020), which enables small scale initiatives (below 200 kW). Which other CE implementation models that will be supported by the legislator will depend on the policy decisions that will be taken in the future steps of the EU Directive implementation process. Whether this will lead to a revival of local, small-scale experiences as those developed in the 2008–2013 period or will reinforce the national paradigm developed by the larger Italian CE initiatives (or a combination of both) is an open question worthy of analysis and discussion in the future.

Author Contributions: Conceptualization, C.C. and G.R.; methodology, C.C.; data curation, C.C. and G.R.; writing—original draft preparation, C.C. and G.R.; writing—review and editing, C.C. and G.R.; funding acquisition: C.C.; supervision, C.C. Both authors have read and agreed to the published version of the manuscript.

Funding: Research funding from the Marie Curie Fellowship LocalEnergy (FP7-PEOPLE-2012-IEF, GA-331081) and from COMETS (COllective action Models for Energy Transition and Social Innovation), Call Reference N°: H2020-LC-SC3-2018-2019-2020, Grant Agreement N°: 837722 are gratefully acknowledged.

Acknowledgments: We also acknowledge the data update effort by Veronica Lupi in Oct–Dec 2019.

Conflicts of Interest: The authors declare no conflict of interest

Appendix A

Table A1. Dynamics of creation and organizational structure.

Project	Project ID	Proponent	Approach	Legal Form	Instrument for Citizens	Ownership Structure	Citizens Ownership (%)	Citizens Involved	Financing Structure
RETENERGIE	1	Community	Bottom up	coop	Equity/Debt	Citizens	100%	915	70% citizens + 30% debt (bank)
DOSSO ENERGIA	2	Mix** (Community & Association)	Bottom up	Ltd company	Equity	Citizens	100%	64	100% equity (citizens)
SOCIETA' LEDRO ENERGIA SO.L.E.	3	Community	Bottom up	coop	NA	Citizens	NA	260	NA
E'NOSTRA	4	Mix (Associations & Companies)	Bottom up	coop	Equity	Citizens + Proponents	80%	300	80% equity (citizens) - 20% (proponents)
MELPIGNANO	5	Municipality	Top down	coop	NA	Citizens	100%	136	100% debt (bank + legacoop)
KENNEDY ENERGIA	6	Municipality	Top down	Ltd company	Equity	Citizens	100%	50	100% equity (citizens)
SOLE PER TUTTI	7	Municipality	Top down	coop	Equity	Citizens	100%	62	40% equity (citizens) + 60% debt (bank)
COMUNITA' ENERGETICA SAN LAZZARO	8	Municipality	Top down	Association	Equity*	Municipality*	100%*	74	100% equity (citizens)
COMUNITA' SOLARE LOCALE	9	Municipality	Top down	Associations	Equity**	Citizens + local ESCO**	0,5%	25	NA
UN ETTARO DI CIELO	10	Municipal Utility	Top down	Ltd company	Bond	Municipal Utility	0%	300	Initially financed by company then opened to citizens. 50% equity (Mun. Utility) + 50% debt (citizens)
IMPIANTO EOLICO MONTE MESA	11	Municipal Utility	Top down	Ltd company	Bond	Municipal Utility	0%	NA	NA
ENERGYLAND	12	Company	Top down	coop	Equity	Citizens + Company	~ 30%	123	Initially financed through private company capital, then opened to citizens
MASSERIA DEL SOLE	13	Company	Top down	coop	Equity	Citizens + Company	~ 90%	187	Initially financed through debt (bank), then equity opened to citizens
FATTORIA DEL SOLE	14	Company	Top down	coop	Equity	Citizens + Company	NA	152	Initially financed through debt (bank), then equity opened to citizens
FATTORIE DEL SALENTO 1	15	Company	Top down	coop	Equity	Citizens + Company	Still Open	175	Initially financed through debt (bank), then equity opened to citizens
FATTORIE DEL SALENTO 2	16	Company	Top down	coop	Equity	Citizens + Company	Still Open	175	Initially financed through debt (bank), then equity opened to citizens
ENERGIA POSITIVA	17	Citizens	Bottom up	coop	Equity	Citizens	100%	304	100% equity (citizens)

* Municipality formally owner of the PV system, but investment financed by citizens association, who manages the project and gets returns out of it. ** Initiative proposed by municipality, PV systems developed by local ESCO which then open ownership to citizens

91

Table A2. Type of activity and timing.

Project	Project ID	Start Date	Primary Activity	Technology	Plant Size (kWp)	Investment Cost (euro)**	Scope
RETENERGIE	1	2008	Mix (Electr production & energy services)	PV	879 kWp (spread over 12 projects)	2.2 Mn (collected from citizens Investments. Cumulated, 2016)	National
DOSSO ENERGIA	2	2010	Electr. Production	PV	74,56 + 29,36 + 5,04 kWp	369 k	Local
SOCIETA' LEDRO ENERGIA SO.L.E.	3	2007	Mix (Electr production & energy services)	PV	40 kWp + 59 kWp	NA	Local
E'NOSTRA	4	2014	Elect. Supply	-	-	NA	National
MELPIGNANO	5	2011	Mix (Electr production & energy services)	PV	180 kWp (33 plants, 4 of them sold to some members, 29 still under the coop ownership)	400 k	Local
KENNEDY ENERGIA	6	2013	Electr. Production	PV	100 kWp	170 k	Local
SOLE PER TUTTI	7	2011	Electr. Production	PV	102 kWp	450 k	Local
COMUNITA' ENERGETICA SAN LAZZARO	8	2011	Electr. Production	PV	20kWp	49 k	Local
COMUNITA' SOLARE LOCALE	9	2011	Mix (Electr production & energy services)	PV	1378kWp (56 plants)	3 M	Local
UN ETTARO DI CIELO	10	2008	Electr. Production	PV	1000 kWp	5 M	Local
IMPIANTO EOLICO MONTE MESA	11	2014	Electr. Production	Wind	8 MW (4 windtowers)	NA	Local
ENERGYLAND	12	2011	Electr. Production	PV	1000 kWp	3.6 M (about 1M allocated to citizens)	Local
MASSERIA DEL SOLE	13	2013	Electr. Production	PV	999 kWp	1 M	National
FATTORIA DEL SOLE	14	2015	Electr. Production	PV	998.4 kWp	1 M	National
FATTORIE DEL SALENTO 1	15	2017	Electr. Production	PV	999,605	NA	National
FATTORIE DEL SALENTO 2	16	2017	Electr. Production	PV	997,92	NA	NationL
ENERGIA POSITIVA	17	2016	Electr. Production	PV, EO, Idro, energy saving	1571.18 kWp (over 12 plants)*	3.3 Mn (Splitted in quotas of 500 € each. Cumulated value, 2019)	National

* Includes cost of roof insulation. **Investment costs are indicated only for initiatives focus on the development of a single electricity production plant. *** Acquisition and refinancing of an already operating ground mounted PV plant

Table A3. Outcomes.

Project	Project ID	Primary Activity	Return on Investment (%)	Other Monetary Benefits (Citizens/Municipality)	Other Energy Social Services
RETENERGIE	1	Mix (Electr production & energy services)	1.5-3	Monetary benefits (in various forms) for citizens providing assets (e.g. schools providing rooftops)	Collective electricity puchasing scheme for: domestic Pv systems, domestic storage, EV and other services (insurance, internet, bank services, editorial). Collective scheme for domestic energy efficiency audit
DOSSO ENERGIA	2	Electr. Production	~6	Municipality get annual rent for school rooftop use	Wider social engagement promoted by pre-exhisting green assosiation
SOCIETA' LEDRO ENERGIA SO.L.E.	3	Mix (Electr production & energy services)	NA	NA	Promoted: local collective electricity purchasing scheme; local electrical bike sharing scheme
E'NOSTRA	4	Elect. Supply	2	None	Working on pilot distribution of smart meters to cooperative members
MELPIGNANO	5	Mix (Electr production & energy services)	Not applicable	None	Electricity bill savings for end users providing assets (citizens). Scheme for water distribution and reduction of plastic bottles use
KENNEDY ENERGIA	6	Electr. Production	~6	Municipality gets value of electricity bill savings	Education activities in schools promoted by people involved in Kennedy energia
SOLE PER TUTTI	7	Electr. Production	~3	None	School providing roof space also gets roof insulation. Some of the electricity bill savings invested in the school activities
COMUNITA' ENERGETICA SAN LAZZARO	8	Electr. Production	NA	Municipality gets value of electricity bill savings	Promotion of energy efficiency schemes on local public buildings
COMUNITA' SOLARE	9	Mix (Electr production & energy services)	~3.5**	Annual electricity bill discount of 50 € for 20years for citizens	Scheme for domestic energy efficiency audit. Collective purchase scheme for: electric bike, EV, energy efficient appliances
UN ETTARO DI CIELO	10	Electr. Production	5,5(7 years bond);6,5(12 years bond)	None	Offered to citizens 25 allotment gardens on the PV gournd mounted plant field
IMPIANTO EOLICO MONTE MESA	11	Electr. Production	6,5 (7 years bond)	Royalties to municipality (~100k€/year)	Education activities (guided tours for schools)
ENERGYLAND	12	Electr. Production	6.5-8.8*	Electricity bill savings for citizens (proportional to quota)	None
MASSERIA DEL SOLE	13	Electr. Production	~8	Electricity bill savings for citizens (proportional to quota)	None
FATTORIA DEL SOLE	14	Electr. Production	NA	Electricity bill savings for citizens (proportional to quota)	None
FATTORIE DEL SALENTO 1	15	Electr. Production	NA	Electricity bill savings for citizens (proportional to quota)	None
FATTORIE DEL SALENTO 2	16	Electr. Production	NA	Electricity bill savings for citizens (proportional to quota)	None
ENERGIA POSITIVA	17	Electr. Production	~9	Electricity bill savings for citizens (proportional to quota)	None

* including value of electricity bill savings for 1,000 kWh per year, per quota. ** including value of electricity bill savings for 50 € per year for 20 years

References

1. EU Commission. European Energy Security Strategy. Communication from the Commission to the European Parlament and the Council. Brussels, COM 330 final. 2014. Available online: http://eur-lex.europa.eu/legal-content/EN/TXT/PDF/?uri=CELEX:52014DC0330&from=EN (accessed on 28 May 2014).
2. EU Commission. Directive 2009/28/EC of the European Parliament and of the Council of 23 April 2009 on the promotion of the use of energy from renewable sources and amending and subsequently repealing Directives 2001/77/EC and 2003/30/EC. Official Journal of the European Union. Available online: https://eur-lex.europa.eu/legal-content/EN/ALL/?uri=CELEX%3A32009L0028 (accessed on 2 April 2020).
3. EU Commission. Directive 2012/27/EU of the European Parliament and of the Council of 25 October 2012 on energy efficiency, amending Directives 2009/125/EC and 2010/30/EU and repealing Directives 2004/8/EC and 2006/32/EC. Official Journal of the European Union. Available online: https://eur-lex.europa.eu/legal-content/en/TXT/?uri=celex%3A32012L0027 (accessed on 2 April 2020).
4. Hall, S.; Foxon, T.J.; Bolton, R. Financing the civic energy sector: How financial institutions affect ownership models in Germany and the United Kingdom. *Energy Res. Soc. Sci.* **2016**, *12*, 5–15. [CrossRef]
5. Kempener, R.; Malhotra, A.; de Vivero, G.; The Age of Renewable Energy Power. Designing National Roadmaps for a Successful Transformation Power. *IRENA Power Sector Transformation report.* Available online: http://www.irena.org/DocumentDownloads/Publications/IRENA_PST_Age_of_Renewable_Power_2015.pdf (accessed on 15 October 2015).
6. IEA-RETD 2014. Residential Prosumers—Drivers and Policy Options. *IEA-RETD Report. (Revised version of June 2014).* Available online: http://iea-retd.org/wp-content/uploads/2014/09/RE-PROSUMERS_IEA-RETD_2014.pdf (accessed on 2 April 2020).
7. Lavrijssen, S.; Arturo, C.P. Radical Prosumer Innovations in the Electricity Sector and the Impact on Prosumer Regulation. *Sustainability* **2017**, *9*, 1207. [CrossRef]
8. Sokołowski, M.M. European Law on the Energy Communities: A Long Way to a Direct Legal Framework. *Eur. Energy Environ. Law Rev.* **2018**, *27*, 60–70.
9. ILO 2013. *Providing clean energy access through cooperatives*; International Labour Office Cooperative Unit Report; Green Jobs Programme: Geneva, Switzerland, 2013; Available online: https://www.ilo.org/global/topics/green-jobs/publications/WCMS_233199/lang--en/index.htm (accessed on 2 April 2020).
10. Van Der Schoor, T.; Scholtens, B. Power to the people: Local community initiatives and the transition to sustainable energy. *Renew. Sustain. Energy Rev.* **2015**, *43*, 666–675. [CrossRef]
11. REN 21 2016. *Renewables 2016 Global Status Report*; REN 21 Report; REN21 SecretariatParis, France, 2016; Available online: http://www.ren21.net/status-of-renewables/global-status-report/ (accessed on 19 July 2016).
12. Creamer, E.; Taylor Aiken, G.; Van Veelen, B.; Walker, G.; Devine Wright, P. Community renewable energy: What does it do? Walker and Devine-Wright (2008) ten years on. *Energy Res. Soc. Sci.* **2019**, *57*, 101223. [CrossRef]
13. Walker, G.; Devine Wright, P. Community renewable energy: What should it mean? *Energy Policy* **2008**, *36*, 497–500. [CrossRef]
14. Yildiz, Ö.; Rommel, J.; Debor, S.; Holstenkamp, L.; Mey, F.; Mullerr, J.R.; Radtke, J.; Rognli, J. Renewable energy cooperatives as gatekeepers or facilitators? Recent developments in Germany and a multidisciplinary research agenda. *Energy Res. Soc. Sci.* **2015**, *6*, 59–73. [CrossRef]
15. Spinicci, F.; Le Cooperative di Utenza in Italia e in Europa. Euricse Research Report, N.2/2011. Available online: http://www.euricse.eu/wp-content/uploads/2015/03/1296748019_n1615.pdf (accessed on 27 July 2016).
16. Capellan-Perez, I.; Campos-Celador, Á.; Teres-Zubiaga, J. Renewable Energy Cooperatives as an instrument towards the energy transition in Spain. *Energy Policy* **2018**, *123*, 215–229. [CrossRef]
17. Yildiz, Ö. Financing renewable energy infrastructures via financial citizen participation—The case of Germany. *Renew. Energy* **2014**, *68*, 677–685. [CrossRef]
18. Rescoop 2012. REScoop Energy Cooperatives Inventory. *Excel Database. Granted by REScoop, European Federation for Renewable Energy Cooperatives.* Available online: https://rescoop.eu/renewable-energy-citizen-initiatives/all?field_location_country=Allfield_membership_tid=All (accessed on 12 September 2019).
19. Bauwens, T.; Gotchev, B.; Holstenkamp, L. What drives the development of community energy in Europe? The case of wind power cooperatives. *Energy Res. Soc. Sci.* **2016**, *13*, 136–147. [CrossRef]

20. Boon, F.P.; Dieperink, C. Local civil society based renewable energy organisations in the Netherlands: Exploring the factors that stimulate their emergence and development. *Energy Policy* **2014**, *69*, 297–307. [CrossRef]
21. Seyfang, G.; Park, J.J.; Smith, A. A thousand flowers blooming? An examination of community energy in the UK. *Energy Policy* **2013**, *61*, 977–989. [CrossRef]
22. Bauwens, T. Analyzing the determinants of the size of investments by community renewable energy members: Findings and policy implications from Flanders. *Energy Policy* **2019**, *129*, 841–852. [CrossRef]
23. Wierling, A.; Schwanitz, V.J.; Zei, J.P.; Bout, C.; Candelise, C.; Gilcrease, W.; Gregg, J.S. Statistical Evidence on the Role of Energy Cooperatives for the Energy Transition in European Countries. *Sustainability* **2018**, *10*, 3339. [CrossRef]
24. MagnaniI, N.; Osti, G. Does civil society matter? Challenges and strategies of grassroots initiatives in Italy's energy transition. *Energy Res. Soc. Sci.* **2016**, *13*, 148–157. [CrossRef]
25. Magnani, N.; Maretti, M.; Salvatore, R.; Scotti, I. Ecopreneurs, rural development and alternative socio-technical arrangements for community renewable energy. *J. Rural Stud.* **2017**, *52*, 33–41. [CrossRef]
26. Magani, N.; Patrucco, D. Le cooperative energetiche rinnovabili in Italia: Tensioni e opportunità in un contesto in trasformazione. In *Energia e innovazione tra flussi globali e circuiti locali*; Pellizzoni, L., Osti, G., Eds.; EUT Edizioni Università di Trieste: Trieste, Italy, 2018.
27. DECC 2014. Community Energy Strategy: People Powering Change. Department of Energy and Climate Change. Available online: https://www.gov.uk/government/publications/community-energy-strategy (accessed on 12 January 2014).
28. Hatzi, S.; Seebauer, S.; Flei, E.; Posch, A. Market-based vs. grassroots citizen participation initiatives in photovoltaics: A qualitative comparison of niche development. *Futures* **2016**, *78*, 57–70. [CrossRef]
29. Seyfang, G.; Hielscher, S.; Hargreaves, T.; Martiskainen, M.; Smith, A. A grassroots sustainable energy niche? Reflections on community energy in the UK. *Environ. Innov. Soc. Transit.* **2014**, *13*, 21–44. [CrossRef]
30. Hain, J.J.; Ault, G.W.; Galloway, S.J.; Cruden, A.; Mcdonald, J.R. Additional renewable energy growth through small-scale community orientated energy policies. *Energy Policy* **2005**, *33*, 1199–1212. [CrossRef]
31. St Denis, G.; Parker, P. Community energy planning in Canada: The role of renewable energy. *Renew. Sustain. Energy Rev.* **2009**, *13*, 2088–2095. [CrossRef]
32. Seyfang, G.; Longhurst, N. Growing green money? Mapping community currencies for sustainable development. *Ecol. Econ.* **2013**, *86*, 65–77. [CrossRef]
33. Goedkoop, F.; Devine-Wright, P. Partnership or placation? The role of trust and justice in the shared ownership of renewable energy projects. *Energy Res. Soc. Sci.* **2016**, *17*, 135–146. [CrossRef]
34. Hannon, M.J.; Bolton, R. UK Local Authority engagement with the Energy Service Company (ESCo) model: Key characteristics, benefits, limitations and considerations. *Energy Policy* **2015**, *78*, 198–212. [CrossRef]
35. Rutherford, J.; Jaglin, S. Introduction to the special issue—Urban energy governance: Local actions, capacities and politics. *Energy Policy* **2015**, *78*, 173–178. [CrossRef]
36. Legambiente 2016. Comuni rinnovabili 2016. La mappatura delle fonti rinnovabili nel territorio italiano. Legambiente Report. Energy and Climate Department. May 2016. Available online: https://www.legambiente. it/sites/default/files/docs/rapporto_comuni_rinnovabili_2016.pdf (accessed on 2 April 2020).
37. Huybrechts, B.; Mertens de Wilmars, S. The relevance of the cooperative model in the field of renewable energy. *Ann. Public Coop. Econ.* **2014**, *85*, 193–212. [CrossRef]
38. Rescoop Best Practices Report. REScoop 20-20-20 Report. Available online: http://rescoop.eu/best-practices (accessed on 12 September 2019).
39. Sagebiel, J.; Muller, J.R.; Rommel, J. Are consumers willing to pay more for electricity from cooperatives? Results from an online Choice Experiment in Germany. *Energy Res. Soc. Sci.* **2014**, *2*, 90–101. [CrossRef]
40. Viardot, E. The role of cooperatives in overcoming the barriers to adoption of renewable energy. *Energy Policy* **2013**, *63*, 756–764. [CrossRef]
41. Garotta, S.; Project manager of Kennedy Energia, Inzago, Italy. Personal Communication, 2015.
42. Morbi, L.; Project manager Dosso Energia, Castelleone, Italy. Personal Communication, 2016.
43. Feltrin, L.; Technical officer of San Lazzaro di Savena municipality, San Lazzaro di Savena, Italy. Personal Communication, 2015.

44. Garotta, S.; Adotta un Pannello. Produrre in comune l'energia per le nostre esigenze comuni. Una via per una partecipazione diretta e attiva al bene comune. Available online: https://www.scuoladellebuonepratiche. it/wordpress/wp-content/uploads/2013/06/Adotta-un-pannello-Inzago.pdf (accessed on 9 April 2020).

45. Nicolis, G.; Zanini, G.; WeForGreen President and WeforGreen board member, Verona, Italy. Personal Communication, 2015.

46. Setti, L.; University of Bologna, Bologna, Italy. Personal Communication, 2016.

47. Antonelli, M.; Desideri, U. The doping effect of Italian feed-in tariffs on the PV market. *Energy Policy* **2014**, *67*, 583–594. [CrossRef]

48. GSE 2011. Rapporto Attività 2011. GSE Report. Available online: http://www.gse.it/it/Dati%20e%20Bilanci/ Rapporti%20delle%20attivit%C3%A0/Pages/default.aspx (accessed on 18 July 2016).

49. GSE 2014. Rapporto Statistico—Energia da Fonti Rinnovabili Anno 2014. Report by Gestore Servizi Elettrici. Available online: http://www.gse.it/it/Statistiche/RapportiStatistici/Pagine/default.aspx (accessed on 22 March 2017).

50. GSE 2016. Energia da Fonti Rinnovabili in Italia—Dati Preliminari 2015. Report by Gestore Servizi Elettrici. Available online: http://www.gse.it/it/Dati%20e%20Bilanci/GSE_Documenti/osservatorio%20statistico/ Stime%20preliminari%202015.pdf (accessed on 22 March 2017).

51. GSE 2019a. GSE, Rapporto Statistico—Energia da Fonti Rinnovabili Anno 2017 Report by Gestore Servizi Elettrici. Available online: https://www.gse.it/documenti_site/Documenti%20GSE/Rapporti%20statistici/ Rapporto%20Statistico%20FER%202017.pdf (accessed on 10 February 2020).

52. GSE 2019b. Energia da Fonti Rinnovabili in Italia—Rapporto Statistico 2018. Report by Gestore Servizi Elettrici. Available online: https://www.gse.it/documenti_site/Documenti%20GSE/Rapporti%20statistici/ GSE%20-%20Rapporto%20Statistico%20FER%202018.pdf (accessed on 10 February 2020).

53. Candelise, C.; Winskel, M.; Gross, R. The dynamics of solar PV costs and prices as a challenge for technology forecasting. *Renew. Sustain. Energy Rev.* **2013**, *26*, 96–107. [CrossRef]

54. Marigo, N.; Candelise, C. What is behind the recent dramatic reductions in photovoltaic prices? The role of China. Economia e Politica Industriale. *J. Ind. Bus. Econ.* **2013**, *40*, 5–41.

55. RETENERGIE 2008. Retenergie Statute. Available online: http://retenergie.pbworks.com/w/page/24763065/ statuto (accessed on 9 April 2020).

56. RETENERGIE 2017. Mappa Dei Progetti. Retenergie Website. Available online: http://www.retenergie.it/ impianti/mappa-degli-impianti (accessed on 2 May 2017).

57. FORGREEN 2010. Press Release. Available online: http://www.forgreen.it/da-multiutility-a-forgreen-nel-segno-del-risparmio-energetico (accessed on 9 January 2010).

58. Zanini, G. Energyland—La Filiera Delle Energie Rinnovabili Cittadini, Imprese e Amministrazioni Insieme per un Nuovo Modello di Sviluppo Economico nel Rispetto del Territorio e dell'Ambiente Presentation at Patto dei Sindaci Verona. Available online: http://portale.provincia.vr.it/uffici/uffici/7/723/documenti/patto-dei-sindaci/patto-dei-sindaci-20-20-20-attivita-anno-2011/convegno-il-patto-dei-sindaci-per-lo-sviluppo-ed-il-miglioramento-delle-politiche-energetiche-locali/energyland-la-filiera-delle-energie-rinnovabili/at_ download/file (accessed on 22 March 2017).

59. Energia Positia 2020. Energia Positiva Website. Available online: https://www.energia-positiva.it/ (accessed on 31 January 2020).

60. Bauwens, T.; Huybrechts, B.; Dufays, F. Understanding the Diverse Scaling Strategies of Social Enterprises as Hybrid Organizations: The Case of Renewable Energy Cooperatives. *Organ. Environ.* **2019**, 1086026619837126. [CrossRef]

Article

Empowering Vulnerable Consumers to Join Renewable Energy Communities—Towards an Inclusive Design of the Clean Energy Package

Florian Hanke * and Jens Lowitzsch

Faculty of Business Administration and Economics, European University Viadrina, 15230 Frankfurt, Germany; Lowitzsch@europa-uni.de
* Correspondence: hanke@europa-uni.de; Tel.: +49-(0)-335-5534-2560

Received: 15 February 2020; Accepted: 25 March 2020; Published: 2 April 2020

Abstract: The unequal distribution of costs and benefits of the energy transition is a challenge for energy justice and energy policy. Although the empowerment of consumers to participate in renewable energy communities (RECs) has great potential for a just energy transition, vulnerable consumers remain underrepresented in RE projects. The recast of the European renewable energy directive obliges the European Member States to facilitate the participation of vulnerable consumers and support their inclusion in its *"enabling framework"* for prosumership. However, the type and specific design of corresponding measures remains unclear. Against this background this article investigates consumer empowerment in a vulnerability context. In particular we stress the need to understand how vulnerability affects participation in RECs to inform both policy makers and practitioners on its specificities and restrictions for the *"enabling framework"*. To prevent the inclusion of vulnerable consumers in RECs from remaining an idea on paper lawmakers need to be made aware of the implications for a consistent *"enabling framework"*. We argue that both individual vulnerable consumers as well as RECs need incentives and support to boost RECs' capacity to include groups that until now remain underrepresented.

Keywords: renewable energy community; vulnerable consumer; consumer empowerment; energy transition; energy justice; Clean Energy Package

---

1. Introduction

Over the past five years, the European Union launched a new design for the Energy Union introducing common rules and new forms of cooperation between the various actors with the adoption of the so-called Clean Energy Package (CEP) in 2018/19 (a package of measures that the European Commission presented on 30 November 2016 to keep the EU competitive as the energy transition changes global energy markets). In the European Green Deal the European Commission (EC) proposes a "new growth strategy that aims to transform the EU into a fair and prosperous society, with a modern, resource-efficient and competitive economy" [1]. To comply with its international commitments under the Paris Agreements the EU promotes a comprehensive energy transition, promoting renewable energy (RE) and entailing a more decentralised energy system with consumers becoming producers of the energy they consume, that is, prosumers. The rationale behind incentivising energy prosumership is triggering private investments for the development of RE installations while assuming more energy efficient behaviour [2,3]. The recast of the renewable energy directive (RED II) [4] as well as the internal electricity market directive (IEMD) [5] promote this new role of consumers as prosumers of (renewable) energy both as (1) individual and/or jointly acting self-consumer, and (2) organised in citizen energy communities (CECs) and renewable energy communities (RECs). To boost the deployment of RE the

RED II further obliges Member States to introduce an *"enabling framework"* to allow RECs competition "on an equal footing" [4] (Recital 26) with established actors on the energy market and enshrines the protection and empowerment of vulnerable energy consumers. Both the European Green Deal as well as RED II promote the inclusion of vulnerable consumers in RECs as a form of consumer empowerment and a means to fight energy poverty. Art 22 para. 4 RED II specifies:

> *"4. Member States shall provide an enabling framework to promote and facilitate the development of renewable energy communities. That framework shall ensure, inter alia, that:*
>
> *(a) unjustified regulatory and administrative barriers to renewable energy communities are removed;*
>
> *. . .*
>
> *(f) the participation in the renewable energy communities is accessible to all consumers, including those in low-income or vulnerable households;*
>
> *(g) tools to facilitate access to finance and information are available;*
>
> *(h) regulatory and capacity-building support is provided to public authorities in enabling and setting up renewable energy communities, and in helping authorities to participate directly;*
>
> *(i) rules to secure the equal and non-discriminatory treatment of consumers that participate in the renewable energy community are in place."*

(emphasis by the authors)

Above catalogue stipulates the minimum requirements of the *"enabling framework"* while EU lawmakers have left room for other support measures (*"inter alia"*). In the National Energy and Climate Plans (NECPs) all European member states outline respective national energy frameworks, policies and measures to comply—inter alia—with the Union Climate and renewable energy (RE) targets. According to RED II Art 22 para. 5, *"(t)he main elements of the enabling framework (...) and of its implementation, shall be part of the updates of the Member States' integrated national energy and climate plans and progress reports (...)"*. The NECPs shall thus include the design of an "enabling framework" inter alia enabling the participation of vulnerable households including possible tax incentives and exemptions from levies. In consequence prosumption should provide tangible benefits in form of lower energy costs, additional revenues and local economic development [6]. In the context of this paper we focus on the design of a socially just energy transition and the promotion of RE and therefore focus on RECs as foreseen in the RED II.

In this context, the European Commission (EC) promotes consumer empowerment by providing access to information and extending consumption options e.g., in form of facilitated supplier switching or engaging in RE prosumption [7]. At the individual level this includes the need to choose among consumption/prosumption options requiring individual cognitive capacity to process relevant information [8]. Moreover, prosumership requires the willingness to take risks, access to financing as well as time and know-how [6]. While national policies promote prosumership as a central element of the energy transition, only those fulfilling these requirements can acquire their own RE installation and thus benefit from the *"enabling framework"* and its subsidy schemes. Those who do not—mainly vulnerable consumers—not only do not benefit from an *"enabling framework"* but carry the increased burden of rising grid tariffs, levies and energy costs [9]. While those well off can benefit from prosumption more than 50 million people in the EU cannot afford an adequate level of energy consumption and live in energy poverty [10] detrimental to their personal wellbeing and to notions of equity and justice [11].

This unequal distribution of costs and benefits of the energy transition has increasingly become subject of debate not only in academic literature with a focus on energy justice [12] and energy poverty [13] but also in energy policymaking at the European level. While the CEP promotes their participation in RECs as a key factor to overcome energy poverty the European legislator does not specify how to achieve this aim. As will be discussed in detail the empowerment of vulnerable consumers to become prosumers is linked to energy poverty mitigation. However, the current member structure of existing RE projects for instance in Germany [14] points at a need to facilitate the empowerment of vulnerable consumers who currently remain underrepresented in RE projects.

2. Research Question and Approach

Against this background this paper seeks to answer the following questions:

(1) How is the empowerment of vulnerable consumers to be conceptualised in the Energy Union and which form should it take? Individual empowerment in form of consumer empowerment has been frequently discussed in contemporary energy policy discourses. The transposition of the *"enabling framework"* in the context of RED II as a central element of the empowerment of vulnerable consumers has great potential yet the question of how to achieve it and with which concrete measures remains open.

(2) What are the main barriers for empowerment in a vulnerability context? To develop a concrete policy approach for collective empowerment of vulnerable consumers in RECs we review existing literature concerned with the dynamics vulnerability and inequality produce especially in the energy context. We discuss energy vulnerability as a multidimensional form of deprivation against the background of discriminating systems. We then link insights from behavioural economics on how vulnerability impacts decision making with the discussion of empowerment and its barriers.

(3) Which of the identified measures in particular with regard to participation of vulnerable consumers in RECs can be included in the RED II *"enabling framework"* when transposing the directive into national law? In the light of the discrepancy between the stated political aims and the lack of concrete policy measures it is crucial to make suggestions to lawmakers for the transposition of RED II into national law until June 2021. Therefore, we briefly revise the current versions of the NECPs as well as policy documents from the European Commission to understand the state of play both at the EU and national level. While it is beyond the scope of this paper to provide a comprehensive policy analysis it outlines central considerations any policy approach must take into consideration to facilitate the inclusion of vulnerable consumers. Taking into consideration the most important restrictions previously identified we formulate policy recommendations. In this way concrete ancillary measures to facilitate the participation of vulnerable consumers in RECs and thus their active inclusion in the energy transition can be developed.

Section 3 firstly gives an overview of EU empowerment policy in the context of prosumption. Secondly, it discusses energy prosumption both as a form of consumer empowerment and a means to mitigate energy poverty. Thirdly, it addresses the challenges (energy) vulnerability poses as a structural context affecting individual behaviour and describes the multidimensional aspects of its context. While we mostly rely on theoretical literature—where suitable—we also refer to practical examples and the experience of the ongoing Horizon 2020 research project SCORE (Supporting Consumer Ownership in Renewables (https://www.score-h2020.eu) which amongst others also pursues the inclusion of vulnerable consumers. In the last part of this section we illustrate the advantage of collective approaches drawing on empirical evidence from the reactivation of unemployed in Spanish Sociedades Laborales. We argue that the participation of vulnerable consumers in RECs as a form of collective empowerment has benefits beyond individual empowerment. Based on Section 3, Section 4 discusses policy proposals for collective empowerment approaches to facilitate the participation of vulnerable consumers in RECs. As the description of the *"enabling framework to promote and facilitate the development of renewable energy communities"* in Art. 22 para 4 RED II that all MS need to implement focusses on financial incentives and obstacles we emphasise these aspects in our discussion. However, the enumeration of the elements of Art. 22 para 4 RED II is not exhaustive and therefore also other possible measures, among which incentives and framing are included. Section 5 concludes.

3. Consumer Empowerment in a Vulnerability Context—Inclusion in RECs

3.1. The EU Energy Context

The word "empowerment" entails "em", a prefix used to form verbs such as "to make" or to "cause to be", and "power". Thus "to empower" is to make or cause power [15]. Hence, empowerment is defined as "the act or action of empowering someone or something: the granting of the power, right, or authority to perform various acts or duties" [16]. Empowerment, for the powerless, involves a bottom up process whereby they transform from passive or reactive subjects to positive actors [15] in their own lives and thus on the energy market [17]. Empowerment plays a role beyond individual participation in that it determines social cohesion and thus peace and prosperity. Through empowerment, the formerly powerless become capable individuals who are willing and able to take ownership and responsibility for their own choices, decisions, and actions in society or on markets. They thus are accorded basic social rights, respect and dignity. Material resources, information and knowledge are developed and made accessible. As a result, individual opportunities increase [15] and vulnerability decreases.

In contrast to this universal view on empowerment as a social process, consumer empowerment at the European level is informed by the slogan "the consumer at the heart of the energy market". It constitutes the EU energy policy approach and enshrines the active role consumers should play on the energy market [7]. Consumer empowerment and protection (firstly introduced with the second energy liberalisation package in 2003) have a prominent place in EU energy policy making. Here, in contrast to the above outlined concept of empowerment as a social process, consumer empowerment is limited to a market-based approach. This entails two elements, that is, related to the process access to relevant information and related to the output a wider choice of consumption options [7]. As consumers are empowered to receive more information for a greater quantity of consumption choices competition between producers on the market increases. This should in turn increase market efficiency and maximise the end consumer's welfare [8]. Thus, access to relevant information becomes the basis for pro-active decision making of (energy) consumers to e.g., switch suppliers which in turn should increase individual utility on the energy market.

This approach of consumer empowerment in form of enabling and incentivising individual consumers to be active on the market becomes problematic when certain groups among which vulnerable consumers remain passive irrespective of existing incentives. Additional protection measures are needed triggering political and social welfare interventions on the market which may be adversely impacting market liberalisation [8]. Consequently, consumer empowerment is in some cases extended to, in other cases in conflict with consumer protection measures. An example are energy bill subsidies to protect vulnerable households from energy poverty, which stand in conflict to the market-oriented approach of competition.

With the CEP, this market-oriented empowerment approach is now deepened to include prosumership in this active consumer role: As part of the CEP, RED II fosters the new role of the consumer in a *"consumer-centred clean energy transition"*. Extending existing rules that are strengthened (e.g., the right to switch providers in the IEMD), RED II promotes the empowerment of *"renewables self-consumers to generate, consume, store, and sell electricity without facing disproportionate burdens"* [4] (Recital 66). This in turn *"provides opportunities for renewable energy communities to advance energy efficiency at house-hold level and helps fight energy poverty through reduced consumption and lower supply tariffs"* [4] (Recital 67). The empowerment of consumers to become prosumers is the outcome of a comprehensive *"enabling framework"* entailing the provision of information as well as facilitative administrative and regulatory elements. With regard to RE in heating and cooling MS shall further *"ensure the accessibility of measures to all consumers, in particular those in low-income or vulnerable households, who would not otherwise possess sufficient up-front capital to benefit."* [4] (Article 23). The underrepresentation and apparent passivity of vulnerable consumers among prosumers is thus a question of the process dimension of consumer empowerment and highlights the need for re-alignment between measure and individual need.

3.2. Prosumption as a Form of (Collective) Consumer Empowerment in RED II

Prosumption in combination with an "enabling framework" (which may include, e.g., simplified administrative and regulatory requirements, lower levies and taxes) reduces the costs for energy consumption and provides an additional source of income through the sale of excess production to the grid [6]. As every kWh not self-consumed is one potentially sold it also has a positive impact on consumer behaviour and increases energy efficiency (EE) [3]. Thus, RE generation for self-consumption through increased EE decreases energy usage [3]. Given that RES have reached grid parity prosumption decreases energy costs which in turn reduces payments for energy potentially positively impacting household income. Prosumership also triggers a learning process and increases knowledge of RE [3]. In addition to these effects, participating in a RE project may provide access to social groups other than one's primary group when enshrined in a collective scheme like, e.g., a RE-cooperative or a consumer stock ownership plan (CSOP). Given that the socio-cultural context shapes among others, habits, values and norms which in turn have an impact on individual behaviour [18] this is of particular importance for vulnerable households to overcome systemic disadvantages (e.g., higher rates of unemployment, lower education) and social isolation but also boosts the mentioned learning process. These interactions are summarised in Figure 1 illustrating how prosumership can contribute to mitigate two of the major challenges vulnerable energy consumers face on a daily basis, that is, low and precarious income and high energy costs.

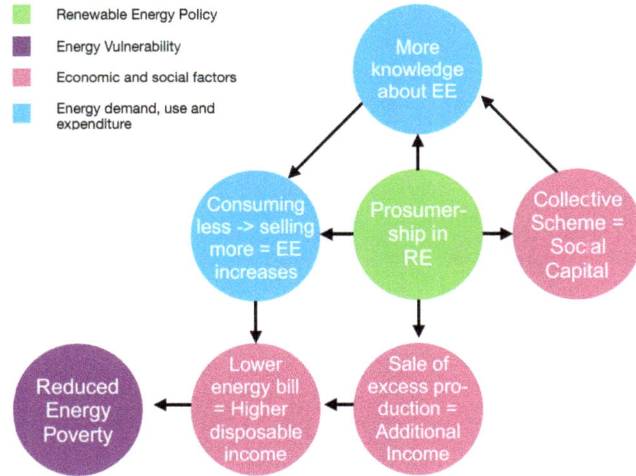

Figure 1. Mitigating Energy Vulnerability (Source: authors' own illustration).

Providing vulnerable consumers with access to the "enabling framework" which includes subsidy schemes to engage in prosumption has energy justice implications as well: The benefits prosumption offers as a result of certain policies should be distributed equally to all social groups in turn increasing social acceptance and political support for the transition itself [19,20]. However, prosumership as mentioned requires a number of prerequisites including access to financing, know-how, a certain willingness to take risks [21]. Especially if one has limited experiences as a (co-)owner the prospect of long-term investments including loans, requirements for maintenance, insecurity in regulatory frameworks (changed in feed-in tariffs) and vague yield expectations in the distant future amount to highly context specific barriers many of which are reinforced by a multidimensional vulnerability. The different dimensions will be discussed in the following sub-section.

3.3. Challenges of Empowerment in a Vulnerability Context

Consumer empowerment intends to enable consumers to exercise full consumer choice to live a life according to their own consumption preferences. According to this logic consumer choice is conceptualised as a form of freedom and may be equated with increased happiness. However, this rational neo-economic approach to individual freedom must be seen critically in that happiness may only be derived though consumption and various studies provide evidence that increases in material satisfaction are not automatically followed by subjective wellbeing [22,23]. Choice as freedom however entails the need to consume because without consumption choice as a form of freedom remains irrelevant [22]. But freedom of choice and material wellbeing is achievable only for those who can participate in the market by gathering information, selling their labour, renting their capital, or otherwise drawing on economic, social and cultural capital [24]. While the fruits of empowerment as proclaimed by the EC [22] are already difficult to reach for the poor the unequal distribution of income and wealth (even in a social welfare state like Germany [25]) reinforces a general vulnerability context. The vulnerability context in turn has different dimensions some of which are of individual nature, others are the outcome of structural dynamics often of inequality. Table 1 provides a non-exhaustive overview of three different vulnerability dimensions. Each vulnerability dimension does not only exist on its own but intersects with the others producing multiple layers of deprivation [26]. The vulnerability context is thus a set of conditions or deprivations (e.g., low education restricting information access) pruning life and consumption choices. These conditions often lead to or recreate circumstances such as poverty or energy poverty [26].

Table 1. General Dimensions of Vulnerability.

(i)	**Individual characteristics** (age, gender, income, health, ethnicity, religion, political orientation)
(ii)	**Discriminating structures** among which housing markets (e.g., vulnerable households have access only to poorly insulated flats) and energy markets (e.g., lack of transparency and complexity of offers and opportunities)
(iii)	**Policy making** (e.g., political underrepresentation of certain social groups, provision of information, compatibilities of different social policies, support programs to not cater for the needs of vulnerable households e.g., credit programs require a certain percentage of equity)

Source: Based on [27].

Understanding the vulnerability context thus entails an understanding of how it creates difficulties to participate and cope with the requirements of the modern society [28] or simply the conditions of a liberalised energy market. For example, individual physical and mental disabilities hinder decision making in that access to information (e.g., internet access, walking to a library, knowledge of ICT infrastructure, utilising a comparison site) is restrained; economic disabilities such as low income and poverty affect decision making in similar ways. Here it is in particular poverty that affects the perception of what is perceived relevant or simply of what is achievable [29]. Market vulnerabilities and discriminating systems are linked to the inability to process relevant information and to get access to certain options. Examples are opacity and inconsistency of information concerning different market offers as well as their high complexity [30].

Social policies often only regard one aspect of vulnerability applying their own form of rationality to solving a problem and sometimes stand in conflict with one another [6,8,27]. One of the most striking experiences from the SCORE project so far is in fact related to these in-compatibilities regarding social policies. On one hand policy makers intent to facilitate the participation of vulnerable consumers in RE projects (which is a form of investment and co-ownership). On the other hand, the requirements to receive social welfare payments (no assets, savings or financial participation in any project) restrain their participation in RE projects. While the (financial) participation in RE projects is beneficial and may even help to overcome structural dis-advantages (comp. previous section), these benefits (e.g., lower energy costs through dividends) only unfold over time. In contrast, the receiving of social welfare payments often conditions the immediate survival of the most vulnerable (paying for rent and food). Their refusal to become prosumer (as experienced in the SCORE project in Italy and Czech Republic) is therefore only logical. This points at a need for increased awareness and re-alignment of existing (social) policies.

The vulnerability context creates context specific challenges for consumer empowerment among which providing the "right" consumption/prosumption choice. In the energy context increased numbers of suppliers and supply tariff options to choose from enable the well-informed consumer to enhance individual utility and is thus a form of empowerment. When these options remain accessible only to those who have the means to know about them (e.g., time and knowhow for market-research and to understand the complexity of market offers) or when other conditions such as a bank account, positive credit history, or the installation of smart meters determine their availability, consumer empowerment leaves out the most vulnerable [31–33]. The same applies to energy prosumption. Simply providing the legal possibility and information about energy prosumption empowers solely those already possessing the means for prosumption. Here the current EU strategy to consumer empowerment experiences its limits. Provision of information about consumption options should lead consumers to the conclusion that e.g., prosumption in the long term is the most cost attractive option and may in some cases mitigate energy poverty. In doing so the legislator follows the logic that consumers—as rational agents maximising their utility—choose the economically most attractive option of e.g., supplier switching or prosumption [7,8].

The challenge of this approach to consumer empowerment in a vulnerability context is however at least twofold. Firstly, human behaviour is not always following a "rational". Human behaviour including energy choices are not exclusively governed by rational thinking but rather by emotional, cognitive and socio-cultural considerations, biases and heuristics [32,34,35]. Decision making is often based on incorrect information and estimations about benefits and costs (biases) and influenced by external factors such as the way information is presented (framing), whereby information that stands out, is novel, or seems relevant is more likely to affect thinking and actions [36]. Secondly as mentioned above vulnerable consumers face particular barriers with respect to consumption choices. Understanding the situation of vulnerable consumers is therefore the first step towards understanding consumer behaviour and choice in a vulnerability context. This in turn informs successful policies for consumer empowerment.

Based on the general vulnerability dimensions described in Table 1 above we can thus summarise the following vulnerability dimensions in the energy prosumption context:

(i)   **Individual characteristics:** Low savings/access to capital; lack of time, experience and knowledge about opportunities to engage in prosumption; limited access to supportive governmental schemes to participate in community energy projects.

(ii) **Discriminating structures:** Complexity of existing opportunities and opacity of the energy markets cause high costs for information gathering to engage in prosumption and often a need for (expensive) legal and economic advice [37]. In this way market inherent complexity discriminates against vulnerable consumers and exacerbates vulnerability.

(iii) **Policy making:** Where supportive policies and programs exist their design often does not consider the specific conditions of vulnerable households and hence remain inaccessible to them [26]. Other policies are mutually exclusive, especially where eligibility for means-tested transfers (e.g., energy or housing subsidies) would be impaired by asset formation effectively preventing participation in RE projects.

Another example for context related challenges is poverty and how it affects decision making unmasking the illusion of rationality [38,39] as the major driver of behaviour (for an overview of how the poverty context impacts energy choices see [6]) or simply the observation that vulnerable consumers in particular cannot make the choices—even if they want to—that would suit their needs best [40]. Even if a household knows that installing a PV installation would, in the long run, drive electricity prices down, without the necessary upfront capital this knowledge remains irrelevant. It is thus often not a question of willingness for example to become (co-)owner of a RE project (which provides tangible benefits such as lower energy costs) but of individual opportunities to receive a loan needed for this investment. The same applies to participation in community energy e.g., in RECs. As exemplified by the SCORE pilot projects vulnerability often translates into financial precarity effectively hindering the participation of vulnerable consumers even under an inclusive and low-thresholding financing approach where the contribution of participating households is small.

3.4. Individual vs. Collective Empowerment

From the discussion so far, it is apparent that empowerment in a vulnerability context must be more than the provision of consumer choices through access to information and consumption options. Understanding and providing access to the prerequisites needed to choose between consumption/prosumption options has to become part of empowerment. Irrespective of the specific context of e.g., promoting prosumption, consumer empowerment in a vulnerability context must be conceptualised as a long-term process [8]. It entails elements beyond consumption choices e.g., in form of self-help for social change [41] including elements such as political participation in form of deliberation [42], financial participation in the energy transition [43] and some form of relief. In addition, given that vulnerable groups often do not have direct control over the social conditions and institutional practices that shape their lives [44] empowerment needs to address the very social dynamics that reproduce social inequality by addressing entire social strata rather than individual consumers.

In general, the empowerment process happens at the interplay of the individual, interpersonal and collective. Unless individuals believe that they can produce desired effects and forestall undesired ones by their actions, they have little incentive to act in the first place [44]. Individuals lacking self-efficacy are convinced not to be in charge of their own destiny, have limited initiative and commitment and as a result tend to engage in passive, unproductive attitudes and behaviours [45]. Applied to the RE context the degree to which individuals engage in environment friendly consumption behaviour (e.g., turning the lights of or investing in RE) is a result of their environmental self-efficacy judgements [46]. For example, the experience of being powerless and at mercy of energy providers who threaten to shut down the power supply due to arrears contributes to a lacking self-efficacy as does the inability to provide desired energy consumption choices for one's children (e.g., leaving the lights on for studying) [47]. Self-efficacy increases when individuals gain a sense of control over their own life *inter alia* in form of increased choice and both material and cognitive means. In the energy context, enabling energy-poor households to become prosumers provides them with control over their own energy supply. Resulting improved EE [3] and reduced energy expenditure [6] allows them to experience a form of self-efficacy. Empowerment of the individual starts by providing education, access to information and processes facilitating the individual experience of self-efficacy [48]. These experiences in turn are not only shaped by the individual but by the respective social environment [44,49].

However, individuals in many cases and specifically in a vulnerability context do not have the control or power over the structural conditions that inhibit individual opportunities and choice [44]. They have thus always turned to others be it their family, neighbours or friends to form collectives to gather necessary competencies, resources and capital (economic, cultural and social [23]). In this way individuals benefit from the power of collective action to overcome individual obstacles [50]. These forms of organisation determine human survival since the beginning of time. Many of which were later enshrined in legal concepts such as the cooperative. The belief in collective agency—people's shared belief in their collective power, the interactive, coordinative, and synergistic dynamics of their transactions [44]—continues to determine empowerment until today. In line with these postulates the UN promotes the cooperative model and builds on individual and collective strengths [51,52] as a form of collective, interpersonal and thus individual empowerment in the global fight against poverty [53,54]. Therefore, we suggest collective empowerment instead of individual consumer empowerment since the need for individual empowerment (in the energy context) is caused by structural and societal dynamics creating systemic disadvantages rather than individual conditions.

3.5. Advantages of Collective Empowerment: The Example of Reactivating the Unemployed in Sociedades Laborales

A prime example in the vulnerability context demonstrating the advantages of collective empowerment is the reactivation of unemployed in the Spanish concept of Sociedades Laborales. A Sociedad Laboral (SL) is a qualified form of conventional corporation, majority-owned by its permanent employees. Since 1985 in lieu of receiving monthly payments, job seekers can choose to capitalise their unemployment benefits into a lump sum in order to establish a new SL or to recapitalise an existing SL by becoming a member. In this way SLs offer unemployed individuals with the right to unemployment benefits to become entrepreneurs creating their own workplace and thus a way out of their precarious situation. What is more, SLs which may be set up by unemployed together with conventional entrepreneurs, or exclusively by either of the two groups appear on the market as regular corporations thus do not bear the stigma of being set up by or with the involvement of formerly unemployed [55]. It is estimated that about one-third of SLs utilise the capitalisation of unemployment benefits at the time of their founding. Between 2006 and 2013 on average 2240 persons capitalized unemployment benefits to set up or join a micro-sized limited liability Sociedad Laboral (SLL) in Spain with an average annual total of around EUR 13,233 per person. It is important to stress that the capitalised unemployment benefit is not a state subsidy but stems from the social security contributions made earlier in times of employment. Similar mechanisms are also available in other members states: in France, unemployed persons can receive up to 50 per cent of their unemployment benefits under the "Aide à la reprise et à la creation d'entreprise" (ARCE) scheme. Portugal introduced the "Support Programme for Entrepreneurship and Self-Employment" or "Programa de Apoio ao Empreendedorismo e à Criação do Próprio Emprego/PAECPE" in 2009, which allows the conversion of unemployment benefits under certain conditions. For example, a full-time job for the unemployed person has to be created and the jobs created must be maintained for at least three years. Finally, in Bulgaria it is possible to receive one's unemployment benefits as a lump sum and to use them as a start-up grant.

This collective reactivation mechanism for unemployed in SLs was compared with individual start-up subsidies to reactivate jobseekers across the European Member States (previously assessed in the European Employment Policy Observatory (EEPO) review [56]) in a 2017 econometric study [55]. Following the EEPO criteria to evaluate the success of start-up subsidies that is, survival rate, access to capital and the capacity to create secondary employment the study on SLs [55] found that in comparison they were superior in all indicators under consideration:

- SLLs generally have higher survival rates than their conventional competitors, surviving long enough to amortise capitalised unemployment benefits: The average paid-out lump sum represents roughly the cost of 1.3 years-worth of unemployment benefits; on average, 88% of all SLs survive this long. Furthermore, in contrast to using up the unemployment benefits month to month both the (formerly unemployed) owner-worker and the SL make social security contributions leading to the accruing of a new expectancy for unemployment benefits from the first day of operation.
- SLLs are set up not only by unemployed persons but also by ordinary entrepreneurs and typically involve external investors which account for 27% of their partners. Unlike conventional start-up subsidies for jobseekers, SLs offer not only access to capital but practical assistance and entrepreneurial advice to an unemployed person joining or setting up an SL.
- With respect to secondary employment according to employment data for 2008–2013, 1.3 additional jobs were created in Spain per founding worker partner. In contrast, the EEPO review concludes that across several studies approximate only 0.2 additional jobs were generated in start-up firms set up under ALMP start-up subsidy programs [56].

These results are a clear indication that with regard to empowerment of formerly unemployed to return back to employment collective schemes like that of SLs are superior to individual schemes.

3.6. Participation in Renewable Energy Communities—Learning from the Sociedades Laborales

Applied to prosumption, access to credit is generally a challenge, but it is even more challenging for people with little business experience and no strong credit track record who bear the social stigma associated with unemployment or other vulnerability characteristics. Collective ventures such as SLs are a means to provide not only financial capital but also social capital and expertise, entrepreneurial experience, training and mentoring. What is more, SLs display the benefits of an efficient alignment between different social policies, i.e., the possibility to capitalise unemployment benefits in form of a lump sum to facilitate entrepreneurship. Applied to energy vulnerability the capitalisation of energy subsidies would facilitate energy prosumption by enabling vulnerable consumers to buy into an existing REC or set up a new one. While energy costs will decrease for the new prosumer, profits from the sale of excess production can be paid out as dividends partly offsetting the energy subsidy no longer available for spending.

Although fully fledged prosumership—that is providing both the possibility to self-consume and sale to the grid or third parties—will be an incentive to become more energy efficient [3], a problem related to vulnerable consumers is underconsumption. Therefore, self-consumption might be larger than the respective share allocated to the individual proportional to his or her investment in the REC bringing energy use to normal levels. Similarly, in the first months of activity of a newly founded SL the new worker-owners may be inclined to pay themselves a lower wage than the market rate. For this reason, the Spanish government and in particular that of the Basque Country additionally subsidise the integration of unemployed persons as worker-owners with non-refundable subsidies paid directly to the SL [55]. This approach could also be applied to RECs which will be discussed in Section 4.3.1. Therefore, when both individual energy-poor consumers and average energy consumers organise themselves in a REC to become prosumers they pull their competencies and resources together and benefit from the exchange with their co-investors. In this way they share their economic, cultural and social capital and increase collective agency as well as individual self-efficacy. Unlike in individual investments, in RECs business decisions need to be taken together, discussing, consulting and justifying them—this facilitates the exchange of experience, a learning process and in the best-case functions as an apprenticeship. Respectively participation in a REC as a form of asset formation becomes an instrument to increase the individual capacity to advance socio-economically beyond the satisfaction of consumption needs [57]. According to Sen (1999) individuals are only capable to shape their own destiny when they have adequate social and economic opportunities to unfold individual capacities [57]. Individual economic capacities in turn are increased through asset formation [57–59]. Participating in a REC not only changes the income side but has an effect on behaviour and attitudes as well [60] and addresses individual disadvantages. What is more, collaborating in a REC breaks up the segregation of disadvantaged communities [61] and affects individuals positively in that the diversity of one's primary group increases with demonstrably positive effects on health and education all the way to career choices re-framing individual self-efficacy beliefs [62]. Thus, the participation in a REC as a co-owner (which entails asset formation) enhances individual capabilities—the core of any empowerment approach.

4. Discussion: Putting Forward Collective Empowerment Strategies for RECs

Today, a wide range of RE projects and organisations already work towards the inclusion of vulnerable groups in the Energy Transition. Examples are Energent in Belgium (http://energent.be), Enercoop in France (http://enercoop.fr) and Energia Positivain in Italy (https://www.energia-positiva.it), as well as numerous projects working on the mitigation of energy poverty such as the EU project STEP (https://www.stepenergy.eu/results/) or the cost action Engager (http://www.engager-energy.net) as well as the Right-to-Energy coalition. The latter unites trade unions, anti-poverty organisations, social housing providers, environmental organisations, health organisations and energy cooperatives under the concrete objective of collaborating on the issue of energy poverty, including measures to alleviate it in the 2030 EU energy package (http://www.righttoenergy.org). However, these initiatives face structural difficulties and stress the importance of an *"enabling framework"* [63–65]. The European legislator acknowledges the potential of RECs to empower vulnerable consumers, introduces a definition for RECs and requires the European member states to ensure that RECs are *"accessible to all consumers, including those in low-income or vulnerable households"* (Art. 22 para. 4 (f)) and to *"assess the possibility to enable participation by households that might otherwise not be able to participate, including vulnerable consumers (...)"* [4] (Recital 67). Yet, while the potential capacity of RECs for the empowerment of vulnerable consumers and the need to include facilitating measures for the participation of vulnerable consumers in RECs in the *"enabling framework"* of RED II are acknowledged a lack of political attention in policy-making for the inclusion of vulnerable consumers in RECs remains.

This is highlighted when looking at the current NECPs: In October 2019 RECs are explicitly mentioned in only 13 out of 28 draft NECPs. The inclusion of vulnerable groups and/or LIHs in RECs or measures to facilitate the participation of these groups was not mentioned in any draft NECP in October 2019. In the analysis of the draft NECPs the EC, therefore, calls on the MS to *"provide additional details and measures on the enabling frameworks for self-consumption and renewable energy communities in line with Articles 21 and 22 of Directive (EU) 2018/2001"* (the EC's recommendations are available at https://ec.europa.eu/energy/topics/energy-strategy/national-energy-climate-plans_en#the-process). However, although until 4th of March 2020 all ten of the final NECPs available in English translation mention RECs, the inclusion of vulnerable households and/or LIHs in RECs is only referred to vaguely in one NECP, namely that of Italy. The remaining nine translated final NECPs (AT, CY, DK, EE, EL, FI, HR, MT, SK) do not mention inclusion of vulnerable households or LIHs. The NECPs of Cyprus, Denmark and Greece mention the preparation of additional legislation as well as the provision of (financial) support schemes for vulnerable households to engage in self-consumption. Such the current shortcomings with respect to addressing the inclusion of vulnerable groups in RECs remain an indicator for a lack of political awareness. A similar knowledge gap with respect to empowerment and inclusion of vulnerable consumers as prosumers in REC appears to prevail in the respective literature until now.

With a focus on the second and third research question outlined in Section 2 we therefore highlight the local conditions for collective empowerment in RECs in Section 4.1. and the implications of behavioural economic for a better understanding of behaviour in the energy vulnerability context in Section 4.2. Section 4.3 discusses access to finance and facilitative ownership and governance models as prerequisites for the successful participation of vulnerable consumer. Section 4.4 discusses the need to understand and frame the relevance of participation for vulnerable consumers. Of course, this selection is not exhaustive and primarily motivated by the postulates of the *"enabling framework"* of Art. 22 para 4 RED II. While we do not provide a comprehensive literature review on all of these elements, we encourage both scholars in the respective fields and policy makers to consider them as a basis for the transposition of the RED II *"enabling framework"* and further research.

4.1. Local Conditions for Collective Empowerment in RECs

Community energy projects in their various forms have been on the research agenda for a while now [66], as has participation in the energy transition and in community energy projects [43]. In different lines of research citizen participation in energy community projects has been linked to aspects such as modes of governance [67], ownership and ownership structures [68], member responsibilities and competences [69], equal opportunities between communities [70], conflicts, trust, and social capital [71], deliberation [72] and power factors [73]. Radtke further mentions relationships and connections to policy makers and the public [74] and network structures linking local communities to energy initiatives [75] which have also been found to effect participation [76].

With respect to success factors of community energy the importance of local factors such factors related to community energy itself (e.g., human capital, skills, access to funds) and factors related to interaction with the local community (i.e., social capital, alignment with community values, attitudes towards RECs, local energy activism) are highlighted [77,78]. These are at least as important as factors related to local and national policies such as fiscal and financial support and planning policies [77,79–81]. Especially in the context of an *"enabling framework"* it must therefore be discussed how national frameworks can be designed to facilitate these local conditions on which RECs depend. Here it proves important to include differences between urban and rural settings and their implications for RECs and inclusive participation in future research.

For individual participation in community energy access to information and knowledge about existing participation opportunities are in turn major drivers of participation [82]. But also additional prerequisites such as financial requirements (acquisition of shares) and the minimum duration of participation determine participation (sometimes indirectly) [82]. While all these factors play a role motives to join energy communities remain diverse [77]. In Germany for instance motives to join an energy community are in tendency less revenue driven and tend to be more ideational (e.g., contributing to the energy transition) for low-income groups and academics than compared to high income groups [82]. With respect to equality of chances, and equity Park (2012) acknowledges differences between communities and their capacity to get involved in renewable energies [70]. Here the importance of practical capacities such as expert knowledge and time (human resources) for administrative, financial and other procedural activities to set up a community, to gain access to grants is highlighted.

However, while it is mentioned that inequity is a consequence of structural and economic factors [83] and therefor results in a "deeply seated problem of involving the most marginalised and deprived (...) communities" [70] the impact of these factors on the individual in a given community remains undiscussed. Although the distribution of prerequisites for individual participation (as mentioned earlier in the text) are equally affected by these inequities. For instance, social capital as the sum of the actual and potential resources embedded within, available through, and derived from the network of relationships possessed by an individual [84] has been linked to aspects such as information gathering [85,86] as a basis for participation [87]. Its role for the participation of underrepresented groups in community energy projects is therefore a promising starting point for further research.

While different authors in community energy research investigate different sets of motives in different settings [68,77,88–91] the discussion of motives for participation of underrepresented social groups such as LIHs remains superficial. Currently the potential motives of those social groups that do not participate in community energy remain unknown. The situation is similar for the respective incentives that could possibly motivate them and the structural conditions that might facilitate their participation. On the other hand, in the energy poverty discourse although different facets of energy vulnerability are already being explored and propositions for both policy making and further research made [92] prosumption is left aside. In a next step, these two strands of investigation—so far unrelated—should be combined and extended to prosumption and the participation in RECs under the lens of energy poverty mitigation.

As is the case for individuals there is a need to better understand the capacities and resources needed for existing RECs to facilitate inclusion. In this line, there has been much interest in capacity-building [73,93,94] and social capital [95–97] as special functioning of communities which facilitates community-based approaches. Existing capacity building approaches for communities to get involved in the production of renewable energy should include the aspect of inclusion. Here, community development literature points at the role of cooperation (e.g., with NGOs working with vulnerable consumers) to share skills, resources and experience to address and include vulnerable individuals pro-actively [83,98]. Following the model of aids for SLs in Spain [55] governments could support capacity building to stimulate the RECs development as required by Article 22 para. 4 (f) and (h) RED II with technical assistance (feasibility studies, auditing, and consulting services) for the RECs and coaching and training (including, e.g., energy efficiency seminars) for individuals.

4.2. Implications of Behavioural Economics

When further assessing the current EU consumer empowerment strategy and its applicability in a vulnerability context, behavioural economics provide valuable insights with respect to the limitations of 'rational consumer behaviour'. Referring to Pollitt and Shaorshadze [99] we distinguish three different areas:

(1) Individuals assess cost and opportunities differently over time [100–102]. Time-varying discount rates not only affect the average consumer by changing consumer preferences depending on the time frame in which financial benefits are received but affect vulnerable consumers in a financial precarious condition in particular [103]. Vulnerable consumers for example perceive financial benefits of prosumption which typically materialise over a timespan of several years as irrelevant if confronted with the immediate need to pay overdue electricity bills to avoid electricity cuts. Here the vulnerability context reinforces this heuristic.

(2) According to the prospect theory consumers tend to make decisions by assessing the extent to which a choice differs from a specific reference point such as the status quo [104]. Rather than assessing costs and benefits of a choice against each other consumers tend to accept more risks to prevent potential losses rather than realise potential gains (status quo bias). As discussed, vulnerable consumers often rely on some form of social transfer payments and are unlikely to engage in any decision that jeopardises their claim for support, e.g., by acquiring assets in a RE project [6]. Empirical evidence confirming this barrier has been gathered in the course of interviews in the SCORE project in the Czech Republic as well as in Germany. Therefore, it is important to understand the prevailing reference point and preferences of different vulnerable households.

(3) The cognitive capacity to process information as a basis for decision-making is limited, a phenomenon that can also be placed under the theory of bounded rationality [40]. Cognitive capacity (sometimes referred to as bandwidth) is utilised by internal processes to derive insights and decisions [105]. Especially under time constraints or in situations where multiple decisions under consideration of their consequences (trade-offs) need to be taken bandwidth is depleted and thus less cognitive capacity remains for other tasks [29]. Moreover, under time pressure rather than engaging in a rational cost-benefit-calculation (utilising and assessing all available information) consumers tend to engage in intuitive judgements and simplified choice strategies [106]. In addition to behavioural economics, an extensive string of social and cognitive psychology investigates the impact vulnerability and here in particular different forms of scarcity (e.g., time, nutrition or financial scarcity) have on the availability and utility of bandwidth. Scarcity as a condition captures one's mind, alters the content of cognition and the perception of options [105]. It adds difficult trade-offs to everyday experiences [107], shifts attention and selects information according to its internal logic to overcome scarcity [105]. As a result, simple activities such as grocery shopping translate to constant and effortful overcoming of buying temptations requiring massive bandwidth [103]. Each of these bandwidth-consuming dynamics alone do not create a burden; cumulative, however, they start to deplete bandwidth restraining its availability for more profound cognitive processes such as those required for efficient economic decision-making.

Consequently, the vulnerability context limits the consideration of alternative options, overshadows possible long-term benefits, depletes willpower necessary to adhere to a long-term objective and makes it more difficult to choose between options or to calculate trade-offs. Understanding the underlying heuristics and biases is the basis of an effective design for a choice architecture that facilitates the empowerment of vulnerable consumers. One of many difficulties here lies in the vast diversity of vulnerability contexts which is likely to render one-size-fits-all approaches ineffective. It is therefore crucial to understand the local context and the reference points and preferences of local vulnerable households. To compensate for these limitations vulnerable consumers need buffers and reserves to be able to consider prosumption and its benefits in the first place [108] and to mitigate the detrimental effects of the vulnerability context: When a single mother spent her monthly salary a week before she receives the next pay-check financial reserves like savings on her bank account can be crucial. Providing financial relief, time for information gathering, trainings and a simple program design may thus facilitate the participation of the most vulnerable [109].

4.3. Providing the Prerequisites for Empowerment of Vulnerable Consumers in RECs

4.3.1. Access to Finance

Traditionally cooperatives and in the past decade RE cooperatives play a prominent role with respect to community approaches and the provision of alternative governance and ownership models in particular compared to national and international commercial RE ventures [110]. However, the inclusion of all consumer groups in cooperatives—although in theory an objective based on the principle of open participation [6]—has so far been difficult to achieve. In Germany, for example, more than 70 percent of RE cooperative members belong to the group of male, high education and high income. Other groups and especially those with low-income are underrepresented [14]. This is first and foremost a question of access to finance: RE projects, especially at the beginning face difficulties to raise sufficient equity; while access to credit is limited due to size, lack of collateral and risk-assessment they often depend on their members to provide required equity [6]. With respect to RECs, a recent study investigating 198 energy communities in nine MS shows that a majority depends on their members equity contribution [37]. Therefore, potential members usually need to buy shares. In Germany the average individual contribution in RE cooperatives amounts to EUR 3899 with an average required minimum contribution of EUR 511 [111]. Such high contributions are a barrier for the participation of vulnerable consumers with limited financial means. This is especially the case since the participation typically does not immediately translate into financial benefits but they only unfold over time (see also time-discounting).

The first obstacle is therefore primarily a financial one: Vulnerable groups as a rule do not have access to finance. This raises a second question: Should vulnerable groups be supported in gaining access to required financing or should RECs receive additional financial means to facilitate the inclusion of vulnerable groups. Ideally the *"enabling framework"* encompasses both dimensions to enable the participation of all consumers *"including those in low-income or vulnerable households"* as postulated by Article 22 para. 4 (f) RED II.

In Germany for instance, cooperative banks could play a major role in facilitating inclusive access to finance as they already have a local network of stakeholders [112] and most RE initiatives are organised as cooperatives [14]. The *"enabling framework"* could facilitate these local networks and provide a structure and incentives for inclusive and green investments in RE as outlined in the European Green Deal [1] and the Sustainable Europe Investment Plan [113]. When transposing the RED II an approach could privilege direct impact investments in those RECs that facilitate the inclusion of vulnerable consumers. Irrespective of direct investments from third parties, the provision of low-interest loans to vulnerable consumers to finance their participation in community energy projects as pioneered, i.a., by the Ærø Windpark on the Danish island of Ærø [114] appears to be a promising starting point. Here a local bank loan system was set up to facilitate access to finance for all residents of the island as a basis for participation in the project. To date such inclusive financing schemes remain local solutions without broader application or support from national policies in RE. In this regard the SCORE project is in contact with institutions like GLS Bank in Germany (https://www.gls.de/) and South Pole in the Netherlands (https://www.southpole.com) to discuss the design of similar financing mechanisms.

As part of the *"enabling framework"* RE projects that fulfil the RED II requirements of RECs in terms of governance and ownership structure should have access to preferential financial support, e.g., in form of zero or low interest rate credit programs and possibly direct subsidies as in the case of SLs specifically to include vulnerable consumers. Additionally, tax exemptions could be granted for those RECs that reach a certain diversity threshold—e.g., 10 percent of members are affected by vulnerability.

Similarly, diversity could be linked to access to preferential treatment in administrative procedures. Articles 15 para. 1 (d) and 16 para. 6 RED II stress the simplification of administrative procedures to facilitate consumer prosumption [4]. In order to further support the development of inclusive RECs establishing an administrative fast track would lead to an additional incentive for RECs to include vulnerable groups.

With respect to measures aiming at the direct empowerment of vulnerable groups, we advocate for a "renewables asset formation agenda for vulnerable consumers" [6] which does not replace social insurance or safety net programs, nor would it contribute to the financialisaton of social policies (e.g., the shift from "defined benefit" to "defined contribution" in social security [115]). Rather it would offer both, financial support and assistance to low-income families, providing a a lever for building up their asset base in a sound economic way by becoming (co-)owners in RECs [116,117]. Similarly, a proposition recently carried forward by the German Green Party envisages citizen funds as a means for old-age provision: Given that for many people shares and real estate are either too expensive or too insecure as a retirement provision, the state should invest for its citizens [118]. Since the financial benefits of RE (co-)ownership materialise in the mid- to long-term, this implies a broadening of the welfare state perspective.

We thus recommend linking existing measures such as energy subsidy schemes with direct empowerment drawing on the above presented best-practice example of SLs in Spain. A similar mechanism could be introduced to energy-vulnerable households. These households usually receive some form of social transfer payment e.g., in form of energy subsidies to pay their energy bills (the Energy Poverty Observatory lists policy measures to mitigate energy poverty here https://www. energypoverty.eu/policies-measures). Providing the possibility to receive a part of their annual social transfers in a lump sum under the condition to invest that money in a local REC would be a cost-effective extension of existing subsidies. One of the major obstacles for the participation of vulnerable consumers namely a lack of access to finance would be overcome while the total of required subsidies would decline over time as more and more households in need would—through their enabled participation in RECs—no longer require public support. In consequence local authorities could use the released budget for the offering of additional (educational) services. However, in doing so the propensity for underconsumption of LIHs described above in Section 3.6 has to be taken into account. Therefore, additional to the capitalisation mechanism governments when transposing the RED II could make available one-time subsidies directly to the REC for each vulnerable consumer becoming a member or shareholder which would be earmarked to compensate for underconsumption. Currently only a few MS provide financial assistance to vulnerable groups to enable them to become prosumers: Cyprus supports vulnerable households to produce RE for self-consumption through net-metering and financial aid to install a PV system, Greece introduced a law on energy communities that promotes energy communities and solidarity in the energy sector, including energy poverty measures. Additional best practise examples remain unknown. Given that this form of empowerment (vulnerable households with no financial assets benefit from a liberalised market through participating in RECs) is a new endeavour for both scholars in the field of participation in the energy transition and policy makers alike, its potential and conditions for implementation as a part of a just energy transition should be further investigated.

4.3.2. Appropriate Ownership and Governance Models

Any type of RE venture is confronted with the assessment of the best-suited ownership and thus governance model—so are RECs. In order to be considered a REC RED II does not proscribe any specific governance model but defines RECs as any *"legal entity (a) which, in accordance with the applicable national law, is based on open and voluntary participation, is autonomous, and is effectively controlled by shareholders or members that are located in the proximity of the renewable energy projects that are owned and developed by that legal entity; (b) the shareholders or members of which are natural persons, SMEs or local authorities, including municipalities"* [4] (Article 2, para. 16). As the choice for the best suited ownership and governance model must confer with the above definition [119] individual ownership and private law partnerships are not eligible; the other of the most common ownership and governance models in RE as listed in Table 2 can be employed. The right choice with regard to the inclusion of vulnerable consumers depends on factors such as the type and scale of investment (smaller PV installations or offshore wind parks) demanding different investment levels and financing mixes, but also required expertise and risk. Limited liability companies and limited partnerships are often used for medium-sized and above all large projects, particularly in the field of wind energy [90]. Cooperatives and trusteed schemes are usually employed for medium-sized and large projects, mainly in the field of photovoltaics and local heating networks.

Table 2. Comparison of different ownership and governance models in RE.

	Individual Ownership	Private Law Partnership	Limited Partnership	Cooperative	Trusteed Scheme e.g., CSOP
Type of investment	PV	PV	Wind	PV and local heating network	PV and local heating network
Size of investment	Small	Medium	Large	Medium–Large	Medium–Large
Financing Mix	High equity ratio	High equity ratio	Low equity ratio	High equity ratio	Low equity ratio
Influence on decision-making	Owner in full control, with oversight from local councils and other such stakeholder groups	Consumer-partners in full control: Voting rights according to contributions/full information rights	Right to demand information; control and veto rights for consumer-shareholders only under exceptional circumstances	Direct: "one member one vote"; general assembly concentrates decision-making power	Indirect: Trustee exercises rights for consumer-shareholders, e.g., participation in management meetings or the right to demand information
Liability	Unlimited personal liability	Unlimited personal liability jointly with partners	No personal liability; liability instead limited to value of share	No personal liability; liability instead limited to value of share	No personal liability; liability instead limited to value of share
Transfer of shares	Not required, unless because of inheritance between individuals	Consent of all consumer-partners needed	Managerial consent needed; entry into the commercial register	Transferable albeit with restrictions; entry into the commercial register	Freely transferable; low transaction cost; only trusteeship agreement is altered
Costs	Low initial setup costs	Low initial setup costs	Higher initial costs to enter commercial register; higher administrative expenses	Higher initial costs to enter commercial register; higher administrative expenses	Expenses of incorporating trusteeship (and holding Ltd. If required due to absence of trust legislation); administrative expenses

Source: Authors' own elaboration based on [19].

For RECs with the objective to facilitate the participation of vulnerable consumers the choice and application of the respective ownership and governance model must take into consideration the characteristics of the vulnerability context discussed in Section 2. Table 3 compares three common models with respect to their ability to cater for the vulnerability context in particular with respect to the requirements of equity contributions being one of the major obstacles for the participation of vulnerable consumers. Even if—as described in the previous section—the provision of access to finance is incorporated in the respective national enabling framework for RECs and the participation of vulnerable consumers therein, ownership and governance models that require low equity contributions still have a central benefit in that both organisation and individual members do not have to spend much time on accessing said financing schemes.

Table 3. Prerequisites for the inclusion of LIHs in RE projects under different business models.

	Limited Partnership	Co-Operative	Trusteed Scheme Like CSOPs
Equity contribution	Moderate; access to credit only against collateral or with guarantor	Moderate; membership shares have to be bought requiring liquidity	Low; future earnings are used to repay acquisition loan
Basic knowledge	Medium; managing partners external management possible	High; setup and management by members; no external management	Low; setup and supervision by trustee; external management possible
Time commitment	Low; involvement limited to control rights	Medium; members expected to be involved in all aspects	Low; involvement limited to crucial decisions; apprenticeship over time
Risk	Low; liability limited to value of share	Low; liability limited to value of share	Low; liability limited to value of share

Source: Modified after [6].

Once more drawing on Germany as an example studies show that irrespective of their heterogeneity RE cooperatives tend to have high equity ratios in contrast to limited partnerships [91]. Here it is especially the prevailing incentive of its members that determines the financing mix: Often yield expectations remain below those of profit-oriented investors [120] and private members of RE cooperatives focus on environmental and social aspects which determine their participation. RECs alike are hybrid organisations in that they entail characteristics of not-for-profit organisations (e.g., governance model) in which however members even if below market levels have a basic yield expectation [90]. Given this lack of a clear focus on profitability and low loan collateral to raise outside capital is difficult [121]. These factors contribute to the mentioned high equity ratio which in turn establishes a barrier for participation of vulnerable consumers.

In addition to pure financial considerations, the degree of involvement necessary to participate in any of the three models consisting of basic knowledge in RE, the (local) opportunity of participation and time commitment determines participation equally. As has been discussed, the vulnerability context puts additional burdens on individuals in form of time and cognitive constraints which should be investigated further. As a hypothesis we assume that the lower the threshold in terms of required knowledge and commitment the more likely participation. This is not to say that vulnerable consumers lack the cognitive capacity to participate but rather that they need to deal with more pressing every-day challenges and need to decide where to spend their resources. The question remains which governance and ownership model facilitates participation and inclusion best. We propose that modernised versions of the traditional cooperative model such as consumer stock ownership plans (CSOPs) enable the participation of vulnerable consumers in that they do not have to pay for their membership up front but repay the acquisition loan of their share from the future earnings of the investment [6].

4.4. Providing Incentives for Vulnerable Consumers to Participate in RECs

4.4.1. Incentives for Participation

As a starting point we suggest the provision of direct benefits for participating in RECs to offset required initial time and monetary investments by vulnerable consumers (necessary to enable their participation in the first place). Under a "renewables asset formation agenda for vulnerable consumers" the decision to participate in a REC should yield immediate benefits [6]. As mentioned above, direct subsidies for vulnerable energy consumer could be tied to membership in RECs immediately increasing household income while providing a strong incentive to participate. And most importantly, investments in RECs should be exempt from necessity to liquidate one's assets when applying for social transfers.

Currently, social policies supporting asset formation at the individual and household level mainly focus on mid to high-income households and appear less inclusive than income-based policies [122]. At the same time, means-tested transfers are a strong disincentive for LIHs to form assets since they typically require liquidation of all assets to become eligible for social transfer payments [60]. Consequently, asset-owning households are supported in further increasing their wealth while poor households are forced to spend down all of their assets, if any, to receive support. This mechanism effectively discourages LIHs from building up assets as every effort to do so directly reduces their eligibility for social transfer payments. This paradox has also been dubbed "dual asset policy" [60]: The same social policy that supports mid and high-income households to form assets and hence increase private wealth disincentivise LIHs to even attempt to increase wealth beyond subsistence. A similar phenomenon is observed with regard to the needs-tested minimum income which takes into account any income received, debiting it from the transfer; any job paying less than the minimum income threshold is thus disincentivised, with the recipient being caught in the "poverty trap". While the latter has been discussed in the literature, we highlight the need to acknowledge welfare-system-inherent conflicts and arising disincentives in the context of energy prosumption and participation in RECs.

On a side note, encouraging vulnerable households to save money to finance their participation in RECs is—at least at the moment—no solution: Putting aside money at this point in time with the European Central Bank continuing its negative interest rate policy [123] is a loss. As a result, the already low saving amounts of LIHs are further reduced. The time it would take a LIH to form sufficient savings to engage in asset formation, e.g., through becoming co-owner in a RE project is too long and not economical. Nudging LIHs to engage in saving behaviour is thus somewhat cynical as it would keep LIHs in the poverty trap. Therefore, providing LIHs with a possibility to form assets that yield a financial return, e.g., participation in a REC, must be detached from accumulating savings in the bank. In addition, although measures like guaranteed feed-in tariffs have in the past successfully enabled private consumers to invest in RE installations [124], LIHs were not addressed as the benefit of feed-in tariffs is linked to the disposability of investment capital. Even though in same cases the legislators introduced low interest rate credit programs for private RE installations, access to these loans is usually still linked to a basic amount of equity capital. For a discussion of feed-in tariffs in the US and the exclusion of LIH see [125]. For Germany see [13].

Finally, what applies to vulnerable consumers must also be considered for RECs. Although many existing RECs want to be inclusive their situation does not always allow for the inclusion of vulnerable consumers (comp. access to financing). We therefore propose to further investigate incentives and conditions that facilitate RECs in their efforts to enable inclusion.

4.4.2. Framing the Participation in RECs

The way how information is presented affects its perception [126]. In consequence, an effective information approach needs to extend beyond consumer choice and include the local community and its decision-making process by framing it around what is perceived relevant and of interest [127,128]. This entails the framing of those incentives discussed as well as of those yet to be identified including aspects such as pro-social and pro-environmental behaviour.

While some approaches address economic decision makers (of a household) others demand for the identification of potential advocates for change at a community and family level. Promoting pro-environmental habits at school through teaching basic sustainable behaviour to school children affects not only the family but the entire community [129]. Once children learn basic energy saving behaviour, they pass this knowledge on to their parents and grandparents. Educating children about the benefits of participation in a local REC is therefore likely to educate their parents as well. Therefore, we suggest that presenting the participation in a REC not only as a consumption or investment choice but as benefitting the future of the family, particularly one's children, the motivation to participate increases. Assuming that, nudged by their child, parents change their behaviour, the child in turn benefits from self-efficacy, an important driver for individual development [48]. The nudging of parents through their children changes the choice architecture in that it alters their behaviour in a predictable—in this case pro-environmental way—without forbidding any options or significantly changing their economic incentives [130]. Nudging has increasingly become an instrument in social policy making especially in the environmental policy domain [131,132]. We, therefore, encourage future research to include this perspective to evaluate the potential and conditions of such framing approaches.

While empowerment in form of more choice can be liberating choice is also disciplining and potentially paralysing [22]. The more choice available the greater the need to gather relevant information increasing the complexity of decision making and the potential for error resulting in a state referred to as "choice paralysis" [133]. When the cost of processing all necessary information to perform the best choice outweighs the benefits of choosing consumers chose to not chose. With regard to framing and communication strategies in general less choice and less complexity helps to engage in particular those already overburdened by the difficult life choices they have to make.

In sum, the way how different choice options are presented by highlighting particular aspects educes information and associations linked to these aspects affecting the perception of a given choice situation [132]. Promoting the participation in RECs among vulnerable consumers calls for fewer but simple choices leading to participation and informing about these choices should be done in a simple and consistent manner. Future research should include these aspects and investigate the potential and limits of such framing approaches in practise.

5. Conclusions

The participation of vulnerable consumers in RECs can make a significant contribution to mitigating energy poverty thus shaping a social and just energy transition. Provided with an *"enabling framework"* (including elements such as tax and levy exemptions) RECs could offer lower energy prices and an extra income from dividends to their vulnerable members (see Figure 1). Prosumption is further linked to EE decreasing individual energy consumption [3]. In this way two major causes of energy poverty can be addressed that is high energy costs and low income. What is more, participation expends the individual social circle and thus helps to overcome social isolation from which many energy poor households reportedly suffer [134].

Moreover, the participation of vulnerable consumers is of particular importance for the overall success of the energy transition as especially this group has so far been underrepresented in RE projects if not entirely excluded. As a form of collective empowerment formerly vulnerable consumers not only become active participants in the energy market—as postulated by the EC as a basis for the most efficient market results—but they are empowered to become self-determining contributing members of society.

Consequently, in the RED II the European legislator obliges the MS to put forward a legally binding *"enabling framework"* for RECs to support and entice their setting up. A connected postulate is to facilitate the participation of vulnerable consumers in RECs. However, if this *"enabling framework"* when transposed into national law is not sufficiently tailored to vulnerable energy consumers' different life situation and behavioural characteristics, the objective of inclusion is unlikely to be achieved. What is more if inclusive RE projects as RECs do not benefit from a robust *"enabling framework"* that encompasses all strata of society, they are likely not to be able to assume their function to increase acceptance of RE since the success of the energy transition depends on the participation of all societal groups.

Therefore, a truly inclusive *"enabling framework"* enables RECs to provide competitive energy prices to their members by removing existing obstacles such as high tax and levy burdens or administrative and regulatory complexities. Simultaneously more knowledge about vulnerable energy consumers and the respective vulnerability context and how it impacts participation in RECs is needed. This serves as a basis for both RECs and policy makers to understand prerequisites, motives and incentives for participation in order to provide those elements and eventually the opportunity to participate.

While research investigating community energy and collective prosumption across Europe exists, research on inclusion and participation of vulnerable consumers in community energy, notably in RECs is scarce. Future research should thus investigate these elements notably the vulnerability context and how it affects participation. This must include the investigation of the role of vulnerable consumers as active consumers in community energy and reasons for why they remain underrepresented. In addition, the participation in RECs could be investigated from the perspective of social innovation and how new transactional modes can be a sustainable alternative to ensure the inclusion of vulnerable households. At the same time, the NECPs which constitute a major step towards the future design of the European Energy Union currently lack the aspect of inclusion and many existing RECs focus on staying operational rather than extending their activities towards inclusion. Policy makers must therefore be informed and made aware of the potential inclusion of (energy) vulnerable households in RECs entails.

In practice projects such as SCORE report barriers with respect to inclusion and energy poverty mitigation that are linked to access to finance and to local eligibility rules for means tested social transfers for vulnerable households which prevent their participation in RE projects as co-owners. Incentives both financial but also ideational linked to pro-environmental behaviour must further be investigated and provided. A first step we propose a "renewables asset formation agenda for LIHs" to enable and support the participation of low-income households in RECs [6]. It is worthwhile to recall the main principles of this agenda and to extend them with particular emphasis on the empowerment of vulnerable consumers while drawing on best practice from Spanish Sociedades Laborales [55] and local banking schemes such as on the Island of Ærø [114]:

(i)   Direct energy subsidies for vulnerable consumers could be tied to membership in a REC; these subsidies then could be capitalised and paid out as a lump sum to join an existing or set up a new REC. Once the REC is operative over time this would increase disposable household income while providing a strong incentive to participate actively in the energy transition. Furthermore, with regard to the acquisition of (co-)ownership in RE to promote prosumership vulnerable consumers receiving subsidies for energy expenditures could be automatically enrolled as (co-)owners in newly founded RECs.

(ii) Investments in RECs should be exempt from necessity to liquidate one's assets when applying for means-tested social transfers; this exemption could have a cap of at least EUR 1000 per person per year which should increase for investments designed to benefit child education and the like.

(iii) An *"enabling framework"* should support capacity building of local municipalities which in turn can then offer coaching and training programs to facilitate the apprenticeship of vulnerable consumers when joining RECs. Of course, they could also provide financial assistance when doing so; this should include easier access to credit, low or no interest loans, credit guarantees and the like. Capacity building would include—building on best practice—the setting up of networks between local banks, impact investors and RECs to provide low interest rate loans to vulnerable households.

Such RECs not only drive forward a truly sustainable energy transition but have the potential to facilitate the inclusion of marginalised and vulnerable households as consumer co-owners providing them with the opportunity to improve their economic situation. Therefore, these issues need to be part of the national *"enabling frameworks"* in the EU Members States and should be accompanied by future research.

Author Contributions: Conceptualization, F.H. and J.L.; methodology, F.H. and J.L.; investigation, F.H.; iginal draft preparation, F.H. and J.L.; writing—review and editing, F.H. and J.L.; visualization, F.H. All authors have read and agreed to the published version of the manuscript.

Funding: The SCORE project that this research is based on has received funding for a coordination and support action from the European Union's Horizon 2020 research and innovation programme under grant agreement No. 784960.

Conflicts of Interest: The authors declare no conflict of interest.

References

1. European Union: European Commission. The European Green Deal—Communication from The Commission to the European Parliament, The European Council, The Council, The European Economic And Social Committee and The Committee of The Regions. COM (2019) 640 Final. 2019. Available online: https://ec.europa.eu/info/sites/info/files/european-green-deal-communication_en.pdf (accessed on 12 March 2020).
2. Lowitzsch, J. (Ed.) *Energy Transition: Financing Consumer Ownership in Renewables*; Palgrave/McMillan: Cham, Switzerland, 2019; ISBN 978-3-319-93517-1.
3. Roth, L.; Lowitzsch, J.; Yildiz, Ö.; Hashani, A. Does (Co-)ownership in renewables matter for an electricity consumer's demand flexibility? Empirical evidence from Germany. *Energy Res. Soc. Sci.* **2018**, *46*, 169–182. [CrossRef]
4. European Union: Council of the European Union. Directive (EU) 2018/2001 of the European Parliament and of the Council of 11 December 2018 on the Promotion of the Use of Energy from Renewable Sources (Recast). OJ L 328, 21.12.2018. 2018. Available online: http://data.europa.eu/eli/dir/2018/2001/oj (accessed on 8 March 2020).
5. European Union: European Parliament; European Council. Directive (EU) 2019/944 of the European Parliament and of the Council of 5 June 2019 on Common Rules for the Internal Market for Electricity and Amending Directive 2012/27/EU (Recast). OJ L 158, 14.6.2019. 2019. Available online: https://eur-lex.europa.eu/legal-content/EN/TXT/?uri=CELEX:32019L0944 (accessed on 8 March 2020).
6. Lowitzsch, J.; Hanke, F. Consumer (Co-)ownership in Renewables, Energy Efficiency and the Fight Against Energy Poverty—A Dilemma of Energy Transitions. *Claeys Casteels Law Publ. Bv* **2019**, *9*, 5–21.
7. European Union: European Commission. Delivering a New Deal for Energy Consumers. COM(2015) 339 Final. 2015. Available online: https://ec.europa.eu/energy/sites/ener/files/documents/1_EN_ACT_part1_v8.pdf (accessed on 8 March 2020).
8. Ioannidou, M. Effective Paths for Consumer Empowerment and Protection in Retail Energy Markets. *J. Consum. Policy* **2018**, *41*, 135–157. [CrossRef]
9. Heindl, P.; Schüßler, R.; Löschel, A. Ist die Energiewende sozial gerecht? *Wirtschaftsdienst* **2014**, *94*, 508–514. [CrossRef]
10. Energy Atlas. *Energy Atlas 2018—Facts and Figures about Renewables in Europe*; Heinrich Böll Foundation; Friends of the Earth Europe; European Renewable Energies Federation; Green European Foundation: Berlin, Germany; Brussels, Belgium; Luxembourg, 2018.
11. Pye, S.; Dobbins, A. Energy Poverty and Vulnerable Consumers in the Energy Sector across the EU: Analysis of Policies and Measures. Policy Report 2. *Insight*. 2015. Available online: https://ec.europa.eu/energy/sites/ener/files/documents/INSIGHT_E_Energy%20Poverty%20-%20Main%20Report_FINAL.pdf (accessed on 8 March 2020).
12. Williams, S.; Doyon, A. Justice in energy transitions. *Environ. Innov. Soc. Transit.* **2019**, *31*, 144–153. [CrossRef]
13. Frondel, M.; Sommer, S.; Vance, C. The burden of Germany's energy transition: An empirical analysis of distributional effects. *Econ. Anal. Policy* **2015**, *45*, 89–99. [CrossRef]
14. Yildiz, Ö.; Rommel, J.; Debor, S.; Holstenkamp, L.; Mey, F.; Müller, J.R.; Radtke, J.; Rognli, J. Renewable energy cooperatives as gatekeepers or facilitators? Recent developments in Germany and a multidisciplinary research agenda. *Energy Res. Soc. Sci.* **2015**, *6*, 59–73. [CrossRef]
15. Staples, L.H. Powerful Ideas About Empowerment. *Adm. Soc. Work* **1990**, *14*, 29–42. [CrossRef]
16. Merriam-Webster Definition of EMPOWERMENT. Available online: https://www.merriam-webster.com/dictionary/empowerment (accessed on 12 March 2020).
17. Freire, P. *Education for Critical Consciousness*; Continuum Impacts; Continuum: London, UK; New York, NY, USA, 2005; ISBN 978-0-8264-7795-8.
18. Bandura, A. Social Cognitive Theory. In *Annals of Child Development*; Vasta, R., Ed.; Vol. 6. Six Theories of Child Development; JAI Press: Greenwich, CT, USA, 1989; pp. 1–60.
19. Jenkins, K.E.H. Energy Justice, Energy Democracy, and Sustainability: Normative Approaches to the Consumer Ownership of Renewables. In *Energy Transition*; Lowitzsch, J., Ed.; Springer International Publishing: Cham, Switzerland, 2019; pp. 79–97. ISBN 978-3-319-93517-1.
20. Sovacool, B.K.; Heffron, R.J.; McCauley, D.; Goldthau, A. Energy decisions reframed as justice and ethical concerns. *Nat. Energy* **2016**, *1*, 16024. [CrossRef]

21. Horstink, L.; Wittmayer, J.M.; Ng, K.; Luz, G.P.; Marín-González, E.; Gährs, S.; Campos, I.; Holstenkamp, L.; Oxenaar, S.; Brown, D. Collective Renewable Energy Prosumers and the Promises of the Energy Union: Taking Stock. *Energies* **2020**, *13*, 421. [CrossRef]

22. Shankar, A.; Cherrier, H.; Canniford, R. Consumer empowerment: A Foucauldian interpretation. *Eur. J. Mark.* **2006**, *40*, 1013–1030. [CrossRef]

23. Diener, E.; Biswas-Diener, R. Will Money Increase Subjective Well-Being? *Soc. Indicators Res.* **2002**, *57*, 119–169. [CrossRef]

24. Kreckel, R. (Ed.) Bourdieu Ökonomisches Kapital, kulturelles Kapital, soziales Kapital. In *Soziale Ungleichheiten. Sonderband 2 Soziale Welt*; Otto Schwartz: Göttingen, Germany, 1983.

25. Grabka, M.; Halbmeier, C. Vermögensungleichheit in Deutschland Bleibt Trotz Deutlich Steigender Nettovermögen Anhaltend Hoch. 2019. Available online: https://www.diw.de/documents/publikationen/73/diw_01.c.679972.de/19-40-1.pdf (accessed on 8 March 2020).

26. Bouzarovski, S.; Petrova, S. A global perspective on domestic energy deprivation: Overcoming the energy poverty–fuel poverty binary. *Energy Res. Soc. Sci.* **2015**, *10*, 31–40. [CrossRef]

27. Großmann, K. Energiearmut als multiple Deprivation vor dem Hintergrund diskriminierender Systeme. In *Energie und Soziale Ungleichheit. Zur Gesellschaftlichen Dimension der Energiewende in Deutschland und Europa*; VS Verlag für Sozialwissenschaften: Wiesbaden, Germany, 2017; pp. 55–78.

28. Lavrijssen, S.A.C.M. The Different Faces of Energy Consumers: Toward A Behavioral Economics Approach. *J. Compet. Law Econ.* **2014**, *10*, 257–291. [CrossRef]

29. Schilbach, F.; Schofield, H.; Mullainathan, S. The Psychological Lives of the Poor. *Am. Econ. Rev.* **2016**, *106*, 435–440. [CrossRef]

30. Parag, Y.; Sovacool, B.K. Electricity market design for the prosumer era. *Nat. Energy* **2016**, *1*, 16023. [CrossRef]

31. Middlemiss, L.; Gillard, R.; Pellicer, V.; Straver, K. Plugging the Gap Between Energy Policy and the Lived Experience of Energy Poverty: Five Principles for a Multidisciplinary Approach. In *Advancing Energy Policy*; Foulds, C., Robison, R., Eds.; Palgrave Pivot: Cham, Switzerland, 2018; pp. 15–29. ISBN 978-3-319-99096-5.

32. OFGEM. What Can Behavioural Economics Say about Gb Energy Consumers? 2011. Available online: https://www.ofgem.gov.uk/sites/default/files/docs/2011/03/behavioural_economics_gbenergy_1.pdf (accessed on 12 March 2020).

33. Allmark, P.; Tod, A.M. Can a nudge keep you warm? Using nudges to reduce excess winter deaths: Insight from the Keeping Warm in Later Life Project (KWILLT). *J. Public Health* **2014**, *36*, 111–116. [CrossRef]

34. Trzaskowski, J. Behavioural Economics, Neuroscience, and the Unfair Commercial Practises Directive. *J. Consum. Policy* **2011**, *34*, 377–392. [CrossRef]

35. Enable.eu Final Comprehensive Literature Review Setting the Scene for the Entire Study. 2017. Available online: http://www.enable-eu.com/wp-content/uploads/2017/08/ENABLE.EU_D2.2.pdf (accessed on 12 March 2020).

36. Dolan, P.; Hallsworth, M.; Halpern, D.; King, D.; Vlaev, I. MINDSPACE: Influencing Behaviour for Public Policy. Available online: http://www.instituteforgovernment.org.uk/publications/ (accessed on 12 March 2020).

37. Horstink, L.; Luz, G.; Soares, M. *Review and Characterisation of Collective Renewable Energy Prosumer Initiatives. PROSEU-Prosumers for the Energy Union: Mainstreaming Active Participation of Citizens in the Energy Transition (Deliverable N°2.1)*; Horizon 2020 (H2020- LCE-2017) Grant Agreement N°764056; University of Porto: Porto, Portugal, 2019.

38. Korobkin, R.B.; Ulen, T.S. Law and Behavioral Science: Removing the Rationality Assumption from Law and Economics. *Calif. Law Rev.* **2000**, *88*, 1051. [CrossRef]

39. Thaler, R.H.; Sunstein, C.R. *Libertarian Paternalism Is Not an Oxymoron*; Theory Working Paper No. 43; University of Chicago Public Law & Legal: Chicago, IL, USA, 2003; p. 47.

40. Kahneman, D. Maps of Bounded Rationality: Psychology for Behavioral Economics. *Am. Econ. Rev.* **2003**, *93*, 1449–1475. [CrossRef]

41. Kahn, A.; Bender, E.I. Self-help groups as a crucible for people empowerment in the context of social development. *Soc. Dev. Issues* **1985**, *9*, 4–13.

42. Römmele, A.; Banthien, H. (Eds.) *Empowering Citizens: Studies in Collaborative Democracy*, Schriftenreihe Kommunikation in Politik und Wirtschaft, 1st ed.; Nomos Verlagsgesellschaft: Baden-Baden, Germany, 2013; ISBN 978-3-8329-7919-5.

43. Holstenkamp, L.; Radtke, J. (Eds.) *Handbuch Energiewende und Partizipation*; Springer Fachmedien Wiesbaden: Wiesbaden, Germany, 2018; ISBN 978-3-658-09415-7.

44. Bandura, A. Exercise of Human Agency Through Collective Efficacy. *Curr. Dir. Psychol. Sci.* **2000**, *9*, 75–78. [CrossRef]

45. Pigg, K.E. Three Faces of Empowerment: Expanding the Theory of Empowerment in Community Development. *J. Community Dev. Soc.* **2002**, *33*, 107–123. [CrossRef]

46. Sawitri, D.R.; Hadiyanto, H.; Hadi, S.P. Pro-environmental Behavior from a SocialCognitive Theory Perspective. *Procedia Environ. Sci.* **2015**, *23*, 27–33. [CrossRef]

47. Middlemiss, L.; Gillard, R. Fuel poverty from the bottom-up: Characterising household energy vulnerability through the lived experience of the fuel poor. *Energy Res. Soc. Sci.* **2015**, *6*, 146–154. [CrossRef]

48. Bandura, A. *Self-Efficacy: The Exercise of Control*; W.H. Freeman: New York, NY, USA, 1997; ISBN 978-0-7167-2626-5.

49. Berger. *Luckmann Die Gesellschaftliche Konstruktion der Wirklichkeit*; Fischer Taschenbuch Verlag: Frankfurt am Main, Germany, 1980.

50. Axelrod, R.; Hamilton, W. The evolution of cooperation. *Science* **1981**, *211*, 1390–1396. [CrossRef]

51. Pinderhughes, E.B. Empowerment for Our Clients and for Ourselves. *Soc. Casework* **1983**, *64*, 331–338. [CrossRef]

52. Rappaport, J.; Swift, C.F.; Hess, R. (Eds.) *Studies in Eempowerment: Steps toward Understanding and Action*; Haworth Press: New York, NY, USA, 1984; ISBN 978-0-86656-283-6.

53. Birchall, J. The potential of co-operatives during the current recession; theorizing comparative advantage. *J. Entrep. Organ. Divers.* **2013**, *2*, 1–22. [CrossRef]

54. Birchall, J.; Organización Internacional del Trabajo; Servicio de Cooperativas. *Rediscovering the Cooperative Advantage: Poverty Reduction Through Self-Help*; Cooperative Branch, International Labour Office: Geneva, Switzerland, 2003; ISBN 978-92-2-113603-3.

55. Lowitzsch, J.; Dunsch, S.; Hashi, I. *Spanish Sociedades Laborales—Activating the Unemployed*; Palgrave/McMillan Pivot: Cham, Switzerland, 2017; ISBN 978-3-319-54869-2.

56. European Commission. European Employment Policy Observatory Review—Activating Jobseekers through Entrepreneurship: Start-up Incentives in Europe. 2014. Available online: https://www.google.com.hk/url?sa=t&rct=j&q=&esrc=s&source=web&cd=3&ved=2ahUKEwjW7MvFk8boAhUGGaYKHfwFDJkQFjACegQIBhAB&url=http%3A%2F%2Fwww.agenziagiovani.it%2Fmedia%2F156405%2Fdgempl_eepo_autumn_review_accessible_v20.pdf&usg=AOvVaw3M2AfVR-KYcn3dh1hTw5CQ (accessed on 8 March 2020).

57. Sen, A. *Development as Freedom*, 1st. ed.; Knopf: New York, NY, USA, 1999; ISBN 978-0-375-40619-5.

58. Nam, Y.; Huang, J.; Sherraden, M. *Assets, Poverty, and Public Policy: Challenges in Definition and Measurement*; Poor Finances: Assets and Low-Income Households; Oxford University Press: New York, NY, USA, 2008; p. 45.

59. Sherraden, M.W. *Assets and the Poor: A New American Welfare Policy*; Routledge: London, UK, 1991; ISBN 978-1-56324-066-9.

60. Sherraden, M.; Johnson, L.; Clancy, M.M.; Beverly, S.G.; Sherraden, M.S.; Schreiner, M.; Elliot, W.; Shanks, T.R.W.; Adams, D.; Curley, J.; et al. *Asset Building Toward Inclusive Policy*; NASW Press and Oxford University Press: Oxford, UK, 2013; Volume 1.

61. Ludwig, J.; Duncan, G.; Gennetian, L.; Katz, L.; Kessler, R.; Kling, J.; Sanbonmatsu, L. *Long-Term Neighborhood Effects on Low-Income Families: Evidence from Moving to Opportunity*; National Bureau of Economic Research: Cambridge, MA, USA, 2013.

62. Lent, R.W.; Brown, S.D.; Hackett, G. Toward a Unifying Social Cognitive Theory of Career and Academic Interest, Choice, and Performance. *J. Vocat. Behav.* **1994**, *45*, 79–122. [CrossRef]

63. Liger, Q.; Stefan, M.; Britton, J. *European Parliament*; Directorate-General for Internal Policies; Policy Department A: Economic and Scientific Policy; European Parliament; Committee on the Internal Market and Consumer Protection Social Economy: Study; European Union: Brussels, Belgium, 2016; ISBN 978-92-823-9058-0.

64. Brummer, V. Community energy—Benefits and barriers: A comparative literature review of Community Energy in the UK, Germany and the USA, the benefits it provides for society and the barriers it faces. *Renew. Sustain. Energy Rev.* **2018**, *94*, 187–196. [CrossRef]

65. Friends of the Earth Europe; Energycities; REScoop.eu. *The New Energy Market Design: How the EU Can Support Energy Communities and Citizens to Participate in the Energy Transition*; Renewable Energy Sources Cooperative, 2018; Available online: https://energy-cities.eu/wp-content/uploads/2018/11/commuity_energy_coalition_pp_trilogues_mdi_final.pdf (accessed on 8 March 2020).

66. Šahović, N.; da Silva, P.P. Community Renewable Energy-Research Perspectives-. *Energy Procedia* **2016**, *106*, 46–58. [CrossRef]

67. Ison, N. From Command and Control to Local Democracy Governance of Community Energy Projects. Independent Research Project, Lancaster Environment Centre, 2010. Available online: https://refubium.fu-berlin.de/bitstream/handle/fub188/19440/Ison-From_command_and_control_to_local_democracy-416.pdf?sequence=1&isAllowed=y (accessed on 8 March 2020).

68. Walker, G. What are the barriers and incentives for community-owned means of energy production and use? *Energy Policy* **2008**, *36*, 4401–4405. [CrossRef]

69. Herbert, S. The Trapdoor of Community. *Ann. Assoc. Am. Geogr.* **2005**, *95*, 850–865. [CrossRef]

70. Park, J.J. Fostering community energy and equal opportunities between communities. *Local Environ.* **2012**, *17*, 387–408. [CrossRef]

71. Walker, G.; Devine-Wright, P.; Hunter, S.; High, H.; Evans, B. Trust and community: Exploring the meanings, contexts and dynamics of community renewable energy. *Energy Policy* **2010**, *38*, 2655–2663. [CrossRef]

72. Fast, S. A Habermasian analysis of local renewable energy deliberations. *J. Rural Stud.* **2013**, *30*, 86–98. [CrossRef]

73. Middlemiss, L.; Parrish, B.D. Building capacity for low-carbon communities: The role of grassroots initiatives. *Energy Policy* **2010**, *38*, 7559–7566. [CrossRef]

74. Cass, N.; Walker, G.; Devine-Wright, P. Good Neighbours, Public Relations and Bribes: The Politics and Perceptions of Community Benefit Provision in Renewable Energy Development in the UK. *J. Environ. Policy Plan.* **2010**, *12*, 255–275. [CrossRef]

75. Parag, Y.; Hamilton, J.; White, V.; Hogan, B. Network approach for local and community governance of energy: The case of Oxfordshire. *Energy Policy* **2013**, *62*, 1064–1077. [CrossRef]

76. Radtke, J. A closer look inside collaborative action: Civic engagement and participation in community energy initiatives. *PeoplePlace Policy Online* **2014**, *8*, 235–248. [CrossRef]

77. Bauwens, T. Explaining the diversity of motivations behind community renewable energy. *Energy Policy* **2016**, *93*, 278–290. [CrossRef]

78. Warbroek, B.; Hoppe, T.; Bressers, H.; Coenen, F. Testing the social, organizational, and governance factors for success in local low carbon energy initiatives. *Energy Res. Soc. Sci.* **2019**, *58*, 101269. [CrossRef]

79. Aguirre, M.; Ibikunle, G. Determinants of renewable energy growth: A global sample analysis. *Energy Policy* **2014**, *69*, 374–384. [CrossRef]

80. Marques, A.C.; Fuinhas, J.A. Are public policies towards renewables successful? Evidence from European countries. *Renew. Energy* **2012**, *44*, 109–118. [CrossRef]

81. Polzin, F.; Migendt, M.; Täube, F.A.; von Flotow, P. Public policy influence on renewable energy investments—A panel data study across OECD countries. *Energy Policy* **2015**, *80*, 98–111. [CrossRef]

82. Radtke, J. *Bürgerenergie in Deutschland*; Springer Fachmedien Wiesbaden: Wiesbaden, Germany, 2016; ISBN 978-3-658-14625-2.

83. Gilchrist, A.; Taylor, M. *The Short Guide to Community Development*, Community Development/Social Studies, 2nd ed.; Policy Press: Bristol Chicago, IL, USA, 2016; ISBN 978-1-4473-2783-7.

84. Nahapiet, J.; Ghoshal, S. Social Capital, Intellectual Capital, and the Organizational Advantage. *Acad. Manag. Rev.* **1998**, *23*, 242. [CrossRef]

85. Lin, N. Building a Network Theory of Social Capital. *Connections* **1999**, *22*, 28–51. Available online: https://www.academia.edu/1033821/Building_a_network_theory_of_social_capital?auto=download (accessed on 8 March 2020).

86. McElroy, M.W.; Jorna, R.J.; van Engelen, J. Rethinking social capital theory: A knowledge management perspective. *J. Knowl. Manag.* **2006**, *10*, 124–136. [CrossRef]

87. Naranjo-Zolotov, M.; Oliveira, T.; Cruz-Jesus, F.; Martins, J.; Gonçalves, R.; Branco, F.; Xavier, N. Examining social capital and individual motivators to explain the adoption of online citizen participation. *Future Gener. Comput. Syst.* **2019**, *92*, 302–311. [CrossRef]

88. Hoffman, S.M.; High-Pippert, A. Community Energy: A Social Architecture for an Alternative Energy Future. *Bull. Sci. Technol. Soc.* **2005**, *25*, 387–401. [CrossRef]

89. Volz, R. *Genossenschaften im Bereich Erneuerbarer Energien: Status quo und Entwicklungsmöglichkeiten Eines Neuen Betätigungsfeldes*; Veröffentlichungen der Forschungsstelle für Genossenschaftswesen an der Universität Hohenheim; Forschungsstelle für Genossenschaftswesen an der Univ. Hohenheim: Hohenheim, Germany, 2012.

90. Holstenkamp, L.; Kahla, F.; Degenhart, H. Finanzwirtschaftliche Annäherungen an das Phänomen Bürgerbeteiligung. In *Handbuch Energiewende und Partizipation*; Holstenkamp, L., Radtke, J., Eds.; Springer Fachmedien Wiesbaden: Wiesbaden, Germany, 2018; pp. 281–301. ISBN 978-3-658-09415-7.

91. Kahla Das Phänomen Bürgerenergie in Deutschland Eine betriebswirtschaftliche Analyse von Bürgergesellschaften im Bereich der Erneu- Erbaren Energien-Produktion. 2018. Available online: https://d-nb.info/1155587189/34 (accessed on 8 March 2020).

92. Thomson, H.; Petrova, S.; Bouzarovski, S.; Simcock, N. *Energy Poverty and Vulnerability: A Global Perspective*; Routledge: Abingdon, UK, 2018.

93. Warburton, D. A Passionate Dialogue: Community and Sustainable Development. In *Community and Sustainable Development*; Warburton, D., Ed.; Routledge, 2018; pp. 1–39. ISBN 978-1-315-07119-0.

94. Craig, G. Community capacity-building: Something old, something new? *Crit. Soc. Policy* **2007**, *27*, 335–359. [CrossRef]

95. Muntaner, C.; Lynch, J.; Smith, G.D. Social capital and the third way in public health. *Crit. Public Health* **2000**, *10*, 107–124. [CrossRef]

96. Rydin, Y.; Holman, N. Re-evaluating the Contribution of Social Capital in Achieving Sustainable Development. *Local Environ.* **2004**, *9*, 117–133. [CrossRef]

97. Andrews, R. Social Capital and Public Service Performance: A Review of the Evidence. *Public Policy Adm.* **2012**, *27*, 49–67. [CrossRef]

98. Purdue, D. *Community Leadership in Area Regeneration*; Policy Press: Bristol, UK, 2000; ISBN 978-1-86134-249-2.

99. Pollitt. Shaorshadze the Role of Behavioural Economics in Energy and Climate Policy. In *Handbook on Energy and Climate Change*; Edward Elgar Publishing: Cheltenham, UK, 2011.

100. Benzion, U.; Rapoport, A.; Yagil, J. Discount Rates Inferred from Decisions: An Experimental Study. *Manag. Sci.* **1989**, *35*, 270–284. [CrossRef]

101. Thaler, R. Some empirical evidence on dynamic inconsistency. *Econ. Lett.* **1981**, *8*, 201–207. [CrossRef]

102. Thaler, R.H. Mental accounting matters. *J. Behav. Decis. Mak.* **1999**, *12*, 183–206. [CrossRef]

103. Spears, D. Economic Decision-Making in Poverty Depletes Behavioral Control. *B.E. J. Econ. Anal. Policy* **2011**, *11*, 1–44. [CrossRef]

104. Kahneman, D.; Knetsch, J.L.; Thaler, R.H. Anomalies: The Endowment Effect, Loss Aversion, and Status Quo Bias. *J. Econ. Perspect.* **1991**, *5*, 193–206. [CrossRef]

105. Mullainathan, S.; Shafir, E. *Scarcity: Why Having too Little Means so Much*, 1st ed.; Times Books; Henry Holt and Company: New York, NY, USA, 2013; ISBN 978-0-8050-9264-6.

106. Tversky, A.; Kahneman, D. Judgment under Uncertainty: Heuristics and Biases. In *Utility, Probability, and Human Decision Making*; Wendt, D., Vlek, C., Eds.; Springer Netherlands: Dordrecht, The Netherlands, 1975; pp. 141–162. ISBN 978-94-010-1836-4.

107. Haushofer, J.; Fehr, E. On the psychology of poverty. *Science* **2014**, *344*, 862–867. [CrossRef]

108. Mullainathan, S.; Shafir, E. *Scarcity: The New Science of Having Less and How It Defines Our Lives*, 1. Picador ed.; Picador/Henry Holt and Co.: New York, NY, USA, 2014; ISBN 978-1-4299-4345-1.

109. Bertrand, M.; Mullainathan, S.; Shafir, E. Behavioral Economics and Marketing in Aid of Decision Making Among the Poor. *J. Public Policy Mark.* **2006**, *25*, 8–23. [CrossRef]

110. Huybrechts, B.; Creupelandt, D.; Vansintjan, D. Networking Renewable Energy Cooperatives—The experience of the European Federation REScoop.eu. In *Handbuch Energiewende und Partizipation*; Holstenkamp, L., Radtke, J., Eds.; Springer Fachmedien Wiesbaden: Wiesbaden, Germany, 2018; pp. 847–858. ISBN 978-3-658-09415-7.

111. DGRV; D.G.R. e.V. *Energiegenossenschaften. Ergebnisse der DGRV-Jahresumfrage 2018*; DGRV: Berlin, Germany, 2018; Available online: https://www.genossenschaften.de/sites/default/files/20190715_DGRV_Umfrage_Energiegenossenschaften_2019_0.pdf (accessed on 8 March 2020).

112. Baumgärtler, T.; Popović, T. *Genossenschaftliche Innovationsökosysteme—Transformation aus der Kraft der Gemeinschaft*; Akademie Deutscher Genossenschaften e.V.: Montabaur, Germany, 2019.
113. European Union: European Commission. Commission Communication on the Sustainable Europe Investment Plan; 14.1.2020 COM(2020) 21 Final. 2020. Available online: https://ec.europa.eu/commission/presscorner/api/files/attachment/860462/Commission%20Communication%20on%20the%20European%20Green%20Deal%20Investment%20Plan_EN.pdf.pdf (accessed on 8 March 2020).
114. Busch, H. Community-Owned Wind Farm on the Island of Ærø, Denmark. 2019. Available online: http://co2mmunity.eu/wp-content/uploads/2019/03/Factsheet-Aerö.pdf (accessed on 8 March 2020).
115. Lavinas, L. New Trends in Inequality: The Financialization of Social Policies. 2015. Available online: https://www.researchgate.net/profile/Lena_Lavinas/publication/268091123_Inancial_Inclusion_As_a_Basic_Human_Right_Reframing_Inequalities_in_the_South/links/5666d02b08ae4931cd628154/Inancial-Inclusion-As-a-Basic-Human-Right-Reframing-Inequalities-in-the-South.pdf (accessed on 8 March 2020).
116. Shiller, R.J. *Finance and the Good Society*; 3. Print., and 1. Paperback Print.; Princeton Univ. Press: Princeton, NJ, USA, 2013; ISBN 978-0-691-15809-9.
117. Zeit Online "Grüne fordern Bürgerfonds zur Altersvorsorge"—The Green Party Demands for a Citizen Fund. 2020. Available online: https://www.zeit.de/wirtschaft/geldanlage/2019-02/rentenreform-buergerfonds-altersvorsorge-gruene (accessed on 26 February 2019).
118. Cramer, R.; Shanks, T.R.W. (Eds.) *The Assets Perspective*; Palgrave Macmillan US: New York, NY, USA, 2014; ISBN 978-1-349-48196-5.
119. Jens Lowitzsch, I. Investing in a Renewable Future—Renewable Energy Communities, Consumer (Co-)Ownership and Energy Sharing in the Clean Energy Package. 2019. Available online: https://www.ingentaconnect.com/content/cclp/relp/2019/00000009/00000002/art00003#Refs (accessed on 9 July 2019).
120. Leuphana Universität Lüneburg; Nestle, U. *Marktrealität von Bürgerenergie und Mögliche Auswirkungen von Regulatorischen Eingriffen in Die Energiewende*; Leuphana University: Lüneburg, Kiel, Germany, 2014; Available online: http://www.energiegenossenschaften-gruenden.de/fileadmin/user_upload/downloads/Bündnis_Bürgerenergie/Studie_Marktrealität_von_Bürgerenergie_und_mögliche_Auswirkungen_von_regulatorischen_Eingriffen.pdf (accessed on 8 March 2020).
121. Vilain, M. *Finanzierungslehre für Nonprofit-Organisationen: Zwischen Auftrag und ökonomischer Notwendigkeit*; 1. Aufl.; VS, Verl. für Sozialwiss: Wiesbaden, Germany, 2006; ISBN 978-3-531-90093-3.
122. Sherraden, M.W. (Ed.) *Inclusion in the American Dream: Assets, Poverty, and Public Policy*; Oxford University Press: New York, NY, USA, 2005; ISBN 978-0-19-516819-8.
123. Dpa Rekordtief: EZB hält Leitzins im Euroraum auf null Prozent. *Die Zeit.* 2019. Available online: https://www.zeit.de/news/2019-03/07/ezb-haelt-leitzins-im-euroraum-auf-null-prozent-190306-99-269340 (accessed on 8 March 2020).
124. Couture, T.; Gagnon, Y. An analysis of feed-in tariff remuneration models: Implications for renewable energy investment. *Energy Policy* **2010**, *38*, 955–965. [CrossRef]
125. Powers, M. An inclusive energy transition: Expanding low income access to clean energy programs. *N.C.J. Law Technol.* **2017**, *18*, 540–564.
126. Baucus, M.S.; Rechner, P.L. International Association for Business and Society Framing and Reframing: A Process Model of Ethical Decision Making. *Proc. Int. Assoc. Bus. Soc.* **1995**, *6*, 1–12. [CrossRef]
127. Rogers, J.C.; Simmons, E.A.; Convery, I.; Weatherall, A. Public perceptions of opportunities for community-based renewable energy projects. *Energy Policy* **2008**, *36*, 4217–4226. [CrossRef]
128. Yates, J.F.; de Oliveira, S. Culture and decision making. *Organ. Behav. Hum. Decis. Process.* **2016**, *136*, 106–118. [CrossRef]
129. Lawson, D.F.; Stevenson, K.T.; Peterson, M.N.; Carrier, S.J.; Strnad, R.L.; Seekamp, E. Children can foster climate change concern among their parents. *Nat. Clim. Chang.* **2019**, *9*, 458–462. [CrossRef]
130. Scheufele, D.A.; Tewksbury, D. Framing, Agenda Setting, and Priming: The Evolution of Three Media Effects Models: Models of Media Effects. *J. Commun.* **2007**, *57*, 9–20. [CrossRef]
131. Hansen, P.G.; Jespersen, A.M. Nudge and the Manipulation of Choice: A Framework for the Responsible Use of the Nudge Approach to Behaviour Change in Public Policy. *Eur. J. Risk Regul.* **2013**, *4*, 3–28. [CrossRef]

132. Michalek, G.; Meran, G.; Schwarze, R.; Ö.Yildiz, Ö. Nudging as a new 'soft' tool in environmental policy—An analysis based on insights from cognitive and social psychology. *ZfU* **2016**, *2–3*, 169–207.
133. Schwartz, B. *The Paradox of Choice: Why More Is Less*, 1st ed.; Ecco: New York, NY, USA, 2004; ISBN 978-0-06-000568-9.
134. Anderson, W.; White, V.; Finney, A. Coping with low incomes and cold homes. *Energy Policy* **2012**, *49*, 40–52. [CrossRef]

Article

The Benefits of Local Cross-Sector Consumer Ownership Models for the Transition to a Renewable Smart Energy System in Denmark. An Exploratory Study

Leire Gorroño-Albizu

Aalborg University, Department of Planning, Rendsburggade 14, 9000 Aalborg, Denmark; lga@plan.aau.dk

Received: 14 February 2020; Accepted: 19 March 2020; Published: 22 March 2020

Abstract: Smart energy systems (SESs), with integrated energy sectors, provide several advantages over single-sector approaches for the development of renewable energy systems. However, cross-sector integration is at an early stage even in areas challenged by the existing high shares of variable renewable energy (VRE). The promotion of cross-sector integration requires institutional incentives and new forms of actor participation and interaction that are suitable to address the organisational challenges of implementing and operating SESs. Taking as the point of departure an empirical case and its institutional context, this article presents an exploratory study of the ability of cross-sector consumer ownership at different locations in the power distribution system to address those challenges in Denmark. The methods comprise interviews of relevant stakeholders and a literature review. The results indicate that distant and local cross-sector integration will be necessary to reduce overinvestments in the grid and that consumer co-ownership of wind turbines and power-to-heat (P2H) units in district heating (DH) systems may provide advantages over common separate ownership with regard to local acceptance and attractiveness of investments. Several possibilities are identified to improve the current institutional incentive system in Denmark. Finally, the results suggest the relevance of analysing the possibility for single-sector energy companies to transition to smart energy companies.

Keywords: smart energy system; renewable energy system; sector integration; consumer ownership; local ownership; prosumer; organisational innovation

---

1. Introduction

A drastic reduction in global CO_2 emissions is crucial to mitigate global warming and its devastating consequences [1]. Therefore, the EU has set the target to reduce its greenhouse gas emissions by 80–95% compared to 1990 by 2050 [2]. Achieving this target requires the substitution of fossil fuels, with significant reductions in energy demand and a high penetration of VRE [2]. This implies fundamental changes in the energy system—most remarkably, the significant loss of flexibility on the production side (previously provided by easily and cheaply storable fossil fuels) and the decentralisation of the energy system (in order to harvest local energy resources with modular/scalable technologies such as wind turbines and solar panels). These changes are not only of a technical nature as they are expected to demand and open up important organisational changes, including new business models and the possible reconfiguration of the energy system's ownership [3–5]. Furthermore, the new EU Renewable Energy Directive and Electricity Market Directive, which include definitions for "renewable energy communities" and "citizen energy communities", respectively, could also foster ownership changes by promoting the implementation of renewable energy projects with open and participatory forms of citizen ownership in the EU Member States.

As the implementation of VRE progresses, these technologies are facing greater local opposition [6], lower market prices due to the merit-order-effect, and curtailment due to electricity grid congestion [7]. Several studies conclude that these organisational challenges could be addressed with cross-sector integration (i.e., by integrating the electricity, heating and cooling (H&C), and transport sectors) [8,9] and local inclusive ownership models [10], such as local consumer cooperatives or local municipal companies [11]. However, the institutional incentive system does not yet promote these solutions to the levels that are necessary to address the above mentioned challenges and to implement a renewable SES [12], not even in countries and local regions already pressed by the high shares of VRE, e.g., Denmark.

Denmark is a frontrunner in wind turbine implementation—wind turbines supplied 46.7% of the final electricity demand in 2019 [13]. Moreover, about 64% of the households in the country are connected to DH systems [14]. However, only 1.1% of the heat demand in DH systems was supplied by heat pumps (HPs) in 2018 [14], which indicates a very low integration of the electricity and H&C sectors, in spite of the existing high potential for it. Denmark is also well known for its significant levels of local and inclusive ownership of the energy system [11,15]. Nevertheless, since the second half of the 1990s there has been a trend for exclusive and distant ownership of wind turbines, which is one of the reasons for the observed increase of local opposition to them [6,11].

The country has the target to reduce its greenhouse gas emissions by 70% by 2030 and to become fossil fuel-free by 2050. The achievement of these ambitious targets requires the improvement of the Danish institutional incentive system and possibly new forms of local and inclusive ownership in order to address the mentioned organisational challenges and foster the implementation of a SES [12]. The current Danish institutional incentives do not differentiate between nearby or distant cross-sector integration [12]. This is seen as problematic given that cross-sector integration is expected to reduce electricity grid costs by reducing congestion issues [8,16], which have a strong locational character [7]. Moreover, the current electricity spot market structure makes it necessary for wind investors to access support schemes or arrange beneficial power-purchase-agreements (PPAs) in order to make wind projects economically attractive [17]. In this respect, the abolition of the feed-in tariff scheme and the introduction of the tender scheme considerably reduces the possibility for wind projects with local inclusive ownership to have access to support schemes and favours large commercial wind investors instead [18,19]. Furthermore, the current institutional incentive system completely fails to promote local acceptance of wind turbines in Denmark—proven by the fact that 305 MW of wind capacity was cancelled in 2017 in the country because of protests [20].

In such a changing and hostile environment, local and inclusive ownership of wind turbines continues to develop in Denmark through innovative forms such as local cross-sector consumer ownership, e.g., in Hvide Sande, where the local DH company has bought the local wind turbines [11]. Hvelplund et al. [12] suggest that such cross-sector consumer ownership models might be advantageous to address the organisational challenges of implementing SESs [12]. However, the idea has not been empirically studied yet and that is what the study presented in this article intends to do.

Taking as a point of departure the case of Hvide Sande, this article presents an exploratory analysis that answers the following research questions:

1. What is the (theoretical) ability of cross-sector consumer ownership at different locations to address the organisational challenges of SESs in Denmark?
2. How does the current Danish institutional incentive system encourage/discourage cross-sector consumer ownership at different locations in the power distribution system?
3. Based on 1 and 2, how could the Danish institutional incentive system be improved to better address the organisational challenges of SESs?
4. What issues regarding ownership and SESs can be identified for further research?

Section 2 presents the theoretical approach and methodology of the study. Section 3 is divided into four sub-sections, which answer research questions 1–4, respectively. Finally, Section 4 discusses the results of the study.

2. Theoretical Approach and Methods

2.1. SESs and their Interrelations

SESs are renewable energy systems that comprise smart electricity, thermal and gas grids and are characterised by integrated electricity, H&C, and transport sectors [16]. Figure 1 presents the interrelations between the implementation and operation of the SES, the incentive system, the political system/process, the available resources, and the cognitive/cultural characteristics. The diagram is an adaptation of those presented by Hvelplund et al. [12] and Gorroño-Albizu et al. [11]. The differentiation of "the technical system" and the "actor participation and interaction" presented in Figure 1 intends to emphasise the need to better understand potential organisational possibilities (including different ownership models) for SESs as well as their interrelation with the technical system and the institutional incentive system. Figure 1 suggests that the characteristics of the technical system could influence which/how actors participate in the implementation and operation of the energy system. Thus, different types of actor participation and interactions could be expected, e.g., for centralised and decentralised energy systems. At the same time, the figure suggests that different actor participation and interactions could lead to implementation and operation of the technical energy system in a different way. In this sense, different implementation and operation behaviours could be expected for VRE and P2H in DH systems, e.g., when being owned by different companies belonging to different sectors (which is currently the norm) or by one single (cross-sector) company (i.e., when being regarded as one single system). This understanding motivates the analysis of the co-ownership of wind turbines and P2H by DH companies presented in this article. Such co-ownership represents a cross-sector ownership model. Finally, Figure 1 suggests that, in order to understand the implementation and operation of SESs, it is important to comprehend how the incentive system influences both the technical system and the actor participation and interaction [11].

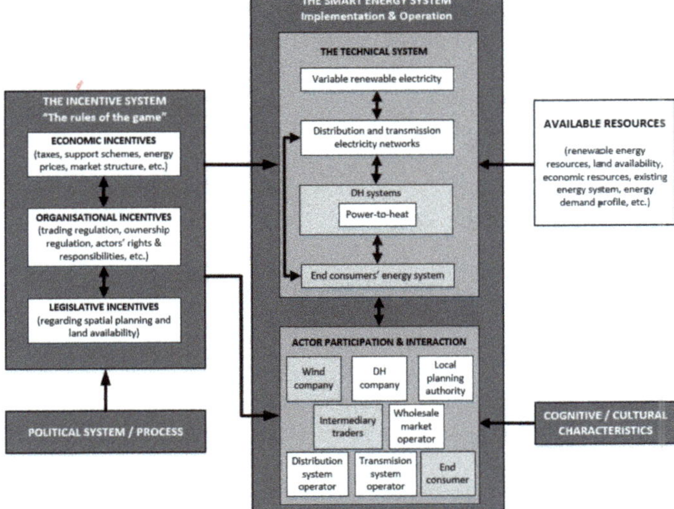

Figure 1. The theoretical approach of the study, inspired by [12] and [11]. The white boxes in the diagram present the elements of SESs included in the scope of the study. The interactions between the actors are not drawn because questioning and analysing those interactions is one of the objectives of the study. DH: district heating.

Figure 1 also presents the delimitations of the study. The study focuses on VRE infrastructure (particularly wind turbines), the electricity grid, and P2H in DH systems. Individual heating systems and other forms of cross-sector integration as well as the end consumers' energy system are outside of the scope. The study analyses the present "rules of the game" but not the political process behind these rules or the end consumers' political influence. However, the study indirectly considers the end consumer through the consumer-owned DH companies and the local community through the local planning authority. Finally, the cognitive/cultural characteristics that have influenced the actor participation and interaction and the dominance of consumer-owned DH companies in Denmark are outside of the scope. The delimitations of the study are defined in line with the problem formulation and the research questions presented in the introduction.

2.2. The Organisational Challenges of SESs, Ownership and Location

The organisational challenges considered in this study are: (1) reduction of overinvestments in the electricity grid, (2) enhancement of local acceptance of wind turbines, and (3) improvement of the economy of wind turbines and P2H in DH systems. In the following it is explained how these three challenges are related to the ownership of the technical SES based on existing literature. Furthermore, the relevance of the location aspect, already mentioned in the introduction, is highlighted. These knowledge forms the theoretical background to answer the research questions.

2.2.1. Reduction of Overinvestments in the Electricity Grid

Increasing flexible demand through cross-sector integration may reduce the need for expanding and reinforcing the electricity transmission grid in order to integrate high shares of VRE [8]. This requires that investments (in VRE, P2H in DH systems and the electricity grid) and operations (of VRE and P2H units) are coordinated with regard to three key aspects: time, size, and location.

The complexity of the necessary coordination to minimise the overall system costs arises from the multiple actors, interests, and institutional incentives that intervene in investment and operations' decision-making, as indicated in Figure 1. Based on a preliminary analysis made for Denmark, Hvelplund and Djørup [3] suggest that local consumer ownership of SESs would facilitate the necessary coordination. In line with that idea, Gill et al. [21] argue that co-ownership is the easiest solution for local coupling of wind power generation and local demand with the purpose of avoiding wind power curtailment due to grid congestions. Moreover, the analysis carried out by Gill et al. indicates that co-ownership reduces the transaction costs of local balancing. The main reason behind those authors' arguments is that, in a co-ownership configuration, there is only one decision-maker for the investments in and operations of the wind turbines, the P2H units, and the rest of the components of the DH system (e.g., thermal storage, solar collectors, combined heat and power (CHP) plants, etc.), which are regarded as parts of the same system [22]. This facilitates the coordination.

2.2.2. Local Opposition to Wind Turbines

Local opposition to wind turbines is a well-documented phenomenon that has caused delays and cancellation of projects. Similar public reactions towards other VRE technologies could be expected as their implementation increases. Reducing the need for wind capacity through, e.g., efficiency measures would result in lower conflicts. Nonetheless, addressing possible local conflicts and enhancing local support is essential. To this end, participatory planning processes and a fair distribution of local impacts and benefits are recommended [23]. In this respect, local and inclusive citizen ownership has proven to be an effective solution [10] as it confers the local community the control over the decisions on the wind turbine project and ensures broad distribution of benefits between the members of the local community. Gorroño-Albizu et al. [11] provide examples of local and inclusive citizen ownership, which include, e.g., local consumer cooperatives and local municipal companies.

2.2.3. Attractiveness of Investments in VRE and P2H in DH Systems

The levelised cost of wind power has decreased significantly in the past decades, reaching similar or even lower levelised costs per MWh than those of fossil fuel technologies [24]. However, the merit order effect and curtailment because of grid congestion reduce the profitability of wind investments as increasing volumes of wind power enter in the electricity system. Increasing flexible demand through cross-sector integration (e.g., P2H) has been presented as a solution to raise wind energy utilisation and, in this way, improve wind economy [9,12]. Therefore, the co-ownership of wind turbines and flexible demand (e.g., P2H in DH systems) could improve the economy of wind turbines—as long as both are located within the same electricity grid congestion node [21].

Some studies have investigated the role of ownership in the attractiveness of investments in wind turbines (see e.g., [25,26]) and in DH systems (see e.g., [27]). The differences in attractiveness of investments is to some extent related to the fact that different types of investors seek different levels of profitability and investment time horizons. The think-tank Grøn Energi [28] argues that shorter or longer time horizons and higher or lower expectations for returns in investments have important implications for the investment choices and future competitiveness of DH systems. In this regard, according to Grøn Energi, long time horizons—which are often preferred by consumer and public investors rather than by commercial investors—will be extremely important to ensure the adoption of more sustainable and flexible technologies (including solar collectors, thermal storage, and HPs). This could imply that wind turbines might also be attractive for consumer-owned DH companies (with P2H units) who seek for long-term return in investments.

2.2.4. SESs, Cross-sector Ownership and Different Location Cases

From the above explanations it may be concluded that the location of the technical system's components and the ownership influence the ability to address the organisational challenges presented in this study. Therefore, this study analyses the co-ownership of wind turbines and P2H by DH companies implemented at different location cases. These are (1) distant, (2) local, and (3) behind-the-meter cross-sector integration, as presented in Figure 2. The location cases should not be regarded as either/or alternatives as they already co-exist and will probably still do so in the future. Nevertheless, they are differentiated in order to assess their influence in the ability of the co-ownership solution to address the organisational challenges of SESs.

2.3. Methodology

Semi-structured interviews and literature review are used to answer the research questions of the study. The interviews were conducted mainly with the stakeholders forming the institutional context in which the case of Hvide Sande is embedded in line with the theoretical approach presented in Figure 1. To the best knowledge of the author, Hvide Sande DH is the only DH company in Denmark that owns wind turbines. Therefore, the input from the stakeholders involved in this case is expected to provide insights about the co-ownership model that other stakeholders might not hold. Moreover, the expertise of Ringkøbing DH, who has explored and discarded the possibility of implementing the co-ownership solution, is also collected. Additionally, a DH consultant has been interviewed to deepen the understanding about the operation of DH systems. Table 1 lists the interviewed stakeholders. Written transcripts were compiled for the interviews.

The objective of the interviews is to understand the (theoretical) ability of the co-ownership model, how the technical system and institutional context are influenced by different ownership models, and how the incentive system encourages/discourages different ownership models. The specific experiences by Hvide Sande DH are out of the scope of the study as they do not answer the research questions. The interviews had different focuses, related to the expertise of the interviewed stakeholder. The questions for grid operators were targeted at understanding the challenges of implementing higher shares of VRE and electrifying the H&C sector as well as the benefits that different cross-sector integration cases, presented in Figure 2, could provide for solving the expected electricity grid issues. The questions for DH companies and the DH consultant were targeted at understanding the operation of the DH system (with and without wind turbines), the economic attractiveness of investing in wind turbines for a DH company under the current institutional incentive system, and the possibility of obtaining local acceptance of wind turbines. Finally, the questions for the local planning authority were targeted at understanding the local energy system and the interactions between the local stakeholders. The interviews also made it possible to capture different opinions on the advantages and disadvantages offered by the different cross-sector integration cases and the consumer cross-sector ownership model under analysis in this study. These opinions are presented as part of the analysis in the results section.

The list of interviewed stakeholders is small and therefore only a limited understanding about stakeholders' expectations regarding cross-sector integration and cross-sector ownership and about relevant research lines regarding ownership and SESs are captured. Nevertheless, this exploratory study advances the existing knowledge about the suitability of institutional incentives and ownership models for SESs and is expected to build a stronger knowledge basis for further research on the topic.

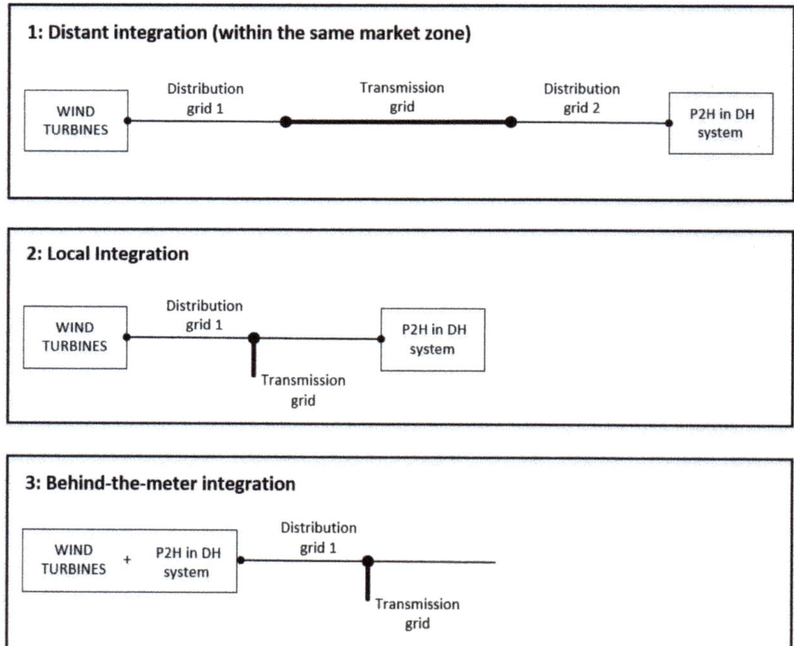

Figure 2. The location cases of cross-sector integration considered for the analysis. The location cases should not be regarded as either/or alternatives as they already co-exist and will probably still do so in the future. VRE: variable renewable energy; P2H: power-to-heat.

Table 1. List of interviewed stakeholders. HP: heat pump; TSO: transmission system operator; DSO: distribution system operator.

Stakeholder	Interviewee	Description
DH Company	Hvide Sande DH [29]	Hvide Sande DH owns wind turbines and an electric boiler in a behind-the-meter solution.
DH Company	Ringkøbing DH [30]	Ringkøbing DH owns an electric boiler and a HP.
Local Planning Authority	Ringkøbing-Skjern Municipality [31]	Hvide Sande DH and Ringkøbing DH are located in this municipality. Ringkøbing-Skjern is a rural municipality with high shares of VRE and ambitious municipal energy targets.
DSO	RAH Net [32]	RAH Net is the local consumer-owned DSO. Hvide Sande DH and Ringkøbing DH are connected to RAH Net's electricity grid.
TSO and Market Operator	Energinet [33]	Energinet is the Danish TSO and market operator.
DH Consultant	EMD International [34]	EMD International is an energy systems software company that provides consultancy to DH companies for the improvement of their operations' strategy.

3. Results

The section is divided into four sub-sections, which answer research questions 1–4, respectively.

3.1. The (Theoretical) Ability of the Consumer Cross-sector Ownership Model in Different Location Cases to Address the Organisational Challenges of SESs

This sub-section answers research question 1. The sub-section starts by explaining the operation of wind turbines and DH systems (i.e., the technical SES) in a co-ownership solution and continues with the analysis of the (theoretical) ability. The analysis builds up on the theoretical approach presented in Section 2.1. and Section 2.2.

3.1.1. The Operation of the Wind Turbines and the DH System in a Co-Ownership Solution

DH companies determine their optimal operational strategy (i.e., the one that meets the heat demand at the lowest possible cost) for the portfolio of technologies available and for every hour [22]. In Denmark, the portfolio may consist of CHP units, boilers, P2H units, solar collectors, waste heat from nearby companies, and thermal storages [14]. Therefore, the calculation of the optimal operational strategy may include production and storage capacities, demand estimation, sun energy resource estimation, and fuel, heat, and electricity prices [22]. In the case of Hvide Sande, where the DH company also owns wind turbines, the calculation also includes wind resource estimation, the market price wind power could get, and the cost of self-consuming the wind power [22]. In this case, the operational strategy defines, among others, when to sell the wind power in the electricity market and when to self-consume it [22,29]. This means that, as suggested by the theoretical approach presented in Figure 1, the operation of the wind turbines and the P2H unit are different in the co-ownership solution implemented in Hvide Sande and the separate ownership solution that is currently the norm.

Hvide Sande DH argues that they—deliberately—built a (smart) energy system that reduces the curtailment of the local wind turbines and the DH system's natural gas consumption while keeping in consideration the need for the wind power in the Danish electricity system. According to the DH company, they self-consume the wind power in periods with low power electricity market prices (i.e., when the demand/value of wind power in the market is low) and they sell it in periods with high electricity market prices (i.e., when the demand for power is high) [29].

EMD International pointed out that the understanding of "high" or "low" power prices by the DH company is subjective as it based on the alternatives that the DH company has to meet the energy demand. During sunny summer days, with low heating demand and high (cheap) heat production from the solar collectors, the DH company may not need (all) the wind production to cover the heat demand and could decide to sell the electricity at lower prices than in winter, when the alternative to self-consumption of wind power could be to operate the (expensive) natural gas boilers. Therefore, EMD International argues that the self-consumption or sale of wind power is not optimised from the Danish electricity system perspective, but from the DH company's perspective; i.e., by not making the wind power available in the Nord Pool market at all times, the co-ownership solution results in "sub-optimisation" of the electricity system [34].

The remark made by EMD International indicates that this stakeholder assumes that the (current) institutional incentive system optimises the operation of the electricity system. This is in line with Energinet's opinion [33]. However, the current institutional incentive system is still strongly influenced by the path dependency of a centralised and fossil fuel energy system with separated energy sectors [12] (as further explained in Section 3.2). One of the consequences is that the current market structure, in combination with electricity grid tariffs and taxes, results in curtailment of wind power (which is assumed not to have any market value) in moments with transmission grid congestions while flexible electricity demand from P2H units in DH systems has not been activated and fossil fuels are being burnt to meet the heat demand of the DH systems [29,30]. This means that the curtailed wind power could have actually had a market value. Therefore, the optimisation of the national electricity (or energy) system through the electricity market and other institutional incentives is also questionable. Moreover, it is not clear if, under the current institutional incentive system, the separate ownership solution results in a better or worse optimisation of the energy resources than the co-ownership solution.

Some of the remarks made by EMD International [34] and Energinet [33] show a rather technocratic approach, where it is assumed that, while keeping the traditional single-sector or separate ownership solution, the right combination of institutional incentives will lead to the optimal operation of the energy system with regard to the political/societal goals. In contrast, scholars of sustainable socio-technical transitions advocate for creating spaces for experimentation and nourishing of niches in order to allow for innovation that could lead to fundamental changes, in this case, in the energy system [35–37]. Therefore, this preliminary study intends to break with the path dependency of the single-sector ownership approach and explore the (theoretical) ability of consumer cross-sector ownership to address the organisational challenges of SESs.

3.1.2. The Location Cases for Cross-Sector Integration and the Reduction of Overinvestments in the Electricity Grid

Grid issues are dependent on the characteristics of the local grid [7]. In the following, a basic technical analysis is provided of what, why, and where electricity grid issues may arise in Denmark as result of the increase of installed VRE capacity in a scenario where no mitigation strategy (e.g., grid reinforcement and expansion or cross-sector integration) is implemented. The technical understanding is essential to discuss the ability of the three different cross-sector integration cases presented in Figure 2 (i.e., distant, local, and behind-the-meter) to address grid issues in Denmark.

The grid issues introduced in the following and in Figure 3 are limited to the scope of the study and the inputs provided by the interviewed grid operators. The grid issues that could arise at transmission and distribution levels due to the implementation of P2H units in DH systems are not discussed in the following. The reason is that, according to the interviewed grid operators, these issues are well-addressed by the current institutional incentive system, which promotes the flexible operation of P2H units in DH systems to avoid grid congestion [32,38].

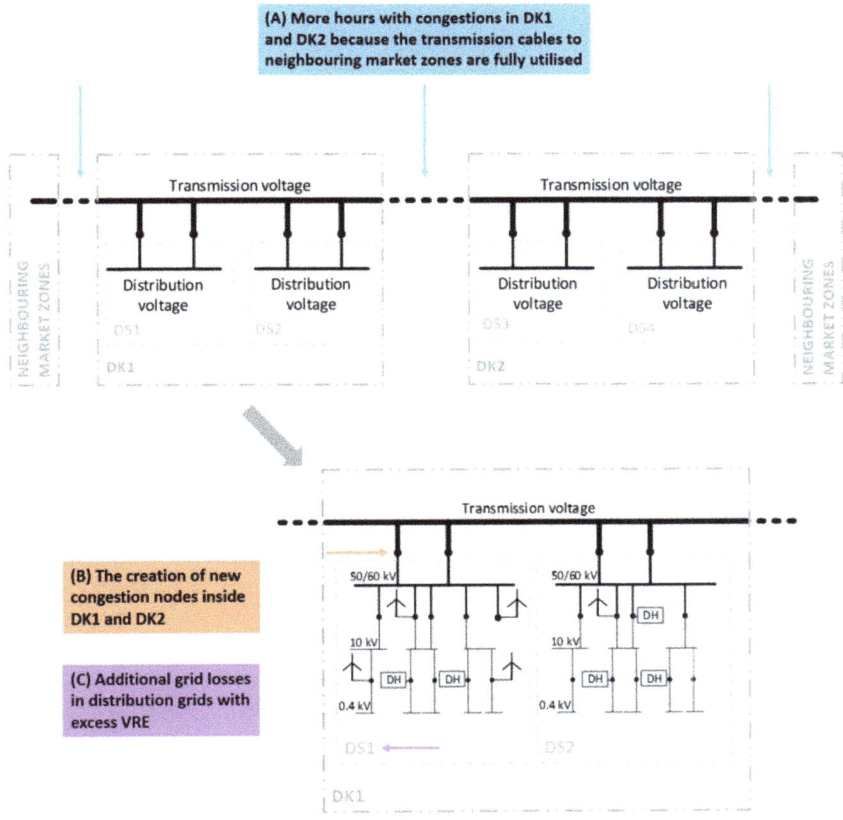

Figure 3. Potential grid issues that could be caused by increasing shares of VRE in a scenario where no mitigation strategy (e.g., grid reinforcement and expansion or cross-sector integration) is implemented illustrated on a schematic representation of the electricity grid in Denmark. DK1 and DK2 are the two electricity market zones in Denmark. DS1 and DS3 represent the distribution system areas with high shares of VRE and DS2 and DS4 the distribution system areas with high electricity and heat demand. The dashed lines represent the transmission voltage connection to other market zones. For a more detailed representation of the Danish electricity grid, see [39] and, e.g., [40]. DS: Distribution system.

Denmark's electricity system is rather decentralised compared to other EU and industrialised countries. About 50% of the electricity generation is directly fed into the electricity distribution grid nowadays, in contrast to 1%–2% in 1980 [41]. Wind turbines, photovoltaic panels, and small-scale CHP plants have been connected to the electricity distribution grids, which has required and resulted in stronger electricity distribution grids than in other EU countries [7]. The total installed electricity generation capacity in 2018 was 15,073 MW, divided into 6121 MW wind (4420 MW onshore and 1701 MW offshore), 5402 MW large-scale power plants (815 MW electricity only and 4586 MW CHP plants), 1904 MW small-scale CHP plants, and 998 MW solar, 9 MW hydro, and 639 MW autoproducers [14]. In addition, Denmark is strongly connected to the neighbouring countries [39,42].

Currently, there is no congestion issue at the electricity distribution level in Denmark; the congestion issues are at the transmission level between market zones [7,33] (see Figure 3). This means that, in moments when the local electricity production e.g., in DS1, exceeds the local electricity demand in DS1, the excess electricity is exported to other parts of the electricity system through the transmission grid, e.g., to DS2 and even to DS4 and neighbouring market zones as long as the transmission grid connecting the different market zones is not congested.

In some Danish municipalities, wind and solar energy produce as much as 500% of the annual electricity demand (these are DS1 and DS3 in Figure 3). In Ringkøbing-Skjern municipality, the share is about 150%. In others, the share is only about 1% (these are DS2 and DS4 in Figure 3) [43]. At the beginning of 2017, the distribution system operator (DSO) NOE Net (which covers fully or partly the municipalities of Holstebro, Lemvig, Struer, and Herning [44]) estimated that there were periods when the exports from their grid were 0.1% of the local electricity demand and expected this number to increase with the connection of the planned new wind turbines [45]. In a scenario where no mitigation strategy (e.g., grid reinforcement and expansion or cross-sector integration) is implemented, the increase of VRE capacity in DK1, DK2, and the neighbouring energy systems could result in:

(A) An increase amount of hours with transmission grid congestions in DK1 and DK2. DK1 and DK2 are the two electricity market zones in Denmark. This problem occurs in moments when the electricity production in DK1 and/or DK2 exceeds the electricity demand in DK1 and/or DK2 and the transmission connections to other market zones are fully utilised. The result is the curtailment of VRE by the power market [7].

(B) The creation of new congestion nodes inside DK1 and DK2. This may occur, e.g., because of congestions in the substations that connect DS1 and DS3 with the transmission voltage cables. Such an issue has already occurred for example in one of the transmission substations in Lolland municipality, where Energinet had to contact the local DSO to achieve down regulation of wind and solar power production. Currently this is only an issue for the transmission system operator (TSO), but it is expected to become a problem for the DSOs as well [46].

(C) Additional electricity grid losses at the distribution level in areas with excess VRE. Grid losses are proportional to the current (i.e., the power flow) and the distance that power is transported. In a centralised fossil fuel energy system (where the electricity is transported from the central power stations to the consumer through the transmission and distribution grids), the power consumption in a given distribution grid can be seen as the cause of the power flow in that given distribution grid and, consequently, of the grid losses in the given distribution grid too. However, in a renewable energy system (where large shares of the power production may be fed directly into the distribution grid and go upstream or downstream) the power flow in a given distribution grid could be caused by power consumption elsewhere in the system. This is the case when the local VRE production exceeds the local power consumption. In this sense, one could say that local excess power production from VRE creates additional grid losses in the local distribution grids where the excess power is produced.

Grid congestions between market zones or inside the market zone may be reduced by reinforcing and expanding the electricity grid and/or increasing flexible demand inside the congestion node through cross-sector integration. Increasing shares of VRE in neighbouring market zones and energy systems reduces the effectiveness of the first two options and demands for more cross-sector integration. Furthermore, cross-sector integration is expected to be strategic to decarbonise the H&C and transport sectors [47]. Therefore, as argued in the theoretical approach, the coordination of investments in and operations of VRE and cross-sector integration infrastructure (e.g., P2H in DH systems)—with regard to time, size, and location—will be essential to reduce unnecessary grid expansion and reinforcement (i.e., overinvestments in the electricity grid). At this point, it is important to highlight the relevance of the location aspect to that end. The congestions between market zones may be reduced by both distant and local cross-sector integration because it does not matter where the VRE production and the P2H demand are located within DK1 and DK2. In contrast, distant cross-sector integration is not suitable to address the congestions inside DK1 and DK2 because it cannot increase the flexible demand within the new congestion zone. To this end, local cross-sector integration would be necessary. Behind-the-meter cross-sector integration may also address the above mentioned congestion issues. However, it is not seen to be strictly necessary given that the congestion issues are not expected to happen at the wind farm connection point. The reason is that the Danish institutional incentive system does not allow DSOs to limit the connection of wind turbines and requires that DSOs make the necessary investments in the grid to enable the connection of new wind farms [32]. This is to avoid the discrimination of power producers.

Regarding the additional grid losses in areas with excess VRE, DSOs have expressed different opinions. In Denmark, there is an "equalization scheme" between all the Danish DSOs that is used to cover the additional grid investments and expenses related the connection of new wind turbines to electricity distribution grids [32,45]. The DSO RAH Net states that the expenses related to the additional grid losses are covered by the equalization scheme [32]. In contrast, the DSO NOE Net argues that the scheme does not adequately cover the additional grid losses [45]. NOE Net adds that additional grid losses have significantly increased in areas with high shares of VRE and raise the electricity bills of the local consumers. NOE Net demands a reform of the scheme [45] and RAH Net points out that avoiding long distance transportation of electricity would reduce grid losses [32]. In this respect, both local and behind-the-meter cross-sector integration could provide a suitable solution to reduce additional grid losses by increasing the local power demand in moments of excess VRE production.

Energinet pointed out that P2H has a strong seasonal profile [33]. Therefore, none of the cross-sector integration cases considered in this study provides a full solution to the grid issues presented in this section. Hence, other integration technologies (such as power-to-gas) are expected to be necessary along with grid expansion and reinforcement [8,12,38].

3.1.3. Consumer Ownership and Local Acceptance of Wind Turbines

The majority of onshore wind turbines in Denmark have citizen ownership, which is very diverse (see [11]). From the middle of the 1990s, a tendency for distant and exclusive commercial and citizen ownership has been observed in the country, which significantly differs from the previous tendency for local and inclusive citizen ownership [11]. The new ownership trend is one of the reasons for the observed increase of local opposition to wind turbines [6,11].

In Denmark, 95% of the DH systems are owned either by a consumer cooperative or a municipal company [11]. The interests of the local DH consumers are strongly represented in the boards of these companies and profits are shared in the form of lower heat bills [27]. The implementation of new turbines or the purchase of existing ones by these DH companies is dependent on a beneficial business economy and the support of the local heat consumers. In the case of Hvide Sande, the purchase of the wind turbines was approved in a general assembly in August 2018 [48]. This means that the ownership of consumer and municipal DH companies in Denmark is local and inclusive [11]. Consequently, based on the theoretical background presented in Section 2.2.2, the ownership of local wind turbines by such local DH companies might bring some advantages with regard to local acceptance compared to the general trend for exclusive and distant ownership observed for the separated ownership solutions [11], where the local community has very limited decision power and access to benefits. The ownership of local wind turbines by distant DH companies would not provide any advantage over the current trend.

When comparing the case of co-ownership with behind-the-meter cross-sector integration and the case of co-ownership with local cross-sector integration, the former has advantages over the latter. In the behind-the-meter case, the closest neighbours to the wind turbines are expected to be connected to the DH system. In contrast, in a local cross-sector integration case, the wind turbines could be placed away from the DH system, probably in the countryside, where the closest neighbours would use individual heating [32]. In this case, the closest neighbours to the wind turbines, i.e., those that will experience the local impacts the most, would not benefit from the ownership of the wind turbines by the DH company. Such local imbalance between benefits and negative impacts should be addressed in order to ensure local acceptance.

3.1.4. The Attractiveness of VRE and P2H for DH Companies

The co-ownership of wind turbines and P2H by DH companies would improve the economy of the wind turbines and the DH company because:

1. Coupling the wind turbines with the flexible demand of the DH system would to some extent resolve the merit-order-effect and the curtailment problems when the wind turbines and the flexible electricity demand are placed in the same congestion node [21].
2. Onshore wind turbines are the cheapest source of electricity in Denmark [49]. Therefore, it would be cheaper for a flexible consumer to self-consume electricity from his wind turbines than buy electricity from a wind power producer (either via the spot market or through peer-to-peer trading). This is because, when buying electricity from a producer, the consumer would have to pay for the cost of producing the wind power and for some benefits for the wind power producer. DH companies who owned wind turbines in windy areas would have an additional advantage because the levelised cost of wind power in these areas is even lower.
3. The DH system could be entitled to a reduction of electricity grid tariffs for the self-consumed electricity based on the advantages it provides for the reduction of overinvestments in the electricity grid expansion and reinforcement (see Section 3.1.2). In this sense, the co-ownership in the behind-the-meter and local cross-sector integration cases would have an economic advantage over the co-ownership in the distant cross-sector integration case.

3.1.5. Summary

This sub-section has analysed the (theoretical) ability of the cross-sector consumer ownership solution implemented at the different location cases presented in Figure 2 to address the organisational challenges of SESs. In the following, a summary of the results is provided.

The results support the argument for the need of coordinating investments (in VRE, P2H in DH systems and the electricity grid) and operations (of VRE and P2H units) with regard to time, size, and location in order to reduce overinvestments in the electricity grid when introducing high shares of VRE. As suggested by the theoretical background, the necessary coordination is expected to be easier in the co-ownership solution than in the separate ownership solution because the wind turbines are regarded as one of the components of the DH system [22,29] and the decisions are made by one single stakeholder, i.e., the DH company. Furthermore, the analysis emphasises the relevance of the location aspects to reduce overinvestments in the grid. Both distant and local cross-sector integration are suitable to reduce congestions in DK1 and DK2 but only local cross-sector integration may address the local grid issues (i.e., the creation of new congestion nodes inside the market zones and the additional grid losses in distribution grids with excess VRE). Behind-the-meter cross-sector integration does contribute to alleviate the above mentioned issues too. However, it is not seen to be strictly necessary given that the congestion issues are not expected to happen at the wind farm connection point.

The ownership of local wind turbines by local consumer- and municipal-owned DH companies may enhance local acceptance as these companies have local and inclusive forms of citizen ownership [11], as recommended by the theoretical background. Besides, the analysis indicates that the behind-the-meter solution is better than the local cross-sector integration solution for local acceptance. In the former, the closest neighbours to the wind turbines are expected be connected to the DH system, whereas in the latter the wind turbines could be out in the countryside where the closest neighbours would use individual heating instead.

Finally, the co-ownership is expected to increase the attractiveness for DH companies to invest in wind turbines and P2H units, as suggested by the theoretical approach. This is particularly so in windy areas, where the levelised cost of wind power is even lower than the country average, and with behind-the-meter or local cross-sector integration solutions, where a higher reduction of electricity grid tariffs for the self-consumed electricity could apply (based on the advantages they provide for the reduction of overinvestments in the electricity grid expansion and reinforcement).

All in all, it may be concluded that especially the local cross-sector integration case with the co-ownership of wind turbines and P2H units by DH companies could (theoretically) provide several benefits for the implementation of SESs in Denmark. These are reduction of overinvestments in grid expansion and reinforcement, improved economic attractiveness of wind turbines and P2H units, improved utilisation of local wind power, reduction of burning of fuels, better economy for the local DH consumers, better economy for the local electricity consumers, and improved local acceptance of wind turbines. Ultimately, this ownership model could have the potential to accelerate the implementation of SESs. Therefore, it is deemed relevant to analyse the possibilities to implement it under the current Danish incentive system.

3.2. The Current Institutional Incentive System for Cross-Sector Consumer Ownership in Denmark

This sub-section answers research question 2. The sub-section analyses how the current Danish institutional incentive system encourages/discourages the different location cases of cross-sector consumer ownership presented in Figure 2. In line with the interrelations presented in Figure 1, some of the institutional incentives introduced in this section directly influence the implementation and operation of the technical SES; others directly influence the actor participation and interaction in the energy system; and, ultimately, all institutional incentives indirectly influence both. The analysis of the institutional incentives is divided into the role and possibilities of electricity grid operators, electricity grid tariffs and taxes, the design of the electricity spot market, and the lack of targeted incentives for local cross-sector integration.

3.2.1. The Role and Possibilities of Electricity Grid Operators

Under the current legislation, grid operators' are responsible for addressing grid congestion issues, such as those that emerge from increasing shares of VRE or the electrification of the H&C and transport sectors. However, the actions they may implement are limited to grid reinforcement and expansion and to introduction of new grid tariffs, which need to be approved by the Danish Utility Regulator. This means that the possibility for grid operators to promote, e.g., the necessary cross-sector integration to reduce grid congestions caused by increasing shares of VRE is very limited and that grid operators might be forced to overinvest in the electricity grid to address congestion issues. This limitation is one of the reasons why the "electricity integration over distance" strategy [38] (i.e., the expansion and reinforcement of the electricity grid) has been (and still is) the main VRE integration strategy implemented in Denmark. This is illustrated, e.g., by the construction of two new transmission connections to the Netherlands and the United Kingdom [50], whereas only 1.1% of the DH demand is supplied by HPs [14].

3.2.2. Electricity Grid Tariffs, and Taxes

In Denmark, the electricity grid costs are distributed among consumers following the waterfall principle. This means that the grid tariff to be paid by the consumer depends on the voltage level of his electricity grid connection and that the consumer pays for the share of the grid expenses he generates in his connection's voltage level and all of the upper voltage levels. This principle was adopted in a fossil fuel energy system, where electricity was produced centrally and transported to the consumption point through transmission lines first and through distribution lines of decreasing voltages afterwards, and where the consumer used all the upper voltage levels of the grid. However, with an increasing share of power production being directly connected to the distribution grid and flowing both downstream and upstream, the waterfall principle might not result in a fair distribution of electricity grid costs for the consumers any longer. Furthermore, the waterfall principle implies that a DH system will have to pay the same grid tariffs when self-consuming or purchasing electricity from a nearby wind farm as when self-consuming or purchasing electricity from a distant wind farm [51]. However, the DH system that consumes electricity from a nearby wind farm in moments of excess electricity is helping to reduce grid costs, as concluded in Section 3.1.2. Hence, the discussion of new cost distribution principles and grid tariffs becomes increasingly relevant.

In Denmark, the electricity taxes and grid tariffs (which, in the case of Hvide Sande, summed approximately 70 EUR/MWh [22,52,53]) have made the private economy of P2H units for DH systems worse than that of other alternatives (e.g., biomass boilers and solar collectors, for which taxes do not apply)—even though the socio-economy of HPs is better [54]. As a result, only a few P2H units were installed in the DH systems of the country up to 2018 [55], and their utilisation has been limited [29,30]. Nevertheless, new economic incentives have been recently introduced to promote P2H in DH systems. These include the reduction of the electricity tax for heating from approximately 41 EUR/MWh to 21 EUR/MWh from 2021 [56]. In addition, in 2017–2018, approximately EUR 6.87 million in subsidies were granted to DH companies with small-scale CHP units to cover up to 15% of the investment in an electric HP. The subsidy was granted through application processes. In total, 29 projects summing 48.8 MW were granted and are supposed to be implemented in 2019–2020. The phasing out of the public service obligation (PSO) electricity tax and the energy companies' energy saving scheme are also expected to improve the economy of DH HPs. [57–59] The new economic incentives are regarded as a positive step towards the promotion of both distant and local cross-sector integration because they are expected to result in several DH companies investing in HPs and a significant increase of the annual hours of operation of the P2H units according to Hvide Sande DH, Ringkøbing DH, and EMD International [29,30,34]. However, the actual positive impacts of the new institutional incentives for P2H in DH systems are still to be seen and their sufficiency to be evaluated.

Regarding the co-ownership of wind turbines and P2H by DH companies, the institutional economic incentives are different for behind-the-meter solutions and any technical solution that utilises the public grid for the electricity consumption (i.e., distant and local cross-sector integration). Currently, no electricity taxes or grid tariffs need to be paid for the wind power that is self-consumed in a behind-the-meter solution [29]. This requires that both the wind turbines, the private cable, and the P2H unit(s) are placed on the same land (with the same cadastral number) [29,51,60]. This requirement significantly limits the potential for behind-the-meter cross-sector integration of wind turbines and P2H in DH systems because wind turbines are usually placed in the countryside, away from the DH systems [29,30,32]. Therefore, in many cases, the implementation of the behind-the-meter solution would require the construction of DH pipelines to the wind farm (where the P2H unit would need to be placed), which could significantly/totally reduce the attractiveness of the investment [30].

Full electricity grid tariffs and taxes need to be paid by the DH company for the electricity that is self-consumed utilising the public grid [29], which completely discourages the co-ownership solutions with distant and local cross-sector integration. Blanco et al. [22] studied the operation and system costs of the DH system in Hvide Sande and assessed the impact of electricity taxes on the rate of wind power that would be self-consumed/sold by the DH company. Their results show that, if the total value of the applicable electricity taxes (about 48 EUR/MWh at that time) had to be paid for the self-consumed electricity, 100% of the wind power production would be sold to the spot market. In other words, it would be more expensive to self-consume wind power than sell the electricity at the Nord Pool market and use the other energy sources and technologies to meet the heat demand instead. The results obtained by Blanco et al. also show that, despite the additional revenue from the sale of wind power, the total system costs would increase significantly compared to the situation where no electricity taxes were paid for the self-consumed wind power.

3.2.3. The Design of the Electricity Spot Market

Denmark is part of the Nord Pool electricity market and has two market zones, DK1 and DK2. Producers and consumers may trade through the Nord Pool market or directly between each other. Peer-to-peer trading is possible as long as the producer and the consumer are placed within the same market zone and the production and consumption occur simultaneously [51].

In Section 3.1.4. it is argued that self-consuming wind power should be cheaper than buying electricity from the Nord Pool market because onshore wind turbines are the cheapest source of electricity in Denmark [49]. However, this does not apply under the current institutional incentive system and with the current spot market design. Energinet [33] and Ringkøbing DH [30] pointed out that the cheapest electricity prices do not necessarily match the periods with local wind resource availability. Ringkøbing DH mentioned: "Other times the wind is not blowing here, but my boiler is running and my price [for purchasing electricity at the Nord Pool market] is −125 DKK [per MWh]"; this is about −17 EUR/MWh [30].

The levelised cost of onshore wind power in Denmark is about 35 EUR/MWh [49]. Figure 4 shows that the number of hours at the Nord Pool with spot market prices below that value has been considerable in DK1 in the last years. Based on a preliminary analysis, the values suggest that it would be unattractive for DH companies to invest in wind turbines, unless a reduction of grid tariffs and taxes would apply for the self-consumed electricity. This was the case in Hvide Sande [29]. At the same time, the values indicate that investments in wind turbines might not be attractive in a separate ownership solution either, unless support schemes or other arrangements such as beneficial PPAs are in place. This idea is supported by the results obtained and the conclusions drawn by Djørup et al. [17], who further argue that "the current electricity market structure is not able to financially sustain the amounts of wind power necessary for the transition to a 100% renewable energy system" (p. 148) in Denmark. This means that, in order to further the implementation of wind power, the market structure will need to be changed so that consumers pay at least the levelised cost of wind power and a reasonable profit for the producer. After such a change, the self-consumed wind power should be cheaper than the purchased wind power.

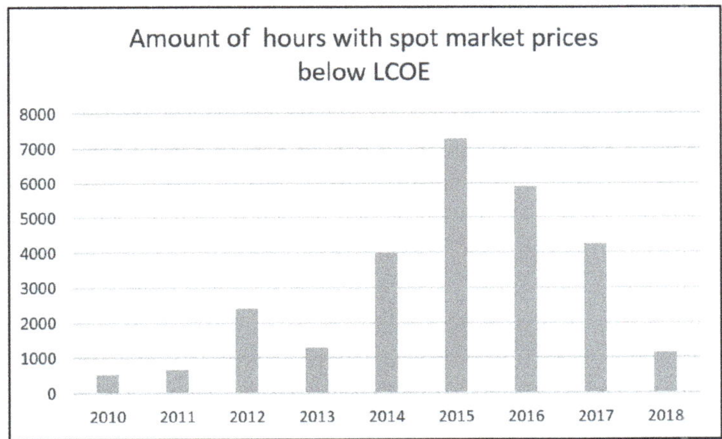

Figure 4. Number of hours per year with spot market prices below the levelised cost of wind power in DK1. Based on data from [61]. LCOE: Levelised cost of energy.

Changes in e.g., ownership regulation, spatial planning regulation, the size of wind projects, and the social normative perception of wind power have been identified as causes for the lowering of wind projects with local and inclusive ownership in Denmark [11,62]. On top of that, the abolition of feed-in tariffs in 2018 and the implementation of the tender scheme introduces an extra burden and increased risk for (small) local initiatives [18,19] and could reduce their number even further. This works against the goal of enhancing local acceptance of wind turbines.

3.2.4. Lack of Specific Incentives for Local Cross-Sector Integration

There are no specifically targeted institutional incentives for local cross-sector integration or local balancing. There is no economic incentive that encourages the activation of the flexible power consumption in a given location within the specific market zone, nor any economic incentive to promote a faster implementation of local cross-sector integration, e.g., in areas with very high/low shares of VRE. This means that the institutional incentive system does not address the issue of the creation of new congestion nodes inside DK1 and DK2 in any way [46], which could lead to overinvestments in the electricity grid. The lack of incentives to address the issue could be explained by the fact that the creation of new congestion nodes is not yet an important technical problem [7]. However, it is expected to become a problem in the near future [46]. In fact, Energinet acknowledges that the distant/nearby location of VRE and flexible electricity demand could have an impact in the grid and argues that the electricity grid should not be "ignored" in the spatial planning of VRE and cross-sector integration technologies [33].

3.2.5. Summary

This sub-section has analysed how the current Danish institutional incentive system encourages/discourages the cross-sector consumer ownership at the different location cases presented in Figure 2 (i.e., distant, local and behind-the-meter). The results presented in this section are in line with the theoretical approach presented in Figure 1, which suggests that the institutional incentive system influences the implementation and operation of SESs by encouraging/discouraging certain technical and actor participation and interaction characteristics, which in turn influence each other.

All in all, it may be concluded that the current institutional incentive system has promoted grid expansion and reinforcement over cross-sector integration for the introduction of VRE in the electricity system [12]. This is seen as problematic given that the potential of this strategy is limited as increasing shares of VRE are implemented in neighbouring market zones and energy systems. Therefore, the cross-sector integration approach is expected to be essential to improve wind power utilisation and business economy [9,12] and reduce overinvestments in the electricity grid [8]. The new incentives for P2H in DH systems are seen as a positive step to promote distant and local cross-sector integration, although the results of the policy are still to be seen and evaluated. However, the lack of institutional incentives that specifically target local cross-sector integration in areas with high shares of VRE is seen as problematic. This should change in order to address local grid issues (i.e., the creation of new congestion nodes inside DK1 and DK2 and the additional grid losses in distribution grids in areas with excess VRE).

The current institutional incentives do encourage to some extent the co-ownership of wind turbines and P2H units by DH systems in a behind-the-meter solution by not applying any electricity grid tariffs and taxes to self-consumed electricity. However, the cases where the behind-the-meter solution may be implemented are very limited because of the requirement of having the wind turbines, the private cable, and the P2H unit on the same piece of land (with the same cadastral number) [29,30]. In contrast, the current institutional incentives completely discourage the co-ownership with local or distant cross-sector integration (i.e., the solutions where the public grid is used) by requiring that the full electricity grid tariffs and taxes are paid for the self-consumed electricity [29].

It may be concluded that the current institutional incentive system seems to block all the benefits that the co-ownership of wind turbines and P2H units by DH companies in a local cross-sector integration solution could (theoretically) offer for the transition to a renewable SES.

3.3. Possibilities for Improving the Current Institutional Incentives

This sub-section answers research question 3. Based on the results presented in Section 3.1. and Section 3.2, this sub-section introduces possible improvements for the current Danish institutional incentive system. Most importantly, local cross-sector integration should be further promoted in order to avoid overinvestments in the electricity grid and improve wind economy. This would reduce the electricity costs to be paid by the consumers. Several institutional incentives could be implemented to that end:

- Geographical bids for the regulating power market in order to enable local balancing, as suggested by Energinet [46] and EMD International [34].
- New (TSO and DSO) grid tariffs to activate flexible demand in a given location in periods of excess electricity generation, as suggested by Ringkøbing DH [30]. "I can prove to them [to the local DSO] mathematically that they would sell more electricity to me [and] make more money and [that] I would make more DH to a lower price on the electrical boiler, if they lowered their price [the electricity distribution tariff]" said Ringkøbing DH [30]. The reconsideration of the waterfall principle for the distribution of grid costs would also be pertinent here.
- Subsidies on investments in VRE and/or P2H in targeted areas. The subsidies could be paid by the savings that would be obtained from not having to reinforce or expand the electricity grid. This could be facilitated by modifying the current legislation so that grid operators could promote the implementation of cross-sector integration technology (on the right time, size and location), when this wasestimated to be a more cost-effective solution from a socio-economic perspective.

Note that no analysis or assessment of the amount and location of the possible new congestion nodes inside DK1 and DK2 has been found, which makes the present understanding of the problem preliminary. Such thorough analysis would be necessary to evaluate which institutional incentive measures would be needed/suitable to tackle the issue. Therefore, the above points should be understood as interim suggestions that require further research in order to define concrete and final policy recommendations.

The promotion of local cross-sector integration could target either the separate ownership solution or the co-ownership solution analysed in this study or both of them. The results of the analysis of the (theoretical) ability of cross-sector consumer ownership to address the challenges of SESs suggests that the co-ownership solution has the additional advantage of improving wind utilisation and economy and of enhancing local acceptance. Therefore, this preliminary analysis suggests that the co-ownership solution should be promoted, e.g., through the reduction of grid tariffs and taxes, as suggested by [22,29,30]. However, it would be advisable to increase the knowledge about the co-ownership solution and its potential implications for the DH and electricity consumers before promoting it. EMD International and Energinet expressed their concern about the fact that the DH company would not pay grid tariffs and taxes [33], which would then need to be covered by other consumers and tax payers [34]. This concern entails an implicit assumption that grid costs would not be reduced by increasing the share of local utilisation of wind power. The results presented in Section 3.1.2. contradict this assumption. In contrast, Hvide Sande DH understands that the production of heat with their own wind turbines is similar to the production of heat with their own solar collectors—they do not have to pay any taxes for the heating they produce from sun energy, nor for the heating they produce from wind energy [29]. Furthermore, DH companies in Denmark do not pay taxes for the biomass they consume either as result of political preferences. The above considerations and mismatch in opinions/perspectives resembles the debates related to individual self-consumption (or individual prosumers) that are taking place in the EU and other industrialised countries, here applied to collective self-consumption (or collective prosumers). As mentioned, a more thorough analysis than the one presented here would be necessary to shed light on this discussion and design adequate institutional incentives for SESs.

Ringkøbing DH also pointed out that "the rules could be better [so that DH companies could own wind turbines and reduce their fuel consumption] but, on the other hand, you know, we are very good at operating the DH system, the HPs and the gas engines, and the CHP and all the pipes in the city and so on. Sometimes when you start thinking on a new market, like you own the wind turbines also, maybe you are not that good at that. Maybe some other guys are better at that" [30]. This comment holds an important observation about the potential need for DH companies to develop new core competences for the transformation of the business model from a DH company into a smart energy company. However, it shall be noted that many DH companies in Denmark already operate CHP units, which means that they do have knowledge about the electricity market. Besides, the planning of the wind turbine project could be done with the help of a specialised consultancy firm, like wind cooperatives have usually done.

What it is clear is that the promotion of the co-ownership model together with the local cross-sector integration solution should also include measures to create benefits for the nearby neighbours who are not consumers of the DH systems. This could be achieved, e.g., by giving them the priority for the purchase of the 20% of the shares that have to be offered to local residents by law [63] or by creating a local wind foundation that would own a part of the wind farm and use the benefits for, e.g., supporting energy renovations and investments in individual HPs outside the DH systems.

3.4. Future Perspectives for Research on Ownership and SESs

This sub-section answers research question 4. The sub-section introduces the issues regarding ownership and SESs that have been identified for further research in this preliminary study.

The suitability of different ownership models to address the organisational challenges of implementing and operating SESs are still rather unknown. This article contributes to build up the knowledge on the topic. However, the results are not conclusive on whether or not local consumer ownership has a higher ability than other ownership models to facilitate and coordinate investments and operations, as suggested by Hvelplund and Djørup [3]. The hypothesis should be tested using a broader scope of the technical (smart) energy system than the one chosen for this exploratory study and in different contexts of actor participation and interaction.

Besides, cross-sector ownership is still rather unknown for medium- and large-scale energy systems. Possibly the only exemption here is CHP plants connected to DH systems. This is not surprising given that cross-sector integration is at early stages even in countries and local regions challenged by the current high shares of VRE. Interesting enough, the idea of cross-sector ownership could have some potential for SESs, as concluded in this article. This opens up for investigating the possibility for energy companies to transition from single-sector energy companies to smart energy companies. Many DH companies in Denmark own CHP plants and, therefore, already provide two energy products, i.e., electricity and heat. In a similar way, the local electricity company or the (current) local wind cooperative could own, e.g., wind turbines, HPs, and electrolysers to supply electricity, heat and hydrogen. It would be relevant to analyse further whether owning a cross-sector energy technology portfolio would provide any competitive advantage in a renewable energy system and under which institutional incentive system. If so, the necessary organisational innovation and strategies to implement it would also be of relevance for further research. This could lead to considerations about bundling of energy sectors and/or (some) services, which would require to assess the implications for energy consumers and study the legal implementation. In this line, it would be advisable to study under which circumstances the smart energy company should or should not be allowed to include a natural monopoly (i.e., electricity, DH, and gas grids) in the portfolio based on the potential implications for the consumers.

4. Discussion

This article presents an exploratory analysis of an interdisciplinary character for which only a few stakeholders have been interviewed. These are the ones that are considered relevant according to the theoretical approach presented in Figure 1 and they are mainly connected to the case of Hvide Sande, which, to the best knowledge of the author, is the only existing case in Denmark for now. Interviewing other DSOs, DH companies, and wind turbine companies could provide a more detailed understanding about the studied issues and show any strong, fair, or weak agreements/disagreements with the views/information collected in this study. Interviewing other DSOs would be necessary, e.g., to find out the general opinion on the equalization scheme and on its adequacy to cover the expenses related to additional distribution grid losses in areas with excess VRE. In addition, it might help to get an idea about where the new congestion nodes inside DK1 and DK2 might appear and the seriousness of this issue. Interviewing other DH companies would be necessary, e.g., to find out their opinion on and interest in investing in wind turbines and their plans for investing in HPs after the new economic incentives. Therefore, a more thorough analysis, based on interviews with a larger number of these stakeholders, would result in better-grounded policy recommendations. Apart from that, interviewing, e.g., DH consumers, the Danish Utility Regulator, the Tax Ministry, or the Energy Ministry would make it possible to collect these stakeholders' opinion on the analysed ownership model and the interim policy recommendations. Furthermore, the study could be extended to include other technologies (e.g., electrolysers) and other countries (e.g., the EU Member States). However, these are out of the scope of the exploratory analysis presented here.

In spite of the limitations of the chosen theoretical approach and methods, this study is sufficient to show that the location of VRE and sector integration infrastructure (e.g., P2H) does matter when it comes to the development of the electricity grid and the attractiveness of investments in VRE. This is in line with knowledge about grid congestion (see, e.g., [7]) and about innovative forms to reduce wind curtailment (see, e.g., [21]), presented in the theoretical approach. The study is also sufficient to show the advantages of the co-ownership with the behind-the-meter solution (when the wind turbines and the DH system are close enough) to enhance the attractiveness of investments in wind turbines and P2H units as well as local acceptance of wind turbines in Denmark. Moreover, the study suggests that, under an improved institutional incentive system, the co-ownership with the local cross-sector integration solution could also provide these benefits. However, further research is necessary to understand the full implications of such ownership model and define suitable institutional incentives to promote local cross-sector integration either with co-ownership, with separate ownership, or both.

The study takes as the point of departure an empirical case in Denmark. Here the material resources, the cultural and cognitive characteristics, the political process and system, and the institutional system may diverge from the conditions in other EU countries and therefore also result in other technical and organisational solutions for the energy system, as indicated by Figure 1 in the theoretical approach. In Denmark, wind power provides almost half of the final annual electricity demand [13] and onshore wind power is the cheapest source of electricity [49]. Denmark has a rather decentralised electricity system (with about 50% of the electricity production directly fed into the distribution grids [40]), which has required and resulted in stronger distribution grids than in other EU countries [7]. Moreover, the Danish electricity system is very well connected to the neighbouring countries through transmission cables [41,42] and has a high integration of DH [14]. Obviously, this puts Denmark in an advantageous position when it comes to handling high shares of VRE compared to other EU countries. Furthermore, the potential within the present Danish technological configuration for the ownership of wind turbines by DH companies is larger than in other countries where, e.g., the shares of DH are still low or the cheapest electricity source is another, e.g., hydropower. Nevertheless, the understanding of how different location cases of cross-sector integration (using different technologies) could reduce electricity grid expansion and reinforcement needs while creating the necessary space for high shares of VRE and the electrification of the H&C and transport sectors is as relevant in other EU countries [8]. One could even think that EU countries with weaker electricity grids could significantly benefit from early considerations regarding locational aspects of cross-sector integration. In this respect, they could find suitable organisational solutions for cross-sector integration that helped avoid overinvestments in the grid. Moreover, a larger deployment of DH systems is recommended to decarbonise the H&C sector [64] and to reduce the costs of integrating high shares of VRE [8] in the EU.

It should also be highlighted that citizen ownership in general and consumer ownership in particular is much more common in Denmark than in many EU countries [65]. As other EU countries advance towards more decentralised energy systems and with EU energy policies that aim at putting the consumer at the centre of the energy transition and at increasing open and participatory forms of citizen ownership for renewable energy projects, increasing shares of consumer ownership could be expected across the EU too. In this regard, advancing the understanding about on-site and off-site collective prosumers and the advantages and disadvantages they offer in a context of developing renewable SESs becomes increasingly important for all EU Member States. On the other hand, it is important to note that the regulation and ownership of DH systems and companies is very diverse in the EU [66,67]. This means that, in other EU countries, DH ownership is not necessarily local and inclusive, i.e., local DH consumers might have little or no power over the decisions of the DH company and the reduction of DH system costs might not be reflected in the heat bills of the consumers. Consequently, the ownership of local wind turbines by the DH company that owns the local DH system might not enhance local acceptance of wind turbines in other EU countries.

Finally, it shall be noted that the current regulation that dictates unbundling of energy sectors/services could be an impediment to implementing cross-sector ownership solutions such as the one presented in this study in other EU countries. On the other hand, this does not mean that the existing regulation should define/limit the organisational solutions of the future if innovative options are proven beneficial to meet the societal goals of the transition to a renewable energy system.

Funding: This research and the APC were funded by European Union's Horizon 2020 research and innovation programme under the Marie Skłodowska-Curie grant agreement number 765515. The article reflects only the author's view and the Research Executive Agency is not responsible for any use that may be made of the information it contains.

Acknowledgments: I thank all the interviewees for their contribution and their time. I also thank Frede Hvelplund and Karl Sperling for their valuable comments to improve the manuscript. Finally, I also thank all those who have helped me understand the technical problems related to the electricity grid.

Abbreviations

CHP	Combined heat and power
DH	District heating
H&C	Heating and cooling
HP	Heat pump
DSO	Distribution system operator
LCOE	Levelised cost of energy
PPA	Power-purchase-agreement
P2H	Power-to-heat
SES	Smart energy system
TSO	Transmission system operator
VRE	Variable renewable energy

References

1. IPCC. *Global Warming of 1.5 °C. Summary for Policymakers*; IPCC: Geneva, Switzerland, 2018.
2. European Commission. *Roadmap 2050*; European Commission: Brussels, Belgium, 2012.
3. Hvelplund, F.; Djørup, S. Consumer ownership, natural monopolies and transition to 100% Renewable Energy Systems. *Energy* **2019**, *181*, 440–449. [CrossRef]
4. *Innovation and Disruption at the Grid's Edge: How Distributed Energy Resources are Disrupting the Utility Business Model*, 1st ed.; Sioshansi, F.P. (Ed.) Academic Press: London, UK, 2017; ISBN 978-0-12-811758-3.
5. Burger, S.P.; Luke, M. Business models for distributed energy resources: A review and empirical analysis. *Energy Policy* **2017**, *109*, 230–248. [CrossRef]
6. Ladenburg, J. Acceptance of Wind Power: An Introduction to Drivers and Solutions. In *Alternative Energy and Shale Gas Encyclopedia*; Lehr, J.H., Keeley, J., Eds.; John Wiley and Sons Inc.: Hoboken, NJ, USA, 2016; pp. 3–9.

7.   Bird, L.; Lew, D.; Milligan, M.; Carlini, E.M.; Estanqueiro, A.; Flynn, D.; Gomez-lazaro, E.; Holttinen, H.; Menemenlis, N.; Orths, A.; et al. Wind and solar energy curtailment: A review of international experience. *Renew. Sustain. Energy Rev.* **2016**, *65*, 577–586. [CrossRef]

8.   Brown, T.; Schlachtberger, D.; Kies, A.; Schramm, S.; Greiner, M. Synergies of sector coupling and transmission reinforcement in a cost- optimised, highly renewable European energy system. *Energy* **2018**, *160*, 720–739. [CrossRef]

9.   Hvelplund, F.; Möller, B.; Sperling, K. Local ownership, smart energy systems and better wind power economy. *Energy Strateg. Rev.* **2013**, *1*, 164–170. [CrossRef]

10.   Berka, A.L.; Creamer, E. Taking stock of the local impacts of community owned renewable energy: A review and research agenda. *Renew. Sustain. Energy Rev.* **2018**, *82*, 3400–3419. [CrossRef]

11.   Gorroño-Albizu, L.; Sperling, K.; Djørup, S. The past, present and uncertain future of community energy in Denmark: Critically reviewing and conceptualising citizen ownership. *Energy Res. Soc. Sci.* **2019**, *57*, 101231. [CrossRef]

12.   Hvelplund, F.; Østergaard, P.A.; Meyer, N.I. Incentives and barriers for wind power expansion and system integration in Denmark. *Energy Policy* **2017**, *107*, 573–584. [CrossRef]

13.   Plechinger, M. Dansk Vind Slog Rekord i 2019. Available online: https://energiwatch.dk/Energinyt/Renewables/article11852237.ece (accessed on 11 February 2020).

14.   Danish Energy Agency. *Energistatistiks 2018*; Danish Energy Agency: Copenhagen, Denmark, 2019.

15.   Rønne, A.; Nielsen, F.G. Consumer (Co-)Ownership in Renewables in Denmark. In *Energy Transition. Financing Consumer Co-ownership in Renewables*; Lowitzsch, J., Ed.; Springer Nature Switzerland AG: Cham, Switzerland, 2019; pp. 223–244.

16.   Lund, H.; Østergaard, P.A.; Connolly, D.; Mathiesen, B.V. Smart energy and smart energy systems. *Energy* **2017**, *137*, 556–565. [CrossRef]

17.   Djørup, S.; Thellufsen, J.Z.; Sorknæs, P. The electricity market in a renewable energy system. *Energy* **2018**, *162*, 148–157. [CrossRef]

18.   Jensen, L.K.; Sperling, K.; Lund, H. Barriers and Recommendations to Innovative Ownership Models for Wind Power. *Energies* **2018**, *11*, 2602.

19.   Grashof, K. Are auctions likely to deter community wind projects? And would this be problematic? *Energy Policy* **2019**, *125*, 20–32. [CrossRef]

20.   Aagaard, L. Kunne du Overveje en Vindmølle i Baghaven? Available online: https://www.danskenergi.dk/nyheder/kunne-du-overveje-vindmolle-baghaven (accessed on 11 January 2020).

21.   Gill, S.; Plecas, M.; Kockar, I. *Coupling Demand and Distributed Generation to Accelerate Renewable Connections*; University of Strathclyde: Glasgow, UK, 2014.

22.   Blanco, I.; Guericke, D.; Andersen, A.N.; Madsen, H. Operational planning and bidding for district heating systems with uncertain renewable energy production. *Energies* **2018**, *11*, 3310. [CrossRef]

23.   Larsen, S.V.; Nielsen, H.; Hansen, A.M.; Lyhne, I.; Clausen, N.-E.; Rudolph, D.; Gorroño-Albizu, L.; Forfang, A.S. *Integrating social consequences in EIA of renewable energy projects: 11 recommendations*; Aalborg University: Aalborg, Denmark, 2017.

24.   IRENA. *Renewable Power Generation Costs in 2018*; IRENA: Abu Dhabi, United Arab Emirates, 2019.

25.   Berka, A.L.; Harnmeijer, J.; Roberts, D.; Phimister, E.; Msika, J. A comparative analysis of the costs of onshore wind energy: Is there a case for community-specific policy support? *Energy Policy* **2017**, *106*, 394–403. [CrossRef]

26.   Yildiz, Ö. Financing renewable energy infrastructures via financial citizen participation - The case of Germany. *Renew. Energy* **2014**, *68*, 677–685. [CrossRef]

27.   Chittum, A.; Østergaard, P.A. How Danish communal heat planning empowers municipalities and benefits individual consumers. *Energy Policy* **2014**, *74*, 465–474. [CrossRef]

28.   Grøn Energi. *Investeringshorisontens indflydelse på fjernvarmesektoren*; Grøn Energi: Kolding, Denmark, 2017.

29.   Hvide Sande DH. Personal Communication, 2019.

30.   Ringkøbing DH. Personal Communication, 2019.

31.   Ringkøbing-Skjern Municipality. Personal Communication, 2019.

32.   RAH Net. Personal Communication, 2019.

33.   Energinet. Personal Communication, 2019.

34.   EMD International A/S. Personal Communication, 2019.

35. Geels, F.W.; Hekkert, M.P.; Jacobsson, S. The dynamics of sustainable innovation journeys. *Technol. Anal. Strateg. Manag.* **2008**, *20*, 521–536. [CrossRef]

36. Seyfang, G.; Smith, A. Grassroots innovations for sustainable development: Towards a new research and policy agenda. *Env. Polit.* **2007**, *16*, 584–603. [CrossRef]

37. Schot, J.; Geels, F.W. Strategic niche management and sustainable innovation journeys: Theory, findings, research agenda, and policy. *Technol. Anal. Strateg. Manag.* **2008**, *20*, 537–554. [CrossRef]

38. Energinet. *Energy System Perspective 2035*; Energinet: Fredericia, Denmark, 2018.

39. Energinet Transmission System Data. Available online: https://en.energinet.dk/Electricity/Energy-data/System-data (accessed on 2 January 2020).

40. Evonet Net Kort. Available online: https://sekort.dk/Html5/Index.html?viewer=SE_NetWeb.SE (accessed on 11 January 2020).

41. State of Green The Future Role of DSOs: Europe's Energy System is Turned Upside-Down. Available online: https://stateofgreen.com/en/partners/state-of-green/news/the-future-role-of-dsos-europes-energy-system-is-turned-upside-down/,2017/ (accessed on 11 January 2020).

42. European Commission Expert Group on electricity interconnection targets. *Towards a Sustainable and Integrated Europe*; European Commission: Brussels, Belgium, 2017.

43. Vittrup, C. Denmark as a Bottom-Up Case. Pitch at the Session: "Decarbonisation of the European Energy System". Available online: https://my.eventbuizz.com/assets/editorImages/1570519706-Workshop1_1150-1300_-_Carsten_Vittrup.pdf (accessed on 2 January 2020).

44. NOE Net Historie om NOE. Available online: https://www.noe.dk/noe_s_historie (accessed on 2 January 2020).

45. Tornbjerg, J. Vestjyske Elkunder Betaler Overpris for Grøn Omstilling. Available online: https://www.danskenergi.dk/nyheder/vestjyske-elkunder-betaler-overpris-gron-omstilling (accessed on 2 January 2020).

46. Energinet. *Ny Viden til Sikker Forsyning. F&I Årsrapport 2018*; Energinet: Fredericia, Denmark, 2019.

47. Mathiesen, B.V.; Lund, H.; Hansen, K.; Skov, I.R.; Djørup, S.R.; Nielsen, S.; Sorknæs, P.; Thellufsen, J.Z.; Grundahl, L.; Lund, R.S.; et al. *IDA's Energy Vision 2050: A Smart Energy System Strategy for 100% Renewable Denmark*; Aalborg University: Aalborg, Denmark, 2015.

48. Danish District Heating Association. Nyheder. Nærdemokrati Hvide Sande køber vindmøller. Available online: https://www.danskfjernvarme.dk/nyheder/presseklip/arkiv/2016/160812naerdemokrati-hvide-sande-koeber-vindmoeller (accessed on 20 March 2020).

49. Energinet. *Systemperspektiv 2035. Baggrundsrapport*; Energinet: Fredericia, Denmark, 2018.

50. Energinet International Infrastructure Projects. Available online: https://en.energinet.dk/Infrastructure-Projects/Projektliste (accessed on 31 January 2020).

51. Energinet. Personal Communication, 2020.

52. RAH Net Gældende nettariffer og abonnementer. Available online: https://rahnet.dk/priser/nettariffer-og-abonnement/ (accessed on 11 February 2020).

53. Energinet TARIFFER OG GEBYRER. Available online: https://energinet.dk/El/Elmarkedet/Tariffer (accessed on 11 February 2020).

54. Djørup, S. Competing Market Regimes: When and where are supply and demand able to shake hands? An exploration of demand side policies in the Danish energy transition. In Proceedings of the SWEDES Conference, Palermo, Italy, September 30–Octomber 4 2018.

55. Smartvarme.dk Homepage. Available online: http://smartvarme.dk/ (accessed on 11 January 2020).

56. Danish Government. *Energiaftale af 29. June 2018*; Danish Government: Copenhagen, Denmark, 2018.

57. Danish Energy Agency Varmepumper—En God Forretning. Available online: https://presse.ens.dk/news/varmepumper-en-god-forretning-268036 (accessed on 17 January 2020).

58. Danish Energy Agency 15 Værker Får Støtte til Store Varmepumper. Available online: https://presse.ens.dk/pressreleases/15-vaerker-faar-stoette-til-store-varmepumper-2562721 (accessed on 17 January 2020).

59. Danish Energy Agency Grundbeløbets Ophør og Grundbeløbsindsatsen. Available online: https://ens.dk/ansvarsomraader/varme/grundbeloebets-ophoer-og-grundbeloebsindsatsen# (accessed on 17 January 2020).

60. SKAT Bindende Svar. Available online: https://www.rksk.dk/Edoc/Dagsordenspublicering/\T1\Okonomi-ogErhvervsudvalget/2016-08-0909.00/Dagsorden/Referat/Hjemmeside/2016-08-0913.10.32/Attachments/1147601-1452497-1.pdf (accessed on 11 January 2020).

61. Energinet Markesdata. Available online: http://osp.energinet.dk/_layouts/Markedsdata/framework/integrations/markedsdatatemplate.aspx (accessed on 11 January 2020).

62. Mey, F.; Diesendorf, M. Who owns an energy transition? Strategic action fields and community wind energy in Denmark. *Energy Res. Soc. Sci.* **2018**, *35*, 108–117. [CrossRef]
63. Danish Government. *Stemmeaftale Mellem Regeringen (Venstre, Liberal Alliance, Det Konservative Folkeparti) og Dansk Folkeparti om ny Støttemodel for Vind og sol i 2018–2019*; Danish Government: Copenhagen, Denmark, 2017.
64. Connolly, D.; Lund, H.; Mathiesen, B.V.; Werner, S.; Möller, B.; Persson, U.; Boermans, T.; Trier, D.; Østergaard, P.A.; Nielsen, S. Heat Roadmap Europe: Combining district heating with heat savings to decarbonise the EU energy system. *Energy Policy* **2014**, *65*, 475–489. [CrossRef]
65. Energy Transition. *Financing Consumer Co-ownership in Renewables*, 1st ed.; Lowitzsch, J., Ed.; Palgrave Macmillan: Cham, Switzerland, 2019; ISBN 978-3-319-93517-1.
66. Zeman, J.; Werner, S. *District Heating System Ownership Guide*; IEE DHCAN project: Watford, UK, 2004.
67. Werner, S. International review of district heating and cooling. *Energy* **2017**, *137*, 617–631. [CrossRef]

 *energies*

Article

Collective Action and Social Innovation in the Energy Sector: A Mobilization Model Perspective

Jay Sterling Gregg [1,*], Sophie Nyborg [2], Meiken Hansen [2], Valeria Jana Schwanitz [3], August Wierling [3], Jan Pedro Zeiss [3], Sarah Delvaux [4], Victor Saenz [4], Lucia Polo-Alvarez [5], Chiara Candelise [6], Winston Gilcrease [7], Osman Arrobbio [7], Alessandro Sciullo [7] and Dario Padovan [7]

[1] Department of Technology, Management and Economics, Technology Transitions and System Innovation Division, UNEP-DTU Partnership, UN City, Marmorvej 51, 2100 Copenhagen Ø, Denmark

[2] Department of Technology, Management and Economics, Innovation Division, DTU-Technical University of Denmark, Akademivej Building 358, 2800 Kongens Lyngby, Denmark; sonyb@dtu.dk (S.N.); meih@dtu.dk (M.H.)

[3] Department of Environmental Sciences,HVL-Western Norway University of Applied Sciences, Postbox 7030, 5020 Bergen, Norway; Valeria.Jana.Schwanitz@hvl.no (V.J.S.); August.Hubert.Wierling@hvl.no (A.W.); Jan.Pedro.Zeiss@hvl.no (J.P.Z.)

[4] VITO-Vlaamse Instelling voor Technologisch Onderzoek, 2400 Mol, Belgium; sarah.delvaux@vito.be (S.D.); victor.sdmp@yahoo.es (V.S.)

[5] TECNALIA-Parque Tecnológico de Bizkaia, Astondo Bidea, Edificio 700, 48160 Derio, Biakaia, Spain; lucia.polo@tecnalia.com

[6] UB–GREEN (Centre for Research in Geography, Resources, Environment, Energy and Networks), ICEPT (Imperial Centre for Energy Policy and Technology), Imperial College London, London SW7 2AZ, UK; chiara.candelise@unibocconi.it

[7] Department of Culture, Politics and Society, UNITO-University of Turin, 10153 Turin, Italy; gregorywinston.gilcrease@unito.it (W.G.); osman.arrobbio@unito.it (O.A.); alessandro.sciullo@unito.it (A.S.); dario.padovan@unito.it (D.P.)

* Correspondence: jsgr@dtu.dk

Received: 19 November 2019; Accepted: 24 January 2020; Published: 4 February 2020

Abstract: This conceptual paper applies a mobilization model to Collective Action Initiatives (CAIs) in the energy sector. The goal is to synthesize aspects of sustainable transition theories with social movement theory to gain insights into how CAIs mobilize to bring about niche-regime change in the context of the sustainable energy transition. First, we demonstrate how energy communities, as a representation of CAIs, relate to social innovation. We then discuss how CAIs in the energy sector are understood within both sustainability transition theory and institutional dynamics theory. While these theories are adept at describing the role energy CAIs have in the energy transition, they do not yet offer much insight concerning the underlying social dimensions for the formation and upscaling of energy CAIs. Therefore, we adapt and apply a mobilization model to gain insight into the dimensions of mobilization and upscaling of CAIs in the energy sector. By doing so we show that the expanding role of CAIs in the energy sector is a function of their power acquisition through mobilization processes. We conclude with a look at future opportunities and challenges of CAIs in the energy transition.

Keywords: collective action; social innovation; mobilization model; energy communities; energy collectives

---

1. Introduction

In this conceptual paper, we develop a framework to better understand mobilization of Collective Action Initiatives (CAIs) within the energy sector. In so doing, we draw from sustainable transition theory and social movement theory to analyze the underlying mechanisms of CAIs in the energy transition.

As part of meeting sustainable development challenges, it is becoming increasingly recognized that the social aspect is an essential, and often overlooked, component to the energy transition. Research has traditionally focused on market-based, technology-driven changes, while the social aspect has traditionally been framed in terms of "social acceptance" (e.g., [1]).

Recognizing the importance of co-evolutionary innovation, this viewpoint is now broadening beyond social acceptance, especially towards "sustainability transitions" [2–4]. This has led to a more nuanced understanding of the dynamics between local communities and the energy transition [5]. The literature on sustainable transitions (e.g., [2,3]) has contributed to a better understanding of the role of citizen initiatives and CAIs in transitioning towards sustainable energy systems [6]. However, a range of prominent transition scholars [7] have called for clarifying how sustainable energy projects, such as urban living labs and other initiatives, can 'scale up' and impact society beyond their initial geographical scope [8]. Furthermore, they encourage inclusion of new theoretical approaches to challenge the current academic socio-technical transition regime, arguing that too little attention is being paid to 'opposition movements' and their effects on sustainability transitions [7].

Transition scholars have emphasized the relevance of social movements in relation to socio-technical transitions [9,10] since socio-technical transitions in the energy sector imply changes in both the social and technical systems. Transition scholars recognize that social movements affect cultural values and beliefs in society [7]. Scholars have argued that social movement theory may be useful in relation to transition studies to investigate a range of topics related to the effectiveness of activists' repertoire of contention, such as how various forms of activism complement or detract from each other, and which technological innovations are socially acceptable [10]. Recent research has highlighted how the energy transition is motivating changes in communities and neighborhoods [11], indicating that disparate movements can be interlinked and synergistic.

Moreover, the exchange of theoretical ideas goes both ways: social movement scholars also see potential in elaborating their theoretical approaches based on transition studies. For example, Törnberg [12] recognizes that socio-technical transition theory may support theory development in the social movement literature. Thus, by combining these perspectives, this allows us to address issues regarding how social movements break through and change social systems, as well as how social innovation can lead to institutional reform. This may be a cyclical process, as other research indicates that institutional reform may play an increasingly important role in motivating the formation of CAIs in a variety of sectors in response to the changing role of the welfare state and privatization of public services [13].

CAIs can take several forms of management and organizational structures—from working groups, grassroots organizations, and foundations to neighborhood associations and cooperatives. These structures provide opportunities for citizens to be more engaged with each other and can offer platforms to be involved in policy-making processes. Tilly [14] created a model for understanding how individuals from the civil society form collectives that mobilize and act as contenders to challenge incumbent regimes.

In particular, the emergence of citizen-led energy CAIs is motivating many municipalities, towns, and villages to create a more low carbon society that involves sustainable energy [11]. In this respect, involving citizens and their local communities in the energy transition is paramount [6]. Self-organization in particular can facilitate socio-institutional practices which link citizen-driven energy projects to local government institutions [15]. This, in turn, can have lasting effects on policy making to support the sustainability and scaling up of CAIs.

This paper augments sustainable transition theory with social movement theory to better understand the role of social innovation and CAIs within the energy sector. We argue that incorporating the concept of mobilization offers a perspective on how change occurs within the energy sector, and thus serves as the structure of this paper. We aim to contribute to a better understanding of civil society's role in sustainability transitions by applying four dimensions that shed more light on mobilization and upscaling of CAIs. With this approach, we suggest a framework that enlightens dimensions of importance for the mobilization and upscaling of CAIs that support social innovation within the energy sector. Moreover, this perspective allows us to discuss the generative and innovative power in mobilizations against the status quo and explore why some CAIs are successful and others are not.

2. Social Innovation in the Energy Sector

2.1. Social Innovation

Mulgan, et al. [16] define social innovation as "innovative activities and services that are motivated by the goal of meeting a social need and that are predominantly developed and diffused through organizations whose primary purposes are social." This is seen in opposition to "business innovation", which is mostly driven by maximization of profit and diffused through organizations in the private industry [16]. Furthermore, the European Union expands the definition of social innovation analogously to the one above: "Social innovations are innovations that are social in both their ends and their means. Specifically, we define social innovations as new ideas (products, services and models) that simultaneously meet social needs (more effectively than alternatives) and create new social relationships or collaborations. They are innovations that are not only good for society but also enhance society's capacity to act" [17]. Notably, social innovation has been linked to critical societal challenges, such as climate change, because of the need for multi-level governance and a coordinated effort to succeed [17]. In the following, we utilize the definition from the European Union, elaborating on the relevance of social innovation for the energy sector.

2.2. Social Innovation and Social Movements

While social innovation can be understood as new processes and practices with social means and ends, social movements, on the other hand, consist of dynamic alliances and complex interactions between social actors and motivations that are not as easy to define. In a literature review, Diani [18] developed the following definition for social movements, as "networks of informal interactions between a plurality of individuals, groups and or organizations, engaged in political or cultural conflicts, on the basis of shared collective identities." We thus understand a social movement as a grouping of civil society actors engaged in a common goal to bring about social change. Consequently, social movements, such as CAIs, can be seen as instantiations and drivers of social innovation. Beyond this, social movements also seek to disrupt and redefine power structures, form new collective identities, and overcome social and structural barriers to change [19]. Social innovation can itself be viewed as a social movement in that its adherents frame the current issues from a social perspective and see social innovation as a promising pathway to achieve solutions to our pressing needs [20] (e.g., the energy transition).

While researchers are increasingly emphasizing the importance of social innovation to the energy transition, it nevertheless still remains unclear as to what extent social movements such as CAIs are contributing to the current energy transition. Additionally, it remains unclear as to how they can better bring the transition to fruition by changing power structures, forming identities around community energy systems, and overcoming political and cultural barriers. Therefore, as we frame the concept of CAIs for the energy transition, it may not only be promoted in terms of instrumental solutions, nor to convince others that such solutions matter, but rather to question technical regime conventions and to debate the critical implications of sustainable energy when understood in new ways [21]. These types of initiatives can be framed theoretically through Critical Theory [22] as far as they activate processes that make apparent the social structure dominating an issue and propose actions to liberate people from such dominance.

2.3. Energy Communities as Social Innovation

The transition towards sustainable energy not only entails a shift from centralized systems of energy provision towards mixed forms, but also a change in the organizational structure, which comes along with new actors who partly replace incumbents in the market. Decentralized, community-based ownership of energy equipment, sources, and distribution systems (i.e., an energy community) is a prominent example of energy generation and distribution under the control of local owners and used by the community members [23] and thus represents a CAI for the energy transition.

Energy communities are typically understood to be locally based, non-commercial, and small enough that they rely on engagement of motivated people with otherwise limited power and resources [24]. Walker and Devine-Wright [25] defined an energy community as entities that have a high degree of citizen ownership and control, that derive collective benefits, and include both supply and demand side energy initiatives. In this way, an energy community is defined by the beneficiaries of an initiative (who the project is for) and who is participating (who the project is by) [25].

The ambitions for local communities to achieve carbon neutrality through self-sufficient, sustainable energy production have also led a trend toward decentralized, renewable energy [11]. CAIs are characterized by local involvement and ownership, grassroots innovation, citizen participation, individual motivations, consumer demand and incentives, and financial and legislative support mechanisms [15]. According to Hielscher et al. [26], community energy projects differ from governmental or private sector projects in three principal ways: (1) community energy projects tend to be multifaceted, in that they tend to address more than one technology, as well as incorporate behavior changes (e.g., energy efficiency measures); (2) they empower communities to collectively change their social, economic, and technical contexts (e.g., energy poverty mitigation); and (3) they enable citizen participation to develop solutions applicable at the local context. Thus, energy communities may go beyond energy to address a wide range of sustainable development issues [6].

Informal networks and social movements are often very important to the development of energy communities (see the case study represented by the Cloughjordan Ecovillage, in [27] (pp. 13–17). These communities may be supported by networks of individuals or by associations which may in turn be supported by local administrators. Social movements and network initiatives in this area are often the result of initiatives undertaken by citizens denouncing problems generated by over-professionalization, privatization, and lack of a real commitment to sustainability from major energy suppliers [28].

There is a myriad potential social impacts from energy community CAIs. For example, local energy communities contribute to local economic development [29,30], address issues of energy poverty [31–33], raise awareness and engagement in sustainable energy [6,34,35], and promote energy justice through grassroots democratic processes [36–40]. Barr and Devine-Wright [40] found that community energy projects help to promote a more sustainable and resilient society while offering communities legitimacy, consensus, and voice. Along these lines, Seyfang et al. [6] highlight that by enabling and empowering communities to collectively change their social, economic, and technical contexts to transition to more sustainable lives, their ideological commitment to sustainability and community energy projects helps groups and individuals to overcome the structural limitations of individualistic measures by bringing communities together with a common purpose.

3. CAIs in the Energy Domain and Sustainable Transition Theory

Current research on citizen initiatives, social innovation and the energy transition has relied on a variety of different analytical and theoretical frameworks. Specifically, the family of sustainable transition theories includes the Multi-Level Perspective (MLP) [2], Strategic Niche Management (SNM) [41,42], Technology Innovation Systems (TIS), and Transition Management

(TM) [3]. Such theories recognize that grand societal challenges, including the energy transition, require more than incremental technological improvements, but also radical transformation of our socio-technical systems [7]. Of these theories, MLP and SNM include significant social aspects and have been applied to understand social innovation within the energy transition. TIS and TM focus on innovation and governance perspectives, respectively, and have not yet been as extensively applied to the role of CAIs in the energy transition. In addition to sustainable transition theory, we also consider theories from institutional dynamics [43,44], which scholars have recently applied to understand CAIs role in the energy transition. Central to all of these perspectives, and the strength of sustainable transition theory in this regard, is the co-evolutionary and multidimensional understanding of transition processes; i.e., how the social and technological aspects of society co-develop. This implies a systemic perspective to capture not only co-evolutionary complexity, but also key phenomena, such as path-dependency, emergence, and non-linear dynamics [7].

3.1. Multi-Level Perspective (MLP)

The MLP approach argues that transitions come about through the interaction between three analytical levels: (1) Niches or niche situations, which are protected spaces and locus for radical innovation (e.g., demonstration projects, social movements, etc.); (2) socio-technical regimes, which represents the institutional structuring of existing systems; and (3) exogenous socio-technical landscape developments (which includes demographic trends, political ideologies, societal values, and macro-economic patterns) [7] (p. 4) [45] (p. 28).

The socio-technical regime is characterized by being locked-in to certain pathways, i.e., it is difficult to change. Multiple dimensions, i.e., rules, practices, and institutions related to the system, co-evolve with each other (hence socio-technical), thus upholding the system: science, technology, industry, politics, markets, user preferences, and cultural meanings.

Niche-innovations may widely break through if landscape developments (e.g., the increasing recognition of effects of climate change) put pressure on the socio-technical regime (e.g., incumbent fossil-based energy system) that leads to tensions in the regime and creates windows of opportunity. Thus, in a transition theory perspective, CAIs occur in the interaction between the niche level and the existing regime. One potential role of CAIs is to take a proactive position in order to shape and moderate a transition (promote an evolution of the regime) and avoid a chaotic regime collapse [5].

Seyfang and Haxeltine [5] emphasize the importance of balancing attention and resource allocation between internal niche-formation and external diffusion, strategically focusing on how group cohesion is maintained, i.e., how identity, belonging, purpose, and community can bolster a CAI and the evolution of its vision.

3.2. Grassroots Innovation and Strategic Niche Management (SNM)

Grassroots innovation takes the perspective of the CAIs as (social) innovation niches. By doing so, the theory seeks to provide insight into the challenges, needs, and potential of grassroots initiatives [46]. When regimes undergo radical change, there is typically an underlying network of organizations and technologies on the margins (i.e., niches). Such niches allow for the development of new ideas and practices (e.g., social innovation) while being shielded to some degree from the processes affecting regime development [41,46–49]. From this perspective, strategic niche management (SNM) is seen as a means to present revolutionary solutions that could not otherwise emerge from the dominant incrementally path-dependent regime [50]. Seyfang and Haxeltine [5] see civil society as an agent of innovative change, able to form new protected niche spaces and develop new practices and ideas.

Hasanov and Zuidema [15] argue that the niche-regime interaction leads to small, adaptive changes that lead to new socio-institutional structures. They focus on how communities self-organize to form new niches, and the value-led features that facilitate that process. They find that energy communities are motivated by collective norms and they are strengthened by sharing a common vision and activities. However, such initiatives typically require intermediaries: e.g., semi-public organizations [15].

Dóci, et al. [51], also find that links with powerful regime actors are a key indicator for the success of a renewable energy community. Additionally, a clear vision and knowledge of the goals, clear structural rules, common events, and networking platforms, and heterogeneity (in actors, motivations, and technologies) of the group are all attributes that contribute to its success [51]. They distinguish between externally oriented niches, which are organized around technological innovations and internally oriented niches, where technology is primarily a tool for some other social goal [51]. As such, renewable energy communities are more about developing new social innovations, such as strategies for managing behaviors and social groups, than they are about promoting a specific technology [51]. These practices strengthen civil society and seek to meet social goals [51].

From a SNM perspective, the important features for contributing to a growing sustainability transition are replication, scaling, and translation [5]. Moreover, a niche development process is more successful if the expectations (i.e., visioning) for the movement are widely shared, specific, realistic, and achievable [5,52]. Additionally, niche experiences are successful when internal and external networks are continually strengthened and when first and second-order learning comes from their activities and experience [5,52].

3.3. Technological Innovation Systems (TIS)

From the TIS perspective, sustainable transitions are linked to innovation processes and national innovation policies. As such, it is rarely and only tangentially applied to the dynamics of energy CAIs. However, there may be a future role for employing TIS to energy CAIs, as the theory does contain aspects of knowledge diffusion, resource mobilization, and support from advocacy coalitions.

For example, Agbemabiese et al. [53] employ TIS to gain insights into how innovation policies can accelerate the energy transition in Africa. Among their recommendations, they suggest that entrepreneurs learn to organize to form alliances so that they can act collectively to influence national policies. Hawkey [54] also used a TIS framework to analyze how regional district heating networks can emerge in the United Kingdom, with a focus on local entrepreneurialism, especially on how they mobilize resources (particularly human resources), and build legitimacy within communities, and ultimately the link between local and national scales.

3.4. Transition Management (TM)

The TM perspective was developed to analyze policy and governance of sustainable development transitions, and address the challenges that arise from disagreements of political priorities, distributed political power, pathway determination and lock-in prevention, and short-term versus long-term planning [55]. TM has been less frequently applied to the energy CAIs, as it typically is applied to understand how top-down national policy can facilitate and steer the co-evolution of technology and society within sustainable transitions [55].

Nevertheless, Kaphengst and Velten [56] utilized TM because the theory allows for some description of the normative aspects involved in prescriptive governance. They employed TM to empirically analyze energy collectives in northern Bavaria, Germany (and augmented their study with established cases in Denmark and Spain), focusing on the role of societal actors- including governments, firms, NGOs, and other organizations, and the networks created between them. Kaphengst and Velten [56] sought to identify the features that contribute to the success of energy CAIs, e.g., bottom-up, user-driven innovation processes (in contrast to top-down), business acumen of the members, financial security (e.g., governmental guarantees) during the transition, and high levels of trust between members of the CAIs. From a TM perspective, these factors could then be translated into policy recommendations, such as crafting a favorable legal framework, providing funding opportunities, backing local front-runners and first-movers, and establishing spaces and capacities for open dialogues between citizens.

Späth and Rohracher [57] also employed TM to explore the role that strategic promotion and institutionalization of energy visions by regional governments can have in the creation and mobilization of regional community energy districts. In their case study in Austria, they found that visions helped create discursive niches that helped align the heterogeneous interests of the actors [57].

3.5. Institutional Dynamics

Moss et al. [58] employ theories of intuitional dynamics. While they acknowledge the linkages to MLP and SNM theories, they argue for a more direct focus on the role of institutions (including organizations, such as governing bodies, as well as CAIs) as constitutive components to socio-technical transitions. As such, institutions are both a medium and product of the transition [58]. Oteman et al. [59] show how the institutional arrangement of the energy sector can either promote or restrain community involvement. In particular, decentralization of energy production and multi-level alignment of government discourse and strategies are important enablers of community involvement, providing some degree of stability and predictability [59].

The institutional configuration of the energy sector is a large factor on how community-led initiatives develop because it influences the available space for new initiatives to emerge from the community [59]. As such, barriers to the success of energy communities can include administrative obstacles as well as socio-economic structures [60]. Thus, CAIs as institutions must have agency and power in order to shape the context of the energy transition [58]. Such agency and power is manifested in their ability to create and promote new ideas [58], i.e., drive social innovation. Finally, the concept of ownership within an institution, such as a CAI, is an important dimension, where ownership is understood in the larger context of ideational ownership, such as in a commons: community control, distributional justice, environmental sustainability, and enhanced participation [61].

4. A Mobilization Framework for CAIs in the Energy Sector

4.1. Sustainable Transitions and Mobilization

Sustainability transition theories (and institutional dynamics) have been an essential field of research for improving our understanding of the energy transition as a socio-technical transition. This is especially relevant to the role energy CAIs have in the energy transition from the perspectives of regime structures, innovative niches, of technological innovation, and strategic management policies. While these theories describe the role of CAIs in the energy transition from the different perspectives and institutional dynamics theory shows the necessity of agency and power for CAIs to impact the socio-technical regime, these theories do not yet offer much insight concerning the underlying social dimensions for the formation and mobilization of energy CAIs and how power and agency are acquired [62].

A mobilization perspective, on the other hand, takes into consideration the power structures and the agents that underlie attempts to transform the energy regime. In contrast, adopting a mobilization model best supports investigating how actors are involved in these dynamics. In this light, mobilization is a precondition for the current studies on community energy [63]. To this end, we adapt the mobilization model developed by Tilly [14] and apply it to energy CAIs as a social movement: "Collective action is joint action in pursuit of common ends" [14] (p. 84). Tilly [14] argues that CAIs depend on shared interests, the identity of their organization, and the mobilization that includes the resources available to the group. Furthermore, this framework can be useful in understanding how energy CAIs can mobilize to challenge the current socio-technical regime of the centralized energy system. In so doing, we aim to also build on Hess [64], who explores how social movements mobilize to challenge resistant incumbents to bring about an industrial transition in the energy system.

4.2. The Mobilization Model

Tilly [14] argues that neglected political issues that exacerbate power disparities serve as an impetus for social movements to arise. Thus, people can be motivated to mobilize and thereby impact the current decision-making regime. These movements are a product of the current socio-political structures, but tend to take on a life of their own as the movements test boundaries, and are shaped by the ensuing response and interactions from the incumbent actors [65].

The mobilization model [14] is largely based on a structuralist ontology, and not on a socio-technical perspective. Nevertheless, the mobilization model shines light on several dimensions, including materialities, structures and individual versus collective interests that are of relevance when analyzing the mobilization and upscaling of CAIs in the energy transition. This lens also provides understanding to the contingency of collective action and illuminates factors that can explain why some CAIs are successful and others are not.

The mobilization model describes the behavior of a single contender through some overall dimensions. A contender, according to Tilly [14] (p. 52) is "any group which, during some specified period, applies pooled resources to influence the government. Contenders include challengers and members of the polity. A member is any contender which has routine, low-cost access to resources controlled by the government; a challenger is any other contender. [The polity] consists of the collective action of the members and the government." The mobilization model has both internal dimensions (Interests, Organization, Resources, and Mobilization), as well as external dimensions (Opportunities and Threats) (Figure 1).

Internal dimensions are those related to agency of a collective movement, what Tilly [14] (p. 6) refers to purposive explanations. The internal domain tends to be more normative and receives little attention in the sustainability transition research. The first part of the internal dimension looks at Interests, which includes the costs and benefits to individuals resulting from their interactions within the group. Organization refers to the aspect of a group's structure, which most directly affects its capacity to act on its interests. Resources refers to endowments and forms that can drive collective action, such as financial resources and knowledge. Mobilization is the process through which both the amount of resources and the collective control of the former by the contender can increase over time.

The external dimension of Opportunities and Threats refers to the relationship between a CAI and the current state of the world around it. The external domain lends itself more easily to causal explanation [14] (p. 6), and tends to be the current focus of the research in energy collectives. The opportunity dimension also includes supporting or deterring reactions to the CAI, which can affect the cost-benefit ratio of the CAI. Changes in these relationships can threaten the group's interests or alternatively provide new chances to act on those interests.

Power, according to Tilly [14] (p. 55), is "the extent to which the outcomes of the population's interactions with other populations favor its interests over those of the others; acquisition of power is an increase in the favorability of such outcomes, loss of power a decline in their favorability; political power refers to the outcomes of interactions with governments." In our adapted framework, power is understood both in relation to processes internal to the CAI (interest alignments, functional organization, and command of resources) as well as the external power asserted over the political and social barriers to gain control over the energy systems, which includes the relations to other actors, including government. In the case of energy CAIs, we aim to show that these internal and external domains are interlinked; as the dimensions of mobilization align, the CAI gains more momentum and a greater degree of agency, and thus is able to exert more influence on external structures to gain control over their energy system and moreover reap the social benefits of doing so.

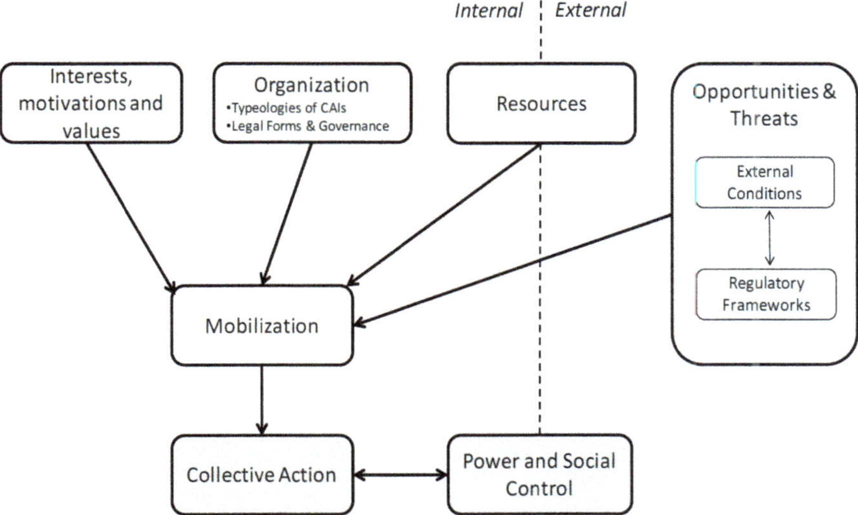

Figure 1. The Mobilization Model, inspired by Tilly [14]. Tilly [14] created a general mobilization model for all forms of collective action from worker strikes to large scale political revolutions. We simplify the model in order to apply it to CAIs in the energy sector. Whereas Tilly [14] describes dimensions of repression and facilitation, we divide this into the two components of external conditions and the regulatory framework, and place this under the opportunities and threats in the external domain. Resources are also not explicitly a component in Tilly's [14] mobilization model, but are an important factor in energy CAIs and this dimension straddles both internal and external domains. Power is also more narrowly defined here than in Tilly's [14] model (which can be applied from striking unions to political revolutions). An energy collective seeks only control over its local energy system with the goal of attaining social benefits that come with this action (e.g., stability, cohesion, identity, etc.). Power internal to the CAI influences its external power, and vice versa. The linkages between these are also simplified, though we acknowledge, similar to Tilly [14], that the processes are complicated and secondary and tertiary links are possible.

5. Dimensions of Collective Action

5.1. Interests, Motivations and Values

Tilly [14] describes the dilemmas of analyzing the interests of groups. The first highlights the contrast between a group's stated interest and their inferred interests from their observable actions, while the second looks at the contrast between individual and collective interests.

When considering the first dilemma with respect to energy CAIs, in order to make energy a collective and common good, a plurality of citizens needs to emerge and claim ownership of the energy system. To claim ownership is not simply a question of defining property rights in the legal sense, but also governs its development, energy production, and consumption within the group, and ultimately, its sustainability. From this perspective, an energy CAI is defined by its interest in liberating a local energy system, specifically electricity, from being a private good—thereby transforming it to a collective good. They are collective goods in the sense that their use value is to a plurality [66]. Thus, on the surface, the response to the first dilemma that Tilly [14] identifies with interests in collective action is straightforward for the case of energy CAIs, since an energy CAI is defined by a goal of developing and managing the community energy system in order to localize its social benefits. However, given the diversity of energy CAIs, recent research has revealed that the picture is more complex. Analyzing a database of German community energy initiatives, Holstenkamp and Kahla [67] found that these entities functioned as "social investments", and in that respect, how the return on that investment was assessed varied significantly across different social settings, geographical locations, and climates.

The second dilemma harkens back to a classical philosophical problem, and is more problematic for the mobilization of energy CAIs. Research in behavioral theory holds that people are driven by a mix of self-regarding motivations as well as pressure to conform to social, cultural, and moral norms [68] (pp. 57, 58). Thus, on the one hand, individuals are more likely to act collectively when there are expectations of some return on the horizon. This return can be determined by an improvement of the current individual or social situation in terms of money (e.g., energy cost savings, return on investment, greater price transparency), self-determination (e.g., greater democratic voice and influence in energy-related decision-making, more control and power over local resources, taking responsibility) and other tangible or intangible goods (e.g., personal environmental consciousness). On the other hand, social cohesion creates a sense of solidarity, equality and participation, and thus shared interests arise. Indeed, studies have found that the so-called "neighbor effect" can have a greater influence on CAI mobilization than information campaigns or financial incentives [69]. In this context, the energy CAIs can be seen as an embodiment of the members' valuing the principle of collective ownership of local energy infrastructure and resources as well as collective governance of the local energy system; i.e., collective autonomy [63]. Some shared vision appears to be necessary in order mobilize collective action and facilitate social resilience within the CAI [70].

Interests and motivation for individuals to form energy CAIs is a topic that is only recently gaining interest in the scientific literature, and is now beginning to be addressed by sustainable transition theory. For example, recent survey-based studies have shed some light on the motivating factors of individuals within CAIs. When Kaphengst and Velten [56] surveyed leaders of energy CAIs, they found that they had a combination of strong social ties, a desire for agency (personal responsibility), and a zest for pioneering new energy technologies. Interestingly, they found that the motivations were more altruistic versus profit-seeking: in their study of Bavarian CAIs, environmental sustainability, followed by a desire to support the local community were the main motivations for members to participate (as opposed to an interest in economic returns on investment) [56]. However, there were differing views among the membership, particularly by age group: investment returns were more important among younger members whereas supporting the community was more important for older members [56]. Bauwens [71] likewise found that members had heterogeneous interests (including monetary, social, and moral) in his survey of members from four community renewable energy initiatives in Belgium. Members' interests were dependent on individuals' psychological, social and moral factors: of interpersonal trust, social identification with the group, and pro-environmental orientation. Moreover, as the collectives evolved from idealistic beginnings to more supply-based initiatives, they attracted members based on different interests, and at different participation levels [71]. With studies such as these, a picture is beginning to emerge that energy CAIs are efforts in aligning and balancing diverse and evolving interests toward a common goal.

5.2. Organizational Perspective

5.2.1. Typologies of CAIs

Tilly [14] (pp. 62–63) describes the organizational perspective as an amalgamation of network connections and categorical similarities (taxonomies) between individual members. Neither network connections, nor categorical similarities on their own, are a sufficient basis for collective action organization. A collection of individuals that are too diverse are unlikely to have shared interests, even if networked; likewise, a collection of individuals with similar categorical characteristics (taxonomy) cannot organize if they are not networked. On the other hand, categorical identities can form the basis for networking.

For example, research has shown that energy CAIs are typically dominated by male participation (e.g., [56]). Gender aspects are rarely considered in research about the determinants of energy communities' development. Among the few examples, a study focused on investments involved in renewable electricity production by citizen participation schemes in Germany [72] revealed differences between men and women in the ownership of citizen participation schemes, in the average investment sum and in the decision-making bodies. Given this situation, a rural Swedish activist Wanja Wallemyr and a small group of nine women, started a women's only collective, Qvinnovindar, to promote sustainable energy, with one of the goals being to shift the gender power balance in the energy sector in rural communities through the economic empowerment of women [27]. Based on this identity, and shared frustration resulting from the reluctance of banks to finance rural female entrepreneurs, the network quickly grew to over 80 members and expanded to a form a second energy collective, Q2 [27]. Kaphengst and Velten [56], found that trust as an essential factor in building networks to grow CAIs, and that furthermore, this trust was built upon the members sharing similar local traditions and having a personal profile and long history within the community.

In other instances, categorical similarities may limit diversity in the network. For example, 50% of the Transition Town members in the UK are aged between 45 and 64, making this group significantly over-represented in the movement compared to the general population (31%) [5]. Time and financial resources (discussed below) may be one possible explanation for why this age-group is over-represented, but another possibility is that other age groups simply lie outside of the network due to the categorical differences from the current members.

Current literature usually refers to CAIs in the energy field as community energy initiatives [6,73–75], which are often organized in the form of energy cooperatives. Individuals in these cooperatives are linked primarily through geographical location, and this aspect is the foundation of the network connections through neighborhood proximity, facilitated by demographic and cultural similarities. Place-based, community identity can thus form the basis for organization, as members have the incentive to join, cooperate, and make sacrifices when they are reassured by the perception that others in the community will act similarly [63].

The International Cooperative Alliance [76] identified a set of principles that are commonly used to characterize the structural properties of cooperatives (the seven principles of the cooperative identity are: voluntary and open membership; democratic member control; member economic participation; autonomy and independence; providing education, training and information; cooperation among cooperatives; concern for the community). Second, most member states of the European Union provide a specific form for cooperatives within their national legislation [77–79]. While there is high variety across legal forms in different countries (discussed below), the most commonly adopted feature is the strong participation of members in decision-making.

The definition of energy communities is far less clear than the energy cooperatives. The heterogeneity in the sector has led to a variety of different definitions: the International Renewable Energy Agency's (IRENA) Coalition for Action, defines an energy community as projects that fulfill two of the following three elements [80]: local stakeholders own the majority or all of a renewable energy project; voting control rests with a community-based organization; the majority of social and economic benefits are distributed locally. Hicks and Ison [81] characterize community energy along different spectra, which include the range of actors and scale of technology, distribution of voting rights, balance of decision-making power, distribution of financial benefits, and the level of community engagement.

Walker and Devine-Wright [25] created a framework, adapted by Candelise and Ruggieri [82], that defines the space of variation of energy communities among the process and outcome dimensions (Figure 2). The process dimension describes the degree to which citizens finance, own, and control the development of an energy initiative, and encapsulates both economic and participatory elements. The outcome dimension describes the degree to which citizens benefit, both monetarily and non-monetarily, from an energy initiative. Examples are given for each of the quadrants within this framework. In the bottom left, a distant and closed energy scenario is represented by a case where a utility external to the community creates a wind farm to add to its energy production portfolio.

No citizen engagement has been responsible for its development, and no citizens directly benefit in any socially defined way; the farm only produces returns for the utility and its shareholders. In the upper left, a distant yet open energy scenario is represented by a utility offering its customers an option to pay a fee to ensure that a greater share of their household electricity consumption is supplied by renewables—so-called green power or renewable energy certificates. In this case, individual citizens are responsible for promoting the shift to renewable energy though their own interests, but there are no direct collective social benefits as there is no organization between the various individuals who purchase the certificates. In the bottom right, a scenario is described where neighboring individual land owners (e.g., farmers) install their own wind turbines. The interests in this scenario are individual, so there is no collective decision to take this action, yet the outcome is nevertheless a community that produces its own renewable energy, thus attaining collective autonomy. The final scenario in the upper right is exemplified by residents of a shared building complex collectively deciding to invest in solar panels on the roof of their building. In this case, both the development process and the benefits are collective. Candelise and Ruggieri [82] expand the upper right quadrant further to include the intersecting dimensions of institutional characteristics (low participation/market logic vs. high participation/community logic) and returns (monetary vs. non-monetary). Alternatively, researchers have also considered degree of integration versus value generation [83]. These participatory and economic elements are the major features that define the overall structure of community energy initiatives, which can be affected by their relative weight by skewing them toward more market- or community-based logics in their dynamics of development and operation [82].

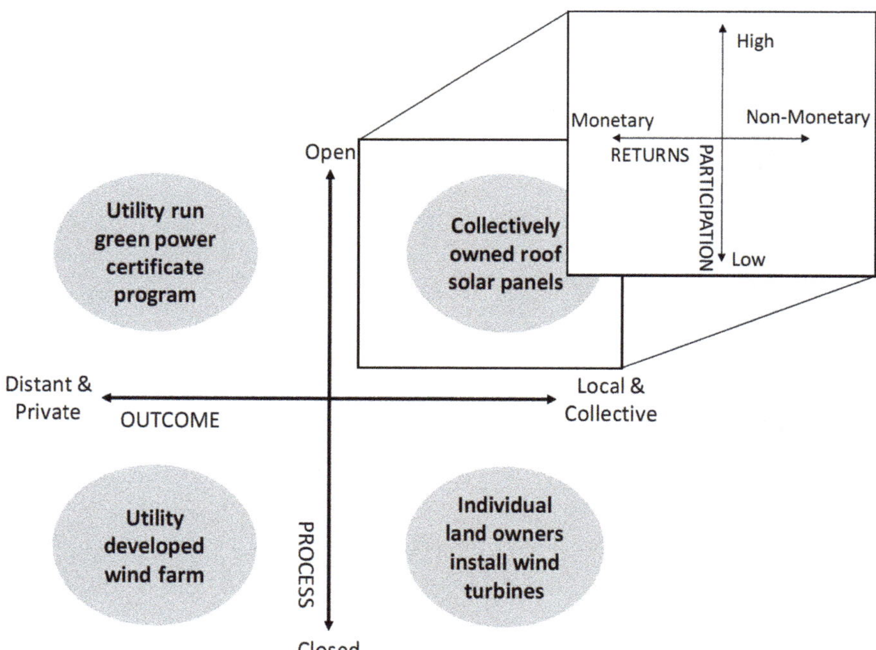

Figure 2. Community energy—two dimensions. Understanding of community renewable energy in relation to project process and outcome dimensions. Adapted from Devine-Wright [25] and Candelise and Ruggieri [82].

There are many alternatives for creating CAI taxonomies. For instance, the Council of European Energy Regulators (CEER) has identified three different types of CAIs with regard to their operational profile, namely: community owned generation assets, virtual sharing over the grid, and sharing of local production through community grids [84]. Moroni et al. [85] suggests a taxonomy for energy communities by intersecting place-based and non-place-based collectives with single and multi-purpose outcomes. Non place-based communities can arise when individuals buy shares into a larger project, for example, Retenergie in Italy [85]. The concept of community can be highly ambiguous [86], and we expect that the concept will continue to evolve. As the digital technologies progress (e.g., blockchain, Carrotmob), the emergence of virtual sharing and thus virtual energy communities becomes a possibility through coordinated consumption infrastructures [87]. This lies at the frontier of research into energy communities, as the members are then no longer geographically localized, and thus the organization is based in virtual spaces of networks of those with similar characteristics. For a successful CAI, the social benefits would also need to apply to the virtual community.

5.2.2. Legal Forms and Governance

In addition to the organizational dynamics of energy communities, it is relevant to highlight how the organizational structure of CAIs is affected by the institutional and legal frameworks of the specific countries in which they are developed. The legal form affects the organization dimension by dictating the available avenues for how CAIs are able to structure themselves. Legal forms can generally be classified into three categories depending on their relevance for renewable energy CAI:

1. Legal forms that require some form of participative decision-making structure, such as the 'one member—one vote' principle (the category with the highest relevance);
2. Legal forms that allow for participative structures (but not required);
3. Legal forms that do not allow participative structures (the least relevant).

Cooperatives are the most common legal form used in the European community energy sector and are generally deemed to provide the best institutional framework for locally owned and participatory approaches to renewable energy projects. They encompass both the social and economic dimension in their scope and are often characterized by a 'one member—one vote' decision making process, thus providing high levels of co-determination [75,88–90].

Depending on the national legal framework, other potentially relevant legal forms are associations, (limited) partnerships and foundations or trusts differently implemented around EU national contexts. Sometimes associations and cooperatives are combined into the same legal form (Denmark and Sweden), while in other cases these are separate legal entities (Germany). Partnerships, on the other hand, are common legal forms for CAIs in Denmark and can be set up by a minimum of two legal or natural persons, while CAIs in the form of trusts can be found in the United Kingdom, as so called 'community development trusts' [6] where the members generally have the right to vote, however this right can be restricted to a specific group of members [91].

The legal form can furthermore provide information on the outcome dimension. Certain legal forms may require the generation and distribution of specific societal benefits such as in the case of Sweden, which differentiates between 'ekonomisk förening' (economic association) and 'ideell förening' (non-profit association) [92,93]. A similar classification can be found in the United Kingdom, where the legislation differentiates between the 'cooperative society (co-op)' and the 'community benefit society (bencom)'. The first, again, focuses on the members' financial interests while the second focuses on community benefits [94].

While the specific governance structure varies depending on national legislation, the participative nature of cooperatives is generally ensured through a general assembly, which convenes at regular intervals. For smaller cooperatives, the general assembly may directly manage its affairs, while larger cooperatives tend to have a board of directors elected by the general assembly [79].

5.3. Resources

Tilly's mobilization model places emphasis on the general idea that the more resources a group has access to, the better the chances are for mobilizing collective action. In Tilly's words, "mobilization refers to the acquisition of collective control over resources" and "contending for power means employing mobilized resources to influence other groups" [14] (p. 78). Tilly and other scholars' focus on resource mobilization has been identified as a separate school of thought in the analysis of social movements as 'Resource Mobilization Theory' [95–97]. This body of work departed from earlier collective behavior studies by moving the focus away from individual participation in social movements and explaining movements as caused by 'individual grievances' and irrational behavior to placing emphasis on the rationality of movement actors. The new focus explained the success of social movements by the ability of movement actors to develop strategies and to arrange and coordinate the use of resources to impact political processes [95,97].

Resources have many definitions—they can be 'internal' (e.g., skills of CAI members) or 'external' (e.g., money) to the group or individual members. Some distinguish between 'tangible' and more 'human' assets [63]. Originally, time and money as well as Tilly's traditional categories of land, labor, capital, and technology have been seen as crucial resources [39], but resources have come to be understood more broadly as "any social, political, economic asset or capacity that can contribute to collective action" [39,98]. Other classifications of resources that have been put forward are socio-organizational resources (e.g., networks, organizations), knowledge resources (e.g., skills, know-how, and technological expertise), material resources (e.g., equipment), symbolic resources (e.g., collective understanding, quest for autonomy, visible and meaningful actions), or structural resources (e.g., investment subsidies or feed-in-tariffs) [39,63]. Bourdieu's [99] notion of economic, social, and cultural capital has also been drawn on by Schwartz [100], who argues that "Bourdieu's view of capitals aligns with resource mobilization perspectives that are open to multiple forms of power resources (not just economic), that can animate social movements." (p. 24). Even Tilly [14] acknowledges the shortcomings of focusing too much on monetary and physical aspects, pointing out that attitudes were more important than any material resource (p. 8).

In an energy transition perspective, it is relevant to consider funding strategies in order to understand monetary resources available to CAIs. For instance, it is self-evident that it is necessary for a cooperative to secure large amounts of capital in order to establish a wind turbine. These require considerable financial commitments and long-term financial planning. In contrast, PV projects are more scalable, and can be completed and expanded with less financial risk. However, with the general increase in the size and complexity of renewable energy projects, the scope for 'traditional' CAIs with direct participation governance and equal rights may be reduced, demanding an adoption of second generation, even more innovative organizational models in the future. In that sense, the collective control over resources in the mobilization model concerning CAIs depends heavily on the chosen energy sources demanding different level of investments. Also, certain types of CAIs do not mobilize separately from existing institutions, e.g., governmental subsidies or other support schemes. The most widely adopted funding strategies for CAIs that are also strongly influenced by the specific national legal frameworks are member-share financing, membership fees, bank loans or community loans, governmental subsidies, tax exemptions and other support-schemes, crowdfunding platforms, refinancing through economic returns, and donations [101].

Other resources that could play a role include, for instance, the extent of communication skills and public awareness, the availability of free technical information and competences, variations in close-knit community spirit [102], and time availability for volunteer-based work (with a side-work for subsistence and family priorities). We consider the issue of time an interesting resource that deserves further attention in the study of energy CAIs. For instance, in the Transition Town movement, many members state they have limited resources of time and money and that this is a significant challenge for wider diffusion of the movement [5]. Internal CAI leadership is another significant resource; a charismatic and trusted initiator (i.e., prime mover) can inspire collective action, particularly in the early stages of formation, as exemplified by Wanja Wallemyr with Qvinnovindar (mentioned above) [27] and Søren Hermansen, with his efforts in establishing an energy collective on Samsø, Denmark, which achieved 100% renewable energy on the island [56].

5.4. Opportunities and Threats

5.4.1. External Conditions

There are many external conditions that influence CAIs mobilization and development. The motivations are diverse and differ across countries, and between regions of the same country. This depends on their specific challenges, such as historical development of national energy markets and other cultural, economic and political factors. The narrow focus on technical and economic aspects of the majority of renewable energy research and policy approaches adopted in this area impedes a deeper understanding of these more nuanced dynamics.

Boon and Dieperink [103] looked at different factors that lead to the formation of energy CAIs in the Netherlands and found both broad national factors: volatile energy prices, societal environmental awareness, inconsistent energy policies, dissatisfaction with national governments' lack of ambition towards environmental targets, desire to reduce dependence on foreign countries for energy; as well as more locate external factors: inspiration from other CAIs, availability of technology suppliers, and support from external parties and institutions. Boon and Dieperink [103] also found symbolic factors, e.g., a green image and enhanced social cohesion could also play a role in the emergence of CAIs. We argue that symbolic factors can also be a strong external driver, because symbols have a shared social understanding, motivate mobilization, and create external visibility. For example, energy memories, the role that the historical process of construction or destabilization of energy cultures has had for a region can be a similar symbolic factor in the creation of CAIs [104]. These key factors of environmental awareness, structural opportunities (e.g., financial incentives, structure of the national energy system) and a presence of a social support system (e.g., in terms of sharing a common identity and ideas) have been observed as positive driving forces for the development of CAIs [105]. A study in Denmark, Germany, and The Netherlands demonstrated that an evolving institutional configuration of the energy sector strongly influences the available space for community initiative development [59]. Another study in Spain concluded that the cooperative tradition is one of the factors that led to the emergence of energy cooperatives in Catalonia, i.e., the Basque Country [106].

External conditions that may play a role as barriers for CAIs' development include the lack of public awareness and sufficient communication resources, the lack of availability of free technical information and competences, limited time availability for volunteer-based work (with a side-work for subsistence and family priorities), lack of close-knit community spirit in big cities and other areas and lack of environmental concerns within influential members of the population. Technological gaps, such as the stabilization of grid infrastructure, may also hamper the establishment of CAIs [105]. Another big issue is connected to legislation and the regulatory framework (see Section 5.4.2), especially legal and regulatory uncertainty for the renewable support schemes evolution, the lack of standardization of such regimes and many other bureaucratic burdens that individuals and collectives face when deciding to start an initiative

5.4.2. Regulatory Frameworks

The structure of the energy sector is a complex structure. It includes the relationships among energy production, energy storage, distribution, energy market and energy demand and consumption. However, traditional European energy systems (in terms of its technical and commercial market design) and its regulatory framework, are organized according to a traditional value chain of production, transport, storage, and distribution of energy. That picture is now far from reality due to the changes produced in the last years [107].

The regulatory framework plays an important role in the creation and development of CAIs. In the last decades, political and legal frameworks in all of Europe have been designed to support an energy system based on centralized production using fossil fuels, in which citizens were passive consumers. The role of consumers has changed and nowadays they are increasingly becoming 'prosumers', broadly 'energy citizens'; drivers of the energy transition [77] to a fairer, democratic, decentralized system with added social benefits [108]. In addition to this, some CAIs not only own the production of energy, but citizens are also now designing creative legal strategies to introduce themselves in the areas of grid ownership and management, and energy supply.

An analysis of community ownership and participation in the production of renewable energy [77] showed that CAIs take many different legal forms. The choice often relates to the goals of the particular community, including tax treatment, profits, or even laws and legal frameworks. Some illustrative examples are shown in this section.

Nevertheless, citizen engagement in renewable energy production only found support in some local and national policies [108]. As the concept of energy communities is varied, the approach and support by the legislative frameworks vary between Member States, e.g., The Netherlands established a regulatory exemption in licensing requirements for new business models, while in Germany there are special rules in action schemes for RES support [109]. This emphasizes that The Netherlands and Germany support a more classical local renewable energy community business models, while the UK, The Netherlands and Poland support more innovative business models. The community energy is less developed in Southern, Central and Eastern Europe, mainly due to the lack of supportive frameworks, or indeed, some abrupt policy changes (withdrawal of support, sometimes retroactively).

After 2000, changes in the EU energy policy provided some opportunities, such as the liberalization of the electricity market. The Clean Energy for all Europeans Package, agreed upon by the EU in 2018, is a significant change. The community energy movement received a boost through the EU's 2030 climate and energy legislative framework that gives more chances for citizens to get involved in the energy transition, allowing communities and individuals the right to generate, store, consume and sell their own energy.

The Package also includes the RED II (The Renewable Energy Directive II, directive 2018/2001/EU). This Directive is important for CAIs in the energy sector because it highlights: (a) citizens and communities are stakeholders in the Energy System; (b) citizens and communities have the right to produce, store, consume and sell renewable energy, and other rights such as consumer's protection or access to all energy markets directly or through a third party; (c) requires Member States produce a National Climate and Energy Plan; (d) it simplifies administration and procedures.

An in-depth assessment of the treatment of energy communities in the 28 draft National Climate and Energy Plans (NECPs) showed that most Member States positively acknowledge renewable energy communities (RECs) in their NECPs and some demonstrate their planned commitment [80]. However, in most cases, this acknowledgment lacks concrete policies or measures. In the analysis of the NECPs, some Member States, like Greece, demonstrate a strong engagement with the role of energy communities in their energy system, whereas others, such as Sweden and Germany, completely ignore this role [110].

Legal frameworks play a role also in shaping the advancement of technologies, telecommunications and data analytics that could provide CAIs with new opportunities. The digitalization of the energy sector gives suppliers the opportunity to have a stronger relationship with consumers, though the security and protection of these data are an increasingly important consideration [111]. Half of all the European Union citizens could be producing their own electricity by 2050 and meeting 45% of the EU's energy demand [112]. This would only be possible assuming that policy and regulatory barriers are removed and national grids, distribution networks, and electricity markets are developed in parallel with the growth of renewable energy production, more storage options, and flexible demand side management.

6. Discussion and Conclusions

6.1. Power and Social Control

The role of CAIs in the energy sector has increased over the last decade. The social relevance of CAIs for the energy transition comes from the fact that it can be the trigger, or at least the accelerator, of the energy transition while creating new conditions for collective and cooperative behavior, thus generating or reinforcing social innovation.

In terms of power, Tilly [14] (p. 55) defines it as both the extent to which a group is able to favor its own interests over those of others, and political power as the outcome of interactions with governmental institutions. We augment this definition by considering internal power as a CAI's ability to align internal interests, build networks, and mobilize resources, and external power as a CAI's ability to seize opportunities and overcome barriers. In our mobilization model, we find that the internal and external domains are linked within energy CAIs. Indeed, a facet of the CAI mobilization process is that it is dynamic and self-reinforcing. As the four dimensions of interest, organization, resources, and opportunities align, and CAIs mobilize, they gain more potential to challenge external barriers to liberate the local energy system from incumbent, centralized energy firms [64]. This process plays a crucial role in building a CAI's identity, including the increase of its networks and membership base, gaining access to more financial and symbolic resources, and imbuing it with greater power to further shape both internal collective interests and external regimes.

Moreover, collective action implies a self-generation of motivations and interests, those that can be linked both to further innovation. The collective creation of guidelines for sustainable consumption of energy services implies a categorical shift in the conceptualization of energy services from a private service to a collective service. In so doing, CAIs provide a structure for fostering sustainability in the energy sector though inherent incentives to develop renewable resources and to promote sustainable consumption patterns. Thus, they garner more attention amongst national and international political and research institutions, and as argued above, this form of social innovation becomes itself a social movement.

6.2. Synthesis of Sustainable Transition Theory and Social Movement Theory

Research into the energy transition is multi-faceted, and many are currently calling for more synthesis between the various perspectives. We have shown here that much can be gained in our understanding of the energy transition by merging aspects of sustainable transition theory and social movement theory. Other researchers are making similar calls for future research to combine understanding of how social movements mobilize in order to shed light upon niche-regime interactions in sustainable transition theory [5]. Moreover, researchers have also argued that studies on energy related behavior change within the energy transition too often focus on the individual energy consumer, and that there is much to be gained from the perspective of researching the community level and viewing individuals as citizens [87]. In this respect, a deeper investigation of the relevance of mobilization in refining motivations and values for people to decide to join and support a collective action (that is tuning, selecting and adapting to the specific context of action) represents a relevant improvement in understanding CAIs' dynamics and performance.

One of the strategies emphasized in the SNM perspective is to support networking activities that involve many stakeholders who can support the growth of a niche by utilizing the available resources of their respective organizations [5]. However, in general, with rare exceptions (e.g., [39,63,113]), sustainable transition theories have not elaborated much on 'available resources' in a social movement perspective as factors that support the upscale of grassroots innovations or mobilization of social movements. There is a focus on resource mobilization within TIS (e.g., [114]), though it does not account well for social movements and collective action. There is also little focus on the other internal dimensions of interests and organization. Instead, sustainable transition theory tends to put more effort in to analyzing the external opportunities and threats to emerging niches. Yet it is highly relevant to gain a deep understanding of how the internal dimensions function in the mobilization of CAIs. This includes how interests are aligned, how organizations form and grow based on taxonomies, networks, and geography, and the how the various types of resources (for instance time, knowledge, skills, money, and materials) and their availability have implications for the mobilization and success of social movements. Thus, the synthesis of transition theory with social movement theory allows for a more nuanced understanding of how energy CAIs form, mobilize, and gain power to affect the transition to sustainable energy.

6.3. Perspectives of CAIs in the Energy Transition

As Tilly [14] (p. 5) notes, the investigation of CAIs is necessarily political and normative, and this can pose challenges for scientific studies that seek to understand CAIs from an objective and non-normative point of view. As such, the new role of CAIs in the political landscape is by no means a one-sided discussion. Undeniably, citizens' engagement with the sustainable energy transition is on the agenda in the EU and other parts of the world, and concepts such as community energy, grassroots innovation, and decentralized energy production are flourishing in academic papers and policy reports. Often the positive merits of these civil society initiatives in terms of pushing the sustainable transition, while ensuring a just and democratic process and distribution of benefits is underlined. Also, citizen engagement and grassroots movements have become increasingly supported politically and institutionally in the first two decades in the new millennium.

However, the expectations of citizens to have an increasingly large role in society concerning, for instance, provision of energy (via energy communities) could also entail a new role of the state: a role where citizens in communities are obliged to 'take care of themselves' and where the service level from the public sector is shrinking [13]. This also relates to discussions in the area of Responsible Research and Innovation where questions of fairness and inclusion are pertinent in relation to projects in the energy sector in the EU. Moreover, many of the imagined energy futures entail investments in advanced technologies (e.g., smart energy systems), which may not be affordable for all. This is compounded by the risk that decentralization may exclude households in very isolated locations, or in areas where CAIs would not prosper, and these would face central system costs less widely socialized than the ones they face today. There is also a risk that the large-scale decentralization may create some negative selection by which the only ones left in the central system to bear most costs are precisely the ones who could not afford to join a CAI.

Outside of these cases, CAIs can be an important form of innovation that produce new types of goods or are able to restore commons that had been monopolized, captured by market forces or privatized. They are a social innovation per se because they counteract privatization and individualization, and because they promote new community interactions and consider a wider definition of social welfare than traditional approaches, which further helps fuel the growth of these initiatives. Here, social innovation, as in the case of innovations that follow from managing new collective goods, is ideally sculpted by principles of environmental and social justice, inclusion, poverty alleviation, and resource sharing as a form of mutual support. All of which generate social welfare.

History has witnessed the transition of how we have conceptualized goods and services, such as energy. One such transition occurred in the 19th century, that of a "moral economy" (one in which the community residents had a right to the resources within the community and the community recognized its obligation to assist resource-less members; i.e., goods and services as collective commons) to that of a "possessive individualism" (where all goods, including labor, should be disposable property and owners had the obligation to use them to their maximum advantage; i.e., goods and services as private) Tilly [14] (p. 4).

Now, as we push against the planet's ecological limits, we are becoming more cognizant of the world's global common resources and the services they provide. Given these constraints, we also reconsider what it means to provide economic wellbeing to citizens and what options are available to promote social cohesion within the civil society. In this light, we may be on the cusp of yet another transition. Because the modern economy is so intrinsically linked to the energy sector, CAIs in the energy sector could be a catalyst to the transformation to a paradigm where sustainability is intrinsically incorporated into our social institutions and technological infrastructure.

Author Contributions: Conceptualization, J.S.G., S.N., M.H., V.J.S., A.W., W.G., O.A., A.S., and D.P.; Formal analysis, J.S.G., S.N., M.H., and D.P.; Funding acquisition, A.S. and D.P.; Methodology, J.S.G., S.N., and M.H.; Project administration, A.S. and D.P.; Writing—original draft, J.S.G., S.N., M.H., V.J.S., A.W., C.C., and D.P.; Writing—review and editing, J.S.G., S.N., M.H., V.J.S., J.P.Z., S.D., V.S., L.P.-A., W.G., O.A., and A.S. All authors have read and agreed to the published version of the manuscript.

Funding: This research was conducted under the COMETS (Collective action Models for Energy Transition and Social Innovation) project, funded by the Horizon 2020 Framework Program of the European Commission, grant number 837722.

Acknowledgments: The authors would like to acknowledge the four anonymous reviewers, whose insightful comments greatly improved the paper.

Conflicts of Interest: The authors declare no conflict of interest. The funders had no role in the design of the study; in the collection, analyses, or interpretation of data; in the writing of the manuscript, or in the decision to publish the results.

References

1. Wüstenhagen, R.; Wolsink, M.; Bürer, M.J. Social acceptance of renewable energy innovation: An introduction to the concept. *Energy Policy* **2007**, *35*, 2683–2691. [CrossRef]
2. Geels, F.W. *Understanding the Dynamics of Technological Transitions: A Co-Evolutionary and Socio-Technical Analysis*; Twente University Press: Enschede, The Netherlands, 2002; p. 426.
3. Markard, J.; Raven, R.; Truffer, B. Sustainability transitions: An emerging field of research and its prospects. *Res. Policy* **2012**, *41*, 955–967. [CrossRef]
4. Grin, J. Understanding transitions from a governance perspective, Part III. In *Transitions to Sustainable Development, New Directions in the Study of Long Term Transformative Change*; Grin, J., Rotmans, J., Schot, J., Eds.; Routledge: London, UK, 2010.
5. Seyfang, G.; Haxeltine, A. Growing Grassroots Innovations: Exploring the Role of Community- Based Initiatives in Governing Sustainable Energy Transitions. *Environ. Plan. C* **2012**, *30*, 381–400. [CrossRef]
6. Seyfang, G.; Park, J.J.; Smith, A. A thousand flowers blooming? An examination of community energy in the UK. *Energy Policy* **2013**, *61*, 977–989. [CrossRef]
7. Köhler, J.; Geels, F.W.; Kern, F.; Markard, J.; Onsongo, E.; Wieczorek, A.; Wells, P. An agenda for sustainability transitions research: State of the art and future directions. *Environ. Innov. Soc. Trans.* **2019**, *31*, 1–32. [CrossRef]
8. Turnheim, B.; Kivimaa, P.; Berkhout, F. *Innovating Climate Governance: Moving Beyond Experiments*; Cambridge University Press: Cambridge, UK, 2018.
9. Foxon, T.J. Transition pathways for a UK low carbon electricity future. *Energy Policy* **2013**, *52*, 10–24. [CrossRef]
10. North, P. The politics of climate activism in the UK: A social movement analysis. *Environ. Plan. A* **2011**, *43*, 1581–1598. [CrossRef]
11. Van Der Schoor, T.; Scholtens, B. Power to the people: Local community initiatives and the transition to sustainable energy. *Renew. Sustain. Energy Rev.* **2014**, *43*, 666–675. [CrossRef]
12. Törnberg, A. Combining transition studies and social movement theory: Towards a new research agenda. *Theory Soc.* **2018**, *47*, 381–408. [CrossRef]
13. Soares da Silva, D.; Horlings, L.; Figueiredo, E. Citizen initiatives in the post-welfare state. *Soc. Sci.* **2018**, *7*, 252. [CrossRef]
14. Tilly, C. *From Mobilization to Revolution*; Addison-Wesley: Reading, MA, USA, 1978.
15. Hasanov, M.; Zuidema, C. The transformative power of self-organization: Towards a conceptual framework for understanding local energy initiatives in The Netherlands. *Energy Res. Soc. Sci.* **2018**, *37*, 85–93. [CrossRef]
16. Mulgan, G.; Tucker, S.; Ali, R.; Sanders, B. Social Innovation: What It Is, Why It Matters and How It Can Be Accelerated. Available online: http://eureka.sbs.ox.ac.uk/761/1/Social_Innovation.pdf (accessed on 31 October 2019).
17. Hubert, A. *Empowering People, Driving Change. Social Innovation in the European Union*; Bureau of European Policy Advisors: Brussel, Belgium, 2010.
18. Diani, M. The concept of social movement. *Sociol. Rev.* **1992**, *40*, 1–25. [CrossRef]
19. Henderson, H. Social innovation and citizen movements. *Futures* **1993**, *25*, 322–338. [CrossRef]

20. Unger, R.M. Conclusion: The task of the social innovation movement. In *New Frontiers in Social Innovation Research*; Nicholls, A., Simon, J., Gabriel, M., Whelan, C., Eds.; Springer: Berlin, Germany, 2015; pp. 233–251.

21. Smith, A.; Hargreaves, T.; Hielscher, S.; Martiskainen, M.; Seyfang, G. Making the most of community energies: Three perspectives on grassroots innovation. *Environ. Plan. A* **2016**, *48*, 407–432. [CrossRef]

22. Feenberg, A. *Transforming Technology: A Critical Theory Revisited*, 2nd ed.; Oxford University Press: Oxford, UK, 2002.

23. Creupelandt, D.; Vansintjan, D. REScoop—Municipality Approach. Deliverable 2.3 of the REScoop MECISE Project. Available online: https://www.dataplan.info/img_upload/c6e3eef692b618867bd4ece4fa16cf48/rescoop-mecise-rescoop-municipality-approach.pdf (accessed on 31 October 2019).

24. Middlemiss, L.; Parrish, B.D. Building capacity for low-carbon communities: The role of grassroots initiatives. *Energy Policy* **2010**, *38*, 7559–7566. [CrossRef]

25. Walker, G.; Devine-Wright, P. Community Renewable Energy: What Should It Mean? *Energy Policy* **2008**, *36*, 497–500. [CrossRef]

26. Hielscher, S.; Seyfang, G.; Smith, A. Grassroots innovations for sustainable energy: Exploring niche development processes among community energy initiatives. In *Innovations in Sustainable Consumption: New Economics, Socio-Technical Transitions, and Social Practices*; Cohen, M., Brown, H., Vergragt, P., Eds.; Edward Elgar: Cheltenham, UK, 2013; pp. 133–158.

27. Ooms, M.; Bijnsdorp, S.; Huygen, A.; Rhomberg, W.; Berger, A. Social Innovation in Energy Supply: Case Study Results. Deliverable 7.3, SI-DRIVE Project. Available online: https://www.si-drive.eu/wp-content/uploads/2017/03/SI-DRIVE-Deliverable-D7_3-Energy-1.pdf (accessed on 31 October 2019).

28. De Moor, T. Homo Cooperans. Institutions for Collective Action and the Compassionate Society. Inaugural Lecture Delivered at the Inauguration of Institutions for Collective Action in Historical Perspective at Utrecht University. Available online: https://dspace.library.uu.nl/handle/1874/349371 (accessed on 31 October 2019).

29. Hoffman, S.M.; High-Pippert, A. From private lives to collective action: Recruitment and participation incentives for a community energy program. *Energy Policy* **2010**, *38*, 7567–7574. [CrossRef]

30. Shaw, S.; Mazzucchelli, P. Evaluating the perspectives for hydrogen energy uptake in communities: Success criteria and their application. *Energy Policy* **2010**, *38*, 5359–5371. [CrossRef]

31. Middlemiss, L.; Ambrosio Albala, P.; Emmel, N.; Gillard, R.; Gilbertson, J.; Hargreaves, T.; Mullen, C.; Tony, R.; Snell, C.; Tod, A. Energy poverty and social relations: A capabilities approach. *Energy Res. Soc. Sci.* **2019**, *55*, 227–235. [CrossRef]

32. Coulon, P.J.; Krieger, K. A Clean Planet. for All? Energy Poverty and Decarbonising Europe's Economy. Available online: http://www.europeanenergyinnovation.eu/Articles/Winter-2018/A-Clean-Planet-for-all-Energy-poverty-and-decarbonising-Europes-economy (accessed on 31 October 2019).

33. O'Brien, S.; Monteiro, C.; Gancheva, M.; Crook, N. *Models of Local Energy Ownership and the Role of Local Energy Communities in Energy Transition in Europe*; European Committee of the Regions: Brussels, Belgium, 2018; Available online: https://cor.europa.eu/en/engage/studies/Documents/local-energy-ownership.pdf (accessed on 31 October 2019).

34. Rogers, J.C.; Simmons, E.A.; Convery, I.; Weatherall, A. Social impacts of community renewable energy projects: Findings from a woodfuel case study. *Energy Policy* **2012**, *42*, 239–247. [CrossRef]

35. Millard, J. *Social Innovation in Poverty Reduction and Sustainable Development*; SI-DRIVE Project: Dortmund, Germany, 2017; pp. 1–10.

36. Hiteva, H.; Sovacool, B. Harnessing social innovation for energy justice: A business model perspective. *Energy Policy* **2017**, *107*, 631–639. [CrossRef]

37. Bianchi, A.; Ginelli, E. The social dimension in energy landscapes. *City Territ. Archit.* **2018**, *5*, 9. [CrossRef]

38. Van Der Schoor, T.; Van Lente, H.; Scholtens, B.; Peine, A. Challenging obduracy: How local communities transform the energy system. *Energy Res. Soc. Sci* **2016**, *13*, 94–105. [CrossRef]

39. Schreuer, A. The establishment of citizen power plants in Austria: A process of empowerment? *Energy Res. Soc. Sci.* **2016**, *13*, 126–135. [CrossRef]

40. Barr, S.; Devine-Wright, P. Resilient communities: Sustainabilities in transition. *Local Environ.* **2012**, *17*, 525–532. [CrossRef]

41. Rip, A.; Kemp, R. Technological change. *Hum. Choice Clim. Chang.* **1998**, *2*, 327–399.

42. Schot, J.; Geels, F.W. Strategic niche management and sustainable innovation journeys: Theory, findings, research agenda, and policy. *Tech. Anal. Strat. Manag.* **2008**, *20*, 537–554. [CrossRef]

43. Radzicki, M.J. Institutional dynamics, deterministic chaos, and self-organizing systems. *J. Econ. Issues* **1990**, *24*, 57–102. [CrossRef]

44. Leach, M.; Mearns, R.; Scoones, I. *Environmental Entitlements: A Framework for Understanding the Institutional Dynamics of Environmental Change*; Discussion Paper 359; Institute of Development Studies: Brighton, UK, 1997.

45. Geels, F.W. The multi-level perspective on sustainability transitions: Responses to seven criticisms. *Environ. Innov. Soc. Transit.* **2011**, *1*, 24–40. [CrossRef]

46. Seyfang, G.; Smith, A. Grassroots innovations for sustainable development: Towards a new research and policy agenda. *Environ. Polit.* **2007**, *16*, 584–603. [CrossRef]

47. Schot, J. The usefulness of evolutionary models for explaining innovation. The case of the Netherlands in the nineteenth century. *Hist. Technol.* **1998**, *14*, 173–200. [CrossRef]

48. Hargreaves, T.; Hielscher, H.; Seyfang, G.; Smith, A. Grassroots innovations in community energy: The role of intermediaries in niche development. *Glob. Environ. Change* **2013**, *23*, 868–880. [CrossRef]

49. Geels, F.W. From sectoral systems of innovation to socio-technical systems: Insights about dynamics and change from sociology and institutional theory. *Res. Policy* **2004**, *33*, 897–920. [CrossRef]

50. Smith, A.; Voß, J.P.; Grin, J. Innovation studies and sustainability transitions: The allure of the multi-level perspective and its challenges. *Res. Policy* **2010**, *39*, 435–448. [CrossRef]

51. Dóci, G.; Vasileiadou, E.; Petersen, A.C. Exploring the transition potential of renewable energy communities. *Futures* **2015**, *66*, 85–95. [CrossRef]

52. Kemp, R.; Schot, J.; Hoogma, R. Regime shifts to sustainability through processes of niche formation: The approach of strategic niche management. *Technol. Anal. Strateg. Manag.* **1998**, *10*, 175–198. [CrossRef]

53. Agbemabiese, L.; Nkomo, J.; Sokona, Y. Enabling innovations in energy access: An African perspective. *Energy Policy* **2012**, *47*, 38–47. [CrossRef]

54. Hawkey, D. District heating in the UK: A Technological Innovation Systems analysis. *Environ. Innov. Soc. Trans.* **2012**, *5*, 19–32. [CrossRef]

55. Kemp, R.; Loorbach, D.; Rotmans, J. Transition management as a model for managing processes of co-evolution towards sustainable development. *Int. J. Sustain. Dev. World* **2007**, *14*, 78–91. [CrossRef]

56. Kaphengst, T.; Velten, E.K. *Energy Transition and Behavioural Change in Rural Areas-The Role of Energy Cooperatives. WWW for Europe Working Paper*; WIFO: Vienna, Austria, 2014; Volume 60.

57. Späth, P.; Rohracher, H. 'Energy regions': The transformative power of regional discourses on socio-technical futures. *Res. Policy* **2010**, *39*, 449–458. [CrossRef]

58. Moss, T.; Becker, S.; Naumann, M. Whose energy transition is it, anyway? Organisation and ownership of the Energiewende in villages, cities and regions. *Local Environ.* **2015**, *20*, 1547–1563. [CrossRef]

59. Oteman, M.; Wiering, M.; Helderman, J.K. The institutional space of community initiatives for renewable energy: A comparative case study of the Netherlands, Germany and Denmark. *Energy Sustain. Soc.* **2014**, *4*, 11. [CrossRef]

60. Magnani, N.; Osti, G. Does civil society matter? Challenges and strategies of grassroots initiatives in Italy's energy transition. *Energy Res. Soc. Sci.* **2016**, *13*, 148–157. [CrossRef]

61. Cumbers, A. *Reclaiming Public Ownership: Making Space for Economic Democracy*; Zed Books Ltd.: London, UK, 2012.

62. Shove, E.; Walker, G. CAUTION! Transitions ahead: Politics, practice, and sustainable transition management. *Environ. Plan. A* **2007**, *39*, 763–770. [CrossRef]

63. Bomberg, E.; McEwan, N. Mobilizing community energy. *Energy Policy* **2012**, *51*, 435–444. [CrossRef]

64. Hess, D.J. Social Movements and Energy Democracy: Types and Processes of Mobilization. *Front. Energy Res.* **2018**, *6*, 1–4. [CrossRef]

65. Tilly, C. *Popular Contention in Great Britain, 1758–1834*; Routledge: New York, NY, USA, 1995.

66. De Angelis, M. *Omnia Sunt Communia. On the Commons and the Transformation to Post-Capitalism*; Zed Books Ltd.: London, UK, 2017.

67. Holstenkamp, L.; Kahla, F. What are community energy companies trying to accomplish? An empirical investigation of investment motives in the German case. *Energy Policy* **2016**, *97*, 112–122. [CrossRef]

68. Bowles, S.; Gintis, H. Beyond Enlightened Self-Interest: Social Norms, Other-Regarding Preferences, and Cooperative Behavior. In *Games, Groups, and the Global Good*; Levin, S.A., Ed.; Springer: Berlin/Heidelberg, Germany, 2009; pp. 57–78.

69. Adil, A.M.; Ko, Y. Socio-technical evolution of decentralized energy systems: A critical review and implication for urban planning and policy. Renew. *Sustain. Energy Rev.* **2016**, *57*, 1025–1037. [CrossRef]

70. Parkhill, K.A.; Shirani, F.; Butler, C.; Henwood, K.L.; Groves, C.; Pidgeon, N.F. 'We are a community [but] that takes a certain amount of energy': Exploring shared visions, social action, and resilience in place-based community-led energy initiatives. *Environ. Sci Policy* **2015**, *53*, 60–69. [CrossRef]

71. Bauwens, T. Explaining the diversity of motivations behind community renewable energy. *Energy Policy* **2016**, *93*, 278–290. [CrossRef]

72. Fraune, C. Gender matters: Women, renewable energy, and citizen participation in Germany. *Energy Res. Soc. Sci.* **2015**, *7*, 55–65. [CrossRef]

73. Walker, G.; Devine-Wright, P.; Hunter, S.; High, H.; Evans, B. Trust and community: Exploring the meanings, contexts and dynamics of community renewable energy. *Energy Policy* **2010**, *38*, 2655–2663. [CrossRef]

74. Wirth, S. Communities matter: Institutional preconditions for community renewable energy. *Energy Policy* **2014**, *70*, 236–246. [CrossRef]

75. Yildiz, Ö.; Rommel, J.; Debor, S.; Holstenkamp, L.; Mey, F.; Müller, J.R.; Radtke, J.; Rognli, J. Renewable energy cooperatives as gatekeepers or facilitators? Recent developments in Germany and a multidisciplinary research agenda. *Energy Res. Soc. Sci.* **2015**, *6*, 59–73. [CrossRef]

76. International Cooperative Alliance (2018). The Cooperative Identity. Available online: https://www.ica.coop/en/cooperatives/cooperative-identity (accessed on 18 July 2019).

77. Roberts, J.; Bodman, F.; Rybski, R. *Community Power; Model Legal Frameworks for Citizen-Owned Renewable Energy*; Client Earth Energy: London, UK, 2014.

78. Cocolina, C.Q. The power of cooperation: Cooperatives Europe Key Figures 2015. Cooperatives Europe, 2016. Available online: https://coopseurope.coop/sites/default/files/The%20power%20of%20Cooperation%20-%20Cooperatives%20Europe%20key%20statistics%202015.pdf (accessed on 31 October 2019).

79. European Parliament. Cooperatives: Characteristics, Activities, Status, Challenges. European Parliamentary Research Service. Briefing 635541, 2019. Available online: http://www.europarl.europa.eu/RegData/etudes/BRIE/2019/635541/EPRS_BRI(2019)635541_EN.pdf (accessed on 31 October 2019).

80. IRENA Coalition. Community Energy: Broadening the Ownership. 2018. Available online: http://irena.org/-/media/Files/IRENA/Agency/Articles/2018/Jan/Coalition-for-Action_Community-Energy_2018.pdf?la=en&hash=CAD4BB4B39A381CC6F712D3A45E56E68CDD63BCD&hash=CAD4BB4B39A381CC6F712D3A45E56E68CDD63BCD (accessed on 31 October 2019).

81. Hicks, J.; Ison, N. An exploration of the boundaries of 'community' in community renewable energy projects: Navigating between motivations and context. *Energy Policy* **2018**, *113*, 523–534. [CrossRef]

82. Candelise, C.; Ruggieri, G. *Community Energy in Italy: Heterogeneous Institutional Characteristics and Citizens Engagement*; IEFE Working Papers; IEFE, Universita' Bocconi: Milano, Italy, 2017; Volume 93, pp. 1–33. Available online: https://ideas.repec.org/p/bcu/iefewp/iefewp93.html (accessed on 31 October 2019).

83. Koirala, B.P.; Koliou, E.; Friege, J.; Hakvoort, R.A.; Herder, P.M. Energetic communities for community energy: A review of key issues and trends shaping integrated community energy systems. *Renew. Sustain. Energy Rev.* **2016**, *56*, 722–744. [CrossRef]

84. Council of European Energy Regulators (CEER). *Regulatory Aspects of Self-Consumption and Energy Communities*; CEER Report; CEER: Brussels, Belgium, 2019; Available online: https://www.ceer.eu/documents/104400/-/-/8ee38e61-a802-bd6f-db27-4fb61aa6eb6a (accessed on 31 October 2019).

85. Moroni, S.; Alberti, V.; Antoniucci, V.; Bisello, A. Energy communities in the transition to a low-carbon future: A taxonomical approach and some policy dilemmas. *J. Environ. Manag.* **2019**, *236*, 45–53. [CrossRef]

86. Rae, C.; Bradley, F. Energy autonomy in sustainable communities. *Renew. Sustain. Energy Rev.* **2012**, *16*, 6497–6506. [CrossRef]

87. Heiskanen, E.; Johnson, M.; Robinson, S.; Vadovics, E.; Saastamoinen, M. Lowcarbon communities as a context for individual behavioural change. *Energy Policy* **2010**, *38*, 7586–7595. [CrossRef]

88. Huybrechts, B.; Mertens, S. The relevance of the cooperative model in the field of renewable energy. *Ann. Public Coop. Econ.* **2014**, *85*, 193–212. [CrossRef]

89. International Labour Organisation (ILO). *Providing Clean Energy and Energy Access Through Cooperatives Providing Clean Energy and Energy Access*; ILO: Geneva, Switzerland, 2013; Available online: https://www.ilo.org/wcmsp5/groups/public/---ed_emp/---emp_ent/documents/publication/wcms_233199.pdf (accessed on 31 October 2019).

90. Viardot, E. The role of cooperatives in overcoming the barriers to adoption of renewable energy. *Energy Policy* **2013**, *63*, 756–764. [CrossRef]

91. Wilcox, D. What is a Development Trust? Available online: http://partnerships.org.uk/AZP/what.html (accessed on 31 October 2019).

92. Bolagsverket, Förening. Available online: https://bolagsverket.se/fo/foreningsformer (accessed on 29 July 2019).

93. Bolagsverket, Företagsform. Available online: https://www.verksamt.se/starta/valj-foretagsform (accessed on 29 July 2019).

94. Alastair, C. *Guidance on Legal Forms and Ownership Models for Business, Policies, Department for Business, Energy & Industrial Strategy*; BIS/11/1399; Department for Business, Innovation and Skills: London, UK, 2011.

95. Jenkins, J.C. Resource Mobilization Theory and the Study of Social Movements. *Annu. Rev. Sociol.* **1983**, *9*, 527–553. [CrossRef]

96. Bate, P.; Bevan, H.; Robert, G. Towards a million change agents. A review of the social movements literature, 2005. Available online: https://discovery.ucl.ac.uk/id/eprint/1133/1/million.pdf (accessed on 7 January 2020).

97. Seyfang, G.; Haxeltine, A.; Hargreaves, T.; Longhurst, N. *Energy and Communities in Transition: Towards a New Research Agenda on Agency and Civil Society in Sustainability Transitions*; CSERGE Working Paper EDM; University of East Anglia, The Centre for Social and Economic Research on the Global Environment (CSERGE): Norwich, UK, 2010; pp. 10–13.

98. Jenkins, J.C. Social Movements: Resource Mobilization Theory. In *International Encyclopedia of the Social and Behavioral Sciences*; Smelser, J., Ed.; Pergamon: Oxford, UK, 2001; pp. 14368–14371.

99. Bourdieu, P. The Forms of Capital. In *Handbook of Theory and Research for the Sociology of Education*; Richardson, J.G., Ed.; Greenwood Press: New York, NY, USA, 1986; pp. 241–258.

100. Swartz, D. *Symbolic Power, Politics and Intellectuals. The Political Sociology of Pierre Bourdieu*; The University of Chicago Press: Chicago, IL, USA, 2013.

101. Wierling, A.; Schwanitz, V.; Zeiß, J.; Bout, C.; Candelise, C.; Gilcrease, W.; Gregg, J. Statistical evidence on the role of energy cooperatives for the energy transition in European countries. *Sustainability* **2018**, *10*, 3339. [CrossRef]

102. Theodori, G.L.; Luloff, A.E. Urbanization and Community Attachment in Rural Areas. *Soc. Nat. Resour.* **2000**, *13*, 399–420.

103. Boon, F.P.; Dieperink, C. Local civil society based renewable energy organisations in The Netherlands: Exploring the factors that stimulate their emergence and development. *Energy Policy* **2014**, *69*, 297–307. [CrossRef]

104. Lettmayer, G.; Schwarzinger, S.; Koksvik, G.; Skjølsvold, T.M.; Velte, D. Energy Memories: A new concept for understanding the impact of historic events on energy choices. In Proceedings of the Behave Conference, Zurich, Switzerland, 5–7 September 2018.

105. Carrus, G.; Tiberio, L.; Panno, A.; Lettmayer, G.; Kaltenegger, I.; Demir, M.H.; Solak, B.; Dimitrova, E.; Tasheva-Petrova, M.; Mutafchiiska, I.; et al. Analysis of Enabling Factors for Consumer Action. ECHOES Project Deliverable 5.3. Available online: https://echoes-project.eu/sites/echoes.drupal.pulsartecnalia.com/files/D5.3.pdf (accessed on 31 October 2019).

106. Jimenez Iturriza, J.; Polo, L.; Beermann, M.; Lettmayer, G.; Carrus, G.; Tiberio, L.; Biresselioglu, M.E.; Demir, M.H.; Solak, B.; Dimitrova, E.; et al. Collective Energy Practices in Europe. ECHOES Project Deliverable 5.4. Available online: https://echoes-project.eu/sites/echoes.drupal.pulsartecnalia.com/files/D5.4.pdf (accessed on 31 October 2019).

107. Hoppe, T.; Butenko, A.; Heldeweg, M. Innovation in the European energy sector and regulatory responses to it: Guest editorial note. *Sustainability* **2018**, *10*, 416. [CrossRef]

108. Friends of the Earth Europe. Unleashing the Power of Community Renewable Energy. Available online: https://energy-cities.eu/unleashing-the-power-of-community-renewable-energy/ (accessed on 31 October 2019).

109. Tounquet, F.; De Vos, L.; Abada, I.; Kiekichowska, I.; Klessmann, C. Energy Communities in the European Union. ASSET Project Final Report, 2019. Available online: https://asset-ec.eu/wp-content/uploads/2019/07/ASSET-Energy-Comminities-Revised-final-report.pdf (accessed on 31 October 2019).

110. Roberts, J.; Gauthier, C. *Energy Communities in the Draft National Energy and Climate Plans: Encouraging but Room for Improvements*; REScoop: Berchem, Belgium, 2019; Available online: https://uploads.strikinglycdn.com/files/ed4a94af-a4ea-458b-a87b-676c6e300ffa/Briefing%20-%20NECPs%20and%20energy%20communities.pdf (accessed on 31 October 2019).

111. Eurelectric. *The Power Sector Goes Digital-Next Generation Data Management for Energy Consumers*; Eurelectric: Brussels, Belgium, 2016; Available online: https://www3.eurelectric.org/media/278067/joint_retail_dso_data_report_final_11may_as-2016-030-0258-01-e.pdf (accessed on 31 October 2019).

112. Kampman, B.; Blommerde, J.; Afman, M. *The Potential of Energy Citizens in the European Union*; Report by CE Delft for Greenpeace European Unit, Friends of the Earth Europe, European Renewable Energy Federation (EREF) and REScoop, Publication code: 16.3J00.75; CE Delft: Delft, The Netherlands, 2016; Available online: https://www.foeeurope.org/sites/default/files/renewable_energy/2016/ce-delft-the-potential-of-energy-citizens-eu.pdf (accessed on 7 January 2020).

113. Verbong, G.; Loorbach, D. *Governing the Energy Transition: Reality, Illusion, or Necessity*; Routledge: New York, NY, USA, 2012; pp. 180–202.

114. Anna Bergek, A.; Jacobsson, S.; Carlsson, B.; Lindmark, S.; Rickne, A. Analyzing the functional dynamics of technological innovation systems: A scheme of Analysis. *Res. Policy* **2008**, *37*, 407–429. [CrossRef]

 *energies*

 MDPI

Article

Participatory Experimentation with Energy Law: Digging in a 'Regulatory Sandbox' for Local Energy Initiatives in the Netherlands

Esther C. van der Waal [1,*], **Alexandra M. Das** [1] **and Tineke van der Schoor** [2]

[1] Science and Society Group, Faculty of Science and Engineering, University of Groningen, P.O. Box 70017, 9704 AA Groningen, The Netherlands; alexandra.das@xs4all.nl

[2] Research Centre for Built Environment NoorderRuimte, Hanze University of Applied Sciences, P.O. Box 3037, 9701 DA Groningen, The Netherlands; c.van.der.schoor@pl.hanze.nl

* Correspondence: e.c.van.der.waal@rug.nl; Tel.: +31-6-2921-0422

Received: 31 October 2019; Accepted: 8 January 2020; Published: 17 January 2020

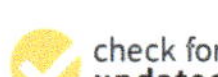

Abstract: To facilitate energy transition, regulators have devised 'regulatory sandboxes' to create a participatory experimentation environment for exploring revision of energy law in several countries. These sandboxes allow for a two-way regulatory dialogue between an experimenter and an approachable regulator to innovate regulation and enable new socio-technical arrangements. However, these experiments do not take place in a vacuum but need to be formulated and implemented in a multi-actor, polycentric decision-making system through collaboration with the regulator but also energy sector incumbents, such as the distribution system operator. Therefore, we are exploring new roles and power division changes in the energy sector as a result of such a regulatory sandbox. We researched the Dutch executive order 'experiments decentralized, sustainable electricity production' (EDSEP) that invites homeowners' associations and energy cooperatives to propose projects that are prohibited by extant regulation. Local experimenters can, for instance, organise peer-to-peer supply and determine their own tariffs for energy transport in order to localize, democratize, and decentralize energy provision. Theoretically, we rely on Ostrom's concept of polycentricity to study the dynamics between actors that are involved in and engaging with the participatory experiments. Empirically, we examine four approved EDSEP experiments through interviews and document analysis. Our conclusions focus on the potential and limitations of bottom-up, participatory innovation in a polycentric system. The most important lessons are that a more holistic approach to experimentation, inter-actor alignment, providing more incentives, and expert and financial support would benefit bottom-up participatory innovation.

Keywords: polycentricity; local energy initiatives; community energy; smart grid; legal innovation; socio-technical innovation; bottom-up

---

1. Introduction

Perhaps one of the most critical issues for the energy transition is matching sustainable energy supply and demand, and especially managing the local peak loads and the influx of prosumer energy since many renewables are intermittent resources. For now, the existing grid is used for balancing, but when renewable electricity production and use further increase, the grid capacity will not be sufficient and reinforcement will be very expensive. New options for grid management that have been explored are smart meters, smart grids, demand response, and storage technologies to reduce peak loads and manage congestion. These technological developments create opportunities for new roles in the energy system, such as aggregators [1–3].

New technological developments are also relevant from a prosumer perspective [4,5]. Until recently, project partners in smart grid projects perceived users primarily as a barrier [6], or as passive subscribers to grid services [7]. Planko et al. show that end-users are scarcely represented in the system-building networks that are active in the development of the Dutch smart grid sector [8]. Yet, times are changing with the increase of local energy initiatives [9], which increasingly broaden their activities that aim to further influence the direction and pace of the energy transition [10]. Potentially, local energy initiatives can extend their role from energy generation to performing active functions within the smart grid. They could 'actively offer services that electric utilities, transmission service operators or other prosumers have to bid for' [11] (p.4), such as offering storage capacity for balancing, or avoiding grid reinforcement through flattening the usage profile and increasing real-time use and local storage [12].

Local energy initiatives or other local actors would need to be enabled to organise a more integrated resource management at the local or regional level to extend and optimise such services. For instance, peer-to-peer supply and flexible tariffs could increase local use.

However, the extant law is sometimes a limiting factor for energy management innovation towards a renewable energy (RE)-based system that needs matching demand and supply both in terms of available energy and grid capacity [3]. For instance, for household consumers, law might need to enable pricing of grid services based on actual loads instead of connection capacity.

Several countries' regulators have devised 'regulatory sandboxes' to create a participatory experimentation environment for exploring revision of energy law to overcome such legal obstacles for energy transition. A main characteristic of these sandboxes is that they allow for a two-way regulatory dialogue between an experimenter and a regulator to innovate regulation and enable new socio-technical arrangements. For instance, in the Netherlands, the executive order 'experiments decentralized, sustainable electricity production' (EDSEP) allows for the implementation of innovative energy services at the local level [13,14]. Another example is the UK, where innovators can get a temporary derogation of some rules in order to run a trial if the proposed product or service is considered to be genuinely innovative and able to deliver consumer benefits [15]. Importantly, new actors, such as local energy initiatives, take centre stage in these sandbox experiments, and they are seen as a locus of agency, in contrast with 'business as usual' in smart grid experiments, as described above [6–8].

What is especially interesting about these experiments is that, while experimenters can take on new roles due to exemptions, they do not operate in a vacuum, but experiments need to be designed and implemented in a multi-actor, multi-centered decision-making system. Such a system was coined by V. Ostrom et al. as a polycentric system [16] and was further elaborated by E. Ostrom [17,18]. In the particular polycentric system in this study, the experimenters need to collaborate with the regulator, but also energy sector incumbents, such as the distribution system operator.

Little is known regarding the functioning and innovative potential of local energy initiatives as experimenters in polycentric actor-constellations [19], while they are earmarked as potential providers of new grid services in such a system by governments creating these experimentation environments [11]. Our central question, therefore, is: What can be learnt about local energy initiatives' bottom-up experimentation with smart grids in a polycentric energy system? By answering this question, we aim to provide policy relevant insights regarding the preconditions for and obstacles to using end-user collectives as innovators informing new energy regulation, which is more facilitative of the integration of renewables within the limits of the grid. Furthermore, we would like to introduce the polycentricity concept to the community energy literature and demonstrate its value to better understand the relationality and interdependencies in governing energy.

To research this, we focus on the aforementioned case of the Dutch EDSEP, which invites homeowners' associations and energy cooperatives to propose projects that are prohibited by extant energy regulation. Local experimenters can, for instance, organize peer-to-peer supply and determine their own tariffs for energy transport in order to localize, democratize, and decentralize sustainable energy provision. We further introduce our case in Section 2. Subsequently, we elaborate on our theoretical framework, in particular the concept polycentricity, in Section 3. In Section 4, we describe the used case study methodology and introduce the four EDSEP projects that are analyzed in-depth. Afterwards, we will describe the polycentric configuration under the EDSEP, and the functioning of the experimenters in this configuration in Section 5. Finally, Section 6 concludes the article with a discussion of our findings in a broader context and the value of the polycentricity literature for studying the potential and limitations of bottom-up, participatory innovation in a polycentric system.

2. Policy Background and Introduction EDSEP

In this section, we introduce the policy developments that led to the EDSEP, and the EDSEP itself.

2.1. Policy Background

The direct reason for the EDSEP is the 2013 Social and Economic Council (SER) energy agreement for sustainable growth between over 40 Dutch organizations and supported by the Dutch national government [13]. In the text of the energy agreement, it is stated that: "To realize the energy transition the legislation needs to be providing a consistent framework to provide investors with long-term security. In addition, the legislation needs to facilitate innovation. This means that the legislation needs to provide sufficient space to enable desired new developments, specifically when it comes to the production of RE. To this end, the Gas and Electricity Acts will be revised" [20]. For the revision, the Dutch government had established the legislative agenda STROOM (abbreviation of streamlining, optimizing and modernizing, in Dutch: STROomlijnen, Optimaliseren en Moderniseren), which had achieving clearer and simpler rules to reduce bureaucracy, streamlining with European legislation, and being facilitative of a competitive economy and transition towards as sustainable energy system as its goal. This legislative proposal offered a merger of the Electricity Act 1998 and the Gas Act [21].

However, instead of waiting for the new Gas and Electricity Act, the parties in favor of local, sustainable energy lobbied to make use of article 7a sub 1 of the Electricity Act 1998. This article states that, through executive order, in accordance with European Union legislation, the Electricity Act can be derogated from by the experiment [22]. The article intends to enable relatively small-scale, localised, RE experiments for which the strictly regulated separation between the commercial activities production and supply, and the publicly managed distribution side of the energy system can be relaxed, to a certain extent under specified conditions for a particular target group of homeowners associations (HOAs) and cooperatives.

Such derogation has to be laid down in an executive order (in Dutch: Algemene Maatregel van Bestuur) and it has taken the shape as the EDSEP, which entered into force on the 28th of February 2015. The objective of the EDSEP is stated in its explanatory memorandum and it is to observe whether it is necessary to strictly apply the rules of the current Electricity Act for decentrally produced renewable electricity.

2.2. Executive Order 'Decentral, Sustainable Electricity Production Experiments'

To informedly revise the Electricity Act, the Dutch government strives to obtain more knowledge regarding grid stabilization by prosumers and obstacles that are created by present regulations. For this reason, the Executive order 'Decentral, sustainable electricity production experiments'(in Dutch: Besluit experimenten decentrale duurzame elektriciteitsopwekking) was designed [23]. The goals of the executive order are stimulation of more renewable energy (RE) at the local level, more efficient use of the existing energy infrastructure, and more involvement of energy consumers with their own energy supply.

It provides energy cooperatives or HOAs the opportunity to get an exemption from the Electricity Act and carry out the functions of the grid operator. The cooperatives and HOAs can carry out two main types of experiments:

-   the project grids up to 500 users. In this case, the grid is owned by the project and has only one connection to the public grid;
-   the larger experiments up to 10,000 users and 5 MW generative capacity, usually in cooperation with the grid operator. The grid operator remains owner of the grid. These experiments are concerned with balancing the electricity grid through peak shaving, and dynamic electricity tariffs.

The size of the experiments is chosen, so that the projects remain manageable and the general security and safety of the electricity provision on the regional grid will be guaranteed. Safeguarding provision within the projects is the responsibility of the participants of the projects. Thus, the protection of the consumer is partly taken care of through the assumed control that the participant can exert in the cooperative or HOA. The members should hold each other accountable for the responsibilities of the local energy initiative regarding production, supply, and transport.

Initiatives that are willing to make use of the EDSEP need to apply at the Netherlands Enterprising Agency (in Dutch: Rijksdienst voor ondernemend Nederland, RVO) for the derogation of the Electricity Act. Yearly, 10 projects of both types could be admitted, but only a total of 18 projects have been approved (see Appendix A), and only few are actually being implemented. The admission started in 2015 and ended in 2018. The experiments will be evaluated in early 2020.

3. EDSEP Experimenters As Decision-Making Unit in a Polycentric System

The EDSEP is designed to identify the obstacles that the extant Electricity Act presents to the development of local collective solutions to the production of more RE and its more efficient use. When experiments receive derogation under the EDSEP, this means that they become part of a system with decision-making units at several levels, with whom they have to cooperate, or by whom they are supervised or even opposed. These include, amongst others, grid operators, energy companies, the Netherlands Authority for Consumers and Markets (ACM), the Ministry of Economic Affairs and Climate and its executive organization RVO.

A polycentric approach is suitable for analysing the functioning of these experiments as part of such a larger system in which decision-making power is distributed [24], and it has been used for previous work on smart grids [25,26]. Polycentricity means that there are "many centres of decision-making which are formally independent of each other" [16], but which in practice often need to collaborate with others to execute what they are formally allowed to do. For instance, in the case of the EDSEP experiments, experimenters pursuing a project grid only formally need to discuss their plans regarding grid design and distribution with the regional distribution system operator (DSO), as they are allowed to take the role of DSO in their mini-grid, but in practice the approval of the regional DSO is important for obtaining the exemption.

Polycentric systems are characterised in the literature as being multi-level, multi-sectoral, multi-functional, and multi-type, as displayed in Table 1 [26,27]. We will use these concepts to describe the polycentric setting in which the experiments operate in Section 4.1, as the authority of a decision-making centre in energy regulations is defined by these characteristics. For instance, a locally functioning energy initiative is a private sector initiative and has therefore previously been excluded from the function grid management, as it was deemed a public good.

We rely on Ostrom et al. [16] for the analysis of the polycentric system, who propose four criteria to evaluate the well-functioning of a polycentric system: control, political representation, efficiency, and local autonomy. We briefly define these criteria in Table 2.

Table 1. Aspects of polycentric constellations based on [26,27].

Aspects	Definition
Multi-level	Geographical level of scale (e.g., local, regional, provincial, national, and global)
Multi-sectoral	Actors are active in different sectors (e.g., public, semi-public, voluntary, community-based, private, and hybrid kinds)
Multi-functional	Different functions are performed by different actors (i.e., specialized units for different functions, such as production, provision, sale, financing, etc.)
Multi-type	Several types of jurisdictions are present at the same time (e.g., territorial jurisdictions: nested, multi-purpose jurisdictions; and organizations with functional jurisdiction: specialized, cross-territorial organizations)

Table 2. Criteria for evaluating the functioning of polycentric decision-making systems [16].

Criteria	Definition
Control	Formal powers of the decision-making unit within the applicable legal frameworks;
Efficiency	Whether the collaboration between the multiple decision-making centres has advantages for getting to the desired outcome;
Local autonomy	The power of local stakeholders to be a decision-making unit;
Political representation	Inclusion of the political interests of the decision-making unit within the decision-making arrangements.

4. Methods

4.1. Case Study

We study the EDSEP as a multiple case study. Case study research allows for in depth analysis of a contemporary phenomenon in a real-life context and the combination of various complementary research methods [28]. Our sample of cases includes four projects that were approved under the EDSEP: Schoonschip, Endona, Collegepark Zwijsen, and Aardehuizen. All of these started relatively early (in 2015 or 2016) and their projects have reached an advanced stage. Two of these are so-called large experiments and two are project nets, so both types of experiments that are possible within the EDSEP are equally well represented.

Many (web-)documents that describe the four cases are available, and for each case the project initiators or other participants heavily involved in the development have been interviewed in a semi-structured face-to-face interview. Although these representatives provided us with key information for this research, we acknowledge that other participants to the experiments could have different perspectives. Furthermore, we conducted interviews with other relevant actors in the polycentric system related to the EDSEP, mostly telephonic. Appendix B presents an overview of the interviewees.

This information has been analysed through reflexive thematic analysis, starting with the criteria indicating the functioning of polycentric systems as analytical framework. The coding has been based on the six-step methodology of Braun et al. [29], which consists of the steps: familiarisation, generating codes, constructing themes, revising themes, defining themes, and writing the report. We used the qualitative data analysis software Atlas.ti for our analysis.

4.2. Cases

Via Tables 3–6, we will shortly introduce all four of our case studies based on their project type, delineation of the experiment, its organization and governance, its energy system, and the use of the EDSEP.

Table 3. Case study description Endona.

Aspects	Description
Project type	Large experiment
Delineation	The first pilot is in the village of Heeten, but eventually Endona wants to supply the medium voltage grid (part Raalte) with locally produced RE as well as increase the region's real-time electricity use.
Organisation and governance	Endona is an energy cooperative, with the board members registered as its members. This structure has been chosen to keep the decision-making with its day-to-day management. Endona has a large portfolio of projects and is part of several collaborations with grid operators, technology developers and knowledge institutions.
Energy system	With some of its partners, Endona installed sea salt batteries. It also implemented household level energy management systems (EMSs) [1] in a neighbourhood with 47 households [30], and an overarching EMS that uses the inputs from these EMS for neighbourhood level optimisation. Furthermore, Endona developed a solar park with 7200 photovoltaic (PV) panels on 3.5 ha of former agricultural land.
Use of EDSEP	The derogation has not yet been effectuated. The cooperative only acts as producer and does balancing experiments that are allowed within the framework of the current Electricity Act. At present, the electricity sale is through a cooperative energy company. Endona has not found a suitable business model for being energy supplier, and is investigating the financial risks. In the long run, it wants to take on this role so both the costs and benefits of the energy system are local, and they can possibly offer a lower price to their users because of the integrated management.

[1] An EMS is a system of computer-aided tools used by to monitor, control, and optimize the performance of the energy system.

Table 4. Case study description Aardehuizen.

Aspects	Description
Project type	Large experiment
Delineation	The location is at the outskirts of the village of Olst, and is situated in a rural landscape. Incidentally, it is near Heeten, where our first case, Endona, is located. Aardehuizen is in contact with Endona. 23 houses have been built, of which 3 rental social houses, and a community house.
Organisation and governance	The project is operated by a HOA and part of a worldwide movement, Earth Ships, which wants to build houses with little environmental impact built from recycled and regionally sourced material. The project's decision -making system is a sociocracy, which means everyone is involved and informed, although decisions are not made by consensus. The occupants of the rental houses are also a member of the HOA.
Energy system	Electricity generation in Aardehuizen is realised by PV-panels on individual houses, while at a later stage collective PV may be placed at a parking lot. The PV panels are privately owned, but the battery will be collectively owned. A collective battery is under investigation, in cooperation with a different higher education institution. No gas connection is present, and because the energy performance coefficient of the buildings is almost zero, the little auxiliary heating that is required is done with heat pumps and wood stoves. Next to the direct current (DC) grid, in the future, an inverter will be placed, to make storage possible. Some of the houses have a private EMS. An investigation is ongoing to place EMSs in all houses, which can be connected to a higher level collectively owned EMS. Not all households are connected yet, because not all participants are certain about their privacy. Smart appliances and smart connectors are under investigation.
Use of EDSEP	At present, the HOA acts as producer. Once the collective smart grid is in place, peer-to-peer supply based on dynamic tariffs is planned. At this moment, every household has its own energy supplier. Later, an external cooperative energy company will buy and sell electricity, and handle the administration of the project. Ownership of the grid was not feasible financially as the grid was already in place and the grid was too expensive compared to the benefits of having Aardehuizen managing it.

Table 5. Case study description Collegepark Zwijsen.

Aspects	Description
Project type	Project grid
Delineation	The project consists of a HOA for 115 apartments built in a monumental, former school building in the village Veghel, in the south of the Netherlands.
Organisation & governance	The derogation for a project grid has been arranged by the project developer before the houses were sold. The HOA has been set up by the developer so that the residents can use it as a vehicle to decide on matters related to their energy system.
Energy system	Collegepark Zwijsen has solar PV and solar collectors. These installations are jointly owned (in Dutch: mandeligheid). All households are connected to one shared large-scale use connection to the national grid. Grid balancing measures will be achieved through individual EMSs for each household. No smart appliances are involved in the project for reasons of privacy. The EMSs, in combination with dynamic tariffs are expected to incentivize the apartment owners to better align demand to supply. Storage will be as heat, not as electricity.
Use of EDSEP	The HOA acts as supplier, producer and distributor, but is not a balance responsible party (BRP). Project grid management, management of the energy technologies and the administration of energy use for billing are done by an external organization affiliated with the project developer. The apartment owners will pay a fee for these services commissioned by the HOA. The initial tariff structure is in place and approved by regulator ACM. The occupants are guaranteed to a 3-year zero energy charge, provided their consumption remains within a certain bandwidth. Later on, grid balancing is seen as a way to negotiate better tariffs, and then the HOA will be involved in deciding upon tariffs and new investments.

Table 6. Case study description Schoonschip.

Aspects	Description
Project type	Project grid
Delineation	Schoonschip is an HOA of the owners of 46 houseboats and one communal boat in the Amsterdam quarter Buiksloterham, which is a city quarter that develops all kinds of sustainable building projects.
Organisation & governance	The project was started by a group of friends, who were later joined by other friends and acquaintances. There are other goals than RE, e.g., wastewater treatment, and the use of recycled building materials. The board of the HOA is responsible for daily decisions. Working groups have been established, e.g., in supervising the building process. These working groups may give presentations about their findings, to keep all members involved. For some decisions it is necessary for all members of the HOA to be present.
Energy system	The boats are all-electric, part of a project grid, and connected to the national grid via one connection. The HOA generates electricity through individually owned solar panels. Batteries are placed on each boat, but collectively owned. Shared electric vehicles are part of future plans. The administration and some of the maintenance are done collectively. A smart grid is in place, and every household has an EMS. The smart grid is part of a project of a consortium with external expertise, which researches the optimization of smart grid technologies and algorithms [31]. Dynamic tariffs are not foreseen as part of demand management. Efficiency should occur through the smart grid: using and storing electricity when production is high. Eventually, the energy management should result in providing electricity to the main grid at the highest price.
Use of EDSEP	The HOA acts as supplier, producer and distributor. The administration of electricity use and supply is outsourced to a commercial electricity company, which acts as BRP and provides electricity when a shortage occurs, and buys surplus electricity.

5. Results

In this section, we first discuss the polycentric constellation of actors that EDSEP experimenters need to function in, and thereafter we analyse the well-functioning of the experiments in this context.

5.1. The Polycentric Constellation of Actors Under the EDSEP

In this section, we will introduce the polycentric energy system that EDSEP experimenters are part of and function within (see Figure 1 for an overview). The selection of the actors that we discuss here is limited to actors that are directly involved in EDSEP experiments, and therefore does not include actors, such as the high voltage system operator.

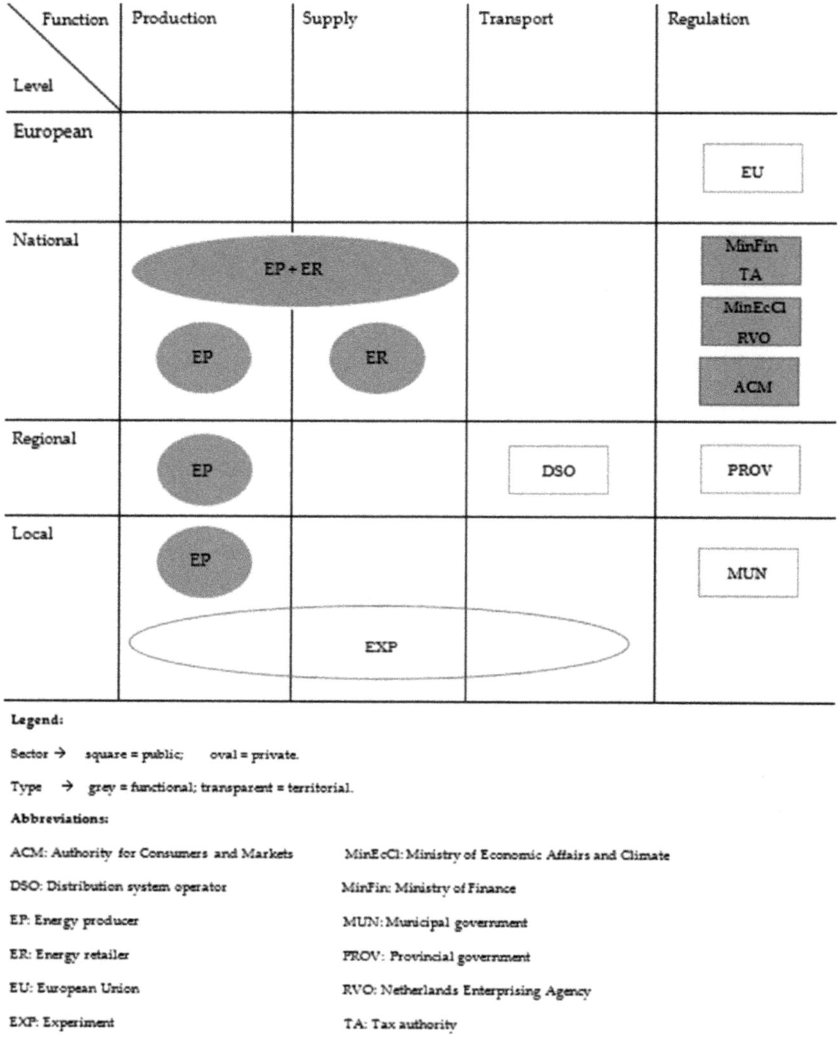

Figure 1. Overview of polycentric energy system under the EDSEP.

- Energy supplier

Dutch energy companies are traditionally large, nationally operating, private companies. In recent years, some cooperative energy companies have been founded that are closely related to local energy initiatives and seek to return at least part of the benefits to the region.

The functional energy companies receive surplus electricity from the projects and deliver electricity when the projects do not meet demand with their own production. They take care of the administration and billing for the electricity produced and consumed.

Energy suppliers that supply to small scale users, such as households, need a supply permit. This permit is given by ACM when the supplier can show amongst others that supply will be reliable, tariffs are reasonable, and the company is financially, organisationally and technically compliant with the conditions of the Electricity Act. Under these conditions, it is not feasible for local energy initiatives to act as the supplier. However, a few cooperative energy suppliers exist that supply energy that is produced by a growing number of local energy initiatives. These suppliers are cooperatives of cooperatives, of which local energy initiatives producing energy are member.

Furthermore, an energy supplier needs to have balance responsibility (in Dutch: programmaverantwoordelijkheid) or have a contract with a balance responsible party (BRP). The BRPs share the responsibility of balancing and they have to inform grid operators about their planned injections, offtakes, and transports. At the moment, experimenters are not able to take up balance responsibility and they rely on the larger national energy companies to provide this function for them.

- DSO

The DSOs in the Netherlands are territorially organized, monopolist utility companies that operate regionally. They are specialised in the transport of electricity and the maintenance and extension of the grid.

As utility companies, they are subject to forms of public control and regulation. The Authority for Consumers and Markets yearly determines the tariffs that the DSOs can charge to their clients to connect them, be connected and transport energy, and how much profit they can make on their investments.

In the large projects, the DSO remains the owner and manager of the grid, but, in the project grids, the grid is part of the project, and is built and maintained by the experimenters. The DSOs are asked by RVO to give a reaction on the project grids, and they try to be involved in the design of these grids. They want to be formally involved in the process towards the derogation.

The DSOs have considerable experience and they are well equipped to build and maintain grids. However, as the regulatory focus in the Netherlands is primarily on the public values of affordability and availability of supply, the safeguarding of sustainability is prioritized at a much lower level [4,32]. While DSOs can benefit from the sustainability experiments, they are concerned about the knowledge that is present among the experimenters to perform DSO tasks. After the 10-year-derogation, the project grid has to be potentially handed over to the DSOs, and they wonder whether the quality of these grids will be sufficient, and who must pay the costs if this is not the case.

- ACM

The ACM is a nationally and functionally operating, independent public organization. It is a business regulation agency, which is charged with competition oversight, sector-specific regulation for several sectors, and enforcement of consumer protection laws. In the context of the EDSEP, the ACM checks the calculation method for the energy and transport tariffs if the energy experiment wants to take over the task of the supplier and the DSO.

- Tax authority

The tax authority is a nationally and functionally operating public organization. It is tasked with the tax collection and customs service of the Dutch government and it is part of the Ministry of Finance. It levies and collects the energy tax on electricity (in Dutch: Energiebelasting elektriciteit). This is a type of environmental tax that disincentivizes use. The energy tax per kWh for 0–10,000 kWh electricity was in 2019 € 0.09863 [33]. This is a large share of the average electricity price in the first quarter of 2019 of € 0.203 per kWh for households using 2.5–5 MWh [34]. In the experiments, it is dependent on the circumstances within each project whether energy tax needs to be paid, and no special conditions exist.

Another tax that needs to be paid is for the storage of renewable energy (in Dutch: Opslag duurzame energie), which is € 0.0189 per kWh until 10,000 kWh [33]. In addition, a payment of 21% VAT is charged over supply costs, transport costs, and levies.

- RVO

RVO is an executive organization of the ministry of Economic Affairs and Climate, which operates nationally in the public domain with a functional agenda targeted at executing policies that support Dutch enterprising. RVO provides the derogation to the projects and supervises its implementation.

Once or twice a year it organizes meet-ups for the experiments, together with the national platform organisation for community energy, Hieropgewekt. Here, projects can create a community of practice and share learning experiences.

The types of experiments under the EDSEP are left rather open to see what kind legal changes are required to facilitate energy transition. This meant that some of the problems that the projects encountered were not foreseen, e.g., whether energy tax needed to be paid was first also not clear to RVO.

- Experimenting HOA or cooperative

The experimenters are locally operating, territorial decision-making units. The HOA's and cooperatives themselves are voluntary bodies, but a hybrid sometimes develops where a private party is the main developer and is either founding the HOA or cooperative, or paid by it to take on an important role in the design of the experiment. The functions that an experiment can fulfil under the EDSEP in the energy system can be any type of activity in the domain of energy production, supply, or grid management for projects grids, whereas large experiments are more constrained (see Section 2.2).

- Municipality, provincial government, and European Union

The governmental bodies are, similarly to the previously described departments of the national government, public, territorial bodies, which operate at their respective scales. In the context of the EDSEP, these governments have played various roles in the polycentric energy system, such as subsidizer and provider of permits. This will be discussed in more detail in Section 5.2.4 regarding political representation.

5.2. The Functioning of the Experiments in Their Polycentric Environment

We will now discuss the functioning of the EDSEP experiments within the afore-described Dutch, polycentric energy system, based on the criteria from our conceptual framework: control, efficiency, local autonomy, and political representation.

5.2.1. Control

Under the EDSEP, experimenters can carry out several tasks that were not permitted under the current Dutch model. Energy transport and grid management are considered to be a public utility, and production and supply are commercial activities. Without the EDSEP, the experiments can only be active in production and supply. However, supply requires a specific permit and it is not feasible for most local energy cooperatives or HOA's due to the required scale of customer base and financial risk. Before 2014, most of the energy cooperatives that acted as supplier sold electricity through energy companies as reseller, while using a so-called white label construction [35]. Others outsource tasks, such as administration and balance responsibility, to a back office of one of these companies while still using their own brand and image [36].

With a derogation, experiments can take over the tasks of both the energy supplier and the DSO, to the extent that they deem to be most beneficial for their projects. Note that derogations only apply to specific articles of the Electricity Act [23]. Other laws and regulations, such as the General Data Protection Regulation, continue to be applicable. In short, the derogation presents the following opportunities to derogate from the Electricity Act

- derogation from the prohibition to carry out DSO tasks;
- derogation from the obligation to have a supply permit;
- freedom to determine grid tariffs, tariff structures, and requirements as set by ACM. ACM only checks the method by which the tariffs are determined, not the tariff itself;
- derogation from certain specific rules that apply to data processing (which are mainly about the requirement to participate in sector-wide discussions to align data related procedures to the benefit of the consumer);

- derogation from certain specific rules regarding transparency and liquidity of the energy market (which are mainly about the right of the government to create additional requirements regarding supply conditions and information provision in case of an illiquid market); and,
- derogation from rules regarding metering device requirements.

There are regulations that limit the control of the experiment. One of these that poses a particular threat to the experimenters is the European Union (EU) legal obligation to provide third-party access to a network whether it is a public or a private network (see article 32 Third electricity directive, 2003/54/EG. Pb EU L 211/55.). This means that participants need to be able to choose another energy supplier. From the perspective of the experimenters, this third-party access is a threat, because it can undermine the business model, as only as much energy is allowed to be generated as the projected use of the participants [23]. Moreover, collective energy management and storage are at risk when the user group decreases. The installations are dimensioned to supply for the initially projected users, and part of the production capacity can potentially not be used anymore if the user number decreases. A reason for this is that the government wants to keep the experiments as self-contained as possible to minimize the risk of blackouts or safety issues in surrounding areas.

Secondly, the prohibition of a flexible transport tariff limits the control of the experiments. Currently, it is only allowed for the DSO to charge a fixed daily transport tariff that is proportional to the capacity of the grid connection [37]. This limits the attractiveness of balancing, as the DSO cannot vary the costs based on the actual used capacity.

Finally, non-energy legislation can also limit the control of experiments over their project. For instance, project grids are only attractive when there is no existing grid and, therefore, go along with the development of houses or apartments. The experimenters then need to obtain a building permit and might need to obtain permission from an aesthetics committee of the built environment. For instance, for Collegepark Zwijsen it was hard to get the design with solar collectors on the façade approved, as it was first deemed to negatively affect spatial quality.

5.2.2. Efficiency

Having an experiment under the EDSEP can lead to a number of cost savings for the participants. We list the most important below [38]:

- Grid connection and DSO transport costs for project grids: A one-time saving on the grid connection costs can be realized. Experimenters that newly construct a grid can save costs, because one high-volume connection to the regional grid is cheaper than the sum total of connections for individual dwellings to the regional grid. This is a financial incentive to balance the energy on project grids, because, the smaller the connection with the regional grid required, the lower the connection costs. Furthermore, the periodical transport costs that need to be paid to the DSO are also lower when the capacity of the connection is lower. This can result in a rather significant saving as the DSO costs are about 1/3 of the total electricity bill.

  To give an example: The total of the DSO tariffs for a household with an average 3×25 A connection at the DSO Stedin € 230.36 (other DSOs do not differ much in their tariffs) [39]. Schoonschip annually pays € 6759.74 according to their business model, which comes down to an average of € 225.32 per dwelling. As this is an all-electric neighborhood, where the electricity consumption is higher, the balancing brings these dwellings back to rather average DSO costs).

  However, if dwellings do not have their own connection to the grid, they miss out on the annual levy rebate for a part of the energy tax.

- DSO transport costs for large grids: the periodical transport costs on a large grid can be reduced by creating a virtual connection through a shared code for a group of participants that cooperate to create balance. The lower the required peak capacity, the lower the transport costs. Additional costs can be saved by helping the DSO to realize a flat usage profile (using the same capacity of the grid throughout the day), because this has value to the DSO. However, sufficiently adjustable capacity is needed for this.

- Energy tariff of large net: If the experiment can realize the aforementioned flat usage profile, it can potentially negotiate a lower tariff for the energy that it does not generate with its own capacity and needs to buy from an external supplier.
- Fixed supplier costs: Most energy suppliers charge a fixed supply tariff. If the experiment (project grid or large experiment) has one connection, these costs are lower than when each individual user would need to conclude a contract with the supplier. However, costs need to be made to measure the usage within the project and bill the participants.

When the EDSEP started, not all of the decision-making units were familiar with the regulation, because RVO did not prepare them for working with the EDSEP. This led to various instances when the experimenters needed to explain the regulations to the DSOs, ACM, and the tax authority. The compartmentalization of DSOs had a negative impact on the progress of projects, because the functioning of decision-making units within DSOs was not always well aligned. Accordingly, after informing and convincing the civil-servants in one unit, experimenters met with resistance of the executive staff, and had to re-explain their plans. RVO has asked organizations that have dealings with EDSEP-experiments to assign a case-manager with whom the projects can communicate at an early stage to improve this situation.

The scale is another efficiency related factor. It is questionable whether the experiments are an interesting party for the DSO to do business with for grid balancing. Grid operators could for example contract experimenters to make use of their storage capacity, or compensate them for the investment costs of grid reinforcement that are avoided by the experiment. However, some grid operators prefer to deal with larger parties and find projects with a size of up to 10,000 households too small and not very interesting to buy flexibility from. The creation of a legal requirement to buy balancing services through tendering could be a solution here, giving priority to small-scale providers. Or oblige DSOs to buy local balancing services for a price that reflects their value. Historically, such a similar obligation has been embedded in the law for DSOs regarding grid connection to make sure energy production and consumption would be accessible at any location in the country.

Furthermore, energy tax needing to be paid twice for stored energy is a major inefficiency [40] (once when the electricity is uploaded in a battery and once when it is taken out again). As the energy tax is a high proportion of the energy price (see footnote 5), this limits experimentation with storage solutions. Unfortunately, alignment between the Ministry of Economic Affairs and Climate and the Ministry of Finance to avoid this double taxation has been lacking. In the near future, this problem will no longer occur, because the EU has adopted the 'Clean Energy for all Europeans' package, which states that owners of storage facilities should not be subject to any double taxation [41].

Additionally, the interpretation of current energy tax rules makes the experiments less efficient. Energy tax can be saved if the ownership structures make sure that there is no supply to third parties, and the participants make use of their own production and distribution capacity. However, a third party is a party with a different real estate valuation tax object, according to the taxation criteria (REV-object, in Dutch: woz-object). Each house or apartment is a REV tax object, and, therefore, energy tax on electricity needs to be paid when a participant uses energy from the production installation of another participant. A possible solution would be for the municipal government to register the houses as one REV-object (this has no consequences for the REV-tax and the procedure is the same as for other REV-tax objects with multiple owners).

Moreover, whilst DSOs embrace the goal of the EDSEP to keep production and consumption local, they fear that private project grids threaten the socialization model that underlies Dutch grid management. The DSOs have the perception that some experiments are motivated by the evasion of the energy tax, as it appeared at first to some participants that this tax would not apply for the experiments.

Last, but certainly not least, the experimenters need to fully comprehend a whole gamut of complicated energy related regulation to be successful. Misinterpretation can lead to a worsening of the business model and can, ultimately, lead to an inviable project. Experimenters progressed slowly despite some support from RVO and Hieropgewekt due to this complexity. Slow progress even led to the strange situation that the government has decided to draft a follow-up EDSEP without waiting for the formal evaluation of the present experiments.

5.2.3. Local Autonomy

Formally, for experimenters, the two structures to self-organize and function as a decision-making unit in the polycentric energy system are HOA and cooperative.

While HOA and cooperative seem to be structures that are explicitly designed for high commitment of the involved households, these do not, per se, imply a high level of participation of all participants. For example, Endona is a cooperative, but only its board members are members to keep decision-making with the daily management. The organizational structure is primarily set up to run the sub-projects efficiently, it is not geared to involve many local participants. A second example is Collegepark Zwijsen, which was designed without input from its future inhabitants. The derogation was applied for by its project developer, but assigned to the HOA, which was not yet in existence at that time. The HOA only started its regular meetings after the residents started living in the apartments. From then on, the autonomy of the HOA will be larger, as it will decide on topics, such as maintenance and tariffs.

The other two HOA's, Aardehuizen and Schoonschip, functioned from the beginning of the projects as decision-making units run by the future inhabitants. Both outsourced tasks to professional parties, but took the decisions about project design themselves. The working groups prepared proposals about e.g., sustainability, but these decisions were then taken collectively.

All of the projects, except Zwijsen, which is entirely professionally developed, mention that working as a HOA or a cooperative with participation based on the input of volunteers, who are mostly not professionals in the field of energy, has made it harder to function as a local decision-making unit, because they need to invent the wheel by themselves and it was not always easy to acquire all of the required information for informed choices. Additionally, in the communication with other decision-making units such as DSOs, the tax service and ACM, the status as cooperative or HOA was by times a disadvantage and they needed to first convince the other parties of their know-how and professionality.

5.2.4. Political Representation

The municipal government was the political body that was most involved in the projects. Sometimes the relationship with the local government depended on the political tide, but most projects had a productive working relationship with the municipality and felt supported. Two projects got a municipal subsidy: Endona for a feasibility study for its solar park, Schoonschip a contribution per household for the high energy efficiency of the houses.

Additionally, motions at the local council functioned as a mechanism to realise political representation of the interests of projects in local politics. Aardehuizen and Collegepark Zwijsen both benefited from political motions. Aardehuizen benefited from a motion about sustainable building prior to the project, which helped to increase the support for the project. The project developer of Zwijsen successfully lobbied for a motion that would reduce the fee for the building permit, which is proportional to the building costs and was high due to the costs of the energy sustainability measures and techniques. The project developer was also successful in lobbying to overrule the negative advice of the aesthetics committee for the built environment, so Zwijsen could have its solar collectors.

Furthermore, Endona, Aardehuizen, and Schoonschip received a provincial subsidy, e.g., to hire an architect or for feasibility studies. Aardehuizen also received a European subsidy for the community building, although this had to be partly paid back, as the building could not be realised in time.

At the national level, no specific representation of the experiments exists. RVO reports on their progress to the ministry of Economic Affairs and Climate, but only from their position as an executive organisation, not as lobbyists. For this reason, it is unlikely that the experiences of the experimenters will be influential in the revision of energy law, especially because the experimenters were not asked for their input during the consultation for the draft of the follow-up executive order.

6. Conclusions and Discussion

We studied the EDSEP as an example of a regulatory sandbox, a participatory experimentation environment for exploring the revision of the Electricity Act. When projects receive a derogation under the EDSEP, they can perform new tasks and combine roles that are otherwise legally separated and thereby deliberately unbundled to protect the consumer and safeguard security of supply, affordability, and safety. On the one hand, the project grids can act at the same time as the supplier, producer, and distributor of energy, managing an own mini grid. On the other hand, the large experiments cooperate with the DSO, while the grid remains owned by the grid operator, and are concerned with flattening the usage profile and balancing supply and demand.

By taking on these tasks, experimenters become part of a polycentric energy system with decision-making units at several levels. Interested in their functioning, we asked ourselves the question: What can be learnt about local energy initiatives' bottom-up experimentation with smart grids in a polycentric energy system? In this section, we conclude on our findings and discuss our conceptual framework, and then put these in a broader perspective of legal innovation for energy transition.

6.1. Lessons Learnt from Participative Experimentation under the EDSEP

For potential experiments, the EDSEP has shown to be a complicated procedure with limited attractiveness for local energy initiatives, which resulted in only 18 experiments of the potential 80 in a four-year period. We want to make four main points, related to the four criteria for the well-functioning of polycentric decision-making structures.

- Efficiency: Combining exemptions with a pro-active nurturing of experimentation

The EDSEP's exemptions should make the integration of RE and grid balancing more attractive, which adds to the overall efficiency of the energy system. The EDSEP enables taking on new roles, but taking on these roles is hardly attractive or facilitated in the polycentric constellation. First, our case studies show that the EDSEP provides only a modest improvement for the business case of smart grids at the project grid level, and that for the large experiments we studied a good business case has not yet been found due to the limited financial attractiveness and the large organizational capacity required for taking on the balancing and supply roles while they come with considerable financial risks.

Second, for developing the experiments, there is no financial support available and, therefore, the experimenters have to rely solely on their own political efficacy and networking capacities to attract subsidies, or partners with knowledge or capital to invest. RVO has an important task to distribute subsidies for energy innovations, especially for innovations in the early stages. Hence, a special fund or subsidy for experiments would fit in seamlessly in the overall aims of the RVO. In addition to this, we suggest that more support should be created to overcome knowledge differences in small-scale volunteer organizations.

Third, alignment between decision-making units, such as the DSOs, ACM, and the experiments, was initially lacking due to poor communication with the other actors about the regulation by RVO, which made it harder to establish a productive collaboration with these decision-making units. This reduced the efficiency of experimentation, as enrolling such established actors in their network is very beneficial for bottom-up technological innovation projects [42].

Hence, our findings suggest that the smart-grid niche that the EDSEP provides lacks sufficient nurturing to function efficiently [43]. Nurturing can take place through assisting learning processes, articulating expectations, and helping networking processes [43]. All of these could be strengthened to increase the efficiency of the polycentric constellation that is created under the EDSEP.

- Control: the benefits and limitations of the new roles

The EDSEP fulfills a need to explore regulation that better facilitates the integration of intermittent resources. By making use of the EDSEP, the experimenters can take on new roles as grid managers (for project grids even the role of grid owners is possible) and as energy suppliers. For project grids, we saw that this incentivizes grid balancing through providing the opportunity to bring down the DSO costs by minimizing the exchange of energy (import or export) between the project grid and the regional grid. Additionally, the exemption from getting a supply permit is used for the project grids, but, in both cases, the administration has been outsourced to either an energy company or a company related to the project developer. These tasks require more time and expertise than the local initiatives could give and, therefore, they chose to outsource the tasks to commercial organizations.

Taking on the roles of supplier and balancing agent is more difficult when it comes to extra control for the large grids. First, when it comes to supply, the customer base is bigger than for the small projects, so the risks of, for instance, late payments are also higher, but the company is still not big enough (or not sure whether it is in the case of Endona) to carry these risks. Second, when it comes to taking the role of grid manager for a larger area, this is complicated due to the fact that for flattening the usage profile, adjustable capacity is required to create a good business case, which is expensive for experimenters, as it has to come largely from storage because they cannot use industrial partners' capacity, as their participants have to be mainly households. Furthermore, as only the local experimenters could experiment with tariff structures and the regional DSOs not, business opportunities regarding balancing are limited. Lastly, the supplier role of the BRP is out of reach for the experimenters, as the software for this is too expensive and the risks too high for the small-scale experiments.

Thus, having the opportunity to take more control over the local energy system from a legal perspective does not always mean that all of this control can be taken over and all new roles can be enacted. Some of the tasks are not (yet) feasible, mostly due to financial, organizational, practical, or sometimes legal constraints. However, despite the fact that experimenters cannot take full control, the EDSEP provides end-user collectives with an incentive to balance their grid, e.g., enabling p-2-p supply without intervention of a DSO.

- Political representation: approach sustainability more holistically in policymaking

Experimentation would have been more effective if the Dutch tax authority was enabled by the ministry of Finance to co-experiment and to, for instance, exempt the experiments from double taxation on storage. However, communication regarding the EDSEP between the ministries of Economic Affairs and Climate and Finance was lacking. Some projects have tried to come up with project designs to pay less energy tax. However, no exceptions or reduced tariffs were granted to these relatively small energy cooperatives, in contrast to the tax rulings for large international companies. Hence, similarly to the work of Kooij et al. on niche–regime interactions between the tax authority and collective PV producers, our case also 'illustrates the political and power-laden nature of sustainability transitions, going beyond the focus on organizational and technological challenges' [44] (p.10). Ultimately, the EDSEP-sandbox shows that an experiment is not always fully a two-way regulatory dialogue between an experimenter and a regulator.

Furthermore, the lack of alignment between ministries shows that the development of policies that affect sustainability evolve in parallel worlds, and a more holistic approach is needed [1]. Stepping away from silo thinking and strengthening inter-ministerial alignment would be helpful in designing effective energy transition policies. Stronger political representation of a lobby organizations or intermediaries [45,46] at the national level would also be useful in this case. For instance, EnergieSamen, a Dutch lobby organization for local energy initiatives, could take on such a role.

- Local autonomy: a legislative balance between self-responsibility and the protection of consumers

The experiments show that, while the HOA and cooperative seem to be structures that are explicitly designed for high commitment of the involved households, these do not, per se, imply a high level of participation by all participants. In the context of smart electricity, energy legislation needs to strike the balance between opportunities for self-responsibility and the protection of consumers [1]. Options for users to shape their own energy system are desirable in the context of energy democracy [26], but consumer protection against high prices could be threatened, e.g., when making tariffs flexible. Therefore, further experimentation with legal innovation should not only explore how legalislation can be facilitative of technological innovation, but also of social innovations to create an energy system that represents the interests of its users and is acceptable to them. Involving local energy initiatives or users cannot function as the sole mechanism of user involvement, because our cases show that such a characteristic does not always guarantee high participation. Furthermore, adequate insight of end users in the experiment necessary to protect their interests might be lacking.

6.2. Theoretical Reflection on Polycentricity

The advantage of the concept of polycentricity is that an actor constellation can be described by four different actor-characteristics (level, type, sector, and function), which provide helpful tools for understanding the context of experimentation. We find that this concept provides more guidance for our study in defining actor roles and their position in the energy system than e.g., the multi-level perspective (MLP), which predominantly focuses on levels and rather general dimensions, such as science, market preferences, technology, socio-cultural, and policy [47]. With the concept of polycentricity, it is easy to see what a nested system of decision-making units looks like and in which ways it is layered, whereas MLP puts more focus on which sectors (market, science, policy, etc.) are represented in a system.

Furthermore, the concepts for evaluating the role of actors in polycentric systems (local autonomy, control, efficiency, and political representation) help to understand what is necessary for a decision-making unit in such a system to function well. They were especially helpful when studying legal innovations due to the inclusion of the concepts of control and political representation. The same goes for studying participative bottom-up innovation due to the inclusion of local autonomy. Lastly, the concept efficiency helps to understand whether the decision-making unit can provide added value to the system, which is a useful indicator in assessing whether sustainability experiments contribute to an efficient progress towards a more sustainable energy system.

However, it needs to be realised that, while using these concepts, the success of the experimenters in the polycentric context does not equal the value of the experiment for legal innovation. When evaluating the experiments, the question should also be whether the experiment has resulted in new insights for guiding energy transition, in this case study for revising energy law, and not only whether the experimentation constellation itself is efficient in providing added value. Learning potential, instead of replication potential, should be central in evaluating experimentation for legal innovation.

Furthermore, the analytical framework is focused on the functioning of the polycentric system, but does not give theoretical guidance on what actors can do to nurture experimentation, or how they can better work together and create alignment in the system. Strategic niche management and actor-network theory may be helpful frameworks to further explore these aspects of innovation management.

6.3. Final Remarks

For the Dutch legislators, learning from the EDSEP experiences is important, because the EDSEP is only the start of experimentation informing revisions of energy law. A follow-up of the EDSEP has already been drafted, being based on the 2018 Law Progress Energy Transition. This executive order expands the size of experiments, experimenting actors, and also enables experiments under the Gas Act. The new regulation has been presented to the parliament in May 2019 and new experiments can apply once the new executive order has received positive advice of the Council of State, which is expected early 2020.

We would like to briefly summarize the conclusions of this study, so they can be taken into account for the evaluation of the EDSEP as well as for future experimentation. Experimentation under the EDSEP shows us that inter-actor alignment was initially lacking and pro-active nurturing would have smoothened the implementation. Furthermore, EDSEP experimenters faced significant constraints, had very limited political representation, and varying representation of the users within the experiment.

As a starting point to improve both the well-functioning of the experiments and the quality of the learning process, an intermediary could be more of a bridge between national and regional actors and the locally operating experimenters, and take a more active role in developing a knowledge base, providing project development support, spreading knowledge in the polycentric experimentation system, and extending the learning community. A first option for this could be an extension of the role of the executive organization, RVO, as it is already involved in the derogation process. In the Scottish context, Community Energy Scotland, which provides such support, also grew from a governmental initiative. Alternatively, the national community energy platform Hieropgewekt could take on this role, or even the regional umbrella organizations for energy cooperatives. Yet, to realize this, such intermediaries should pro-actively follow developments in energy legislation relevant for local energy initiatives and attract or train expert staff that can assist experimenters with their project development. As many of such organizations do not have the financial means for this, a government that truly wants to support inclusive innovation and transition processes should allocate budget to them for staff time.

Thus far, a lot has been expected from the experimenters without much active facilitation. Resultantly, the distribution between the risks of and incentives for experimentation is rather uneven and, therefore, it could have been expected that experimenters' progress was relatively slow and interest in new roles limited. This decreased the potential of the sandbox for generating lessons for revising energy regulation to facilitate energy transition. A more holistic approach, inter-actor alignment, the availability of expert support by an intermediary, and facilitation of a more close-knit learning community would bring benefits to the bottom-up participatory innovation.

Author Contributions: Conceptualization, A.M.D., T.v.d.S. and E.C.v.d.W.; Methodology, A.M.D. and E.C.v.d.W.; Formal Analysis, A.M.D.; Investigation, A.M.D. and E.C.v.d.W.; Data Curation, E.C.v.d.W.; Writing – Original Draft Preparation, V.d. S., T. and V.d. W., E.C.; Writing – Review & Editing, T.v.d.S. and E.C.v.d.W.; Project Administration, E.C.v.d.W. All authors have read and agreed to the published version of the manuscript.

Funding: This research and the APC were funded by The Netherlands Organisation for Scientific Research (NWO) grant number [313-99-304].

Acknowledgments: We would like to thank all the interviewees for generously sharing their time, experiences, and knowledge. We would also like to thank Henny van der Windt for his constructive feedback. Furthermore, we are grateful for the constructive feedback of two anonymous reviewers during the publication process.

Appendix A

Table A1 displays an overview of EDSEP projects.

Table A1. Overview of EDSEP experiments.

Year	Name	Type (Project/Large)	Legal Entity	Project Goals	Scale
2015	Parq Green	P	HOA	Collective PV, sustainable heat	292 recreational houses
	Black Jack/Withdrawn	L			
	Experiment DDE Collegepark Zwijsen	P	HOA	Generations, EMS, tariff differentiation, cogeneration	115 apartments in renovated school
	Endona EXP	L	Coop	Generation, cooperation with biodigester, supply to members, increasing direct usage, EMS, storage.	47 with EMS and towards 5000 members in 10 years
2016	Schoonschip	P	HOA	EMS, generation, batteries, heat pumps, heat storage in buffer and smart appliances	46 water houses
	Noordstraat 111 Tilburg	P	HOA	EMS, generation, smaller grid connection	3 houses in old office (owned)
	Villa de Verademing	P	Coop	Insulation, generation, smart grid connected to the neighborhood, storage.	18 apartments and 1 city residence
	Groot Experiment Aardehuizen e.o.	L	HOA	Community battery, EVs, EMS, generation, no gas, smart software dynamic electricity tariffs and demand response, p-2-p.	3 rental and 20 owned
	Kringloopgemeenschap Bodegraven-Reeuwijk	L	Coop	Generation and determining own tariff	2500 households
2017	Republica Papaverweg	P	Coop	EMS, generation, own grid, smart charging with EVs, thermal storage and batteries	Newly built housing block with various accommodations
	Micro Energy Trading Eemnes	L	Coop	P-2-P, EVs, blockchain, storage, generation, smart software.	100–200 social houses; scaling up to 1500
	Micro Energy Trading Amersfoort	L	Coop	P-2-P, smart software and block chain	400–600 social houses
2018	Duurzame Wijkenergiecentrale Trudo	L	Coop	Generation, EMS, batteries, EV chargers, and tariff differentiation	260 apartments in old industrial building(owned/rental)
	Smart Grid Groene Mient	L	HOA	Generation, heat pumps, no gas, battery and EVs	33 newly built houses (2017) with communal garden
	Zeuven heuvels Wezep	P	Coop	EMS, generation, no gas, own grid.	57 newly built houses
	Smart energy grid Bajeskwartier	L	Coop	Generation, neighborhood battery, EVs, heat pumps and thermal storage, smart grid software platform EMS	950 apartments, school, hotel, 340 student houses and various other services
	Kleine Duinvallei Katwijk/ Gave Buren	P	Coop	Balancing, joint electricity purchase and distribution, generation.	80 ecological houses
	Shared energy-mobility community Amersfoort	L	Coop	P-2-P, car sharing with EVs	400–800 houses of housing cooperation
2019	Cooperatie zonnepark Bad Noordzee U.A.	P	Coop	Heat pumps, P-2-P, PV, battery storage.	322 recreational houses and a few large use connections

Appendix B

Table A2 displays an overview of interviewed actors.

Table A2. Overview of interviewed actors.

Interviewed Actor	Type of Interview
Resident of case Schoonschip	Face-to-face
Resident of case Aardehuizen	Face-to-face
Project developer of case Collegepark Zwijsen	Face-to-face
Resident board member and advisor of case Endona	Face-to-face
Grid operators from the different territorial jurisdictions, who engage with experiments (3)	Phone (all 3) and one also face-to-face
Energy company staff member: EnergieVanOns & Nuts&co. (2)	Phone
RVO	Phone
Policy maker ministry of Economic Affairs	Phone
Tax authority staff member	Phone
Consultant in legal, technical and fiscal aspects of renewable energy and energy efficiency. Focus on complex projects and political processes.	Face-to-face
Employee regional umbrella cooperative for supporting local energy cooperatives	Phone
Management, ICT, energy and sustainability advisor, creator of web environment with information overview for EDSEP experimenters	Face-to-face

References

1. Diestelmeier, L. Changing power: Shifting the role of electricity consumers with blockchain technology–Policy implications for EU electricity law. *Energy Policy* **2019**, *128*, 189–196. [CrossRef]
2. Pallesen, T.; Jacobsen, P.H. Solving infrastructural concerns through a market reorganization: A case study of a Danish smart grid demonstration. *Energy Res. Soc. Sci.* **2018**, *41*, 80–88. [CrossRef]
3. Lavrijssen, S.; Parra, A.C. Radical prosumer innovations in the electricity sector and the impact on prosumer regulation. *Sustainability* **2017**, *9*, 1207. [CrossRef]
4. Edens, M.G.; Lavrijssen, S.A.C.M. Balancing public values during the energy transition–How can German and Dutch DSOs safeguard sustainability? *Energy Policy* **2019**, *128*, 57–65. [CrossRef]
5. Parra, D.; Swierczynski, M.; Stroe, D.I.; Norman, S.A.; Abdon, A.; Worlitschek, J.; O'Doherty, T.; Rodrigues, L.; Gillott, M.; Zhang, X.; et al. An interdisciplinary review of energy storage for communities: Challenges and perspectives. *Renew. Sustain. Energy Rev.* **2017**, *79*, 730–749. [CrossRef]
6. Verbong, G.P.J.; Beemsterboer, S.; Sengers, F. Smart grids or smart users? Involving users in developing a low carbon electricity economy. *Energy Policy* **2013**, *52*, 117–125. [CrossRef]
7. Schick, L.; Winthereik, B.R. Innovating relations - Or why smart grid is not too complex for the public. *Sci. Technol. Stud.* **2013**, *26*, 82–102.
8. Planko, J.; Chappin, M.M.H.; Cramer, J.M.; Hekkert, M.P. Managing strategic system-building networks in emerging business fields: A case study of the Dutch smart grid sector. *Ind. Mark. Manag.* **2017**, *67*, 37–51. [CrossRef]
9. Oteman, M.; Kooij, H.J.; Wiering, M.A. Pioneering renewable energy in an economic energy policy system: The history and development of dutch grassroots initiatives. *Sustainability* **2017**, *9*, 550. [CrossRef]
10. Proka, A.; Loorbach, D.; Hisschemöller, M. Leading from the Niche: Insights from a Strategic Dialogue of Renewable Energy Cooperatives in The Netherlands. *Sustainability* **2018**, *10*, 4106. [CrossRef]
11. Parag, Y.; Sovacool, B.K. Electricity market design for the prosumer era. *Nat. Energy* **2016**, *1*, 16032. [CrossRef]
12. Koirala, B.P.; Hakvoort, R.A.; Van Oost, E.C.J.; Van der Windt, H.J. Community energy storage: Governance and business models. In *Consumer, Prosumer, Prosumager: How Service Innovations Will Disrupt the Utility Business Model*; Sioshansi, F.P., Ed.; Academic Press: London, UK, 2019; pp. 209–234. ISBN 9780128168363.
13. Koenders, J.F.; Pipping, S.L. Het Besluit experimenten decentrale duurzame elektriciteitsopwekking doorgelicht. *Ned. Tijdschr. Voor Energier* **2016**, *4*, 146–155.
14. Heldeweg, M.A. Normative alignment, institutional resilience and shifts in legal governance of the energy transition. *Sustainability* **2017**, *9*, 1273. [CrossRef]
15. OFGEM Insights from Running the Regulatory Sandbox|Ofgem. Available online: https://www.ofgem.gov.uk/publications-and-updates/insights-running-regulatory-sandbox (accessed on 31 October 2019).
16. Ostrom, V.; Tiebout, C.M.; Warren, R. The Organization of Government in Metropolitan Areas: A Theoretical Inquiry. *Am. Polit. Sci. Rev.* **1961**, *55*, 831–842. [CrossRef]
17. Ostrom, E. *Understanding Institutional Diversity*; Princeton University Press: Princeton, NJ, USA, 2005; ISBN 9780691122380.
18. Ostrom, E. Polycentric systems for coping with collective action and global environmental change. *Glob. Environ. Chang.* **2010**, *20*, 550–557. [CrossRef]
19. Van der Schoor, T.; Scholtens, B. *Scientific Approaches of Community Energy, a Literature Review*; CEER: Groningen, The Netherlands, 2019.
20. SER Energieakkoord voor Duurzame Groei. Available online: https://www.energieakkoordser.nl/energieakkoord.aspx (accessed on 13 August 2019).
21. Ministerie van Economische Zaken en Klimaat. *Voorstel van wet Houdende Regels Met Betrekking Tot de Productie, het Transport, de Handel en de Levering van Elektriciteit en gas (Elektriciteits- en Gaswet)*; Ministerie van Economische Zaken en Klimaat: The Hague, The Netherlands, 2014.
22. Artikel 7a Elektriciteitswet 1998. Available online: https://maxius.nl/elektriciteitswet-1998/artikel7a/ (accessed on 28 September 2019).
23. Wetten.nl Regeling: Besluit Experimenten Decentrale Duurzame Elektriciteitsopwekking - BWBR0036385. Available online: https://wetten.overheid.nl/BWBR0036385/2015-04-01 (accessed on 6 September 2019).

24. Goldthau, A. Rethinking the governance of energy infrastructure: Scale, decentralization and polycentrism. *Energy Res. Soc. Sci.* **2014**, *1*, 134–140. [CrossRef]

25. Lammers, I.; Diestelmeier, L. Experimenting with Law and Governance for Decentralized Electricity Systems: Adjusting Regulation to Reality? *Sustainability* **2017**, *9*, 212. [CrossRef]

26. Lammers, I.; Arentsen, M. Polycentrisme in lokale besluitvorming over duurzame energie: De casus slimme netten. *Bestuurswetenschappen* **2016**, *70*, 20–31. [CrossRef]

27. McGinnis, M.D. An introduction to IAD and the language of the Ostrom workshop: A simple guide to a complex framework. *Policy Stud. J.* **2011**, *39*, 169–183. [CrossRef]

28. Yin, R.K. *Case Study Research: Design and Methods*, 5th ed.; Sage: Thousand Oaks, CA, USA, 2013; ISBN 1483322246.

29. Braun, V.; Clarke, V.; Hayfield, N.; Terry, G. Thematic analysis. In *Handbook of Research Methods in Health Social Sciences*; Liamputtong, P., Ed.; Springer Nature Singapore Pte Ltd.: Singapore, 2019; pp. 843–860. ISBN 9789811052514.

30. GridFlex Initiatief Gridflex Heeten: GridFlex. Available online: https://gridflex.nl/over-gridflex/ (accessed on 8 October 2019).

31. CWI Grid Friends–Demand response for grid-friendly quasi-autark energy cooperatives. Available online: http://www.grid-friends.com/ (accessed on 8 October 2019).

32. Zomerman, E.; Van Der Windt, H.; Moll, H. The distribution systems operator's role in energy transition: Options for change. *WIT Trans. Ecol. Environ.* **2018**, *217*, 411–422.

33. Belastingdienst Tabellen Tarieven Milieubelastingen. Available online: https://www.belastingdienst.nl/wps/wcm/connect/bldcontentnl/belastingdienst/zakelijk/overige_belastingen/belastingen_op_milieugrondslag/tarieven_milieubelastingen/tabellen_tarieven_milieubelastingen?projectid=6750bae7-383b-4c97-bc7a-802790bd1110 (accessed on 28 September 2019).

34. CBS StatLine - Aardgas en Elektriciteit, Gemiddelde Prijzen van Eindverbruikers. Available online: https://opendata.cbs.nl/statline/#/CBS/nl/dataset/81309NED/table?fromstatweb (accessed on 28 September 2019).

35. Van Der Schoor, T.; Van Lente, H.; Scholtens, B.; Peine, A. Challenging obduracy: How local communities transform the energy system. *Energy Res. Soc. Sci.* **2016**, *13*, 94–105. [CrossRef]

36. Hieropgewekt Wat Kan Een Energiebedrijf Voor Een Initiatief Betekenen? Ruggensteun Energiebedrijven Een Zegen Voor Lokale Initiatieven? Available online: https://www.hieropgewekt.nl/sites/default/files/documents/Watkaneenenergiebedrijfvooreeninitiatiefbetekenen.pdf (accessed on 6 September 2019).

37. Enexis Uitleg Tariefopbouw Netwerkkosten. Available online: https://www.enexis.nl/consument/aansluiting-en-meter/tarief/uitleg-tariefopbouw (accessed on 13 January 2020).

38. HIER Opgewekt Waarom Meedoen Met de Experimentenregeling? Available online: https://www.hieropgewekt.nl/kennisdossiers/waarom-meedoen-met-experimentenregeling (accessed on 6 September 2019).

39. Stedin Tarieven. Available online: https://www.stedin.net/tarieven#3x25 (accessed on 7 January 2020).

40. Smart Storage Magazine Staatssecretaris Snel: Problematiek Dubbele Belasting bij Energieopslag per 2021 Oplossen Door Wetswijziging. Available online: https://smartstoragemagazine.nl/nieuws/i18940/staatssecretaris-snel-problematiek-dubbele-belasting-bij-energieopslag-per-2021-oplossen-door-wetswijziging (accessed on 28 September 2019).

41. European Parliament Texts Adopted - Common Rules for the Internal Market for Electricity. Available online: http://www.europarl.europa.eu/doceo/document/TA-8-2019-0226_EN.html (accessed on 22 October 2019).

42. Van der Waal, E.C.; van der Windt, H.J.; van Oost, E.C.J. How local energy initiatives develop technological innovations: Growing an actor network. *Sustainability* **2018**, *10*, 4577. [CrossRef]

43. Smith, A.; Raven, R. What is protective space? Reconsidering niches in transitions to sustainability. *Res. Policy* **2012**, *41*, 1025–1036. [CrossRef]

44. Kooij, H.-J.; Lagendijk, A.; Oteman, M. Who Beats the Dutch Tax Department? Tracing 20 Years of Niche–Regime Interactions on Collective Solar PV Production in The Netherlands. *Sustainability* **2018**, *10*, 2807. [CrossRef]

45. Hargreaves, T.; Hielscher, S.; Seyfang, G.; Smith, A. Grassroots innovations in community energy: The role of intermediaries in niche development. *Glob. Environ. Chang.* **2013**, *23*, 868–880. [CrossRef]

46. Hielscher, S.; Seyfang, G.; Smith, A. *Community Innovation for Sustainable Energy*; CSERGE Working Paper; Centre for Social and Economic Research on the Global Environment (CSERGE): Norwich, UK, 2011.
47. Geels, F.W. The multi-level perspective on sustainability transitions: Responses to seven criticisms. *Environ. Innov. Soc. Transit.* **2011**, *1*, 24–40. [CrossRef]

 *energies*

Article

Collective Renewable Energy Prosumers and the Promises of the Energy Union: Taking Stock

Lanka Horstink [1,2,*], Julia M. Wittmayer [3], Kiat Ng [1,4], Guilherme Pontes Luz [5], Esther Marín-González [5], Swantje Gährs [6], Inês Campos [5], Lars Holstenkamp [7], Sem Oxenaar [3] and Donal Brown [8]

1   Faculty of Engineering, University of Porto (FEUP), Rua Dr. Roberto Frias, 4200-465 Porto, Portugal; kiatng@fe.up.pt
2   Institute of Social Sciences, University of Lisbon (ICS-UL), Av. Professor Aníbal de Bettencourt, 9, 1600-189 Lisboa, Portugal
3   DRIFT—Dutch Research Institute for Transitions, Erasmus University Rotterdam, Postbus 1738, 3000 DR Rotterdam, The Netherlands; j.m.wittmayer@drift.eur.nl (J.M.W.); oxenaar@drift.eur.nl (S.O.)
4   CIIMAR-Interdisciplinary Centre of Marine and Environmental Research, University of Porto, Terminal de Cruzeiros de Matosinhos, Av. General Norton de Matos s/n, 4450-208 Matosinhos, Portugal
5   Centre for Ecology, Evolution and Environmental Changes (CE3C), Faculty of Sciences of Lisbon University, Campo Grande, 1749-016 Lisboa, Portugal; gpluz@fc.ul.pt (G.P.L.); emgonzalez@fc.ul.pt (E.M.-G.); iscampos@fc.ul.pt (I.C.)
6   Institute for Ecological Economy Research (IÖW), Potsdamer Str. 105, 10785 Berlin, Germany; swantje.gaehrs@ioew.de
7   Institute of Banking, Finance and New Venture Management, Leuphana University of Lüneburg, Universitätsallee 1, 21335 Lüneburg, Germany; holstenkamp@uni.leuphana.de
8   Sustainability Research Institute, University of Leeds, Leeds LS2 9JT, UK; D.L.Brown@leeds.ac.uk
*   Correspondence: lanka.horstink@ics.ulisboa.pt; Tel.: +351-919852781

Received: 1 November 2019; Accepted: 30 December 2019; Published: 15 January 2020

Abstract: A key strategy in the European Union's ambition to establish an 'Energy Union' that is not just clean, but also fair, consists of empowering citizens to actively interact with the energy market as self-consumers or prosumers. Although renewable energy sources (RES) prosumerism has been growing for at least a decade, two new EU directives are intended to legitimise and facilitate its expansion. However, little is known about the full range of prosumers against which to measure policy effectiveness. We carried out a documentary study and an online survey in nine EU countries to shed light on the demographics, use of technology, organisation, financing, and motivation as well as perceived hindering and facilitating factors for collective prosumers. We identified several internal and external obstacles to the successful mainstreaming of RES prosumerism, among them a mismatch of policies with the needs of different RES prosumer types, potential organisational weaknesses as well as slow progress in essential reforms such as decentralising energy infrastructures. Our baseline results offer recommendations for the transposition of EU directives into national legislations and suggest avenues for future research in the fields of social, governance, policy, technology, and business models.

Keywords: renewable energy prosumer; energy transition; collective prosumer; energy union; community energy

---

1. Introduction

The European Commission (EC) is spearheading the EU's plan to 'lead the clean energy transition, not only adapt to it' [1]. In 2016, the EC started developing a 'Clean Energy Package' that has now been finalised (the eight legislative acts that compose the Clean Energy package were recently concluded with the adoption of the recast of the Renewables Directive (RED II), the new Governance Regulation of the Energy Union and Climate Action, the new Energy Efficiency Directive as well as the recast of the Electricity Directive [2]), completely overhauling the EU's energy policy framework, with the objectives of reducing CO_2 emissions by 40%, increasing the share of renewable energy sources to 32%, and improving energy efficiency by 32.5% by 2030. The vision of an 'Energy Union'—providing all EU consumers with secure, sustainable, competitive, and affordable energy—includes an appropriate regulatory framework, strategic investments to innovate the EU's energy system, and an integrated multi-level energy governance framework. Having promised safe, viable, and accessible energy supply for all, the EC and EU countries are keen on embedding fairness, inclusiveness, local economy stimuli, and job growth in the transition toward a climate-neutral energy system [3].

The Energy Union aims to stimulate the involvement of energy consumers in the energy market 'to generate electricity for their own consumption, store it, share it, consume it or sell it back to the market' [4]. At the very least, citizens are expected to be 'active customers' (i.e., not merely buying electricity, but participating either in energy production, demand-response, or energy efficiency schemes: see the new recast EU Electricity Directive [3]), at best they will become what the EC is now calling 'renewables self-consumers', who generate, store, and/or sell self-generated electricity from renewable energy sources (RES) as per the RED II Directive [5], and which in the scientific literature is also referred to as an 'energy prosumer' [6]. By placing citizens at the centre of the Energy Union, and giving them the right to produce, store, or sell their own energy, whether individually or collectively, EU institutions are betting on a more rapid take-up of renewables in the energy system [2,5].

With EU countries being encouraged to support decentralised renewable energy through the relaxing of rules and/or the offering of incentives for RES self-consumption, the development of energy cooperatives and energy communities is accelerating all across Europe [7]. Representing approximately one million citizens, the European Federation of renewable energy cooperatives (REScoop.eu), established as recently as 2013, has rapidly grown to a network of 1500 renewable energy cooperatives and energy communities [8]. The pace at which the adoption of renewable energy has spread through Europe, additionally facilitated by the unexpected drop in prices of a number of RES technologies [9], has taken legislators and policy-makers by surprise, creating a fertile ground for ad hoc rather than strategic responses [10]. Important dimensions of prosumerism, such as the development of technology, the choice of organisational models, and innovation in funding solutions are still a long way from stabilising [7].

A number of promising case studies on community energy initiatives (e.g., [11–13]) as well as helpful analyses of the mitigating factors at work (see for instance [14–16]) support the claim that placing citizens at the core of a clean and fair energy transition is key to its success. There are, however, no reviews of the full range of collective prosumers (i.e., non-household) beyond the better-known energy cooperatives: who are they, what are their characteristics, behaviour, needs, and socio-economic impact? How does one collective prosumer initiative differ from others? Which of these initiatives should be incentivised and how? This gap in our knowledge makes it difficult to assess RES prosumerism's contribution to an energy transition that is expected to meet ambitious social, economic, and ecological objectives as well as measure the effectiveness of the policies being put in place to stimulate the prosumer phenomenon. These issues are especially salient when considering the accelerated timeframe of the Energy Union and the expected growth-spurt in prosumer initiatives once the Clean Energy Package is in place. This article aims to address this gap in the literature by providing a much-needed overview of the diversity of collective RES prosumer initiatives as well as a stock-take of the demographic, technological, organisational, financial, motivational, and hindering/facilitating factors that characterise them, and assess how the state of the art aligns with current energy policies and incentives. Our research, part of a larger project aiming to provide a framework of incentive structures for collective prosumers, is guided by the following question:

> What is the current state of play for collective forms of RES prosumerism in Europe considering the demands and promises of the Energy Union?

The article is structured as follows: in Section 2, we will embed our research within its scope. We will then, in Section 3, present the methodology employed to survey a diversity of RES prosumer initiatives in nine countries in Europe: Belgium (BE), Croatia (HR), France (FR), Germany (DE), Italy (IT), the Netherlands (NL), Portugal (PT), Spain (ES), and the United Kingdom (UK). In Section 4, we present the results of our collective RES prosumer characterisation, the most significant of which are subsequently discussed in light of their policy implications in Section 5. In Section 6, we sum up our key conclusions and make some policy recommendations to support the continued growth of RES prosumerism in the EU, while safeguarding the vision of the Energy Union.

2. Background Review of Renewable Energy Sources (RES) Prosumerism

Reviewing nine EU countries as well as the EU as a whole, we found that differences in the take-up of RES prosumerism can be attributed among others to the varying investment in RES [17], energy path dependencies related to the natural resources available in the different countries (see for example [18]), as well as cultural factors (e.g., [19,20]). Of the countries studied by us, only Portugal and Croatia approached the mark of a 30% share of renewable energy sources in gross final consumption of energy, while France, Spain, and Germany scored around the EU average (17.53%), and the United Kingdom, Italy, the Netherlands, and Belgium scored closer to the 9% mark [21]. Each country's context is reflected in the overall energy forms that it consumes (i.e., countries with good hydric conditions have a high production of hydro-electricity, countries that (historically) have access to natural gas (NL, UK), have gas-driven heating systems, France is largely dependent on its nuclear energy production, etc). Overall, in the EU at the RES level, hydro continues as a leading energy technology, with wind energy coming in either in first or in second place in terms of production capacity. Solar-powered electricity, meanwhile, is growing fast in most countries including the more northern countries [21].

Despite the advances made in restructuring the legislative and policy framework to prepare the EU's clean energy transition, Campos et al. [22] highlighted considerable disparities in legislative and policy support for RES prosumerism in different EU countries, resulting in varying levels of prosumer development. The Clean Energy Package is intended to homogenise the attitude of EU countries toward prosumerism, but it presents several challenges. A key challenge is the imposition of new definitions and rules for individuals as well as collective forms of RES prosumerism that, besides falling short of representing the full diversity of prosumer initiatives sprouting up [10,23–25], is prone to different interpretations in the subsequent transposition to national legislations, a process that must mandatorily be concluded in 2021.

Several reports, reviews, and case studies have tried to produce insight into the drivers, facilitating factors as well as barriers for energy cooperatives and communities. In the sub-sections below, we aim to summarise the most recent and relevant conclusions available from the literature.

2.1. Sociocultural and Socio-Economic Factors of Prosumerism

- The institutional features of communities that decide to self-produce will influence/facilitate the process (e.g., whether there is a tradition of cooperatives and/or of collective ownership, how strong is the sense of responsibility for community, etc.) [19,20];
- Social drivers tend to be predominant in community initiatives including in the area of energy (e.g., the need to respond to societal challenges or local social demands) [26]. Bauwens' studies point to a strong desire of energy community initiatives to oversee the (clean) energy supply for their community [27,28];
- A recent review of community energy initiatives found that the latest wave of prosumer initiatives was less tied to advances in renewable energy technology or changes in legislation than to the desire to democratise and decentralise energy [29];
- Aside from responding to societal challenges, energy community initiatives can also bring financial benefits for the community engaged [30,31];

- Members of energy community initiatives report satisfaction in being part of a community experiment [30] and they tend to be more norm-driven as well as more positive toward an energy system change when part of collective, participative energy projects [28,30,32];
- There are a number of cultural and socio-political barriers to the further development of RES prosumerism, among them the lack of (technical) knowledge in the prosumer initiatives (which are more than often run by volunteers), the spread of misinformation about renewable energy alternatives, and the lack of legitimacy attributed to the cooperative model [30,31];
- A gender imbalance has been posited in energy prosumerism: a relationship has been established between gender and risk of energy poverty [33] as well as between the different ways in which women and men participate in the energy sector and in energy policy decisions, with women generally being highly under-represented in both [34,35].

2.2. Technical Factors

- Energy communities are strongly connected to the use of renewable energy [30];
- Collective energy initiatives provide an opportunity for bottom-up innovation in energy efficiency and production and for innovative business models: this topic has been well-researched, not only in terms of grassroots innovation in energy, but also in other key areas of human production (see among others [14,36,37]);
- RES production is increasingly attractive as well as accessible, considering the rapidly falling costs of solar photovoltaic (PV) installations and batteries for storage [31].

2.3. Financial Factors

- There is an increasing incentive to self-consume rather than sell to the grid, with feed-in-tariff rates dropping or being abolished, such as in the United Kingdom [31];
- Despite falling costs for RES technologies and accessories, installations can still command considerable investment, in particular, wind parks [30,31].

2.4. Political Factors

- There are considerable legal-political implications from relocating control over such a crucial resource as energy to emerging new actors, such as prosumers and prosuming energy communities [9,38]. This may cause governments to hesitate, delay, or stall the development of prosumers. For example, Germany, having 100 years of experience with electricity cooperatives and hundreds of small grid operators, has been under pressure for years by the EU to minimise its number of grid operators [31];
- Policy advocates for RES prosumerism complain that on the one hand, energy infrastructures are insufficiently digitalised, and on the other, existing digital systems (such as trading and billing) are still in the hands of large energy companies [7];
- EU countries are being very slow in phasing out subsidies for fossil fuels, which is creating overcapacity in the energy market, with the EU barely keeping up with the growth of RES [7] (p. 6);
- Due to the liberalisation of energy markets, there is a growing number of purely commercial RES initiatives set up by project developers and incumbent energy companies, in some cases creating legal structures that appear collaborative (i.e., the cooperative), but are not de facto in citizens' hands [31];
- The federation of RES cooperatives and communities complain of rigorous lobbying by large energy companies to reign in the amount of control RES initiatives may command over the energy system [31];
- Energy cooperatives and communities report increasing bureaucratic and regulatory hurdles for starting and running a prosumer initiative, and the current political and legal framework is unstable, with a tapering off of RES prosumer incentives [29–31];

Our review reveals several opportunities for the thriving of RES prosumerism, but also an alarming number of barriers and legal/policy contradictions. One concern raised by a number of scientists, policy-makers, as well as representatives of energy communities and cooperatives themselves [31] is how to ensure a more inclusive and democratically-run energy transition such as that being promoted by the EC as a cornerstone of the Energy Union [13,37,39]. For example, should civic-focussed renewable energy initiatives be treated differently than self-interest/profit-focussed initiatives, and who should run the transmission and distribution networks [31]? These are sensitive and under-discussed topics that will influence the pathway of collective RES prosumerism.

In the next section, we present the methodology of our study of collective forms of RES prosumerism.

3. Methodology

To elucidate the current state of play for collective (i.e., non-household) prosumer initiatives in Europe, we drew upon an interdisciplinary mix of qualitative and quantitative methods, used in different iterations. Our review was conducted for the whole of the EU as well as zooming in on nine EU countries: BE, HR, FR, DE, IT, NL, PT, ES, and UK. Our survey process included: (i) content preparation; (ii) sampling strategy; (iii) survey administration; (iv) data processing; and (v) data analysis. Our main objective was to obtain an overall picture of the profiles of collective RES prosumer initiatives and the context(s) in which they are developing.

The survey form was designed using a collaborative and iterative approach, drawing on the pooled information needs from the multi-disciplinary research team as well as the knowledge acquired in previous, similar surveys. The survey questionnaire was designed and programmed by us to be answered online and covered six categories, each corresponding to a different information need, with a total of 32 questions (see Appendix A for a full list of the questions).

The main categories were:

1. Control questions (e.g., name and whether the initiative produces/will produce RES);
2. General demographics of collective RES prosumers (e.g., legal form, founding date, location);
3. Use of technology by collective RES prosumers (e.g., energy needs, technologies used);
4. Governance/organisation of collective RES prosumers (e.g., staff characteristics, decision-making mechanisms);
5. Motivation/ambition of the RES prosumer initiative (e.g., reasons to start the initiative); and
6. Hindering and facilitating factors as perceived by collective RES prosumers.

Due to the ambitious nature of our information needs—implying a longer questionnaire—the survey was set up as a multiple case-study. The online, user-friendly survey form was made available in the respondent's own language (a total of eight languages), and its launch was, for most countries, combined with a soft-push approach in two or three steps (telephone calls to leaders of the initiatives, an explanatory email with a link to the survey, and a follow-up email or phone call, as needed).

The final questionnaire is publicly available [23] (pp. 90–116), and has also been submitted as Supplementary Material (Document S1), while examples of questions can be found in Appendix B.

The sample for our self-administered survey was drawn from the nine countries. We included countries with fertile environments for RES prosumerism (DE, UK, NL); two countries with a long history of self-consumption either at an industry or at the regional level but where new prosumer initiatives encounter significant challenges (BE, IT); and four countries where RES prosumerism has only just been legalised: two small countries (HR, PT), and two large ones (FR, ES).

Since there is no established overview of RES prosumer initiatives across Europe, we took an iterative approach to respondent identification. Research teams in the different countries were asked to build exploratory databases of collective RES prosumers in several steps, each being subjected to database analysis to improve these exploratory actor types. In the first iteration, it became clear that our collective forms of RES prosumers were not easily categorised, with attribute overlaps existing between the exploratory types found. In a next step, we decided to distinguish between those actors actually prosuming (i.e., producing and consuming energy from renewable energy sources, as an entity or through its members) and those actors influencing RES prosumerism (i.e., facilitating, promoting, financing, supporting, benefitting from, or even hindering) (see Table 1).

Table 1. Working definitions of collective renewable energy sources (RES) prosumers and RES prosumer stakeholders. Source: [23] (pp. 24–25).

Broad Actor Type	Working Definition
Collective RES prosumer	A collective energy actor that produces energy from renewable sources with the primary objective of providing in its own energy needs and/or those of its members, and in some cases selling excess energy to clients, thereby actively participating in the energy markets.
RES prosumer stakeholders	Organisations, institutions, or collectives—or their representatives—that influence, facilitate, benefit from, and/or may hinder the development and evolution of RES prosumer initiatives, in particular, its collective form.

After a few iterations, our exploratory database analysis identified six broad categories of collective RES prosumer actors capable of describing the RES prosumer initiatives and stakeholders that were collected in our databases across nine EU Member States (Table 2).

Table 2. Key categories of collective RES prosumer actors. Source: [23] (pp. 25–27).

Type of Collective RES Prosumer Actor	Examples	Notes
1. Energy cooperatives	Wind energy cooperatives; local energy cooperatives; regional energy cooperatives; cooperatives of cooperatives; dispersed-site cooperatives.	Cooperatives come in many shapes and forms, may have a local or broader focus, and may be for-profit or not-for-profit. They may even technically function as a utility.
2. Renewable energy communities	Partnerships between municipalities and local organisations and/or citizens; village energy communities; neighbourhood initiatives; informal collectives for RES prosuming; other forms of partnerships with a community focus.	Under the new EC definition these communities will have to have a legal entity (which may be a cooperative) running the initiative and have a clear local as well as a not-for-profit focus. Virtual communities as well as informal communities are not officially recognised.
3. Organisational prosumers	Public institutions (city council, school, retirement home); not-for-profit organisations such as NGOs and associations; businesses from different sectors (farming, services, sales).	Many of these organisations will behave as large households, bringing them closer to residential prosumers. Nevertheless, their motivations and ambition may vary significantly.
4. Property-sector prosumers	Social real-estate projects; home owner associations; municipal real estate schemes; district heating schemes.	Although technically this is a sub-sector of the previous category 3, organisational prosumers, this is a special case where business or public sector interests meet community interests.
5. RES prosumer-focussed initiatives	P2P energy trading platforms; Other energy aggregators; Energy developers; ESCOs.	These are not prosumer initiatives, but provide services to them or benefit in some other way from them.
6. Other RES and RES prosumer stakeholders	Municipal, regional or NGO campaigns that promote CO_2 neutrality, energy efficiency, green mobility, greener housing, or more generally 'sustainability' in their territory; EU governments; energy agencies; the EC; conventional energy companies; RES utilities.	Their campaigns may promote prosumerism, but they do not engage in it. Other stakeholders may influence the RES prosumerism phenomenon negatively or positively or may even compete with prosumers.

Our final sample population included close to 1000 RES prosumer initiatives. Each country research team had an objective according to the size of their prosumer population (i.e., countries such as Germany, the Netherlands, and the UK contacted hundreds of prosumer initiatives, whereas countries such as Croatia and Portugal did not have a population larger than 20 to 30 initiatives). Sampling was adjusted dynamically according to the type of respondents that answered our survey. A snowball technique was attempted to capture initiatives beyond our sample and countries of focus, but the lack of a personalised approach proved to be less successful, and only two additional initiatives, from Denmark and Finland, respectively, responded to this method.

Upon conclusion of the survey, we reclassified the dataset considering the distribution of respondents across countries and the high diversity of legal forms of the initiatives surveyed (see Section 4.1), after which the data were cleaned, tested, treated, plotted, and analysed, using the computational programme MATLAB, complemented with Excel, for our statistical analysis and generating graphs. The full details of how we proceeded can be found in [23]. Our dataset has not been made publicly available, since it contains sensitive information that would identify the initiatives that participated, to which we have promised full anonymity.

In the next section, we present and discuss our main results grouped as follows: distribution of the final dataset; general demographics and operational information; organisational structure; and the key drivers as well as perceived hindering/facilitating factors for developing a RES prosumer initiative.

4. Results

4.1. Distribution of the Final Dataset

Despite the challenges of an online questionnaire and indirect contact with respondents, the average response rate was 21.8%, corresponding to 198 initiatives that concluded our questionnaire. The number of respondents per country followed the anticipated trend and contacting strategies, with countries with longer histories of prosumerism achieving higher numbers (NL, DE, UK, FR, respectively). Smaller countries and/or countries where RES prosumerism is a more recent phenomenon achieved smaller numbers. With respondents from several countries (UK, NL, DE) warning us about survey fatigue, especially among energy cooperatives, Belgium provided less respondents than would have been expected when looking at the history of prosumerism in that country, while the Netherlands provided more.

We plotted all the answers for the whole dataset, for each of the countries, as well as for the top four legal forms encountered. Overall, and as expected, most of our respondents were energy cooperatives (60%, $n = 119$). Their spread across countries more or less followed the trends documented by the few statistical overviews that are available of energy cooperatives, which state that countries like Denmark lead with over 1,000 cooperatives and other northern European countries such as Germany, the UK, and Austria each count hundreds [40], whereas in the south of Europe, the numbers tend not to exceed two dozen [41,42]. There were three other main organisational forms: the for-profit company (14.5%); the public institution (9%); and the private not-for-profit organisation (8%). The prevalence of other legal forms was too residual to draw conclusions on correlations.

We also found and included in our analysis three types of initiatives that come close to a more direct form of energy community: public–private partnerships, partnerships between organisations and/or collectives, and informal civil society initiatives or collectives. Finally, we found RES prosumer initiatives that were run as projects by organisations or collectives (for example, a store that puts a RES installation on its roof as a stand-alone project, or when RES production is just one activity within an organisation promoting sustainable development). In total, we registered around 50 legal forms in the nine countries—of which many were similar, such as the legal form of the association or NGO as well as limited companies and corporations, but some were also quite different, such as the many 'sociétés' in France and the community societies in the UK. In consultation with the research teams in the different countries, we reclassified the legal forms, which resulted in a more manageable list of 10 legal forms (Table 3).

Table 3. Reclassification of 50+ legal and organisational forms reported by survey respondents. Source: [23] (p. 54).

Legal or other Organisational Forms Given by Respondents (in Original Language)	Reclassification for Data Analysis
SAS (Société par actions simplifiée) cooperative, SAS d'interêt collectif, Community Benefit Society, Societé cooperative à resp. limitée, eingetragene Genossenschaft (eG), CVBA, Community Development Trust, Cooperativa, Industrial Provident Society	Cooperative
Societé à resp. limitée, Privatno firma, Malo poduzece, S.A., ESCo, GmbH & Co. KG (Kommanditgesellschaft), Gesellschaft mit beschränkter Haftung (GmbH), Aktiengesellschaft (AG), Besloten Vennootschap (BV), Limited (Ltd)	Company (for-profit)
Publieke organisatie, Staatliche Behörde, Kommune, Overheids orgaan, Körperschaft des öffentlichen Rechts, Gemeente, Municipalidade, Gebietskörperschaft, Escola pública	Public Institution (incl. local authorities)
Association (ex: of homeowners, sports, …), Stichting, associação, Gesellschaft bürgerlichen Rechts (GbR), associazione privativa	Private not-for-profit organisations (e.g., NGO, association, foundation …)
Social purpose business, Empresa de no lucro, Community Interest Company	Social Enterprise (for-profit as well as a social objective)
Private project	Project run by an organisation or collective (i.e., not a legal form)
Partnership between family farms and a town community, partnership between cooperatives, partnership between companies and community interest companies	Partnership between private organisations and/or collectives
Unincorporated community group, informal association	Informal collective or community
Partnership between a GmbH & Co.KG, partnership between municipality and other organisations	Public–Private-Partnership
Other	Other

4.2. Key Demographics and Operational Information on Collective RES Prosumers

Figure 1 presents the distribution of the top four legal forms across the top four countries in terms of sample size as well as the average distribution for the remaining countries.

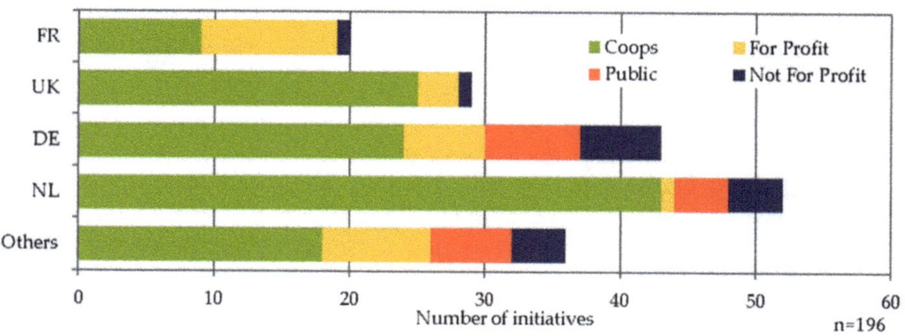

Figure 1. Distribution of legal forms.

While the cooperative was clearly the preferred form for prosumer initiatives in our dataset, nevertheless, in some countries the balance was different than expected, such as in France, which had a higher percentage of initiatives opting for the for-profit/company form, or Croatia, where prosuming initiatives were almost exclusively companies, and finally Spain, where almost half of the respondents were public institutions. Even though a considerable number of informal prosumer collectives were contacted, the response rate was low.

Since respondents had space to comment in most of the questions of the survey, we analysed the additional information, which provided more evidence of the apparent mismatch between the legal form and organisational mission:

- Many initiatives highlighted that, independently of their legal form, they considered themselves an energy community (commenting for example: 'we are a community interest association and not-for-profit'). In several cases, instead of giving a straightforward answer to the question about their legal form, respondents would state that 'we are a municipality working with local organisations', 'we are a company/association but run as a cooperative', 'we are a citizens' cooperative', etc. One respondent did not identify as the company it clearly was, instead called their organisation a 'project developer'. As an example of the 'legal form dilemma', see [43] on legal forms chosen by early community energy initiatives in Germany.
- Larger cooperatives, but also municipalities in NL, UK, and BE, reported that increasingly they were opting to create energy companies that will mediate for them in the energy market. They feel that they can move quicker and comply better with legislation by using a company. In this way, cooperatives are hybridising, but keeping their different missions separate, at least legally.
- There were a number of interesting outliers: an association that represents firms located on the same grounds that wish to aggregate their RES production; farming cooperatives that also wish to be prosumers, but not become energy cooperatives; energy suppliers that enable individuals as well as organisations to prosume and buy up the excess energy; and companies taking advantage of pro-renewable energy legislation to set up for-profit RES initiatives that buy up energy from others (an example is Croatia, where biogas is obtained from farmers by companies and then resold).

The growth trend of RES prosumer initiatives (Figure 2) shows a slow growth period until 2010, a period of acceleration followed by a slowing down of growth in the period that the Energy Union and its pillars were debated as well as questioned, a period that starts in 2014, and a possible new growth spurt starting from 2017, with new countries joining the RES prosumerism phenomenon. However, over 12% of our dataset had not started producing yet, with quite a few initiatives complaining of excessive and complex bureaucracy and/or strict urban planning regulations, some even stating that they had given up on producing due to the above-mentioned barriers compounded with high investment requirements for some of the RES technologies (in particular, wind energy). These initiatives are now focussing on energy advice services and promoting energy efficiency or (e.g., in the UK) considering developing their own RES-ready housing.

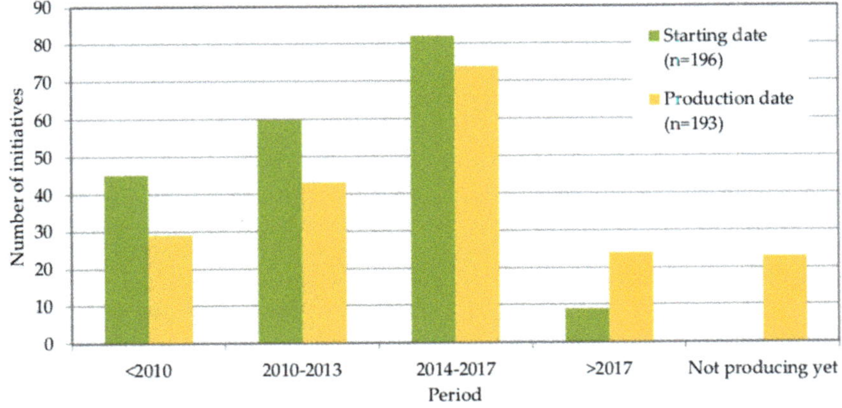

Figure 2. Starting dates of the initiatives and of production.

In terms of the scale of the initiatives in our dataset, we found 80% to have a local focus (i.e., town, city, municipality), with 16% having a regional focus, while a residual percentage had a national focus (3%). Of those that operated locally, 12% also reported on regional engagement.

We inquired as to the energy needs addressed by the RES prosumer initiatives in our dataset, and found that RES-powered electricity significantly took the lead, followed by heating and mobility, with cooling appearing in last place (Figure 3). Practically half of our respondents only focussed on producing electricity, while a bit less than half addressed several energy needs simultaneously, the combination of electricity and heating being most popular. The only legal form that did not follow this trend was that of the public institution, where cooling came in third place, and mobility in last. This may be linked to the fact that public institutions most often manage large(r) buildings. In terms of country trends, Germany stood out, with initiatives attributing almost equal importance to electricity and heating, which practically shared first place. The initiatives from the other countries followed the main trend and appeared to have heating trailing significantly behind electricity. This is despite the fact that half of the countries surveyed had a considerable share of heating and cooling from RES sources as a percentage of their total RES consumption: HR (36.6%); PT (34.4%); FR (21.4%); and IT (20.1%), with Spain (17.5%) and Germany (13.4%) scoring in the mid-range, and all remaining countries having a negligible share of RES in their heating and cooling energy use [21].

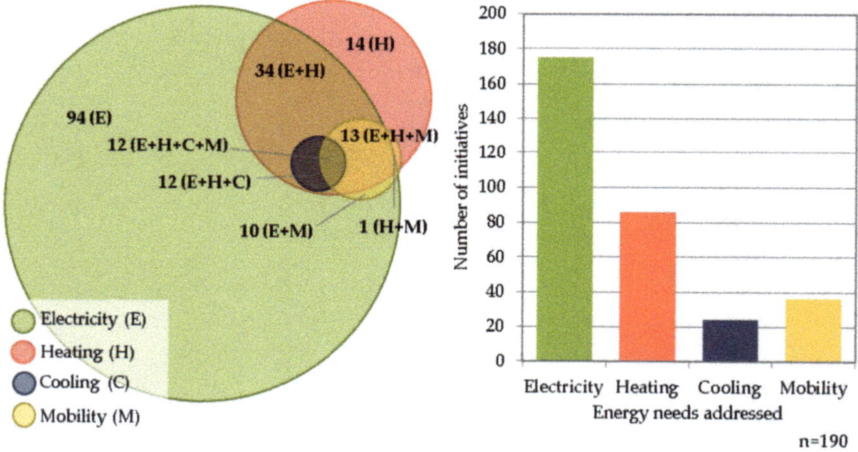

Figure 3. Energy needs addressed by initiative including combined needs.

The overwhelming majority (90%) of our respondents were producing energy from, or planning to install, solar PV (Figure 4). Nevertheless, one third of respondents were producing energy from wind, which came second in terms of popularity, followed by biomass, storage in batteries, biogas, and solar thermal, respectively. More than half of the prosumer initiatives invested in more than two technologies, with a considerable percentage (20%) investing in more than four technologies.

Country differences were quite relevant: Belgian respondents reported wind energy as their leading technology, while Croatian respondents focussed on biogas, biomass, and co-generation. The Italian initiatives invested in the highest number of different technologies, each reporting that they were using on average about five to six technologies. The 'big four' from our sample (NL, DE, UK, and FR) as well as Spain reported solar PV as their main technology, but choices for secondary technologies varied significantly: biomass and storage in Germany, wind and storage in the UK, whereas in our French dataset, we found almost no experimentation with energy storage. Even though few initiatives were actively investing in mobility options, several mentioned that they were planning to invest in storage in future. From the trend observed in our dataset, we expect both storage and clean mobility to become more significant.

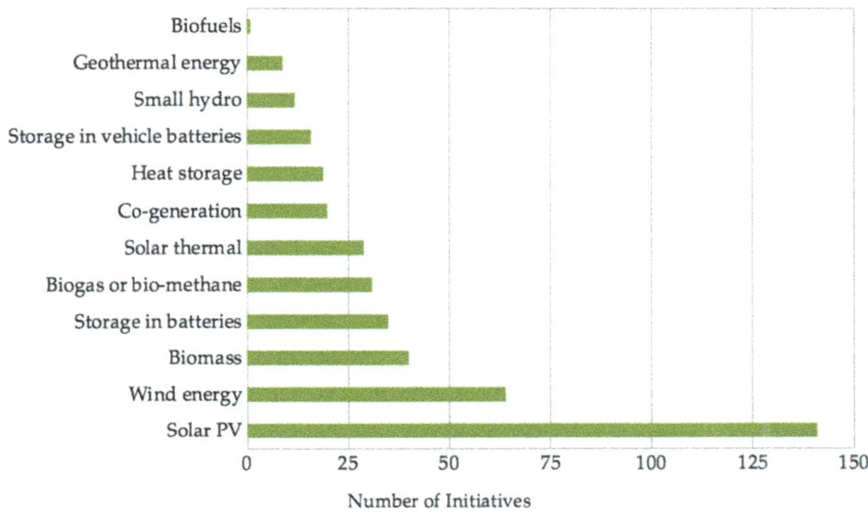

Figure 4. Renewable energy technologies used.

To gauge the different sizes of the RES prosumer initiatives in our dataset, we inquired about the number of members, and client base as well as the number of staff. Our dataset had a predominance of middle-sized initiatives (more than half of the respondents reported having between 51 and 500 members) (Figure 5). Having members, mandatory in most cases, was the norm for cooperatives and the not-for-profit sector (NGOs, associations, foundations, and informal collectives). About half of the cooperatives and the not-for-profit initiatives reported having direct clients besides members. A surprisingly high number of companies reported not having direct clients, but when verifying their websites, we concluded that there may have been a miscommunication due to our use of the term 'direct client', since they did report having clients on their websites. In terms of staff size, while the average number of staff members was low (13, with a median of eight), when compared to the member sizes of the initiatives, there were extreme outliers (an overall range from 1 to 150 staff members) as well as differences between companies, cooperatives, and the not-for-profit sector, not to mention between countries. Unsurprisingly perhaps, considering their for-profit nature, private companies had the highest average number of staff (although they have the same median), while cooperatives on average had a little over half as many. The not-for-profit sector reports the lowest number of total staff, which may reflect their size and/or limited financing options, while their focus may also not be exclusively on producing RES. The public sector's sample size was too small to make a definitive observation. Initiatives from NL, UK, and ES (and to a certain extent IT) reported the highest average number of staff.

We also collected information on the financing strategies of the RES prosumer initiatives in our dataset, which we correlated with their legal form and country of origin. Most respondents indicated more than one form of financing. The top choice in terms of financing (Figure 6), whether correlated by country or by legal form, was through member contributions and/or the founders of the initiative (reflecting the high representation of cooperatives in our dataset). This was followed by public funding, whether regional, national, or from the EU, and then by bank loans, whether traditional or ethical/non-traditional. The latter was a financing form par excellence for those investing in (typically expensive) wind energy projects. More alternative forms of financing, such as collecting single donations from individual citizens and crowdfunding, tended to be residual choices, as reported by less than 10% of our dataset, while these forms of financing were completely absent from the German sample. Almost half of the initiatives stated that they had to borrow more than €150,000 to kick-start their initiative, with another significant number (27%) claiming that they did not need to borrow any money. Most of the 'larger' investors were cooperatives, including all of the wind cooperatives, a few homeowner associations and other initiatives that invested in a heating system, and half of the Croatian companies. Among those that did not borrow any capital were most of the public institutions in our dataset, several local cooperatives, associations with a local focus (32 initiatives), and all the informal collectives.

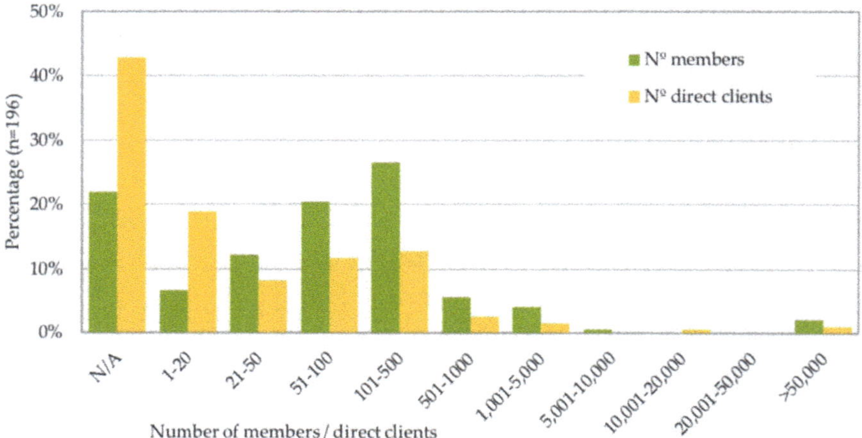

Figure 5. Number of members and/or direct clients of initiatives.

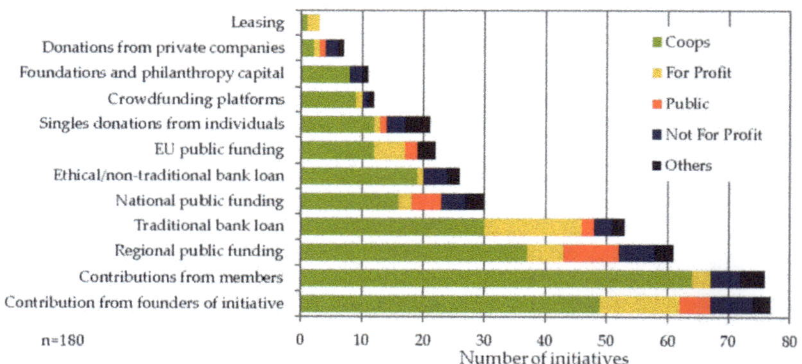

Figure 6. Financing strategies of the initiatives.

Regardless of the initial form of financing, two-thirds of the RES prosumer initiatives from our survey ended up owning their RES installation. In half of the remaining initiatives, the founding or supporting organisation owned the installation. This was the case of the initiatives founded by another cooperative or by an NGO, or that varied their partnerships according to each project (letting the partner own the equipment). The remaining options (such as co-owning with a utility or even the possibility of each member owning an installation) were very residual.

4.3. Organisational Structure of Collective RES Prosumers

As stated earlier, most of the initiatives with members in our dataset were mid-sized, with an average staff of 13, and a median staff of eight, meaning that the teams responsible for running these operations are generally on the smaller side (146 initiatives had less than 15 people involved in running the initiative, and half of those that have members reported between 50–150 members).

There was a significant gender imbalance in most of the surveyed initiatives, as illustrated by Figure 7. Most people working in the prosumer initiatives were male (72% overall), with the biggest imbalance found in German initiatives, where 80% of staff was reportedly male. Only in 17% of the 153 initiatives that answered these questions were there more women active in the initiative than men. Overall, these numbers resemble the gender distribution reported in the literature [44,45]. There were also three examples in the sample that were run by women only and 23 examples that were run by men only. The picture improves slightly when we move from management staff to non-management staff, especially in the public sector; however, on average, the change was only ~10%.

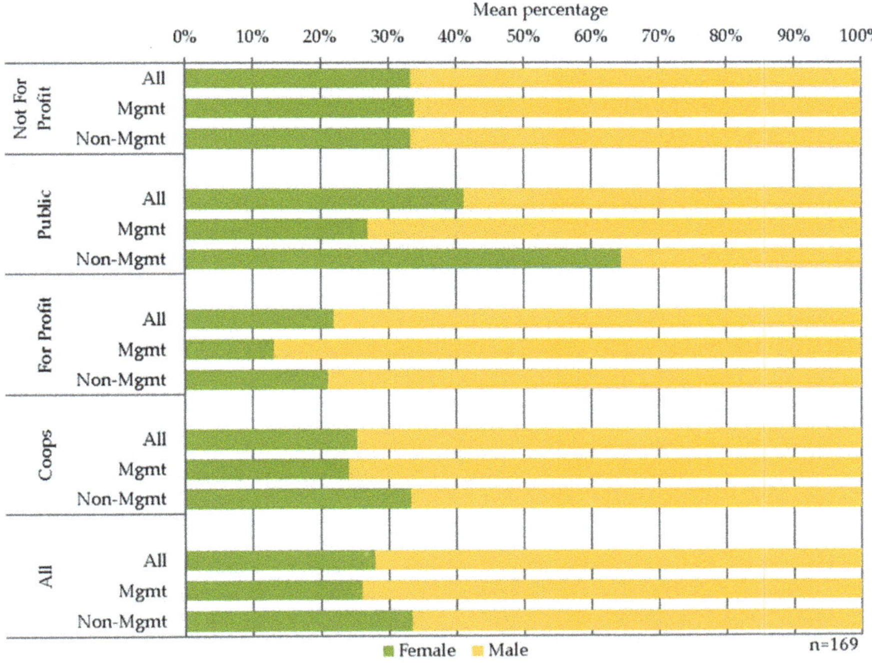

Figure 7. Balance of female vs male staff according to the type of position and top four legal forms.

The differences between legal forms and between countries when it comes to the balance between paid staff and volunteer staff were even more significant (Figure 8). On average, cooperatives depend on volunteers for more than two-thirds of their staff positions, the not-for-profit sector is almost exclusively dependent on volunteers (82% on average), whereas this balance inverts when we look at the other top legal forms. Looking at the different countries, we found that the Belgian and Dutch initiatives were the most dependent on volunteers (89% and 81%, respectively), with German, UK, and French initiatives also showing a high dependence (between 72–75%). The Spanish, Italian, and Croatian initiatives showed the opposite trend: they paid between 71% to 100% of their staff (Croatian initiatives reported 100%, but as mentioned, these were practically all companies), while the Italian cooperatives had been in general established much earlier and often functioned as utilities for their region [19], which may explain their ability to pay their staff. The Spanish initiatives that responded to our survey were highly diverse, with no obvious factor explaining why these were outliers. We will discuss the implications of the dependence on volunteering in the next section.

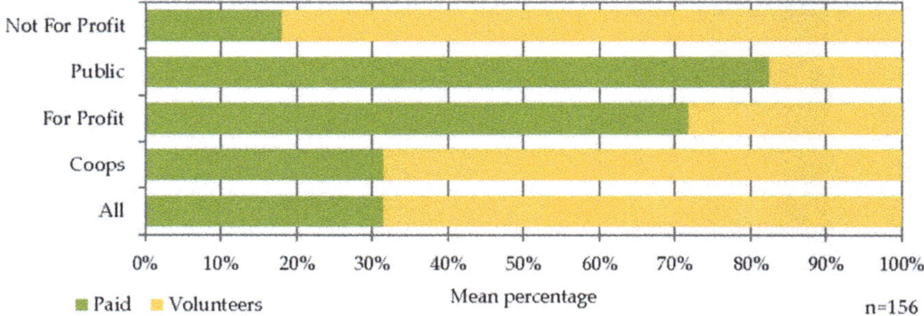

Figure 8. Balance of paid staff vs volunteers overall and for to the top four legal forms.

In terms of governance models, our survey measured how major (strategic) decisions were taken by the initiatives as an indicator of the degree of participation and inclusiveness [46]. Figure 9 presents the decision-making style at three levels of decision-making (founders, core team, and general assembly) as well as the level of involvement of staff in strategic decision-making, ranging from not informing staff, offering lip service to involvement (i.e., simply informing of decisions), asking for opinions, asking for actual input, and involvement in the discussion and analysis resulting in decisions, to fully including those that will be impacted or have the relevant experience in decision-making, and, finally, taking all strategic decisions together with all staff. We offered three forms of decision-making to choose from: majority vote, consensus, and consent, where we defined consensus as a decision on which everyone, without exception, agrees; whereas consent is a decision that not everyone may agree with but that all can live with.

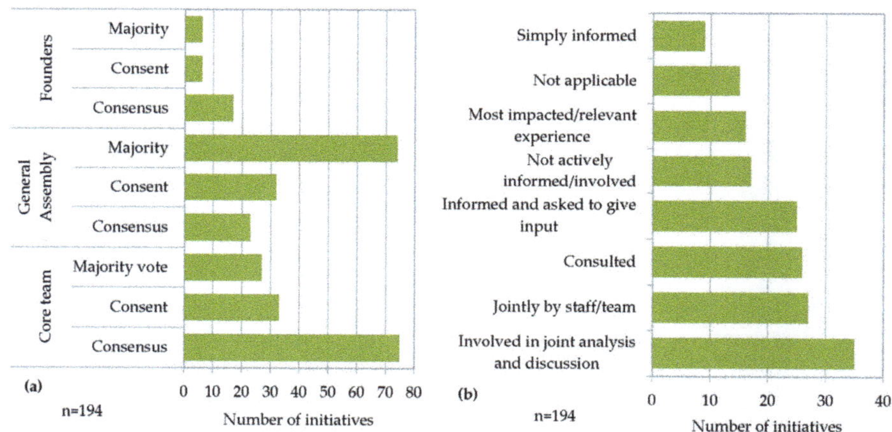

Figure 9. Strategic decision-making at the initiative: (**a**) at three levels of decision-making; (**b**) according to degree of involvement of staff.

About half of the cooperatives and not-for-profits reported that they decided by majority vote at the level of the general assembly. This is a common finding for the functioning of cooperatives and associations at this level, since they are legally obliged to hold at least one general assembly a year. About a quarter of cooperative and not-for-profit initiatives use the consent form of decision-making, with consensus (the most demanding form of decision-making) coming in last place. In contrast, the picture was inverted when it came to making important decisions at the level of the management team/core team and/or the founders. The favoured form here was decision-making by consensus (reported by 76 initiatives at the level of management), followed by a shared second place between consent and majority vote. These results point to issues of trust: in smaller, self-selected groups (i.e., founders, core teams), trust tends to be higher and consensus becomes a non-threatening decision-making tool to use. There were some relevant outliers, which could be further investigated: French initiatives on average reported that they did not use consensus in decision-making at the assembly level, whereas, in contrast, proportionally more of them opted for consent-based decision-making in the core team than the overall average. Other outliers were the initiatives from the UK, which tended to make major decisions primarily at the core-team level by consensus, and those from the Netherlands, where the use of consensus in assemblies was higher than that of consent (but still lower than the majority vote), compared to the overall average.

In terms of involving staff, the RES prosumer initiatives in our dataset clearly appear to favour the more participative/inclusive forms of decision-making. The top most participative forms of staff involvement were also the most reported by our respondents (close to 60% chose these). Importantly, most initiatives that reported the more participative forms of decision-making were cooperatives, thus staying true to the spirit of this organisational form. Of our 'big four' prosumer countries, only the UK initiatives deviated from the trend, converging on less participatory forms. The public sector initiatives, albeit a small sample, showed a clear trend in their answers: they either did not know if staff were involved, or stated that these were not informed. Since public sector organisations tend to follow stricter and multi-levelled hierarchies, this was not a surprising outcome.

Another measure we used to gauge inclusiveness in RES prosumer initiatives was the type of criteria for joining the initiative (Figure 10). The most popular answer was the absence of criteria for joining, followed by the need to be a local resident, and by the impossibility of joining (mostly the case of public institutions and companies). Although small in number (15), it is worth mentioning that some initiatives stated that a mandatory investment was their main criterium. This category would be larger if initiatives had not been forced to choose one answer option, since it is quite likely that most of the prosumer initiatives with members will require them to contribute upon joining. Finally, several initiatives made a point of mentioning that newcomers should agree with the initiative's principles and/or goals, highlighting a desire to create a sense of community. Again, the exact number could be higher if initiatives had not been forced to only choose one answer. The UK initiatives once again represented an outlier, with mandatory investment coming up as the second most important criterium, after 'no criteria'.

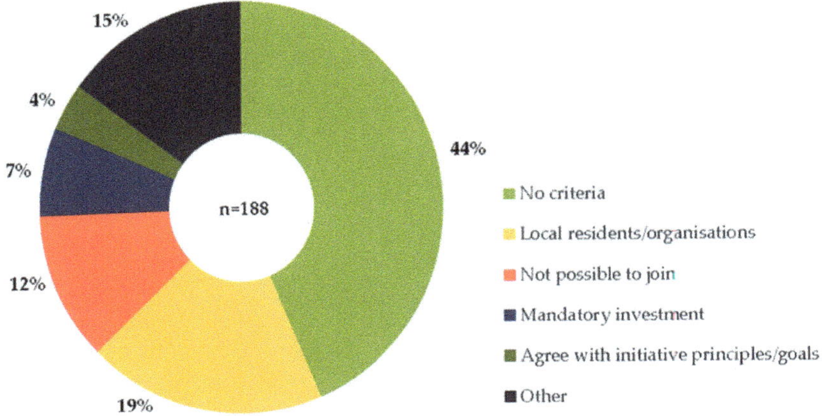

Figure 10. Criteria for joining the initiative.

We further asked respondents to indicate with which stakeholders (of a list of 15) they tended to collaborate and for what purposes (of a total of five types of relationships: 'knowledge sharing', 'self-promotion', 'access to funding', 'access to human resources', and 'access to material resources'). Their answers are shown in Figure 11 for the cooperatives, our largest sub-set, whereas answers for the top legal forms can be found in the Supplementary Materials (Document S2). Overall, the most popular stakeholder group with which prosumer initiatives engaged was that of communities/collectives and/or cooperatives for the purpose of knowledge sharing, followed by engagement with citizens/households for purposes of self-promotion, and finally, contact with civil society organisations and other prosumers, again for the purpose of knowledge sharing. All legal forms, except for public institutions, additionally showed an interest in engaging with local and regional government for the purpose of self-promotion. For-profit companies differed from other forms in that they engaged more with regional and national government for purposes of fund raising than any of the other forms of relationships. Public institutions slightly favoured self-promotion over knowledge sharing with citizens/households and civil society organisations, which is unsurprising considering their function is often to regulate and mediate, while outright cooperation with citizens and civic organisations is a more recent phenomenon. Nevertheless, although the sample was small, public institutions were the only legal form to almost equitably try to engage with all other stakeholders, reinforcing the idea of public authorities as mediators or hubs for energy transition. The for-profit sector also tended to favour self-promotion over knowledge sharing, a trait common to the business sector. Finally, all legal forms, except for the for-profit sector, showed considerable interest in engaging with national networks, interest organisations, or social movements. The opportunities mentioned by some respondents included the building of synergies between RES prosumerism and other climate-friendly activities, such as energy efficiency measures and awareness creation.

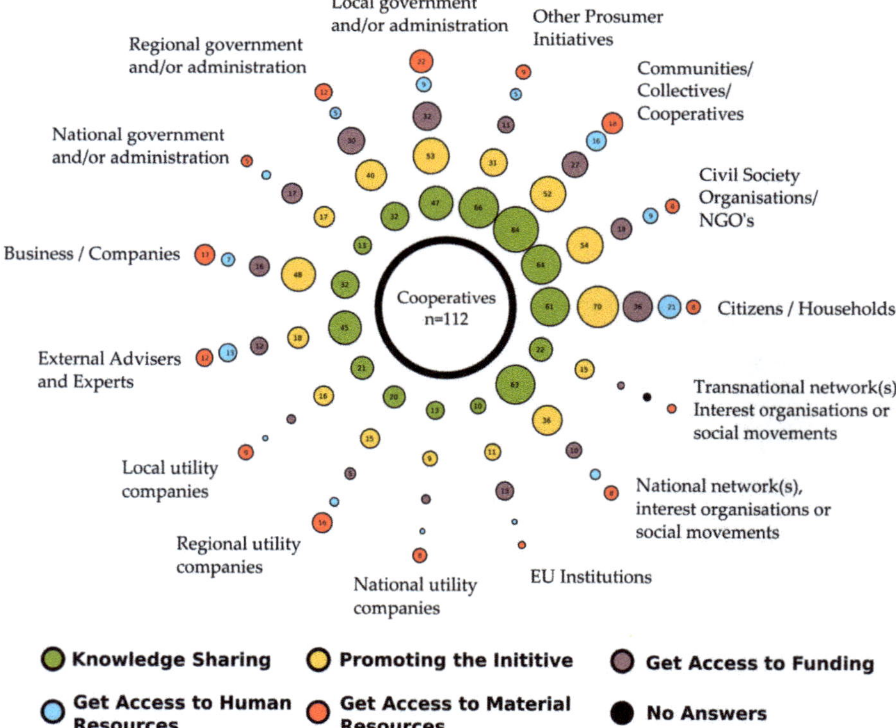

Figure 11. Key networking relationships cultivated by initiatives (please consult Document S2 for plots of top legal forms).

4.4. Key Drivers and Perceived Hindering and Success Factors

One of our main objectives was to understand the diversity of drivers behind the development of collective RES prosumer initiatives as well as hear from them what they perceive to be the key success factors and barriers to their development.

We asked respondents to grade, on a Likert scale, the degree to which certain drivers motivated them to start the initiative (spider graph in Figure 12). The outcome was quite unequivocal. Over 60% attributed the highest score to the driver 'tackling the climate change problem', followed by 'being part of the clean and low carbon transition', 'decentralising production', and finally the possibility to 'create a sense of community', closely followed by 'taking advantage of new RES technologies'. This overall trend was mirrored by the energy cooperatives in our dataset, except for the latter motivation, which was replaced by 'reducing the environmental impact of existing activities of the organisation/collective/community'. The for-profit sector appeared to be divided about their key motivation between 'responding to local demand/needs' and 'tackling the climate change problem', with the second place also divided, this time between 'being part of the low-carbon energy transition' and 'decentralising energy production'. The public sector differed from the previously mentioned legal forms by electing 'reducing the environmental impact of existing activities of the organisation/collective/community' as their second choice, while it did not appear to value 'creating a sense of community'. Finally, the not-for-profit sector placed 'reducing energy costs' in second place, before 'being part of the clean and low carbon transition', and relegated 'decentralising production' to fifth place.

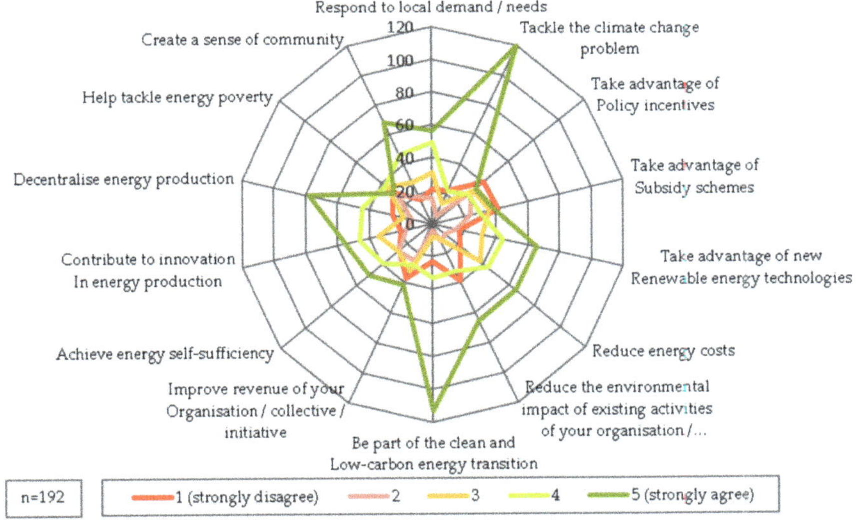

Figure 12. Main reasons for starting a prosumer initiative.

As an indirect measure of motivation, we asked RES prosumer initiatives about any additional services they might offer (Figure 13) and obtained mixed results. A third of initiatives focussed exclusively on self-production and consumption, and offered no other services. About half of these were cooperatives, while the other half was made up of public institutions, some smaller associations, and two of the Croatian aggregator companies. When additional services were offered, energy efficiency advice took first place, followed by community-focussed services, such as community organising. Energy storage appeared here as an upcoming technology as much as an additional service, with several initiatives contemplating offering this service in the near future.

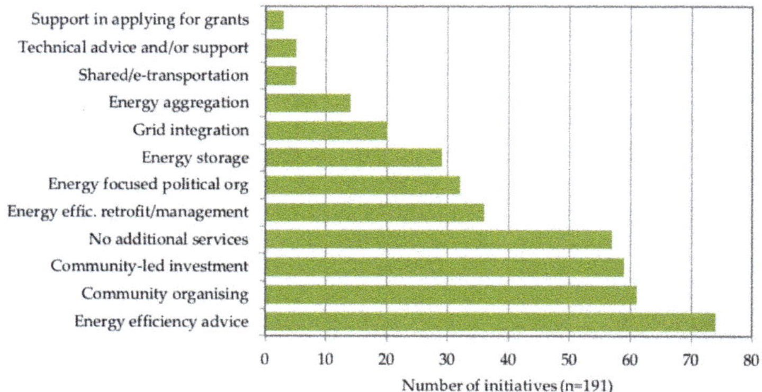

Figure 13. Additional services offered by prosumer initiatives.

Besides the key drivers, the key facilitating and hindering factors as perceived by our respondents were among our most elucidating results. Respondents were asked to choose the top three factors that most facilitated as well as those that most slowed down the development of their initiative, in their opinion. The results are presented in the spider graph in Figure 14 below.

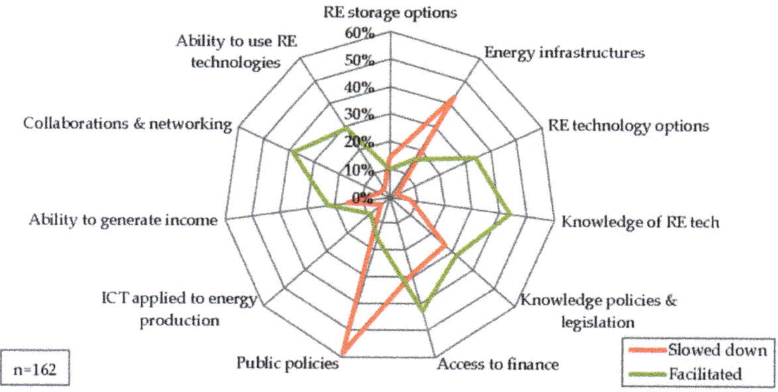

Figure 14. Main hindering (slowing down) and facilitating factors for the development of a prosumer initiative.

We found that the top four factors perceived as most facilitating by our respondents were:

- Knowledge of renewable energy technologies.
- Access to finance, subsidies or grants.
- Collaborating and networking with others.
- Renewable energy technology options available.

These factors were consistent across the countries as well as the legal forms, except for initiatives in France and other countries newer to RES prosumerism (PT, ES). These tended to rank the factor of 'ability to use RES technology' above that of 'renewable energy technology options available'.

The top four factors perceived as most hindering by our survey respondents were:

- Public policies and legislation for renewable energy initiatives.
- Energy infrastructures (e.g., grid, meter).
- Access to finance, subsidies, or grants.
- Knowledge of policies and legislation in RES production.

These 'negative' factors were equally consistent across the countries as well as the legal forms. Curiously, there were a couple of factors that received a high classification in both categories, meaning that, depending on whether the initiative can harness the factor in question, it will be either a facilitating factor or a barrier. Such is the case for 'access to finance, subsidies or grants', which is considered to be critical to the initiative's development as well as a contributor to its potential failure. This was also true for 'knowledge of policies and legislation in RES production', which was considered as important for both the successful development (rated in fifth place) as for its failure when absent. It is illustrative of the importance of networking that the socio-political factor of 'collaborating and networking with others' was deemed as important as having the appropriate RES technology.

Several specific complaints were volunteered by respondents from different countries. German, Dutch, and Belgian respondents complained about complex bureaucracies amidst inconsistent laws and rules, and about the contradictory attitudes of authorities at different levels (regional vs national). French initiatives complained that they were not allowed to consume what they produced, while UK initiatives expressed their apprehension about the end of FiT. In smaller countries such as the Netherlands and Belgium, a lack of space for RES installations was reported as another barrier.

5. Discussion

Taking stock of the survey results and comparing them to our review of the context for RES prosumerism, we found that true collective prosumers (in the sense of simultaneous self-production and -consumption in a collective context) were hard to find. In some countries, of which France is an example, energy may be produced and sold, but not self-consumed. In other countries (Croatia and Germany), it is quite easy to self-consume, but very hard to sell (a license is needed). The nine countries under study, as explained in detail in [10], varied significantly as to whether they recognised energy communities, allowed neighbours in the same building or apartment block to self-consume collectively, allowed energy communities to share electricity among members, or whether a supplier license was needed, to name but a few legislative features.

Our results open up several avenues for future research. Concerning legal forms, given the variety of forms that prosumer initiatives can choose from (see Table 3), we were expecting to find a high number of energy cooperatives, a legal form that allows for the hybridisation of socio-ecological objectives and the ability to make profit as well as share the latter among 'shareholders' (i.e., the members of the cooperatives). We were not expecting to find so many other hybrid forms such as public–private partnerships, other formal and informal partnerships, and the choice of the Ltd. or associative form to represent energy communities, which the EC so far does not recognise as a legal entity in itself (see Clean Energy Package Directives [3,5]). The occurrence and implications of hybrid organisational forms in the field of energy, particularly the cases of energy cooperatives and social enterprises, is becoming a topic in itself, and has been discussed by a number of researchers: Raven [47], Huybrechts and Haugh [8], and Bauwens et al. [48]. Although the cooperative form appears to offer initiatives that identify as energy communities a satisfying legal entity, it is also patent in our survey results that this choice is not always possible or ideal. In the Netherlands, to qualify for the so-called 'PostCodeRoos' incentive, you need to be a cooperative, whereas in France, you can run a for-profit organisation such as a SAS (Société par actions simplifiée) as a cooperative, reaping benefits from both organisational forms. More qualitative or in-depth research among RES prosumer initiatives operating under different legal forms may explain better what is happening.

As for the energy needs addressed by RES prosumer initiatives, electricity clearly stands out. Factors influencing the choice of energy needs that prosumer initiatives wish to address still require further research. Among these are the fact that many RES subsidies are for electricity production and that RES electricity is easier to share collectively, whereas a heating system requires very specific conditions (e.g., the need to refit entire blocks, neighbourhoods, or districts, a higher entry barrier, a significant change in basic grid infrastructure).

Although we showed earlier (in Section 2) that official country numbers for the shares of renewable energy in gross final energy consumption maintain wind energy as the leading RES technology, the trend encountered in our results—with photovoltaic PV leading the RES technology choices—may be explained by the fact that the growth of wind energy has occurred largely independently from prosumerism, with projects mostly developed by energy companies [49]. Additionally, wind energy projects are very costly [31]. On the other hand, solar-powered technology is growing fast, and it has proven easier to build and set up. It is therefore expected to take the lead from wind energy in the upcoming years [50].

Gender is without a doubt an important crosscutting topic for research on RES prosumers. The extremely low overall involvement of women in energy initiatives has already been mentioned in a number of studies and reviews [34,35]. Some explanations for this, in the case of Germany, can be found in [44], whereas French data indicate that gender differences are not just linked to individual preferences and investment attitudes, but also significantly influenced by cultural, social, and political factors [33]. There were interesting country differences regarding the gender balance between our initiatives. Whereas the average percentage of women in the initiatives in our dataset did not exceed 30%, Portuguese and Spanish initiatives reported on average close to 50% women in management positions, and 60% in non-management positions. Spain and Portugal happen to be countries that have been flagged for low levels of overall volunteer participation, which is why, despite the small sample size for these countries, finding more female than male volunteers merits further investigation [51]. Between the legal forms, no significant differences can be reported as to the female/male balance, apart from the public sector, the only type of organisation with high numbers of female staff, albeit mostly in non-management positions.

Regarding the financing choices of our respondents, we found that most of those that did not borrow any capital were public institutions, aside from several local cooperatives, associations with a local focus (32 initiatives), and all of the informal collectives. These patterns closely resembled those obtained for community energy in Germany [52]. Kahla, for instance, reported an 'inverted leverage effect' that has been observed for social enterprises in general: high equity ratios may be explained by lower costs of equity for some initiatives compared with costs of debt (for more also see [53,54]). Another observation that we can make is that the United Kingdom, one of the countries with the broadest mix of financing in our dataset, is also the country with the least initiatives opting for a traditional bank loan (9%), whereas Germany and France have quite a high share (46 and 78%, respectively), reflecting different banking cultures and/or systems in these countries. UK initiatives appeared on average to be more professional and better financed, but the recent abolishment of the Feed-in-Tariff may turn this panorama upside down [23].

The extreme dependence of cooperatives on volunteer labour constitutes one of their key weaknesses, as reported by our respondents: one representative from a Dutch cooperative put it this way: 'We need to move from hobby to lobby, from volunteer organisation to professionalisation' (translated from the original Dutch). This fragility is also something Brummer discusses in his comparative review of community energy in the USA, UK, and Germany [30]. He found that a reliance on volunteers was a barrier in the USA and Germany, but not in the UK. Although volunteers are generally highly motivated, their lack of expertise in certain areas, and possible limitations in terms of dedication, means that cooperatives and energy communities will need to spend money on qualified consultants. Research on non-profit organisations highlights professionalisation as a stage of development in the life-cycle of non-profits [55,56], which respond to external pressures—'isomorphism' [57]—or rely on government grants and trading [58,59]. As a reaction, non-profits have taken different paths [60]. However, research into not-for-profit organisations also documents some flipsides of professionalisation: the danger of 'mission drift', less engagement by volunteers, or diminishing capacity of social capital production [61–63]. In part, these negative effects seem to rest on whether members or external staff are employed [59–61]. Against this background, current EU policy and transposition of directives into national law may create space for the professionalisation of some collective prosumer initiatives, but will most probably also lead to hybridisation and organisational differentiation in this subsector. The heavy reliance on volunteers in the largest sub-sector of collective RES prosumerism may constitute an underappreciated barrier that current EU policy is not addressing.

Our study confirmed the predominance of social drivers over financial and more inward-focussed motivations for collective forms of RES prosumer initiatives. Over 60% of respondents attributed the highest importance to tackling the climate change problem and being part of the clean and low carbon transition, while about half also attached importance to the decentralisation of production and to creating a sense of community. While the nuanced differences in terms of motivation found in the for-profit sector and public institutions can be explained by their respective missions (see Section 4.4.), it remains unexplained—but does not speak to a favourable policy landscape—why over 20% of all legal forms, except for the for-profit sector, had a surprisingly negative view of the possibility of 'taking advantage of policy incentives' or 'subsidy schemes'. Whether this is a result of the difficulty in obtaining said support or whether they find this an unsound reason to start an initiative, is not clear. In contrast, the Croatian initiatives gave high scores to taking advantage of policy incentives as well as subsidy schemes, while 'reducing energy costs' and 'improving revenues of their organisation' also received high votes. Since the legislation in Croatia favours larger, profit-oriented set-ups [10], there is a lack of incentive and support for other forms of prosumer initiatives, particularly more decentralised ones.

Among our strongest findings was the perceived hindering vs facilitating factors reported by our dataset. Consistent across countries as well as legal forms (with the exception of France and recent prosumer countries such as Portugal and Spain, which were more concerned with the ability to be able to use RES technology), we found that initiatives listed as their top four facilitating factors the knowledge of renewable energy technologies, access to finance/subsidies/grants, collaborating and networking with others, and the availability of renewable energy technology options. However, initiatives in our survey felt that having good RES technology options was not sufficient if relevant energy infrastructures were not in place. In an overwhelming first place, existing public policies and legislation for RES initiatives were perceived as a key barrier. This reinforces our findings that the fact that legislation in the surveyed nine countries is currently either being revised, or likely to be revised after the new EU directives come into effect, creates an unstable and uncertain environment for RES prosumerism to flourish. Around 60% of our respondents signalled RES public policies and legislation as the top hindering factor to their development. Indicative of a potential crucial barrier to prosumerism development, access to finance appears as both a facilitating and a hindering factor. Furthermore, the importance attributed to collaboration speaks to its multi-functional aspect, allowing initiatives to join others (strength in numbers), learn from others as well as learn together, promote themselves and their common cause, and share resources. Cooperation of different sorts allows the prosumer initiatives to build up know-how beyond what they would have achieved alone, a function that has also been observed for cooperation among municipal utilities [64,65]. Free knowledge sharing and the use of open source tools are characteristic of the cooperation between collectives that do not have profit as their primary objective, and can jumpstart collaborative economies.

Finally, the initiatives in our dataset showed considerable interest in engaging with national networks, interest organisations, or social movements, a finding that illuminates patterns of cooperation in a sub-sector that shares some similarities with the ones found among (local) municipal utilities [66].

6. Conclusions

This paper examined a wide spectrum of collective prosumers beyond the better-known forms of energy cooperatives. The aim was to establish the current state of play for collective forms of RES prosumerism in Europe considering the demands and promises of the Energy Union. Our documentary review and survey across nine EU countries revealed key differences, challenges, and needs across different types of collective prosumers and across national contexts, and can therefore inform the design of an incentive system supporting clean, fair, and sustainable energy transition pathways. Our research established a comprehensive baseline and a broad cross-section of the diverse profiles of the RES prosumer energy actors, raising several red flags, such as the persistence of an uncertain political and legislative setting; the challenges of volunteer-run structures; the lack of tailor-made policies for collective RES prosumer initiatives—namely those with a civic focus; slow progress in terms of the democratisation of critical energy infrastructures—in particular, the digital infrastructure; the difficulties of accessing finance for RES prosumer initiatives; and the need to more widely share knowledge of RES technology options as well as of how to implement and run RES installations. It also pointed out opportunities and new pathways including the chance to create synergies between RES prosumerism and other climate friendly activities (e.g., complementing prosumerism with energy efficiency measures or awareness creation); the possibility to improve collaborations and knowledge-building between different stakeholders; or the ability for RES prosumers to also become energy suppliers.

Based on these results, we highlight in the following a number of dilemmas for RES prosumer collectives and provide recommendations for national legislators in the transposition of EU directives as well as for future research.

The first dilemma relates to the way the collectives are internally organised: two main types can be distinguished, those collectives relying on volunteer work and thus civic activism, and those relying on paid staff and thus a more commercial and/or bureaucratic attitude. With regard to making energy transitions more inclusive, the—on the surface more accessible—civic activism holds a risk of exclusion as it means that only those who have the time and resources can afford to volunteer. A related concern in this dilemma is the chronic under-representation of women in energy initiatives.

The second dilemma relates to the choice of an appropriate legal form/organisational structure for a RES prosumer initiative: this is a less straightforward choice than one would expect since it is tied to a number of conditions and factors, such as the availability of different legal forms in different countries, or how specific legal forms are tied to specific support and subsidy schemes, and whether there is an obligation to apply for a production license. The reality of legal forms leads to potential conflicts when it presupposes a certain value orientation (such as a for-profit orientation), while the collective may be aspiring to combine such for-profit goals with social goals and thus a more civic-oriented role in the energy transition.

A third dilemma relates to the further formalisation of prosumer initiatives through the advent of the Clean Energy Package. While this promises clarity and support, it also forces such initiatives to formalise. The fact that the newly coined EU concepts for collective RES prosumers—'renewable energy community' and 'jointly acting renewables self-consumers'—imply that the collective must choose one or other legal form to run the community rather than being able to register as such, could limit rather than stimulate the expansion of the more civic-inspired prosumer initiatives. Informal groups or partnerships will not be able to qualify as an energy community. With this limitation comes the risk of hindering the decentralisation of the energy system and the uptake of RES. It is now up to national governments to pick up this challenge as they implement the new EU Directives, a challenge that implies more diverse interpretations and treatment of prosumers of different types in different EU countries, to the benefit of an inclusive, clean, fair, democratic, but also rapid energy transition.

We propose several recommendations for policy-makers from Member States involved in the transposition of the Directives from the Clean Energy package:

- Develop supporting legal and institutional frameworks for collective forms of RES prosumerism that recognise and support a range of organisational forms;
- Work with national regulators and network operators to ensure fair, open, and transparent access for prosumers to the electricity network infrastructure;
- Develop long-term and consistent approaches to financial support for prosumerism, avoiding cliff edges and uncertainty;
- Harness local and bottom-up solutions to solve energy system challenges, recognising the social/non-financial value that is created by these solutions;

- Ensure and support fair, open, and inclusive participation in the prosumer energy transition; especially for marginalised groups, such as women, ethnic minorities, and those with limited material resources.

Within the methodological limitations of a survey study, we have achieved a broad range of cross-cutting aspects of the collective forms of RES prosumers and provided a baseline overview of 'what is'. Now, ongoing and future avenues of research can build on these exploratory findings and examine them more in-depth, across different types of prosumer collectives as well as across cultures and countries in Europe. The most relevant avenues to further explore, in our view, are:

- The implications and policy recommendations stemming from the mismatch between an initiative's legal/organisational form and its mission;
- The key factors driving renewable energy technology needs and choices;
- The reasons behind and fixes for the differences in female/male balance in terms of participation;
- The challenges and fragilities of volunteer-run organisations, and how to overcome these;
- The significance of different internal decision-making and governance styles as well as viable business and financing models that support RES prosumerism.

Supplementary Materials: The following are available online at http://www.mdpi.com/1996-1073/13/2/421/s1, Document S1: Survey form 'New Energy for Europe'; Document S2: Key networking relationships for key RES prosumer legal forms.

Author Contributions: Conceptualisation, L.H. (Lanka Horstink), J.M.W., K.N. and I.C.; methodology: L.H. (Lanka Horstink), J.M.W., K.N., G.P.L., E.M.-G., S.G., I.C. and L.H. (Lars Holstenkamp); software, G.P.L.; validation, E.M.-G., S.G. and G.P.L.; formal analysis, E.M.-G., G.P.L., S.G. and L.H. (Lanka Horstink); investigation, L.H. (Lanka Horstink), J.M.W., G.P.L., E.M.-G., S.G., I.C., S.O., D.B. and K.N.; data curation, E.M.-G., S.G. and G.P.L.; writing—original draft preparation, L.H. (Lanka Horstink), and J.M.W.; writing—review and editing, J.M.W., K.N., I.C., S.O., L.H. (Lars Holstenkamp), E.M.-G., S.G. and D.B.; visualisation, E.M.-G., S.G. and G.P.L.; supervision, K.N., L.H. (Lanka Horstink); project administration, I.C. and L.H. (Lanka Horstink). All authors have read and agreed to the published version of the manuscript.

Funding: The research leading to the results presented in this article has received funding from the European Union's Horizon 2020 research and innovation programme under grant agreement N° 764056. K.N. was supported by the Fundação para a Ciência e a Tecnologia (FCT) postdoctoral fellowship grant (SFRH/BPD/120394/2016).

Acknowledgments: The authors wish to thank the following researchers for their contribution to the data collection and to the discussion that formed the basis of the present analysis: Mark Soares (UPORTO); Jeremie Fosse (ECO-UNION); Kristian Petrick (ECO-UNION); Mireia Reus (ECO-UNION); Thijs Scholten (CE DELFT); Bettina Kampman (CE DELFT); Mark Davis (ULeeds); Stephen Hall (ULeeds); Arthur Hinsch (ICLEI); Marta Toporek (CLIENTEARTH); Michele Zuin (ICLEI); Moritz Ehrtmann (LEUPHANA); Tomislav Novosel (UNIZAG); Tomislav Puksec (UNIZAG); and Ana Lovrak (UNIZAG).

Conflicts of Interest: The authors declare no conflicts of interest. The funders had no role in the design of the study; in the collection, analyses, or interpretation of data; in the writing of the manuscript; or in the decision to publish the results.

Appendix A

Table A1. Survey questionnaire: distribution of questions according to category. Adapted from: [23] (pp. 56–57).

1. Control questions
Q1 (name of initiative)
Q2 (consume RES yes/no)
Q4 (job title of respondent)
Q32 (additional information the initiative might want to give)

2. General demographics of collective RES prosumers
Q3 (legal form)
Q5 (starting date)
Q6 (location)
Q7 (scale)
Q8 (energy needs addressed)
Q20 (N° of members)
Q21 (N° of direct clients)

Table A1. Cont.

---

3. Use of technology by collective RES prosumers
Q9a (which technologies are used or planned)
Q9b (installed capacity of these technologies)
Q10 (is the initiative connected to the grid)
Q11 (when did they start producing/plan producing)
Q12 (energy produced in 2017 for each technology)

---

4. Governance/organisation of collective RES prosumers
Q15–19 (staff characteristics: total number, women/men proportion, volunteer/paid staff proportion)
Q26 (decision-making style in executive organs)
Q27 (involvement/participation of staff/non-management teams in decision-making)
Q28 (networking, openness to others)

---

5. Financing of the initiative
Q22 (who owns the RES equipment)
Q23 (how are initiative activities financed)
Q24 (how much capital was borrowed, if any)
Q25 (what are the 4 largest income generators)

---

6. Motivation/ambition of the initiative
Q13 (whom is energy produced for?)
Q14 (any additional services that are offered)
Q30 (Likert scale (1-5) of reasons to start the initiative)

---

7. Hindering and facilitating factors as perceived by collective RES prosumers
Q31 (which 3 factors have most slowed down and which 3 factors have most facilitated the development of the initiative)

Appendix B

Excerpts from the survey form for collective RES prosumers.

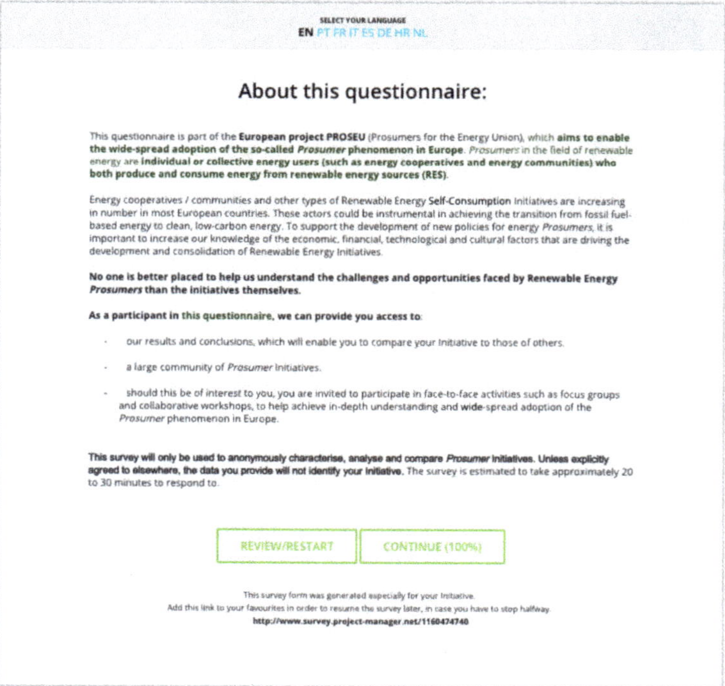

Figure A1. Front page of the survey form.

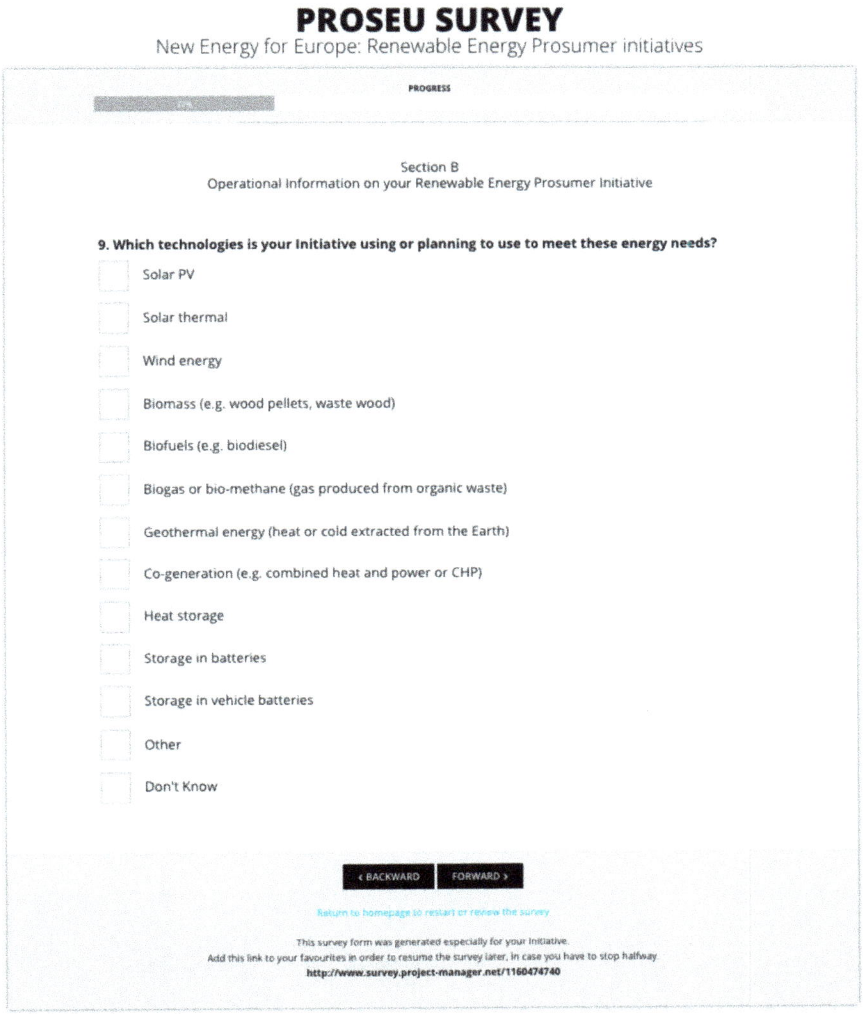

PROSEU SURVEY
New Energy for Europe: Renewable Energy Prosumer initiatives

PROGRESS

Section B
Operational Information on your Renewable Energy Prosumer Initiative

9. Which technologies is your Initiative using or planning to use to meet these energy needs?

- Solar PV
- Solar thermal
- Wind energy
- Biomass (e.g. wood pellets, waste wood)
- Biofuels (e.g. biodiesel)
- Biogas or bio-methane (gas produced from organic waste)
- Geothermal energy (heat or cold extracted from the Earth)
- Co-generation (e.g. combined heat and power or CHP)
- Heat storage
- Storage in batteries
- Storage in vehicle batteries
- Other
- Don't Know

‹ BACKWARD   FORWARD ›

Return to homepage to restart or renew the survey

This survey form was generated especially for your Initiative.
Add this link to your favourites in order to resume the survey later, in case you have to stop halfway.
http://www.survey.project-manager.net/1160474740

Figure A2. Question 9 of the survey form.

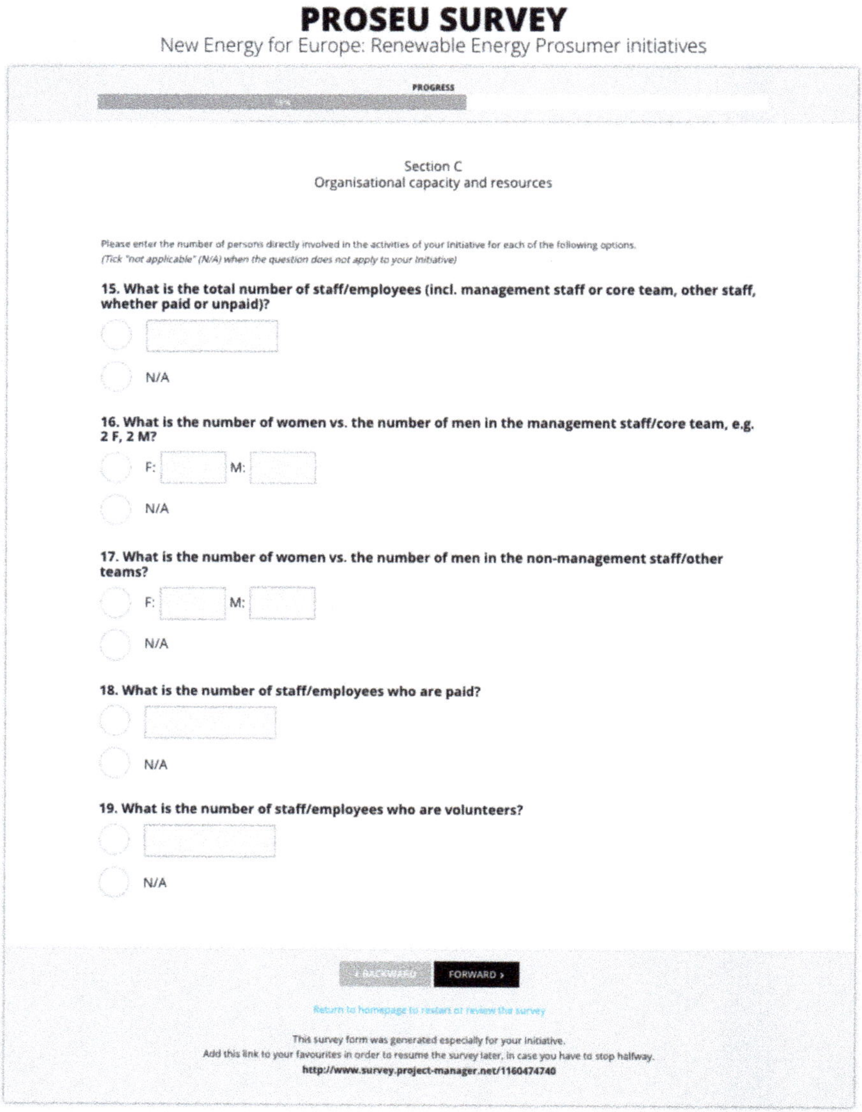

Figure A3. Questions 15–19 of the survey form.

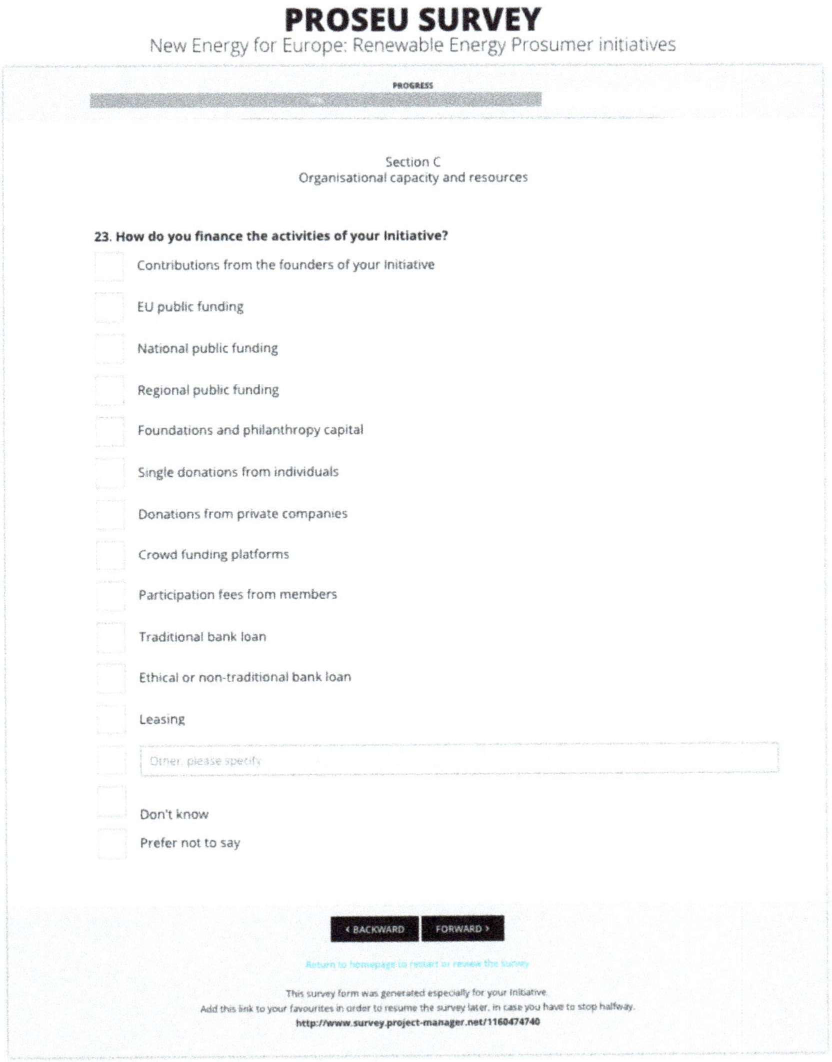

PROSEU SURVEY
New Energy for Europe: Renewable Energy Prosumer initiatives

PROGRESS

Section C
Organisational capacity and resources

23. How do you finance the activities of your Initiative?

- Contributions from the founders of your Initiative
- EU public funding
- National public funding
- Regional public funding
- Foundations and philanthropy capital
- Single donations from individuals
- Donations from private companies
- Crowd funding platforms
- Participation fees from members
- Traditional bank loan
- Ethical or non-traditional bank loan
- Leasing
- Other, please specify
- Don't know
- Prefer not to say

< BACKWARD   FORWARD >

Return to homepage to restart or renew the survey

This survey form was generated especially for your Initiative.
Add this link to your favourites in order to resume the survey later, in case you have to stop halfway.
http://www.survey.project-manager.net/1160474740

Figure A4. Question 23 of the survey form.

References

1. European Commission Commission Proposes New Rules for Consumer Centred Clean Energy Transition. Available online: https://ec.europa.eu/energy/en/news/commission-proposes-new-rules-consumer-centred-clean-energy-transition (accessed on 25 April 2019).
2. European Commission Clean Energy for All Europeans Package Completed: Good for Consumers, Good for Growth and Jobs, and Good for the Planet. Available online: https://ec.europa.eu/info/news/clean-energy-all-europeans-package-completed-good-consumers-good-growth-and-jobs-and-good-planet-2019-may-22_en (accessed on 28 May 2019).
3. European Parliament and Council of the European Union DIRECTIVE (EU) 2019/944 OF THE EUROPEAN PARLIAMENT AND OF THE COUNCIL of 5 June 2019 on Common Rules for the Internal Market for Electricity and Amending Directive 2012/27/EU (Recast). Available online: https://eur-lex.europa.eu/legal-content/EN/TXT/?qid=1570285915486&uri=CELEX:32019L0944 (accessed on 5 October 2019).

4. European Commission. *The State of the Energy Union Explained*; Fact sheet; European Commission: Brussels, Belgium, 2019; p. 5.

5. European Parliament and Council Directive (EU) 2018/2001 of the European Parliament and of the Council of 11 December 2018 on the Promotion of the Use of Energy from Renewable Sources. Available online: https://eur-lex.europa.eu/legal-content/EN/TXT/?uri=uriserv%3AOJ.L_.2018.328.01.0082.01.ENG& toc=OJ%3AL%3A2018%3A328%3ATOC (accessed on 13 March 2019).

6. Ford, R.; Stephenson, J.; Whitaker, J. *Prosumer Collectives: A Review*; New Zealand's Smart Grid Forum, Centre for Sustainability; University of Otago: Dunedin, New Zealand, 2016; pp. 1–28.

7. Bertram, R.; Primova, R. (Eds.) *ENERGY ATLAS 2018: Facts and Figures About Renewables in Europe*; Heinrich Böll Foundation, FOEE, EREF, GEF: Luxembourg, 2018; p. 56.

8. Huybrechts, B.; Haugh, H. The Roles of Networks in Institutionalizing New Hybrid Organizational Forms: Insights from the European Renewable Energy Cooperative Network. *Organ. Stud.* **2018**, *39*, 1085–1108. [CrossRef]

9. Lavrijssen, S.; Carrillo Parra, A. Radical Prosumer Innovations in the Electricity Sector and the Impact on Prosumer Regulation. *Sustainability* **2017**, *9*, 1207. [CrossRef]

10. Toporek, M.; Campos, I.S. *Assessment of Existing EU-Wide and Member State-Specific Regulatory and Policy Frameworks of RES Prosumers (Deliverable N°3.1)*; ClientEarth: Brussels, Belgium, 2019; p. 128.

11. Haggett, C.; Creamer, E.; Harnmeijer, J.; Parsons, M.; Bomberg, E. *Community energy in Scotland: The Social Factors for Success*; University of Edinburgh: Edinburgh, UK, 2013; p. 25.

12. Dóci, G.K. Renewable Energy Communities. A Comprehensive Study of Local Energy Initiatives in the Netherlands and Germany. Ph.D. Thesis, Vrije Universiteit Amsterdam, Amsterdam, The Netherlands, 2017.

13. Dóci, G.; Vasileiadou, E.; Petersen, A.C. Exploring the transition potential of renewable energy communities. *Futures* **2015**, *66*, 85–95. [CrossRef]

14. Smith, A.; Hargreaves, T.; Hielscher, S.; Martiskainen, M.; Seyfang, G. Making the most of community energies: Three perspectives on grassroots innovation. *Environ. Plan A* **2016**, *48*, 407–432. [CrossRef]

15. Walker, G.; Devine-Wright, P.; Hunter, S.; High, H.; Evans, B. Trust and community: Exploring the meanings, contexts and dynamics of community renewable energy. *Energy Policy* **2010**, *38*, 2655–2663. [CrossRef]

16. Lacey-Barnacle, M.; Bird, C.M. Intermediating energy justice? The role of intermediaries in the civic energy sector in a time of austerity. *Appl. Energy* **2018**, *226*, 71–81. [CrossRef]

17. Soares, M.; Brown, D.; Gährs, S.; Horstink, L.; Novosel, T.; Oxenaar, S.; Petrick, K.; Puksec, T.; Toporek, M. *PROSEU Renewable Energy Country Fact Sheets*; University of Porto: Porto, Portugal, 2019; p. 20. Available online: https://zenodo.org/record/3247376#.XfZ9yOvgp0w (accessed on 5 October 2019).

18. Gross, R.; Hanna, R. Path dependency in provision of domestic heating. *Nat. Energy* **2019**, *4*, 358–364. [CrossRef]

19. Wirth, S. Communities matter: Institutional preconditions for community renewable energy. *Energy Policy* **2014**, *70*, 236–246. [CrossRef]

20. Romero-Rubio, C.; de Andrés Díaz, J.R. Sustainable energy communities: A study contrasting Spain and Germany. *Energy Policy* **2015**, *85*, 397–409. [CrossRef]

21. Statistical Office of the European Union. *EUROSTAT: SHARES 2017—Short Assessment of Renewable Energy Sources*; Eurostat: Brussels, Belgium, 2019.

22. Campos, I.; Pontes Luz, G.; Marín-González, E.; Gährs, S.; Hall, S.; Holstenkamp, L. Regulatory challenges and opportunities for collective renewable energy prosumers in the EU. *Energy Policy* **2019**, in press.

23. Horstink, L.; Luz, G.P.; Soares, M.; Ng, K. *Review and Characterisation of Collective Renewable Energy Prosumer Initiatives*; University of Porto: Porto, Portugal, 2019; p. 156.

24. Butenko, A.; Cseres, K. *The Regulatory Consumer: Prosumer-Driven Local Energy Production Initiatives*; Social Science Research Network: Rochester, NY, USA, 2015.

25. Butenko, A. *User-Centered Innovation and Regulatory Framework: Energy Prosumers' Market Access in EU Regulation*; Social Science Research Network: Rochester, NY, USA, 2016.

26. Howaldt, J.; Schröder, A.; Kaletka, C.; Rehfeld, D.; Terstriep, J. *Mapping the World of Social Innovation: A Global Comparative Analysis Across Sectors and World Regions*; Technische Universität Dortmund: Dortmund, Germany, 2016.

27. Bauwens, T.; Gotchev, B.; Holstenkamp, L. What drives the development of community energy in Europe? The case of wind power cooperatives. *Energy Res. Soc. Sci.* **2016**, *13*, 136–147. [CrossRef]

28. Bauwens, T. Explaining the diversity of motivations behind community renewable energy. *Energy Policy* **2016**, *93*, 278–290. [CrossRef]

29. Hewitt, R.J.; Bradley, N.; Baggio Compagnucci, A.; Barlagne, C.; Ceglarz, A.; Cremades, R.; McKeen, M.; Otto, I.M.; Slee, B. Social Innovation in Community Energy in Europe: A Review of the Evidence. *Front. Energy Res.* **2019**, *7*, 31. [CrossRef]

30. Brummer, V. Community energy–benefits and barriers: A comparative literature review of Community Energy in the UK, Germany and the USA, the benefits it provides for society and the barriers it faces. *Renew. Sustain. Energy Rev.* **2018**, *94*, 187–196. [CrossRef]

31. Vansintjan, D. *The Energy Transition to Energy Democracy—Power to the People*; REScoop.eu: Antwerp, Belgium, 2015; p. 76.

32. Bauwens, T.; Devine-Wright, P. Positive energies? An empirical study of community energy participation and attitudes to renewable energy. *Energy Policy* **2018**, *118*, 612–625. [CrossRef]

33. Clancy, J.; Daskalova, V.; Feenstra, M.; Franceschelli, N. *Gender Perspective on Access to Energy in the EU*; European Union: Brussels, Belgium, 2017; p. 106.

34. European Institute for Gender Equality. *Gender and Energy*; Publications Office of the European Union: Luxembourg, 2016.

35. Mort, H. *A Review of Energy and Gender Research in the Global North*; Technical University of Vienna: Vienna, Austria, 2019.

36. Seyfang, G.; Smith, A. Grassroots innovations for sustainable development: Towards a new research and policy agenda. *Environ. Politics* **2007**, *16*, 584–603. [CrossRef]

37. Seyfang, G.; Hielscher, S.; Hargreaves, T.; Martiskainen, M.; Smith, A. A grassroots sustainable energy niche? Reflections on community energy in the UK. *Environ. Innov. Soc. Transit.* **2014**, *13*, 21–44. [CrossRef]

38. Koirala, B.P.; Koliou, E.; Friege, J.; Hakvoort, R.A.; Herder, P.M. Energetic communities for community energy: A review of key issues and trends shaping integrated community energy systems. *Renew. Sustain. Energy Rev.* **2016**, *56*, 722–744. [CrossRef]

39. Parag, Y.; Sovacool, B.K. Electricity market design for the prosumer era. *Nat. Energy* **2016**, *1*, 16032. [CrossRef]

40. Wierling, A.; Schwanitz, V.J.; Zeiß, J.P.; Bout, C.; Candelise, C.; Gilcrease, W.; Gregg, J.S. Statistical Evidence on the Role of Energy Cooperatives for the Energy Transition in European Countries. *Sustainability* **2018**, *10*, 3339. [CrossRef]

41. Soeiro, S.; Ferreira Dias, M. Energy cooperatives in southern European countries: Are they relevant for sustainability targets? *Energy Rep.* **2019**, in press. [CrossRef]

42. Mapa Interactivo-Socio Unión Renovables Coop. Unión Renovables. Available online: http://www.unionrenovables.coop/mapa-interactivo/ (accessed on 15 December 2019).

43. Mautz, R.; Byzio, A.; Rosenbaum, W. *Auf dem Weg zur Energiewende: Die Entwicklung der Stromproduktion aus erneuerbaren Energien in Deutschland; eine Studie aus dem Soziologischen Forschungsinstitut Göttingen (SOFI)*; Universitätsverlag Göttingen: Göttingen, Germany, 2008; ISBN 978-3-938616-98-7.

44. Fraune, C. Gender matters: Women, renewable energy, and citizen participation in Germany. *Energy Res. Soc. Sci.* **2015**, *7*, 55–65. [CrossRef]

45. Bauwens, T.; Eyre, N. Exploring the links between community-based governance and sustainable energy use: Quantitative evidence from Flanders. *Ecol. Econ.* **2017**, *137*, 163–172. [CrossRef]

46. Pahl-Wostl, C. Participative and Stakeholder-Based Policy Design, Evaluation and Modeling Processes. *Integr. Assess.* **2002**, *3*, 3–14. [CrossRef]

47. Raven, R. Niche accumulation and hybridisation strategies in transition processes towards a sustainable energy system: An assessment of differences and pitfalls. *Energy Policy* **2007**, *35*, 2390–2400. [CrossRef]

48. Bauwens, T.; Huybrechts, B.; Dufays, F. Understanding the Diverse Scaling Strategies of Social Enterprises as Hybrid Organizations: The Case of Renewable Energy Cooperatives. *Organ. Environ.* **2019**. [CrossRef]

49. Blanco, M.I. The economics of wind energy. *Renew. Sustain. Energy Rev.* **2009**, *13*, 1372–1382. [CrossRef]

50. Malinowski, M.; Leon, J.I.; Abu-Rub, H. Solar Photovoltaic and Thermal Energy Systems: Current Technology and Future Trends. *Proc. IEEE* **2017**, *105*, 2132–2146. [CrossRef]

51. Gil-Lacruz, A.I.; Marcuello, C.; Saz-Gil, M.I. Gender differences in European volunteer rates. *J. Gend. Stud.* **2019**, *28*, 127–144. [CrossRef]

52. Kahla, F. Die Kapitalstruktur von Bürgerenergiegesellschaften unter dem rechtlichen Rahmenwerk der deutschen Energiewende. *Zeitschrift Umweltpolitik Umweltrecht* **2019**, *2019*, 94–123.

53. Holstenkamp, L.; Kahla, F.; Degenhart, H. Finanzwirtschaftliche Annäherungen an das Phänomen Bürgerbeteiligung. In *Handbuch Energiewende und Partizipation*; Holstenkamp, L., Radtke, J., Eds.; Springer Fachmedien: Wiesbaden, Germany, 2018; pp. 281–301. ISBN 978-3-658-09416-4.

54. Holstenkamp, L. Financing Consumer (Co-) Ownership of Renewable Energy Sources. In *Energy Transition: Financing Consumer Co-Ownership in Renewables*; Lowitzsch, J., Ed.; Springer International Publishing: Cham, Switzerland, 2019; pp. 115–138. ISBN 978-3-319-93518-8.

55. Valeau, P.J. Stages and Pathways of Development of Nonprofit Organizations: An Integrative Model. *Voluntas* **2015**, *26*, 1894–1919. [CrossRef]

56. Sundblom, D.; Smith, D.H.; Selle, P.; Dansac, C.; Jensen, C. Life Cycles of Individual Associations. In *The Palgrave Handbook of Volunteering, Civic Participation, and Nonprofit Associations*; Smith, D.H., Stebbins, R.A., Grotz, J., Eds.; Palgrave Macmillan: London, UK, 2016; pp. 950–971. ISBN 978-1-137-26317-9.

57. DiMaggio, P.J.; Powell, W.W. The Iron Cage Revisited: Institutional Isomorphism and Collective Rationality in Organizational Fields. *Am. Sociol. Rev.* **1983**, *48*, 147–160. [CrossRef]

58. Anheier, H.K. *Nonprofit Organizations: Theory, Management, Policy*, 2nd ed.; Routledge: Abingdon, UK, 2014; ISBN 978-0-415-55046-8.

59. Billis, D. Hybrid Associations and Blurred Sector Boundaries. In *The Palgrave Handbook of Volunteering, Civic Participation, and Nonprofit Associations*; Smith, D.H., Stebbins, R.A., Eds.; Palgrave Macmillan: London, UK, 2016; pp. 206–220. ISBN 978-1-137-26317-9.

60. Marquez, L.M.M. The Relevance of Organizational Structure to NGOs' Approaches to the Policy Process. *Voluntas* **2016**, *27*, 465–486. [CrossRef]

61. Markowitz, L.; Tice, K.W. Paradoxes of Professionalization: Parallel Dilemmas in Women's Organizations in the Americas. *Gend. Soc.* **2002**, *16*, 941–958. [CrossRef]

62. Cumming, G.D. French NGOs in the Global Era: Professionalization "Without Borders"? *Voluntas* **2008**, *19*, 372–394. [CrossRef]

63. Rothschild, J.; Chen, K.K.; Smith, D.H.; Kristmundsson, O. Avoiding Bureaucratization and Mission Drift in Associations. In *The Palgrave Handbook of Volunteering, Civic Participation, and Nonprofit Associations*; Smith, D.H., Stebbins, R.A., Grotz, J., Eds.; Palgrave Macmillan: London, UK, 2016; pp. 1007–1024. ISBN 978-1-137-26317-9.

64. Sander, C. Kooperationen in der Energiewirtschaft. Eine empirische Analyse Kommunaler Energieversorgungsunternehmen [Cooperations in the Power Industry. An Empirical Analysis of Municipal Utilities]. Ph.D. Thesis, University of Münster, Münster, Germany, 2011.

65. Lütjen, H.; Tietze, F.; Nuske, T. *Innovationskooperationen von Stadtwerken: Eine empirische Untersuchung von Treibern und Barrieren*; BoD—Books on Demand: Norderstedt, Germany, 2014; ISBN 978-3-7357-4710-5.

66. Jansen, D.; Heidler, R. Shareholding and Cooperation among Local Utilities: Driving Factors and Effects. In *Sustainability Innovations in the Electricity Sector*; Jansen, D., Ostertag, K., Walz, R., Eds.; Sustainability and Innovation; Physica-Verlag HD: Heidelberg, Germany, 2012; pp. 57–81. ISBN 978-3-7908-2730-9.

Article

Consumer Stock Ownership Plans (CSOPs)—The Prototype Business Model for Renewable Energy Communities

Jens Lowitzsch

East European Business Law and European Legal Policy at the Faculty of Business Administration and Economics, Europa Universität Viadrina, Große Scharrnstraße 59, D-15230 Frankfurt, Germany; lowitzsch@europa-uni.de

Received: 31 October 2019; Accepted: 13 December 2019; Published: 25 December 2019

Abstract: The 2018 recast of the Renewable Energy Directive (RED II) defines "renewable energy communities" (RECs), introducing a new governance model and the possibility of energy sharing for them. It has to be transposed into national law by all European Union Member States until June 2021. This article introduces consumer stock ownership plans (CSOPs) as the prototype business model for RECs. Based on the analysis of a dataset of 67 best-practice cases of consumer (co-) ownership from 18 countries it demonstrates the importance of flexibility of business models to include heterogeneous co-investors for meeting the requirements of the RED II and that of RE clusters. It is shown that CSOPs—designed to facilitate scalable investments in utilities—facilitate co-investments by municipalities, SMEs, plant engineers or energy suppliers. A low-threshold financing method, they enable individuals, in particular low-income households, to invest in renewable projects. Employing one bank loan instead of many micro loans, CSOPs reduce transaction costs and enable consumers to acquire productive capital, providing them with an additional source of income. Stressing the importance of a holistic approach including the governance and the technical side for the acceptance of RECs on the energy markets recommendations for the transposition are formulated.

Keywords: Renewable energy communities; renewable energy directive; prosumership; decentralised energy production; energy clusters; European Union; consumer (co-)ownership.

---

1. Introduction

A consumer stock ownership plan (CSOP) is a financing technique that employs an intermediary corporate vehicle and facilitates the involvement of individual investors through a trusteeship. It is a type of investment transaction that may use external financing, thereby achieving the benefit of financial leverage. The CSOP was applied for the first time in 1958 with spectacular success in the U.S. by its innovator, Louis O. Kelso, a business and financial lawyer turning 4,580 farmers into (co-)owners of the new fertilizer manufacturer Valley Nitrogen Producers, Inc. This involved an investment of USD 120 million which today inflation adjusted would equal around EUR 915 million. It is related to Kelso's best-known financial innovation, the employee stock ownership plan (ESOP), that enabled millions of American workers to become (co-)owners of their employer companies. Both plans repay the acquisition loan not from wages or savings but from the future earnings of the shares acquired. Today the ESOP is an integral part of American corporate finance with around 6,660 ESOPs and a little under 3,000 ESOP-like plans in the USA, about 14.2 million participating employees holding around USD 1.4 trillion in assets as of 2016 [1]. Applied to the energy context as CSOP can buy into an existing or invest in a new renewable energy (RE) plant. Designed to facilitate scalable investments in utilities, it is open to co-investments by municipalities, plant engineers, energy suppliers or other strategic partners.

Moreover, as a low-threshold financing method, it enables individuals to invest in RE projects [2]. The renewable energy consumer stock ownership plan (RE-CSOP) as an alternative financing source for sustainable investments is of particular importance for municipalities that are charged with fulfilling energy efficiency (EE) and climate policy goals but have limited budgets and often lack the funding to make these investments. An objective of this contractual model is, above all, to facilitate single-source financing (i.e., employing one bank loan instead of many micro loans), thus reducing transaction costs. At the same time, individual liability of consumers is avoided, while participating consumers are able to acquire capital ownership, providing them with an additional source of income. Other important issues are easy tradability of shares, deferred taxation for consumer-shareholders and pooling of voting rights.

Especially, low-income households who usually do not dispose of savings necessary for conventional investment schemes are enabled to repay their share of the acquisition loan from the future earnings of the investment: A fiduciary entity that is set up by the local community and managed by an independent director is authorized to take on a bank loan to acquire shares in the RE plant on behalf of the consumers. The shares are allocated among the consumer-beneficiaries in proportion to their respective energy purchases. Monies saved by self-consumption and increased EE as well as revenues from the sale of the excess energy production are used to repay the acquisition loan. After amortisation of this debt, profits are distributed to the consumer-beneficiaries.

In 2018 the European Union has introduced a legal framework for renewable energy communities (RECs) that will have to prove its success in the years to come. A crucial element for the acceptance of RECs by the energy markets will be the underlying business model. This article introduces RE-CSOPs as the prototype business model for RECs. In the limited time since the entering into force of the new rules only very few articles, for the most part policy papers of the different interest groups, have been published. Therefore, the focus lies on the conceptual side of this business model omitting a review of the literature.

1.1. Prosumership in the 2018/19 EU Clean Energy Package

Consumer (co-)ownership in RE is one essential cornerstone of the overall success of energy transition. Marshall McLuhan and Barrington Nevitt as early as 1972 suggested in their book Take Today [3] that technological progress would transform the consumer into a producer of electricity. When consumers acquire ownership in RE, they can become prosumers (Alvin Toffler probably first introduced the artificial word stemming from the Latin in his book The Third Wave [4]), generating a part of the energy they consume, thus reducing their overall expenditure for energy, while at the same time having a second source of income from the sale of excess production. The European Union agreed on a corresponding legal framework as part of a recast of the Renewable Energy Directive (RED II) [5], which entered into force in December 2018:

- Consumers, as prosumers, will have the right to consume, store or sell RE generated on their premises, (1) individually (Art. 21 RED II), that is, households and non-energy small and medium sized enterprises (SMEs); and collectively, for example, in tenant electricity projects, or (2) as part of RECs (Art. 22 RED II) organised as independent legal entities.
- Transposing the RED II into national law, Member States—amongst others—have until June 2021 to adopt an "enabling framework" for prosumership and, in particular, for RECs. The Directive defines citizen's rights and duties and links prosumership to such different topics as increasing acceptance, fostering local development, fighting energy poverty, and incentivising demand-flexibility.

The RED II is part of the "Clean Energy for all Europeans Package" of the European Union, a package of measures that the European Commission presented on 30 November 2016 to keep the EU competitive as the energy transition changes global energy markets; this legislative initiative has four main goals, that is, energy efficiency, global leadership in RE, a fair deal for consumers and a redesign of the internal electricity market. The RED II rules are embedded in those of the 2019 Internal Electricity Market Directive (IEMD) [6] and Regulation (IEMR) [7]. The transposition of these comprehensive rules—in particular those on energy communities—requires developing, implementing and rolling out business models that broaden the capital participation of consumers in all Member States [8].

RED II introduced RECs as a new Europe-wide governance model for RE projects and defined them in Art. 2 as a legal entity:

- *"which, according to applicable national law, is based on open and voluntary participation, is autonomous, and is effectively controlled by shareholders or members that are located in the proximity of the renewable energy projects owned and developed by that community;*
- *whose shareholders or members are natural persons, local authorities, including municipalities, or SMEs;*
- *whose primary purpose is to provide environmental, economic or social community benefits for its members/the local areas where it operates rather than financial profits."*

Complying with the prerequisites for RECs, a corresponding business model needs to have the capability of involving heterogeneous co-investors, that is, local citizens, municipalities, SMEs but possibly also commercial investors in RE projects. Other than bringing together the interests of local citizens and their municipalities, this is an important prerequisite for preferential conditions under the "enabling framework" for RECs, as defined in Art. 22 RED II. This approach facilitates the involvement of municipalities who need to respect the typical prerequisites of municipal law for participation in RE projects, i.e., public purpose, capacities for the investment, subsidiarity, appropriate representation as pacemakers of the energy transition. (Optional) minority stakes for commercial investors is itself nothing new, as citizens' energy models in the wind sector often include professional partners as members of limited partnerships [9]. Depending on the type of project and the underlying technology, it may be useful to include them as operation and maintenance of infrastructure in RE projects can be very complex; this concerns, for example, not only wind energy and bioenergy, but also energy cluster projects aiming at sector coupling that may involve electricity sharing, storage, e-mobility, cogeneration, and the like [10,11].

1.2. Research Questions and Approach

Conventional business models for consumer ownership may not always allow for the combination of different types of co-investors. With regard to cooperatives [12], for example, the one-member one-vote principle is often an obstacle to partnering with SMEs and commercial investors, since these parties will prefer voting rights proportional to their shareholding. Furthermore, municipal co-investments are hindered by the necessity of representation on management and supervisory bodies, as cooperative law does not acknowledge a right of delegation similar to legislation applicable to joint stock companies. Cooperative projects often set up special purpose vehicles (usually a privately held corporation with limited liability) to avoid this problem [13]. The RE-CSOP involves such a standard special purpose vehicle, but with a defined governance structure allowing for the direct involvement of municipalities and strategic partners while safeguarding the interests of the local partners. Unlike cooperatives, where all management and board positions are reserved for members and representation by third parties is not permitted [14], a CSOP may hire external management. Thus, it avoids obstacles related to the principle of self-governance and ensures the representation of municipalities on the board. At the same time members of an energy cooperative can participate in a RE-CSOP, together with strategic partners, when expanding an existing RE plant together with strategic partners.

With regard to the RED II requirements for RECs and the necessary contractual arrangements, this article seeks to answer the following questions:

1. To what extent does the governance model for energy communities stipulated by the Clean Energy Package actually meet the needs of practice?
2. Can the RE-CSOP and similar business models provide attractive conditions respecting both the RED II prerequisites for RECs as well as the individual needs of different types of co-investors?

As the novelty legislation is not broadly known yet, Section 2 on theory first lays out the new legal framework for energy communities with a focus on the governance model for RECs and their importance for RE clusters. Reflecting on available empirical evidence, Section 3 draws on the experience of already existing best practice energy communities in the field of RE, assessing how many involve heterogeneous partners, and in those that do, their relationship to each other with regard to ownership structure and governance. To identify these patterns, the analysis [15] of a dataset of 67 best practice cases from 18 countries covering Europe, North and South America and Asia [16] is referred to asking: (a) Whether they are open to different actors (i.e., the heterogeneity of members or shareholders); and (b) if so, what their governance and ownership structure was. In the light of these empirical findings, Section 4 presents the RE-CSOP putting forward a proposal for future practice using a modular approach: (a) Three levels for co-investments are identified; and (b) the RE-CSOP is adapted to each of these levels describing how it reflects the needs of the different co-investors. Section 5 then discuss specific aspects of this business model, namely, how to convey individual consumers' shareholding, the financing of the investment, and its taxation. Section 6 concludes and formulates policy recommendations with a view to the pending transposition of the RED II. The glossary provides definitions.

2. Theory

Energy communities are mentioned and defined in both the RED II and the IEMD. While the recast of the renewables directive focuses on the promotion of RE and thus speaks of "renewable energy communities" (RECs), the directive on the internal electricity market of the European Union as the more general legal act addresses "citizen energy communities" (CECs) [17]. This raises the question of the relationship between these two types of energy communities and, more generally, the relationship between these two legal acts. Furthermore, the Clean Energy Package introduces a new Europe-wide governance model for RECs and CECs to foster environmental, economic or social community benefits. These benefits are of particular importance for the development of the energy systems of tomorrow, that is, RE clusters that further support the deployment of renewable energy sources (RES) and provide stability of the grid and energy supply in energy markets increasingly characterised by volatility of production [15]. Flexibility [18], bi-directionality, interconnectivity [19] and complementarity [20] are prerequisites to these RE clusters that; however, require an active involvement of all actors involved, including consumers.

2.1. Relation of Electricity Market Directive/Regulation and Renewable Energy Directive

While the purpose of IEMD/R is the completion of the internal market in electricity that has progressively been implemented since 1999, that of RED II on the other hand is to specifically support the deployment of RES for energy production, including electricity, and to foster acceptance for renewables among the Europeans. Both directives expressly see the consumer *"at the heart of the energy markets"*, defining him or her—individually or jointly—respectively as *"active consumer"* (IEMD) and *"renewable self-consumer"* (RED II). With regard to energy communities, the IEMD mainly concerns the horizontal level, that is, their rights and obligations towards public authorities, other electricity enterprises and consumers. This design is also reflected in recital 2 IEMR on the aim of the internal market in electricity *"to deliver a real choice for all consumers in the Union, both citizens and businesses, new business opportunities and more cross-border trade, so as to achieve efficiency gains, competitive prices and higher standards of service, and to contribute to security of supply and sustainability"*. Amongst other issues the IEMD provides energy communities with a level playing field vis-a-vis other market participants (see Art. 65 IEMD). RED II, on the other hand, additionally ensures that RECs can compete for support *"on an equal footing with other market participants"* and calls on the Member States to *"take into account specificities of renewable energy communities when designing support schemes"* (Art. 22 para 7 RED II).

While the framework under IEMD is primarily a *"regulatory framework"* (see Art. 16 para. 1, sentence 1), that of RED II has the explicit aim *"to promote and facilitate the development of RECs"* (see

Art. 22 para. 4, sentence 1), including preferential conditions or incentives. However, the above distinction is not always sharp since the IEMR/D also contain elements that support the deployment of RES. Recital 3a IEMR stipulates as an explicit aim *"to ensure the functioning of the internal energy market while integrating requirements related to the development of renewable forms of energy and environmental policy, in particular specific rules for certain renewable power generating facilities, concerning balancing responsibility, dispatch and redispatch as well as a threshold for CO2 emissions of new generation capacity where it is subject to a capacity mechanism"*. As enshrined in Art. 11, for example, the IEMR defines the principle of priority dispatch for RE plants with an installed electricity capacity of less than 400 kW (for RE plants commissioned after 1 January 2026 less than 200 kW) and for *"demonstration projects for innovative technologies"*. RE-plants that concluded contracts before the entering into force of the IEMR continue to benefit from priority dispatch. Furthermore, with regard to RECs Art 7 para. 3 IEMR stipulates that *"Nominated electricity market operators shall provide products for trading in day-ahead and intraday markets which are sufficiently small in size, with minimum bid sizes of 500 Kilowatt or less, to allow for the effective participation of demand-side response, energy storage and small-scale renewables including directly by customers"*. Figure 1 illustrates the relation of the RED II and the IEMD/R.

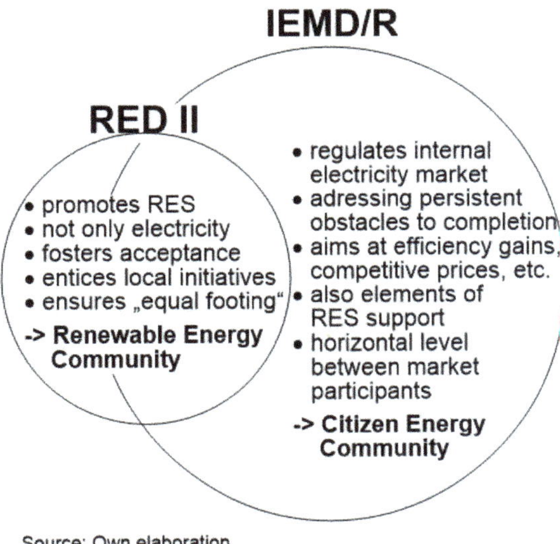

Figure 1. Relation of the Renewable Energy Directive (RED II) and the Internal Electricity Market Directive/Regulation (IEMD/R).

In sum, generally speaking, RECs are a specific form of CECs that benefit from an "enabling framework" promoting and facilitate their development. However, they have their own area of operations not falling under the IEMD/R as far as other types of energy (i.e., not electricity) are concerned. In this regard, the possibility of benefitting from conventional small-scale back-up generation is an important element for REC's micro-grid solutions, be it on- or off-grid. Most importantly, unlike CECs they benefit from the preferential conditions of the "enabling framework".

2.2. The New Governance Model and its Importance for RE Clusters

With regard to energy communities, of course, European energy law does not rule out other private law citizens' or consumer-oriented initiatives facilitated by and implemented with the participation of the public administration in the Member States [17]. However, such initiatives would benefit neither from the possibility of electricity/energy sharing nor from the preferential conditions and incentives foreseen in the "enabling framework" to promote and facilitate the development of RECs under the

RED II. Therefore, the new Europe-wide governance model for energy communities is a determining factor for the choice of business models applied [21]. Both types of energy communities focus more on environmental, economic or social community benefits than on profits and both limit the effective control of the community to their beneficiaries; however, whereas RECs do this by tying control to the criteria of locality and geographic proximity, CECs limit it by the size of the shareholders and their commercial activity, excluding those for which energy constitutes the primary area of activity. An overview is provided in Table 1.

Table 1. The new governance model for energy communities under Renewable Energy Directive (RED) II and Internal Electricity Market Directive (IEMD). Source: Modified after Lowitzsch, Hoicka, van Tulder 2019.

Criteria	Renewable Energy Communities (RECs) Pursuant to Arts. 2 (16), 22 RED II	Citizen Energy Communities (CECs) as Defined in Arts. 2 (11), 16 IEMD
Eligibility	<ul><li>Natural persons,</li><li>Small and medium sized enterprises,</li><li>Local authorities, incl. municipalities;</li></ul>	In principle open to all types of entities;
Primary Purpose	*"environmental, economic or social community benefits for its shareholders / members or for local areas where it operates, rather than financial profits";*	
Membership	Voluntary participation open to all potential local members based on non-discriminatory criteria;	Voluntary participation open to all potential members based on non-discriminatory criteria;
Ownership and control	<ul><li>Effectively controlled by shareholders or members that are located in the proximity of the RE project;</li><li>Is autonomous (no individual shareholder may own more than 33% of the stock).</li></ul>	<ul><li>Effectively controlled by shareholders or members of the project;</li><li>limitation for firms included in shareholders Controlling entity to those of small/micro size (not medium);</li><li>Shareholders engaged in large scale commercial activity and for which energy constitutes primary area of activity excluded from control.</li></ul>
Advantages to qualify as REC or CEC	<ul><li>Preferential conditions defined in the "Enabling framework" to promote and facilitate the development of RECs;</li><li>Energy sharing within the REC.</li></ul>	<ul><li>Level playing field;</li><li>Electricity sharing within the CEC.</li></ul>

With regard to RE, the two crucial consequences of this governance model for the CSOP—as well as for any other business model—are that a REC according to Art. 22 RED II:

1. Must be autonomous and independent of other RES project partners. "Autonomy" in this context should be understood as a 33% ceiling for ownership stakes of individual shareholders or members; recital 71 RED II stipulates that *"REC should be capable of remaining autonomous from individual members and other traditional market actors that participate in the community as members or shareholders, or who cooperate through other means such as investment"*.

2. In addition, *"is effectively controlled by shareholders or members that are located in the proximity of the renewable energy projects owned and developed by that community"* result in a ceiling for the strategic investor's participation of 49% (see the requirements of the definition of Art. 2 of the RED II in Section 1.1 above) or at least binding contractual arrangements that confer decisive influence on

the composition, voting or decisions. Art. 2 pt. (56) IEMD defines "control" as *"rights, contracts or other means which, either separately or in combination and having regard to the considerations of fact or law involved, confer the possibility of exercising decisive influence on an undertaking, in particular by: (a) ownership or the right to use all or part of the assets of an undertaking; (b) rights or contracts which confer decisive influence on the composition, voting or decisions of the organs of an undertaking"*.

These new rules for the lawful control over and administration of (local) energy generation, supply and management concern also the fair distribution of responsibilities and benefits and are the governance side of the technical solutions for the Energy Transition. Energy communities; thus, are the mirror image of energy clusters; the former concern the governance, the latter the technological side of the (renewable) energy systems of the future, entailing flexibility, bi-directionality and interconnectivity options between prosumers and producers of energy and the market [15]. Most importantly they allow energy sharing of a portfolio of RES, that can enhance complementarity, lower energy costs for prosumers [10] and, through (co-)ownership in RES, increase social acceptance of the architecture and logic of a RE future [22].

3. Empirical Evidence: Material, Methods and Results

To cast a light on available empirical information on the structure of renewable energy communities the results of an analysis [15] of a dataset of 67 best-practice examples of consumer (co-)ownership reported in the Palgrave Macmillan publication "Energy Transition: Financing Consumer Co-Ownership in Renewables" [16] are briefly summarised in this section. The notion of (co-)ownership is used not in the technical sense of joint ownership but to indicate that there may be other owners next to the consumers amongst the shareholders such as municipalities or conventional investors. The cases are from 18 countries covering Europe, North and South America and Asia, that is, CZ, DK, FR, DE, IT, NL, PL, ENG, SCT, ES, CH, CAL, CAD, BR, CL, IND, PAK, JAP; these countries were analysed following a consistent pattern including the energy mix, policies supporting consumer (co-)ownership, energy poverty, the regulatory framework, best practice, financing conditions, obstacles and perspectives to enable a like-to-like comparison. In light of the potential for replication of the regulatory framework beyond Europe, and to confirm the existence of projects that fit the criteria elsewhere, the extra-European cases present in the dataset were included in the analysis. The definition of consumer (co-)ownership as *"participation schemes that (a) confer ownership rights in RE projects (b) to consumers (c) in a local or regional area"* [23] (pp. 7–8) is followed in this article.

As mentioned, eligible members for RECs are natural persons, SMEs and local authorities, while CECs are, in principle, open to all entities. Both the IEMD and the RED II; thus, support heterogeneity of members, which follows from the purpose and guiding principle for both types of energy communities "to provide environmental, economic or social community benefits for its shareholders or members or for the local areas where it operates, rather than financial profits". However, with a view to the legislative process it remained unclear whether these guiding principles and in particular the emphasis on local and diverse co-investors originated from political desiderata or practical experience of already operating energy communities. Similar doubts arose with regard to the RED II prerequisites that to qualify as a REC, (a) the effective control should be held by members based in the proximity of the RE installations, and (b) its autonomy from single shareholders is to be upheld by the principle that no single shareholder owns a controlling stake. The IEMD contains a comparable but fairly milder restriction in precluding entities engaged in large-scale commercial activity and for which energy constitutes the primary activity as well as medium and large-sized enterprises from the shareholders effectively controlling the CEC.

The resulting limitations for enterprises which are either not local, too large or dominant in the energy sector with regard to control and size of their shareholding in energy communities may hamper their participation in RECs; together with those stemming from the business models prevalent to date risk to render RECs unattractive for these potential co-investors [21]. While good legislative intentions can lead to over-complex regulations that may actually hinder project implementation, a lot depends on how existing best practice deals with such problems [15]. Amongst other issues Lowitzsch, Hoicka and van Tulder investigated the diversity of co-investors and the prevalent governance structures, testing the dataset for the two following criteria: (a) Heterogeneity of members and (b) governance and ownership. The results of the analyses for these two criteria can be summarised as follows:

- They show that in the evaluation of the 67 cases, 37 had co-investors as envisioned by the RED II for the future RECs. Although these numbers seem low they are nevertheless unsurprising as energy communities operating exclusively in RE are a recent phenomenon not yet widely implemented.
- What is more surprising is, that only 9 projects already meet RE cluster requirements while merely 22 have RE cluster potential. Many projects are of small size and do not or only to a limited extend involve flexibility, bi-directionality, interconnectivity and complementarity; but this is a condition to become fully fledged RECs that will also be able to benefit from energy sharing [15].
- Only in 20 of the 37 cases this involved genuinely heterogeneous co-investors although not all of them comply with the governance structure required by the RED II. Some projects are solely owned by one shareholder; other projects, although showing heterogeneous co-investors are dominated by commercial actors not based in the proximity of the RE installations; in yet other projects a large energy firm has a majority ownership stake violating the autonomy criterion.
- Of the remaining 17 cases that only formally comply with the heterogeneity criterion of the RED II some cases were either cooperatives exclusively with citizens as members or municipal projects without other co-investors.
- Furthermore, geographic and cultural diversity of RE projects even within a given country lead to complexities that do not permit "one size fits all" solutions. While identities and interests are often deeply rooted in geographies and cultures, organizational and contractual arrangements are a more flexible factor that can be adapted to the former two [24].

Against the background of these empirical findings the question which business model is best suited for the RECs of the future becomes even more important. Only a sufficiently flexible business model like the RE-CSOP will be able to fulfil the necessary functions of RE clusters and allow truly heterogeneous partnerships for investment.

4. Presentation of the Renewable Energy Consumer Stock Ownership Plan

The modular approach of the RE-CSOP (see Figure 2) and the structure for each level of co-investment as described in this section is conceived under the assumption of complying with the new RED II governance model in order to benefit from the preferential conditions or incentives foreseen "enabling framework" to facilitate setting up RECs. Therefore, Figures 3–5 emphasise the role of the controlling members of RECs. As a rule, prosumers (households and non-energy small and medium sized enterprises) will hold between 33% and 51% of the shares in the corporation operating the RE-facility (Operating Company) and, together with the municipality, will have a majority interest. However, the CSOP conveys individual shareholding of the participating consumers through a trusteeship. Regarding the exercise of consumer's voting rights, the model offers flexibility: The fiduciary arrangements stipulate which matters are to be decided by the trustee or the managing director of the fiduciary entity (e.g., day-to-day business) and which will be voted on by CSOP-members (e.g., strategic decisions). It is; thus, the consumers themselves that determine the extent of their involvement, thus facilitating a process of apprenticeship. Finally, as the CSOP business model uses the borrowing power of a corporation, it enables the participation of vulnerable consumers that are underrepresented so far.

4.1. The Modular CSOP Approach

In practice, CSOP financing is based on a modular approach, starting with a "base model" and extending to higher levels, depending on the type of different co-investors involved, their investment horizons, needs and aims (see Figure 2a–c).

Level I: The base model is composed of two closely held corporations with limited liability, the fiduciary entity (Trusteeship) and the CSOP operating company (Operating Company). The fiduciary entity can also be a limited partnership or a RE-cooperative already in place which; however, this would have implications for the taxation of individual consumer (co-)owners and their corporate rights. This structure corresponds to a situation where a strategic co-investor has a local long-term interest (e.g., acceptance of a wind project) and does not mind burdening the Operating Company with a capital acquisition loan for consumers; all shareholders are proportionally liable for the debt.

Level II: A more complex structure results when the strategic investor, for example, has a short-term interest and will not engage in the project if his shareholding would be burdened with the acquisition loan that facilitates the consumer shareholding; in this situation the Operating Company stands next to a Holding (again a closely held corporation with limited liability) with only the latter being liable for the acquisition loan. Of course, the Operating Company will still provide security for the loan pledging part of the assets of the RE installation.

Level III: When upscaling and pooling more than one CSOP investment, the structure is still more complex: The Operating Company runs X number of RE projects, while separate Asset Companies own the RE installations of various RE-CSOPs. Strategic investors with differing short- or long-term interest (such as management, capital investment, electricity storage, aggregation and demand response) or a distribution system operator of a micro grid, for example, can invest at different levels accordingly.

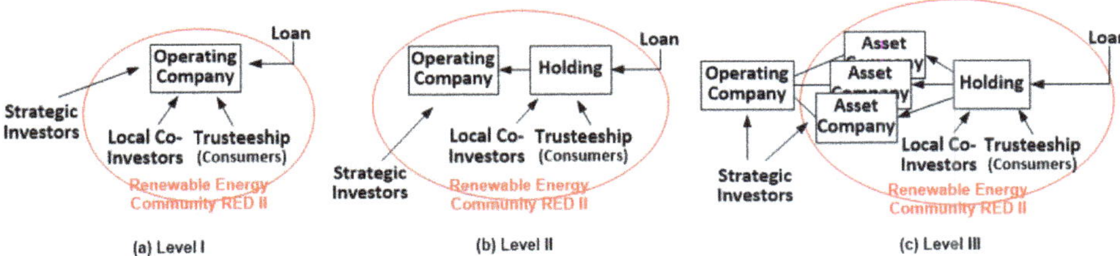

Figure 2. Co-investors in renewable energy communities (RECs). (Source: Own elaboration).

To sum up, compatibility with conventional investments together with the potential of scalability, gives the RE-CSOP the advantage of avoiding concerns of market fragmentation [23]. Sub-scale investments can be eschewed, local projects pooled and partnerships with municipalities set up, thus advancing to economies of scale while retaining the benefits of individual consumer participation. Other than qualifying as a RECs and thus benefitting from the RED II "enabling framework" the RE-CSOP at the same time provides a business model flexible enough to allow for the cooperation with professional energy companies (see in particular Level III).

Against this background, RE-CSOPs can be an important "bridge technology" in financing citizen energy projects while extending the advantages of RE-cooperatives where projects involve heterogeneous co-investors, or where the cooperative model is not feasible for other reasons [12]. This is especially the case in Eastern Europe where citizen energy projects are still rare and where the cooperative model is associated with the socialist past. Furthermore, the flexible governance structure of CSOPs offers the advantage of combining RE projects with active citizen participation, both in financial returns and in decision-making, while also allowing for the participation of commercial investors. Especially in RE clusters that target sector coupling and may involve electricity sharing, storage, e-mobility, cogeneration, etc., including professional operators will become increasingly important as the operation and maintenance of the infrastructure of RE projects becomes more complex [15]. Here the RE-CSOP provides a standard governance model that safeguards the interests of local partners vis-à-vis their co-investors.

4.2. Level I—Key Elements of the Base Model (Leveraged or not)

The first element of the RE-CSOP structure is the RE installation that is operated and managed by the Operating Company. The Operating Company is set up as a closely held limited liability corporation which is the best solution with regard to the functionality of the whole structure as well as with regard to the optimisation of taxation (for example, under Polish law a "spółka z ograniczoną odpowiedzialnością", under Italian law a "societá a responsabilita limitata", under Czech law a "společnost s ručením omezeným" and under U.S. American law a "closely held corporation").

Variant A—A new Operating Company is set up as a special purpose vehicle specifically for the new consumer co-investment: The consumers involved become (co-)owners of the RE installation by themselves or in partnership with other local public partners (e.g., a municipality, entity of local self-administration, public law corporation or a municipal enterprise) and possibly with local private investors such as SMEs.

Variant B—An existing Operating Company is running and managing an existing RE installation: It is taken over partly or entirely by another legal subject assuming control on behalf of the consumers and the other co-investors of the local RE community pursuant Art. 22 RED II.

As the ultimate goal of creating the overall structure is to grant corporate rights to the consumers, it is necessary to answer the question, how will they be included in this plan? This concerns in particular what kind of legal, corporate and property ties will connect the consumers of the RE installation with the Operational Company (independently of the contractual relationship for the supply of energy, of course). On the one hand, consumers could be direct shareholders of the Operating Company, but from a functional perspective this is not a desirable solution. Another component of the RE-CSOP; therefore, is a fiduciary entity. It is this fiduciary entity that on behalf of the consumer-shareholders, together with the other local shareholders, effectively controls the Operating Company (running the RE plant). The legal form of the intermediary entity administering the CSOP shares in the CSOP model for continental Europe, is derived from the Anglo-American Common Law of trusts [25]. In the absence of genuine trust legislation, this requires a two-tier structure (i.e., a closely held corporation with limited liability as fiduciary entity (Trusteeship) that holds consumer's shares in a closely held corporation with limited liability that operates the RE plant (Operating Company)). Figure 3 gives an overview of the financing structure and the key elements of the base model (Level I).

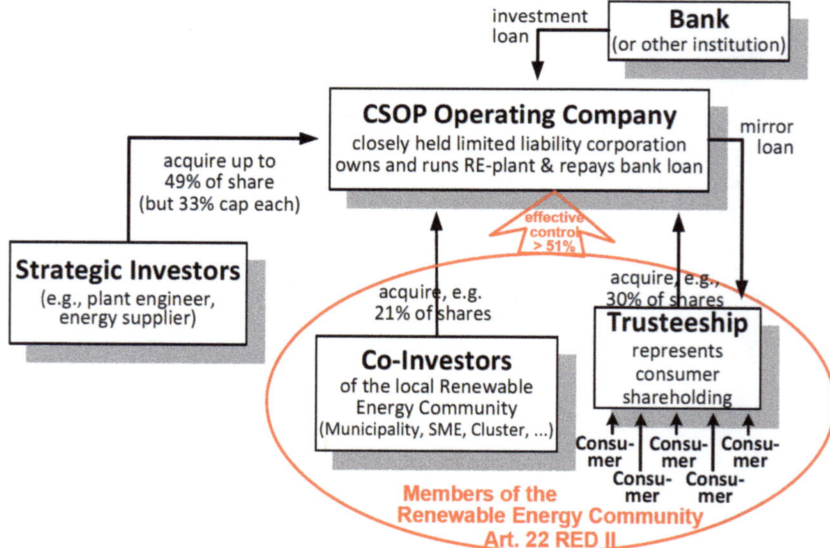

Figure 3. Key elements of the RE-CSOP financing structure in the base model for a REC. Source: Own elaboration.

As mentioned earlier, a RE-CSOP can use a bank loan to leverage the acquisition of shares in a RE project for consumers that have neither savings nor access to capital credit. National company and tax law permitting, using corporate credit to guarantee the loan that funds the acquisition of consumer shares by the CSOP, reduces the financing costs. If the Trusteeship borrows money to buy shares, the Operating Company repays the loan through periodic contributions (however, financing costs will not be tax-deductible) and dividends paid on the shares the fiduciary entity holds in trust for the consumer-shareholders. As the loan is retired, paid-up shares are allocated to individual consumer accounts, usually on the basis of relative energy consumption.

In a variation of the above described loan structure, the lender often prefers to make the loan directly to the Operating Company, followed by a second "mirror loan" from the Operating Company to the Trusteeship. The tax results will be better than in the case of a direct loan to the fiduciary entity. The interest repayments—national company and tax law permitting—will be a deductible expense from taxable corporate income as financing costs of the RE-investment. However, the Operating Company has to make annual contributions to the Trusteeship in amounts sufficient to amortise the internal loan from the Operating Company to the Trusteeship. The amounts paid by the fiduciary entity to the Operating Company to amortise the internal loan will as a rule constitute tax-free loan repayments and will be used by the Operating Company in turn to amortise the external loan. The "mirror loan" structure provides the lender with a stronger security interest in the assets pledged as collateral for the loan [26]. The lender will be in a better position to defend against claims of fraudulent conveyance in the case of default if collateral is taken directly from the borrower rather than from a guarantor of the loan. This should also lower the financing cost for the leveraged transaction significantly.

However, to use this structure the other shareholders of the Operating Company that do not directly benefit from the leveraged transaction must agree to assume the risk associated with financing the acquisition of shares by the Trusteeship with a bank loan. This may be acceptable if these shareholders are all members of the REC and share a genuine interest in involving the consumers. However, in situations where either the interests of the members of the REC are too heterogeneous or where external co-investors are involved, such co-investors may object to the mirror loan structure. In these situations, it may be necessary to set up a Holding Company, as described in the next section.

4.3. Level II—Leveraged RE-CSOP with External Strategic Investor

The following alternative structure of the RE-CSOP model employs a Holding Company which obtains external financing both for the consumers and for the other members of the local REC (i.e., taking on a loan or credit and then investing it in the Operating Company (Variant A); or acquiring the shares from the current owner(s) (Variant B)). The justification for this structure is the diversity of interests of the potential co-investors.

The Holding Company is again a closely held corporation with limited liability which, at the same time, may facilitate the functioning of the entire structure from the viewpoint of tax optimization. The investment or acquisition is financed from external sources, with the loan/credit being repaid from the future profits of the RE installation run by the Operating Company (with such profits coming from the sale of electricity to consumers or to the grid and from the difference in price of the energy provided to the prosumers). National tax law permitting, the Operating Company and the Holding Company may establish a capital tax group (see 5.2. below). In the case of such a structure, profits, losses and, what is most important here, costs, are calculated for tax purposes jointly for the combined tax group. As a result, in practice, financing costs (especially interest) can be deducted from the tax base of the Operating Company. Such a solution has many advantages, including the following:

- Consumers are still not direct shareholders in the dominating Holding Company (in the case of a Holding Company whose direct shareholders would be supplied by the dependent Operating Company, problems relating company law institutions could arise, such as actions to exclude a shareholder, the increase and decrease of share capital, organization of shareholders' meetings, change of statutes, etc.).
- The division of the shareholding between the members of the REC (i.e., municipality, SMEs, and other local co-investors on the one hand, and the consumers represented by the fiduciary entity (Trusteeship) on the other hand) is flexible and reflects the respective contributions and roles, as

long as they together have effective control of the operating company by keeping at least 51% of its shares.

- External strategic investors can buy into the project without being burdened by the leveraged transaction that enables consumers without significant savings to participate.

Thus, at Level II there are three entities in this structure—an Operating Company running the RE installation, a CSOP Holding (dominating company) and a Trusteeship, being the sole shareholder or the co-owner of the Holding, and thus indirectly controlling the Operating Company. Figure 4 shows the advanced scheme of the RE-CSOP model of Level II.

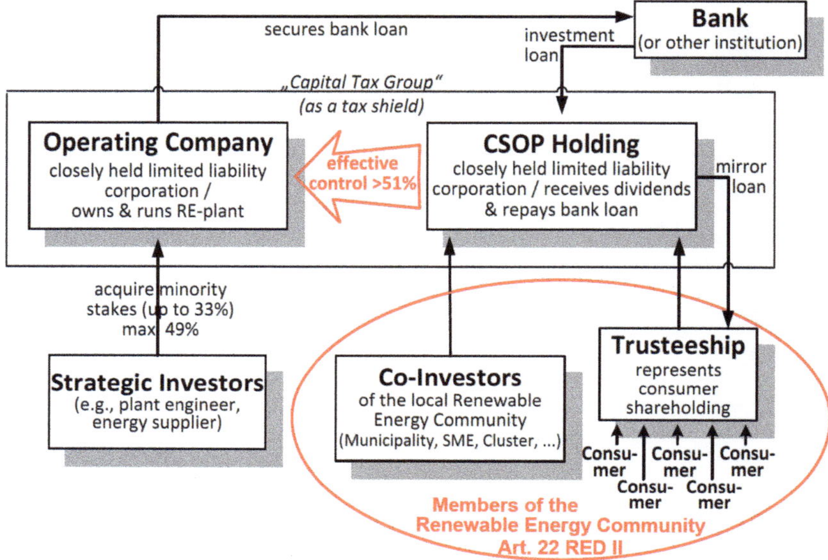

Figure 4. Key elements of the leveraged RE-CSOP with strategic investors for a REC at Level II. (Source: Own elaboration).

In summary, this solution offers two opportunities for co-investments at Level II:

(1) Leveraged investments financed by an investment loan taken on by the Holding Company. The target groups for this type of co-investment are, above all, local co-investors belonging to the REC pursuant to Art. 22 RED II as, for example, a municipality, a small or medium enterprise, members of a RE cluster, etc. They all have in common that their investment horizon is long- to mid-term and that, as a rule, they will have difficulties in obtaining financing individually, or, at least will incur higher financing cost [27], than when benefitting from the borrowing power of the Holding that pledges its shares in the Operating Company to secure the repayment of the investment loan.

(2) Non-leveraged investments financed by a strategic investor in the Operating Company. The target group for this type of co-investment is, generally speaking, external strategic investors that either do not qualify as members of a REC pursuant to Art. 22 RED II and/or have different motivations for their engagement in the project. They typically will have a short- or mid-term investment horizon with preferences for liquidity and a clear exit strategy. Examples are, on the one hand, shareholders engaged in large scale commercial activity for which energy constitutes a primary area of activity (e.g., an energy supplier), or, on the other hand, an external investor with a specific temporary investment interest, as, for example, a plant engineer that seeks acceptance for RE project [28].

4.4. Level III—Upscaling and Pooling RE-CSOP Investments

When RECs reach more complexity both with regard to the technical aspects of energy generation, use or transfer and with regard to the variety of heterogeneous co-investors involved, a need for upscaling and pooling of several RE-CSOP projects will arise. This is, in particular, the case with RE clusters emerging in the Energy Transition [15]. The needs that these RECs will depend on a number of factors that can be grouped into two categories:

1. Technical or engineering requirements [10,11]:

   - The variety of renewable sources (wind, PV, biomass, etc. and their complementarity) or other energy sources (fossils as back-up but also those not easily to divest from);
   - The specific combination of different energy sources where energy production is not the primary aim of economic activity (e.g., cogeneration, waste, biomass, etc.).

2. Management and governance requirements [29]:

   - More than one RE-CSOP project organised in various asset companies with majority ownership stakes of the members of the REC but managed by one operating company in which a professional energy company may have a majority interest;
   - The operating company is run by a third party with expertise in installation and operation, including metering and maintenance, but such third party remains subject to the RECs instructions.

In all the different combinations of scenarios resulting from the factors enumerated above, it will be important to have the possibility to separate the ownership of production assets from their management as illustrated in Figure 5 for Level III.

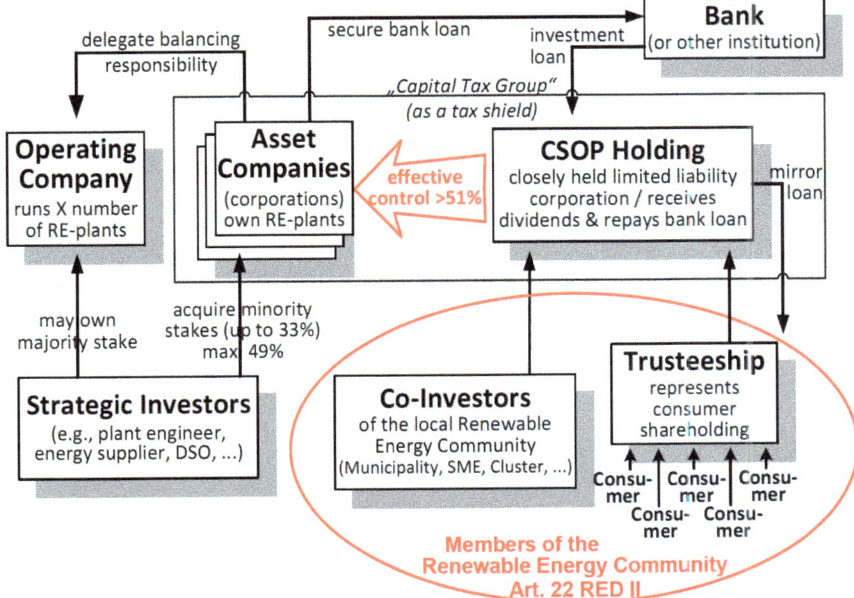

Figure 5. Key elements of pooled/upscaled RE-CSOP investments with strategic investors for a REC at Level III. (Source: Own elaboration).

This is of particular importance, as it will also allow the involvement of strategic investors with majority interest in the Operating Company, which in this case may also be an already existing daughter of a professional energy company; such a strategic partner [29] can be delegated by the REC to provide a variety of services, such as balancing responsibilities, coordination and settlement between REC participants or the implementation of a virtual power plant. Consequently, Art. 21 para 5 RED II foresees that Member States allow the possibility that prosumer's installations are owned by a third party but remain under the direction of self-consumers as Art 16 IEMD permits that Member States allow CECs to own, establish, purchase or lease distribution networks and to manage them or delegate management to third parties. The fact that RECs will bundle functions ranging from generation to distribution and sale is de facto an exception from the unbundling rules for energy markets implemented over the last decades, and again may make them attractive to strategic investors with regard to aggregation, demand flexibility, etc.

5. Discussion of the Key Elements of the RE-CSOP as Applied at Levels I–III

5.1. Indirect Consumer-Shareholding in the Capital of the Operating Company (or Holding)

(Co-)ownership resulting from consumer investment leads in practice to a situation where consumers have influence on the management of the company. From the point of view of co-investors—internal or external—such influence is problematic in terms of predictability and steering of the dynamics in decision-making processes [9]. First, it is highly undesirable that a co-investor would have to interact constantly with all consumer-shareholders, which easily can be hundreds in large CSOPs. Second, with regard to the question of how participating consumers vote their shares, it is undesirable that every consumer takes individual decisions without coordination with the others making it difficult for the remaining shareholders to understand and forecast their voting behaviour and interests. At the same time avoiding fragmentation of their ownership stake ensures that consumers voice has an appropriate weight vis-à-vis that of their co-investors [30]. Therefore, it is desirable that consumer-shareholders take a common position after an informed decision-making process.

5.1.1. Conveying Individual Share Ownership through a Trusteeship

Against this background, the CSOP model conveys individual shareholding of the participating consumers through a Trusteeship, which also—if desired—enables a cautious and gradual transfer of involvement in management decisions; the responsibility for day-to-day decisions of business operations stays with skilled management [31]. The vehicle of a fiduciary entity is a tool for professionalization of decision-making processes on the part of consumers, which at the same time ensures that consumers vote their shares together (en block) after an internal consultation advised by an expert. The fiduciary entity typically takes the form of a closely held corporation with limited liability (however, it could also be, for example, a limited partnership) administered by a managing director [25]. The fiduciary entity has only one shareholder (i.e., its founder; usually the initiator of the RE-project), shown in the list of shareholders at the registry court, with its sole purpose to represent the shareholding of the consumer-shareholders in the operating company. The establishment of the trust follows the conclusion of fiduciary contracts between the trustors and the managing director representing the Trusteeship. From a tax point of view the fiduciary entity is transparent as it is the consumer-shareholders who are the economic owners of the shares.

Instead of direct shareholding in the operating company the RE-CSOP, thus, involves a fiduciary entity that conveys the capital participation of the consumer-shareholders. A (fiduciary, fully fledged) Trusteeship of a shareholding occurs when a shareholder (here the fiduciary entity = trustee) owns the shareholding for the account of one or more other entities (here individual consumer-shareholders = trustors) in the sense that she is entitled to the rights arising from the shareholding only in accordance with a fiduciary contract concluded with the trustors [32]. Unlike in the case of an "authorisation trust" or the "power of attorney trust" in this case the separation of the trustee's external legal competence from his internal fiduciary duty is purely accomplished. The trustee (fiduciary entity) has a dual role: in relation to the other shareholders (e.g., municipality, strategic investor) she is the holder of the shareholder rights and in relation to the settlors she is entitled and obliged to exercise these rights for the account of the settlors (i.e., the participating consumers). The settlors can be described as holders of shareholder rights merely in the economic sense of the term. The trustee is in every respect carrier of the membership (i.e., shareholder) and, consequently, it is the fiduciary entity that is shown in the list of shareholders of the operating company (here a closely held corporation with limited liability).

5.1.2. Core Issues to be Considered for all RE-CSOP Models (Levels I–III)

In the context of enabling consumers to purchase shares, three key aspects need to be considered: (a) Securing the transferability of shares; (b) minimizing the cost of changes of ownership within the consumer-shareholders; and (c) granting corporate rights to the consumers.

Transferability of shares—The rules for changes of ownership among the consumer-shareholders represented by the managing director of the fiduciary entity are enshrined in statutes of the Trusteeship (and will be mirrored in the individual Investment Agreements that the consumer-shareholders conclude with the fiduciary entity):

- Exit of a consumer-shareholder with simultaneous transfer of the capital participation to a new CSOP participant only requiring a change of the party of the fiduciary contract (Investment Agreement).
- Exit of a consumer-shareholder with sale of the capital participation to the Operating Company which holds the share(s) until a new CSOP participant buys into the RE-CSOP. The Operating Company "warehouses" the shares, while at the same time creating a market place of these shares between the CSOP participants; this requires a definition of the legitimate motives to exit and of the period to announce this leave, as well as that of the instalment period for the cashing-out to avoid haemorrhaging of liquidity for CSOP.
- Exclusion of "bad leavers" (e.g., where consumer-shareholders obstruct decision-making within the fiduciary entity (Trusteeship), violate the supply contract substantially, etc). Here a cancellation of shares may occur with a subsequent transfer of monies from the Trusteeship.
- Exit following the death of a consumer-shareholder, which requires rules concerning the transfer by inheritance.

Minimizing the cost of changes amongst the consumer-shareholder—Pooling consumers' ownership rights in a fiduciary entity reduces transaction cost of share transfers between participating individuals (e.g., when CSOP participants move away from the region and transfer their share to new residents). At the same time facilitating consumer (co-)ownership through a fiduciary entity also ensures easy tradability of the shares. "Brokering" consumer shareholding in the Operating Company by the Trusteeship is sufficient to render consumer shares fungible and only requires a fiduciary contract (here Investment Agreement) between the consumers and the Trusteeship: It is the fiduciary entity represented by its managing director that—entering into the Investment Agreement with the consumer-trustors—now holds the shares of the Operating Company on behalf of the consumers. When consumer-shareholders change, the buyer or heir simply steps into the Investment Agreement in lieu of the former trustor. Changes of shareholders need not be registered—as would be the case for direct shareholding in the Operating Company—and the amount of participation held by the Trusteeship can fluctuate making administration easy. The basic mechanism is a fiduciary contract as is used in other investment settings. This structure is a standard solution in Germany tested many times by so-called public companies ("Publikumsgesellschaften" [33]) in real estate investments, who face a similar problem: A very large number of investors is intended to participate in the equity of a company where every change in ownership, whether it be due to death, sale of shares, or seizure has to be signed into the commercial register following the relevant formal procedures. Whether or not the transfer of capital participation from one consumer to another requires notarisation depends on the type of trusteed entity and national company law. For example, in Germany this would be the case for a closely held corporation with limited liability but not for a limited partnership, which in the latter case would have the advantage of lowering the transaction costs of transfers of capital participation from one consumer-shareholder to another. In contrast, the transfer of shares of an Italian closely held corporation with limited liability, following a 2019 reform of company law, does not require notarisation any longer. Depending on national tax and company law the advantages and disadvantages of the different legal design options, therefore, must be weighed against each other.

Granting corporate rights to consumer-shareholders—The statues of the trusteed entity, which as a rule will be a closely held corporation with limited liability, will include a catalogue of decision that can be taken only after a vote among the consumer-shareholders. This leads to a two-tier structure for the decision-making process with regard to representation and control and especially voting rights distinguishing between:

- Decisions concerning the day-to-day business of the Operating Company (or respectively of the CSOP Holding) that the Trusteeship represented by its managing director is authorized to take on behalf of the consumer-shareholders as trustors.

- Decisions of strategic importance (e.g., change of range of activity or business purpose, change of management, and those decisions involving financial commitments above a specific threshold; for example, EUR 50,000 requiring a vote of all trustors).

In this way, as mentioned above, the Trusteeship is also a tool for professionalization of the decision-making processes in the Operating Company while at the same time ensuring that:

- They have the possibility for an internal consultation advised by an expert (the managing director of the Trusteeship should have appropriate qualifications or access to expertise).

- They vote their shares together (in a block proportional to the Trusteeship's share in the Operating Company's or CSOP Holding's capital).

5.2. Financing the Consumer-Investment in the Operating Company

The CSOP is a type of leveraged investment (or buyout) transaction that uses external financing (debt), thereby achieving the benefit of financial leverage [9]. The cost of raising capital, as well as the repayment method, and, above all, the repayment period of the entire debt is all of key importance for the success and efficiency of this type of transaction. This section presents several legal and economic ways to shorten the debt repayment period or reduce the cost of financing and thus increase the effectiveness of financing RE-CSOP transactions.

The basic variable to be analysed is the debt repayment period. This is the period during which the CSOP Holding repays the debt using funds from the profits of the RE installation (the Operating Company). On the Holding or the Operating Company's balance sheet, liabilities from loans taken will gradually decrease in favour of equity. After the repayment period, the debt liabilities will be paid off, which means that external lenders no longer have any claims against the acquirer. In a simplified manner, it can be said that in such a situation the CSOP Holding (and indirectly the consumers) becomes the "full" economic owner of the RE installation (the Operating Company).

The repayment period is influenced by several factors. Determinants can be divided into two groups. The first group are economic factors of a more external nature, one being the size of the debt incurred, measured as the relation between equity and liabilities—the larger the percentage of the CSOP Holding's or Operating Company's assets financed from external funds, the longer the debt repayment period. Another factor is the profitability of the RE installation, that is, of the Operating Company measured by the return on equity ratio (ROE)—the higher the profit generated by the RE installation, the faster the repayment period. The second group of factors affecting the repayment period are legal and economic factors used in a specific transaction. This category includes, among others: (a) funds contributed by consumers; (b) tax optimization; or (c) a preferential loan granted by a public partner.

Contributions by the consumers—The application of CSOP financing in the context of Local Energy Communities according to Art. 22 RED II brings benefits to all parties, especially to the consumers. Therefore, it is justified that consumers make determined financial contributions to the RE-CSOP, which will help to increase its economic efficiency. However, against the background of the principle of proportional participation of CSOP participants depending on consumption (and not on financial strength), a limit is the average income of citizens and their access to savings. The amount of consumer contributions and their importance for the overall project depends on the size of the projected RE installation and the number of consumers supplied, the average purchasing power parity and, above all, the part of the income allocated for contributions. From experience in the U.S., it seems right to limit individual consumer contributions to a maximum of 10% of their respective earnings to avoid risk concentration [31].

Furthermore, it has to be taken into account that there may be changes after the initial allocation of shares to the individuals proportional to the households' respective energy purchases. In order not to incentivise increased energy use by a strict coupling of the acquisition of shares to consumption, a correcting factor should reward increased EE measured by a decrease in consumption per household member. Rewarding consumer-shareholders for reducing their consumption is also justified by the accelerated amortisation of the bank loan, as this will result in an earlier point in time that dividends are paid out.

Capital Tax Group—An important solution may be the creation of a tax capital group [34], which includes the Operating Company (running the RE installation) and the Holding Company. In this way, financing costs (interest) can be deducted from the tax base, which translates into a higher net profit of the entire capital group and enables the use of the so-called tax shield effect. Repayment of debt using a capital tax group can be made using:

- **Fixed capital and interest instalments**—in certain periods additional financial resources will be generated that can be allocated to reserves to ensure timely repayment in the event of an economic downturn or the payment of funds to consumers due to resignation from the plan;
- Or **variable capital and interest instalments**—allocation of a fixed percentage of the net profit for this purpose in each period.

Thus, the setting up of a capital tax group is desirable and—provided that it is permissible under the relevant national taxation legislation—should be considered during the creation of CSOP structures. However, restrictions with regard to the effective control of the two entities may occur. For example, under Polish tax law creating a capital tax group requires that the Holding Company has a 75% majority interest, thus lowering the ceiling for strategic investors' share to 25%.

Preferential conditions, subsidies or loans—Some of the solutions aimed at shortening the debt repayment period and thus improving the efficiency of the entire undertaking, are preferential conditions for land use, public subsidies or, if available, preferential loans from a public partner who owns part of the infrastructure where the investments take place [35]. In the case of a municipality, these may be buildings on which RE installations are constructed. Thus, a part of the funds for RE investments could come from one of the REC's partners according to Art. 22 RED II. This solution facilitates obtaining external financing and reducing the costs of the entire project. In addition, the public partner earns a higher interest rate than is earned on the funds invested in the capital market. Under this method, there are two options for debt repayment:

- **Deferment of the repayment to the public partner until the loan is repaid in relation to the bank:** The public partner agrees to subordinate its loan repayment to the investment loan, and agrees to postpone of its repayment period until other creditors, in particular those of the co-financing bank, have been repaid.
- **Parallel repayment of the bank and the public partner.**

5.3. Taxation of the RE-CSOP and its Consumer-Shareholders

Deferred taxation for consumer-shareholders—Under continental European tax law, the Trusteeship is treated as "transparent" [32] (i.e., the shares of the Operating Company are deemed to be owned by the consumer-shareholders) as beneficial owners (or economic owners) of the Operating Company. However, the standard Investment Agreement of the RE-CSOP (fiduciary contract) stipulates that a consumer-shareholder cannot dispose of his or her share(s) held in trust until fully paid for and until the CSOP participant decides to leave the plan. In this way, deferred taxation of the appreciation of their investment is guaranteed as taxation does not occur until the shares are eligible to be distributed from the Trusteeship and the consumers are actually able to economically dispose thereof. The parallel structure of the Operating Company and the Trusteeship (pooling the shares of the consumer-shareholders) ensures that only dividends paid out are taxed at the level of the consumer-shareholders.

Tax treatment of profits at the level of the Operating Company—In the form of a privately held corporation with limited liability, the Operating Company is, tax-wise, not transparent and with regard to profits incurred at the level of the Operating Company shelters the consumer-shareholders [36]:

- When leveraged, the transaction is financed, if possible, by loans from state development banks with low interest rates under programs specifically promoting RE.
- As a rule, the Operating Company—due to the financing cost of the leveraged transaction—will make losses or, in the best case, very small profits during the first years.
- In the case that the CSOP invests in a new RE plant pro rata profits/losses are allocated directly to the Operating Company; when it invests in an existing incorporated utility, they are allocated indirectly through dividend payments/depreciation of shares. In both cases taxation of profits occurs only once at the level of the Operating Company.

Tax treatment of the financing cost—Usually, the project vehicle will be set up and capitalized as a new Operating Company since buying into an existing utility will be the exception for RE projects. When leveraging the CSOP investment, it is important that the bank loan be taken directly at the level of the Operating Company that is operating the project (e.g., a wind turbine (mirror loan, see above 4.2.)) and that it is the Operating Company that repays the loan from its profits. Only after the bank loan is repaid will profits be paid out to plan participants. Building and running the newly installed facility, profits/losses accrue directly with the Operating Company. Therefore, both deduction of interest payments, as well as depreciation and carry forward of losses, lower the tax burden, increase liquidity and thus accelerate principal payments [36].

The treatment of interest payments is less advantageous in the event of a leveraged investment in an existing incorporated utility. Interest payments incur for the Operating Company rather than at the level of the utility where they would lower the tax burden and thus generate additional liquidity to repay principal. Usually, during the first years the Operating Company will incur losses or, if at all, very small profits as the deductible financing cost, that is the interest on the bank loan, is offset by any taxable income. Of course, the Operating Company must generate enough income to cover the cost of financing servicing both interest and principal of the bank loan. Although, as a rule, double taxation is avoided and the Operating Company in the form of a privately held corporation with limited liability shields the consumer-shareholders taxwise, the benefits are limited under this scenario. Nevertheless, acceleration of principal payments as under the first scenario could be achieved by a debt-push-down through a merger of the Operating Company with the RE utility as target.

6. Conclusions and Policy Recommendations

With regard to energy communities, European energy law does not rule out other private law citizens' or consumer-oriented initiatives than RECs which may be supported by and implemented with the participation of municipalities in the Member States [17] (p. 30). Such projects, while not complying with the RED II / IEMD governance model, would, of course, not benefit from the privilege of energy sharing of IEMD, and in particular the preferential conditions and incentives foreseen in the "enabling framework" under RED II. However, such initiatives could be led and controlled by professional actors on the energy markets who in RECs would be constraint to remain external investors or minority shareholders. The question whether such professional actors will accept the new governance model and decide to join RECs will depend on two factors:

1. The attractiveness and coherence of the RED II "enabling framework";
2. The flexibility of the underlying business model allowing for an adequate division of responsibilities and benefits between the different co-investors according to their expertise and contributions.

The legislative instrument to advance RECs by tying the benefits of the "enabling framework" to the compliance with the governance model can be described as an opt-in mechanism [37] aiming at creating peer-pressure: With a rising number of RECs operating successfully in European municipalities, this new business model will also become increasingly attractive to the incumbents; at the same time the underlying governance model, with its emphasis on the prosumer and the active consumer, will become more acceptable. However, the number of RECs set up in turn will depend on their ability to involve heterogenous co-investors which, as the empirical evidence discussed in Section 3 shows, is key to the success of RE clusters. Here trusteed investment models and in particular the RE-CSOP, introduced in Section 4 as a flexible low-threshold financing method, can play an important role as a bridge technology. The capability to align the interests of municipal, individual and commercial investors, while mitigating the frictions stemming from inherent limitations of conventional approaches make the RE-CSOP the prototype business model for RECs, as has been argued in Section 5.

6.1. Recognising the Challenges of RE Clusters in the Energy Systems of Tomorrow

Against this background, a holistic approach is key to the success of RECs. This has to include not only the governance but also the technical side. The best legislative intentions may lead to over-complexity in one field, while having unintended consequences in another, if not thought through consistently in an interdisciplinary approach. Notwithstanding, the RED II and, to a lesser extent, the IEMD focus on governance issues without providing details on the incentives that make a cooperation let alone partnership of RECs with professional energy companies in RE clusters [15] economically attractive. Therefore, four issues require specific attention:

- With decreasing cost of energy storage and increasing demand for local flexibility, community energy storage systems will become increasingly important for the energy transition as such and, consequently, for RECs. The challenge of integrating community storage in the energy system that presently is still largely centralized demands for socio-technical innovation [38].
- Apart from concerns that the new European regulatory framework does not sufficiently encourage, or in places even inadvertently discourages, complementarity between RES [15], the RED II does not adequately answer the question how energy sharing between local partners within RECs and with the possible involvement of professional energy companies can be facilitated.
- The question of operating and managing electricity networks and especially grid ownership of energy communities both RECs and CECs remains a thorny issue since regulators and the incumbent DSOs are inclined to opposition [29]. Although optional for Member States, it should be supported for RE clusters depending on their complexity and incentivised in a targeted way, in particular during the pioneering period to foster RE deployment.
- Inclusion of low-income households and vulnerable consumers is an important cornerstone in the fight against energy poverty and a postulate of energy justice [39] taken up both in RED II and IEMD. However, although prosumership reduces households' overall expenditure for energy and provides a second source of income through the sale of excess production [40], we observed a lack of concrete proposals in view to facilitate their participation.

Again, a lot will depend on the underlying business models and their capacity to provide flexible solutions that meet the different needs of the diverse actors. To test and demonstrate their potential RE-CSOPs are currently being implemented in the Horizon 2020 project SCORE, which runs from 2018 to 2021 in three pilot regions and in cities across Europe following these pilot projects [41,42]. During implementation, SCORE puts an emphasis on vulnerable groups affected by fuel poverty as a rule excluded from RE investments.

6.2. Spelling Out the "Enabling Framework" for RECs

The provisions on energy communities of the RED II and the IEMD remain relatively open to interpretation and leave the national lawmakers with room to manoeuvre. The transposition into national law until June 2021 is an opportunity to fine-tune and adapt the RED II rules to the needs of RE clusters and to formulate appropriate incentives supporting the underlying business models, like the RE-CSOP. In particular, during this period, the challenge is to overcome obstacles stemming from a lack of compatibility both with the existing regulatory frameworks and the national idiosyncrasies in order not to discourage national legislators. Without going into detail, four general aspects are key to successful transposition:

- Elasticity with regard to the eligibility requirements of proximity of shareholders is important in order not to unintentionally hinder the realisation of more complex RECs, namely fully fledged RE clusters. This is particularly important in view of their impact on complementarity of RES in urban settings [15].
- Where it is expected to delegate the balancing responsibility to professional partners or to pool it for more than one REC, the incentive system of the "enabling framework" should take into account the increased costs of pioneering RE clusters in the still largely centralized present energy systems.
- Energy sharing in RECs is highly sensitive to national regulation, especially when using the public grid, as value creation depends on the ability of its members to sell electricity to each other or make use of offsetting mechanisms of the electricity meters [28]. Network fees should be reduced in proportion to the actual distances in order to maintain the benefits of prosumership in RECs.
- To this end, a real-world testing environment, operated for a limited period of time, also dubbed "regulatory sandboxes" [43], should allow for the testing of incentives for RECs. This would allow to better tailor the "enabling framework" to the most suited business models, proving to meet, in particular, the challenges of RE clusters. Identified best practise could then be supported in a more targeted manner.

Funding: The SCORE project that this research is based on has received funding for a coordination and support action from the European Union's Horizon 2020 research and innovation programme under grant agreement No 784960.

Acknowledgments: The author is indebted to the Kelso Institute for the study of economic systems to have facilitated fruitful exchange of ideas with experts in the field at the 2019 Procida Symposium. The experience from implementing the Horizon 2020 project "SCORE" had a valuable impact on the reflections in this article and publishing it open access was possible thanks to its funding.

Conflicts of Interest: The author declares no conflict of interest.

Glossary

Autonomy of a REC	Recital 71 RED II stipulates the capability "of remaining autonomous from individual members and other traditional market actors that participate in the community as members or shareholders, or who cooperate through other means such as investment".
Capital Tax Group	Corporate structure that permits to calculate profits, losses and, what is most important here, costs, for tax purposes jointly for the combined tax group.

Clean Energy for All Europeans Package of the European Union	A package of measures that the European Commission presented on 30 November 2016 to keep the EU competitive as the energy transition changes global energy markets; this legislative initiative has four main goals, that is, energy efficiency, global leadership in RE, a fair deal for consumers and a redesign of the internal electricity market.
Citizen Energy Communities (CECs)	Defined in Art. 2 (11) of the IEMD as a legal entity that "(a) is based on voluntary and open participation and is effectively controlled by members or shareholders that are natural persons, local authorities, including municipalities, or small enterprises; (b) has for its primary purpose to provide environmental, economic or social community benefits to its members or shareholders or to the local areas where it operates rather than to generate financial profits; and (c) may engage in generation, including from renewable sources, distribution, supply, consumption, aggregation, energy storage, energy efficiency services or charging services for electric vehicles or provide other energy services to its members or shareholders".
Consumer Stock Ownership Plan (CSOP)	A financing technique that employs an intermediary corporate vehicle, facilitates the involvement of individual investors through a trusteeship and may use external financing, thereby achieving the benefit of financial leverage.
Demonstration Projects for Innovative Technologies	Defined in Art. 2 para. 2 (x) of the IEMR as "a project demonstrating a technology as a first of its kind in the Union and representing a significant innovation that goes well beyond the state of the art".
Effective control of RECs and CECs	Defined in Art. 2 pt. (56) IEMD as "rights, contracts or other means which, either separately or in combination and having regard to the considerations of fact or law involved, confer the possibility of exercising decisive influence on an undertaking, in particular by (a) ownership or the right to use all or part of the assets of an undertaking; (b) rights or contracts which confer decisive influence on the composition, voting or decisions of the organs of an undertaking".
Electricity/Energy Sharing (incl. (virtual) net-metering)	Recital (46) IEMD stipulates: "Electricity sharing enables members or shareholders to be supplied with electricity from the generation installations within the community without being in direct physical proximity to the generating installation and without being behind a single metering point". In the context of RECs, this is extended in Recital (71) and Art. 21 para. 6 to energy sharing.
Employee Stock Ownership Plan (ESOP)	An ESOP can use leverage and enables workers to acquire shares of their employer corporations, repaying the acquisition loan not from their wages but from the future earnings of their shares in the company.
Enabling Framework	Art. 22 para. 4 RED II foresees an enabling framework "to promote and facilitate the development of RECs"; furthermore, Art. 21 para. 6. foresees an enabling framework "to promote and facilitate the development of renewables self-consumption".
Fiduciary Trusteeship	A fiduciary, fully fledged Trusteeship of a shareholding occurs when a shareholder (here the fiduciary entity = trustee) owns the shareholding for the account of one or more other entities (here individual consumer-shareholders = trustors) in the sense that she is entitled to the rights arising from the shareholding only in accordance with a fiduciary contract concluded with the trustors.
Internal Electricity Market Directive (IEMD)	Defines amongst others "citizen energy communities" (CECs), introducing in Art. 16 a new governance model and the possibility of energy sharing for them.
Internal Electricity Market Regulation (IEMR)	Mainly focussing on the completion of the internal market in electricity that has progressively been implemented since 1999.
Investment Agreements	In the RE-CSOP these are concluded between CSOP participants and the Trusteeship and stipulate the fiduciary relationship including rights and obligations of both parties.
Leveraged investment	Financing transaction that uses external financing (debt), thereby achieving the benefit of financial leverage.

Mirror Loan	Structure of capital acquisition loan in a CSOP directly to the Operating Company and then in a second "mirror loan" to the Trusteeship resulting in favourable taxation and a stronger position of the lender
Renewable Energy Cluster	(Renewable) energy systems of the future, entailing flexibility, bi-directionality and interconnectivity options between prosumers and producers of energy and the market.
Renewable Energy Community (REC)	Defined in Art. 2 (16) the RED II as a legal entity: "(a) which, according to applicable national law, is based on open and voluntary participation, is autonomous, and is effectively controlled by shareholders or members that are located in the proximity of the renewable energy projects owned and developed by that community; (b) whose shareholders or members are natural persons, local authorities, including municipalities, or SMEs; (c) whose primary purpose is to provide environmental, economic or social community benefits for its members/the local areas where it operates rather than financial profits".
Renewable Energy Directive (RED II)	Defines amongst others "renewable energy communities" (RECs) introducing a new governance model and in Art. 22 the possibility of energy sharing for them, while providing them with an enabling framework.
Trusteeship	Contractual arrangement with a fiduciary (as a rule a legal entity but also a physical person) to facilitate individual shareholding of the participating consumers in a CSOP.

References

1. NCEO. NCEO Statistics. Available online: http://www.nceo.org/articles/esops-by-the-numbers (accessed on 31 May 2019).
2. Lowitzsch, J. The CSOP financing technique—Origins, legal concept and implementation. In *Energy Transition—Financing Consumer Co-Ownership in Renewables*; Lowitzsch, J., Ed.; Palgrave/McMillan: London, UK, 2019.
3. McLuhan, M.; Nevitt, B. *Take Today*; Harcourt College: San Diego, CA, USA, 1972.
4. Toffler, A. *The Third Wave*; Bantam: New York City, NY, USA, 1980.
5. Official Journal of the European Union L 328/82. Directive (EU) 2018/2001 of the European Parliament and of the Council of 11 December 2018 on the promotion of the use of energy from renewable sources (recast). 2018.
6. Official Journal of the European Union L 158/125. Directive (EU) 2019/944 of the European Parliament and of the Council of 5 June 2019 on common rules for the internal market for electricity and amending Directive 2012/27/EU (recast). 2019; to be transposed into national Law until June 2021.
7. Official Journal of the European Union L 158/54. Regulation (EU) 2019/943 of the European Parliament and of the Council 5 June 2019 on the internal market for electricity (recast). 2019; as a Regulation to enter into force on 1 January 2020 without the need for transposition.
8. Gauthier, C.; Lowitzsch, J. Outlook: Energy Transition and Regulatory Framework 2.0: Insights from the European Union. In *Energy Transition-Financing Consumer Co-Ownership in Renewables*; Lowitzsch, J., Ed.; Palgrave Macmillan: London, UK, 2019.
9. Holstenkamp, L. Financing Consumer Co-Ownership of Renewable Energy Sources. In *Energy Transition-Financing Consumer Co-Ownership in Renewables*; Lowitzsch, J., Ed.; Palgrave Macmillan: London, UK, 2019.
10. Ramirez Camargo, L.; Gruber, K.; Nitsch, F.; Dorner, W. Hybrid renewable energy systems to supply electricity self-sufficient residential buildings in Central Europe. *Energy Procedia* **2019**, *158*, 321–326. [CrossRef]
11. Mancarella, P. MES (multi-energy systems): An overview of concepts and evaluation models. *Energy* **2014**, *65*, 1–17. [CrossRef]
12. Lowitzsch, J.; Hanke, F. Renewable Energy Cooperatives. In *Energy Transition-Financing Consumer Co-Ownership in Renewables*; Lowitzsch, J., Ed.; Palgrave Macmillan: London, UK, 2019.
13. REScoop. Mobilising European Citizens to Invest in Sustainable Energy—Final Results Oriented Report of the REScoop MECISE Horizon 2020 Project. 2019. Available online: https://www.rescoop-mecise.eu/ (accessed on 5 December 2019).

14. Fici, A. An Introduction to Cooperative Law. In *International Handbook of Cooperative Law*; Cracogna, D., Fici, A., Henrÿ, H., Eds.; Springer: Berlin/Heidelberg, Germany, 2013; pp. 3–62. [CrossRef]
15. Lowitzsch, J.; Hoicka, C.; van Tulder, F. Renewable Energy Communities under the 2019 European Clean Energy Package—Governance Model for the Energy Clusters of the Future? *Renew. Sustain. Energy Rev.* **2019**. forthcoming.
16. Lowitzsch, J.; van Tulder, F. Overview of the Examples of Consumer (Co-)Ownership from the Country Chapters. In *Energy Transition-Financing Consumer Co-Ownership in Renewables*; Lowitzsch, J., Ed.; Palgrave Macmillan: London, UK, 2019.
17. Jasiak, M. Energy Communities in the Clean Energy Package. *Eur. Energy J.* **2018**, *8*, 29–39.
18. Martinot, E. Grid Integration of Renewable Energy: Flexibility, Innovation, and Experience. *Annu. Rev. Environ. Resour.* **2016**, *41*, 223–251. [CrossRef]
19. Rezaie, B.; Rosen, M.A. District heating and cooling: Review of technology and potential enhancements. *Appl. Energy* **2012**, *93*, 2–10. [CrossRef]
20. Risso, A.; Beluco, A.; De Cássia Marques Alves, R. Complementarity roses evaluating spatial complementarity in time between energy resources. *Energies* **2018**, *11*, 1918. [CrossRef]
21. Lowitzsch, J. Investing in a Renewable Future—Renewable Energy Communities, Consumer (Co-) Ownership and Energy Sharing in the Clean Energy Package. *Renew. Energy Law Policy Rev.* **2019**, *9*, 14–36.
22. Musall, F.; Kuik, O. Local acceptance of renewable energy—A case study from southeast Germany. *Energy Policy* **2011**, *39*, 3252–3260. [CrossRef]
23. Lowitzsch, J. Introduction: The Challenge of Achieving the Energy Transition. In *Energy Transition-Financing Consumer Co-Ownership in Renewables*; Lowitzsch, J., Ed.; Palgrave Macmillan: London, UK, 2019.
24. Baigorrotegui, G.; Lowitzsch, J. Institutional Aspects of Consumer (Co-)Ownership in RE Energy Communities. In *Energy Transition-Financing Consumer Co-Ownership in Renewables*; Lowitzsch, J., Ed.; Palgrave Macmillan: London, UK, 2019.
25. Lowitzsch, J.; Kudert, S.; Neusel, T. *Legal Opinion on the German Trust Model*; Viadrina working paper; European University Viadrina: Frankfurt, Germany, 2012.
26. Ackermann, D. *How to Cash Out Tax-Free, Yet Keep Your Business … ESOPs—A Practical Guide for Business Owners and their Advisors*; Conference Paper; National Center for Employee Ownership: San Francisco, CA, USA, 2002.
27. Kahla, F. *Die Kapitalstruktur von Bürgerenergiegesellschaften unter dem rechtlichen Rahmenwerk der deutschen Energiewende*; Zeitschrift für Umweltpolitik & Umweltrecht (ZfU): Frankfurt, Germany, 2019; pp. 94–123.
28. Tounquet, F.; De Vos, L.; Abada, I.; Kielichowska, I.; Klessmann, C. Energy Communities in the European Union. In *Revised Final Report of the ASSET Project (Advanced System Studies for Energy Transition)*; Energy Communities in the European Union: Brussels, Belgium, 2019.
29. Council of European Energy Regulators. *Regulatory Aspects of Self-Consumption and Energy Communities—CEER Report*; Ref: C18-CRM9_DS7-05-03; Council of European Energy Regulators: Brussels, Belgium, 2019.
30. Jenkins, K. Energy Justice, Energy Democracy, and Sustainability: Normative Approaches to the Consumer Ownership of Renewables. In *Energy Transition-Financing Consumer Co-Ownership in Renewables*; Lowitzsch, J., Ed.; Palgrave Macmillan: London, UK, 2019.
31. Kelso, L.O.; Hetter-Kelso, P. *Democracy and Economic Power: Extending the ESOP Revolution through Binary Economics*; University Press of America: Lanham, MA, USA, 1991.
32. Criddle, E.; Miller, P.; Sitkoff, R. *The Oxford Handbook of Fiduciary Law*; Oxford University Press: Oxford, UK, 2019.
33. Schmidt, K. Handelsgesetzbuch (HGB). In *Münchener Kommentar HGB*; C.H. Beck: München, Germany, 2012; 230p.
34. Oestreicher, A.; Koch, R. *The Determinants of Opting for the German Group Taxation Regime with Regard to Taxes on Corporate Profits*; Springer: Berlin, Germany, 2009.
35. Hunkin, S.; Krell, K. Renewable Energy Communities—A Policy Brief from the Policy Learning Platform on Low-carbon economy. Interreg Europe, August 2018. Available online: https://interregeurope.eu (accessed on 5 December 2019).
36. Thuronyi, V.; Brooks, K. *Comparative Tax Law*; Kluwer Law International B.V.: Alphen aan den Rijn, The Netherlands, 2016.

37. Heeter, J.; McLaren, J. *Innovations in Voluntary Renewable Energy Procurement: Methods for Expanding Access and Lowering Cost for Communities, Governments, and Businesses*; Technical Report; National Renewable Energy Lab.(NREL): Golden, CO, USA, 2012.
38. Koirala, B.; van Oost, E.; van der Windt, H. Community energy storage: A responsible innovation towards a sustainable energy system? *Appl. Energy* **2018**, *231*, 570–585. [CrossRef]
39. Sovacool, B.K.; Burke, M.; Baker, L.; Kotikalapudi, C.K.; Wlokas, H. New frontiers and conceptual frameworks for energy justice. *Energy Policy* **2017**, *105*, 677–691. [CrossRef]
40. Lowitzsch, J.; Hanke, F. Consumer (Co-)ownership in RE, EE & the fight against energy poverty—A dilemma of energy transitions. *Renew. Energy Law Policy* **2019**, *9*, 5–22.
41. SCORE. Horizon 2020 SCORE (Grant Agreement N° 784960) n.d. Available online: https://www.score-h2020.eu (accessed on 5 December 2019).
42. SCORE. CSOP-Financing. 2018. Available online: https://www.score-h2020.eu/?id=16883 (accessed on 5 December 2019).
43. Heldeweg, M.A. Legal regimes for experimenting with cleaner production—Especially in sustainable energy. *J. Clean. Prod.* **2017**, *169*, 48–60. [CrossRef]

Article

Technologies of Engagement: How Battery Storage Technologies Shape Householder Participation in Energy Transitions

Sanneke Kloppenburg [1,*], **Robin Smale** [1] and **Nick Verkade** [2]

1   Wageningen University, Environmental Policy Group, Leeuwenborch, Hollandseweg 1,
     6706 KN Wageningen, The Netherlands; robin.smale@wur.nl
2   Eindhoven University of Technology, School of Industrial Engineering and Innovation Sciences, Room 2.04,
     PO Box 513, 1600 MB Eindhoven, The Netherlands; n.verkade@tue.nl
*   Correspondence: sanneke.kloppenburg@wur.nl

Received: 30 September 2019; Accepted: 14 November 2019; Published: 18 November 2019   

Abstract: The transition to a low-carbon energy system goes along with changing roles for citizens in energy production and consumption. In this paper we focus on how residential energy storage technologies can enable householders to contribute to the energy transition. Drawing on literature that understands energy systems as sociotechnical configurations and the theory of 'material participation', we examine how the introduction of home batteries affords new roles and energy practices for householders. We present qualitative findings from interviews with householders and other key stakeholders engaged in using or implementing battery storage at household and community level. Our results point to five emerging storage modes in which householders can play a role: individual energy autonomy; local energy community; smart grid integration; virtual energy community; and electricity market integration. We argue that for householders, these storage modes facilitate new energy practices such as providing grid services, trading, self-consumption, and sharing of energy. Several of the storage modes enable the formation of prosumer collectives and change relationships with other actors in the energy system. We conclude by discussing how householders also face new dependencies on information technologies and intermediary actors to organize the multi-directional energy flows which battery systems unleash. With energy storage projects currently being provider-driven, we argue that more space should be given to experimentation with (mixed modes of) energy storage that both empower householders and communities in the pursuit of their own sustainability aspirations and serve the needs of emerging renewable energy-based energy systems.

Keywords: battery storage technologies; energy practices; public participation; householders; socio-technical transitions

---

1. Introduction

In Europe and elsewhere, there is an increase in renewable energy generation at domestic and community level. By installing solar panels, more and more householders are becoming prosumers and take responsibility for the decarbonization of the electricity system. However, for the grid, the uptake of solar poses challenges to the balancing of supply and demand of electricity and to grid management. Solar panels only generate energy during day time, whereas a peak in domestic electricity consumption takes place in the evening. Moreover, there are seasonal differences in the hours of day light and in weather conditions. Storage of renewable energy near to their decentralized sources, at the domestic or local level, is increasingly seen as a solution to this problem. Rapid developments in battery technologies have even led some to claim that we are at the brink of a 'storage revolution' [1] that may change the way householders and institutional actors engage with energy in fundamental ways. In addition to promises about the potential of storage for decarbonization and decentralization of the energy system, storage features in discourses about the empowerment of householders and communities to take more control over their energy use and become more independent from energy suppliers [2,3].

Despite the view of storage as a potential enabler of the energy transition, not much is known about the role that householders play, or are imagined to play, in energy systems that include distributed storage [4]. Yet, home batteries open up a range of possible roles and practices for householders. They enable householders to store their energy for use at a later time, but are also an important element in enabling new energy practices such as sharing and trading energy. These new energy practices place householders in a different relationship with the energy system and its key actors, such as energy suppliers. For example, the use of residential energy storage can help householders to become (more) autonomous in their energy supply, but domestic storage may also be used for Demand Response to help stabilize the grid [5].

In this article our aim is to explore potential ways in which home batteries can enable householders to become engaged in the transition towards low-carbon energy systems. Departing from the idea that energy systems are socio-technical configurations, we identify different ways in which householders and communities can become involved in low-carbon energy systems with storage. We link this idea of socio-technical configurations, or storage modes, to theories of 'material participation' [6,7] that argue that through everyday interactions with (energy) technologies, people can express concerns and 'intervene' in the energy system. Our theoretical argument then is that the ways in which householders (are enabled to) engage with energy storage technologies in an everyday, practical sense at the same time shape their participation in wider energy systems and their transitions.

In the following sections, we first explain our approach to energy storage as a technology of engagement, and the way we conducted our research. Next, we distinguish five different socio-technical configurations -or storage modes- in which householders can play a role. We identify how each mode affords specific energy practices for householders, such as storing, trading, or exchanging energy, and how the performance of these practices implies a particular distribution of tasks and responsibilities between householders and others energy system actors. In discussing the wider potential implications of these new types of engagements, we reflect on how energy storage may foster new collective energy practices and engagements that challenge our traditional understanding of energy communities, but also how these new energy practices often imply automation and reliance on intermediaries.

2. Renewable Energy Technologies as Technologies of Engagement

Literature in science and technology studies (STS) views technologies not just as material objects, but argues that the social and the technical are co-dependent and co-evolving [8]. This field stresses that technology and its social context mutually shape each other. Societal values such as sustainability, and ideas about the roles of users shape the technology, and at the same time technologies are constitutive of the social, in the sense that they actively shape their own context of use. Renewable energy technologies too have been approached as configurations of the technical and the social [9–11]. Walker and Cass use of the term 'mode' to understand renewable energy technologies as configurations of technology and social organization [9]. By social organization they refer to the ways technology is 'utilized and given purpose and meaning' [9]. They distinguish for example the traditional 'public utility mode' from modes that have emerged more recently, such as a the 'private supplier', 'community' and 'household mode'. Walker and Cass seek to understand how different modes embed within them different roles for publics in renewable energy deployment. They characterize these roles in terms of people's spatial proximity to the technology and their level of awareness and active engagement with renewable energy. For example, what they call the 'captive consumer' role entails a consumer who is distanced from the sources of renewable energy generation and consumes green energy passively. In an 'energy producer' role, on the other hand, people own and operate their own green energy generation technologies, for example via solar panels on their roofs, and are necessarily active and aware [9]. This approach thus recognizes the variety of roles and engagements of publics that emerge in relation to different renewable energy configurations.

While Walker and Cass discuss a wide range of technologies, from micro to macro scale, other studies have characterized and categorized different sociotechnical configurations around one particular technology. For community energy storage, Koirala et al. [2] have identified three configurations, namely shared residential energy storage, shared local energy storage, and shared virtual energy storage. This allows them to analyze the various ways in which local communities can use energy storage. Parra et al. [12] describe four categories defined by scale and application: single home storage, community storage, grid storage, and bulk storage. We take a different approach in basing our categorization of different storage modes on the question of how householders can become involved in and use energy storage.

Storage Modes and New energy Practices for Householders

Rather than understanding engagement in terms of general 'roles' for 'publics' or positioning ourselves in emerging research on public perceptions of energy storage [13,14], our aim is to examine what these 'publics', as householders who have installed energy storage devices, can do. In other words, we unpack the roles and forms of engagement by focusing on the (new) *energy practices* that become possible in different storage modes. Here we build on the work of Noortje Marres [6], who calls for an appreciation of everyday material practices as forms of participation. She views people's everyday use of energy technologies such as smart meters, as possibilities for public engagements in environmental issues. As she argues, everyday material actions can enable 'practical or physical interventions in current states of affair' [6]. Such an understanding of 'material participation' acknowledges the ways in which people are engaged in sustainable energy transitions through their everyday practices with household and energy devices.

Building on Marres' work, Throndsen and Ryghaug [15] apply the concept of material participation to assess the character of householder engagement in the case of smart grids. They conclude that householders, as 'material publics', articulate widely ranging (and politically engaged) smart grid enactments. Ryghaug et al. [7] argue the introduction of novel energy technologies in householders' everyday lives, such as solar panels, the electric car, and the smart meter, may create new forms of (materially based) energy citizenship. They give the example of the smart meter that through near real-time measurement and visualization of energy consumption makes energy visible in the household. This may result in the articulation of the issue of energy efficiency, and new forms of (practical) participation such as time-shifting of energy consumption, or replacing existing electric appliances with more efficient ones. The theory of material participation thereby challenges the dominant but narrow understanding of participation as involvement in decision-making. Instead, participation also takes the form of households interacting with energy systems through their everyday use of energy technologies in domestic settings, because in these everyday practices, issues around sustainability and climate change are articulated, and energy decisions are taken [7].

We draw upon the theory of material participation to explore how interactions with home batteries can engage people in the energy systems in different ways. For example, through installing a home battery, people can express their concern for climate change. At the same time, the use of batteries can also make them aware of new issues, such as the rhythms of domestic energy production and consumption and the systemic problem of grid balance. Finally, batteries also enable people to intervene in energy systems in a very concrete and physical sense, because batteries allow the redirecting of energy flows between the household and the wider grid. These examples illustrate energy storage devices as 'objects of participation and engagement' [7] in energy systems. Conceptualizing residential energy storage as a technology of engagement thereby allows us to examine not only how different modes imply different roles and energy practices for householders, but also how each mode at the same time shapes householders' participation in the transition to low-carbon energy systems in distinct ways.

In analyzing which different storage modes are emerging and what forms of engagement they imply, we follow a four-step approach (see Figure 1). Our first step is to examine how storage is viewed as a 'solution' to a particular problem, and whose problem this is (or is made to be). Different problematizations of electricity production and consumption entail specific ways of thinking about storage in home batteries as a solution. Some storage rationalities are more directly linked with householders experiences and practices as solar PV owners, while others start from the problems grid operators face in the context of a changing energy system. Starting from these diverse problem-solution sets we then describe the variety of (new) roles and energy practices for householders that are made available. Next, we analyze the distribution of tasks in these practices. It is important to discuss not only the (new) practices that emerge for householders, but also how and with whom these practices are being carried out, as some of these activities and choices may be delegated to technologies or providers and intermediaries. Finally, we examine the storage modes in relation to the wider energy system (outer circle of the figure). Everyday material practices of storing energy in household batteries enable interventions in the direction and management of (green) energy flows within household and between households, but also in the wider energy infrastructures. As such, these practices represent a rather 'direct' form of engaging with, and potentially reshaping the energy system. Our approach therefore also pays attention to the potential implications on the relationships between householders, providers, and technologies in low-carbon energy systems.

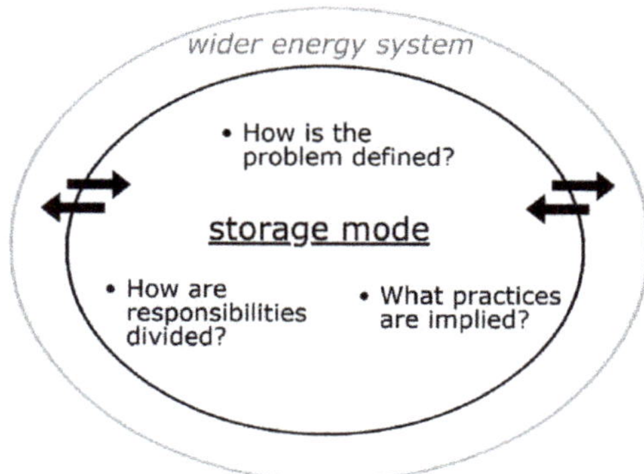

Figure 1. Analytical framework for identifying and analyzing storage modes.

3. Materials and Methods

This paper builds on empirical data that was collected at different moments and sites in the context of a research project on emerging energy practices in the smart grid (2014–2019). The data are qualitative and consists of interviews with different stakeholders in the energy system in the Netherlands, and to a lesser extent Germany and the United Kingdom. We conducted 14 interviews with providers of home battery systems and services, energy storage experts, NGO's and local governments involved in storage pilot projects. The bulk of the data, however, comes from interviews and observations with householders who were involved in storage pilot projects, or who had installed home batteries themselves. In the fall of 2016, 6 interviews were held with householders in Germany who had installed batteries for individual self-consumption, of which a few also participated in a virtual energy community called SonnenCommunity. In the Netherlands we conducted longer term fieldwork in the context of two demonstration projects. Here 14 interviews were held with householders engaged in the pilot project Jouw Energie Moment ('Your Energy Moment') in which home batteries were used for grid balancing. Furthermore, 30 interviews were conducted in the City-zen pilot project, where householders with batteries engaged in wholesale energy trading. A shortcoming is that we were unable to conduct interviews with local communities who owned and operated storage collectively, because there are relatively few real-world examples of this (but see the Feldheim case reported in [2]).

To gather information about community-owned storage, we therefore relied on interviews with storage providers and document study. Due to the variety of research material and differential access to cases, this research has an exploratory character. Hence, we use the real-world cases to conceptualize and identify the different forms of engagement that the use of home batteries may foster, rather than to systematically evaluate the extent to which new energy practices around storage already result in (new forms of) participation.

4. Results

Below we draw out five different socio-technical configurations around home batteries: individual energy autonomy; local energy community; smart grid integration; virtual energy community; and smart grid integration.

4.1. Mode 1: Individual Energy Autonomy

In the first mode, Individual energy autonomy (Figure 2), individual households deploy domestic energy storage for the purposes of using (more) self-generated solar energy. The rationality of this mode is optimizing self-consumption of electricity produced by PV panels. Self-consumption itself is a gratifying project for many PV panel owners. As one of the interviewed householders put it, 'I can use the energy, it gives a good feeling to me. To produce it and to use it'. Beyond this, two main motivations are at play here: (long-term) economic reasoning, and desire for autonomy or self-sufficiency. Self-consumption of solar power with domestic storage emerges as an alternative 'business model' for PV owners, as there is a common expectation that in the near future feeding back electricity into the grid will become less financially attractive. Secondly, domestic batteries appeal to householders who wish to become more energy autonomous, and less dependent on subsidies and energy providers. Here, different levels of energy autonomy may be pursued, ranging from going off-grid, to being self-sufficient during a black-out (back-up power), to remaining connected to the grid but relying on it as little as possible. As one of the householders argued: 'Somewhere the subsidies will stop and then you have a big advantage when owning a battery, then you are independent'.

Figure 2. Individual energy autonomy.

Home batteries for self-consumption often come along with an app or display on the device itself which enables householders to develop monitoring practices. One of the householders described it as a 'little pleasure' when he uses his app and sees 'that the sun shines and that you can see the battery charging'. Several respondents planned energy-intensive activities, like laundering and dishwashing, in such a way that solar (battery) power is used.

Domestic batteries used for the purpose of enhancing self-consumption place ownership in the hands of householders. However, this does not mean that individual householders can operate their batteries directly. The battery installer can translate the wishes of the householder into the learning algorithms which subsequently govern operation of the battery. As one householder put it: 'With the installer you can configure the battery and optimize everything so that it is attuned to the household. What could the customer do herself? Not so much.'

4.2. Mode 2: Local Energy Community

In the local energy community mode (Figure 3), both problem and solution are defined at the community level or within a local area. Local communities cannot always use their locally produced energy within the community itself. For distribution system operators (DSOs), the renewable energy generated by 'green communities' places local pressure on the distribution grid. To both communities and DSOs, an attractive solution is optimizing the local use of locally produced renewable energy. In terms of infrastructures, this mode can either consist of a local community connected to a larger 'neighborhood battery' or be formed by connecting distributed domestic batteries in a local setting. This mode comprises a range of variants from fully self-sufficient off-grid communities to local communities who are sharing energy via the public grid.

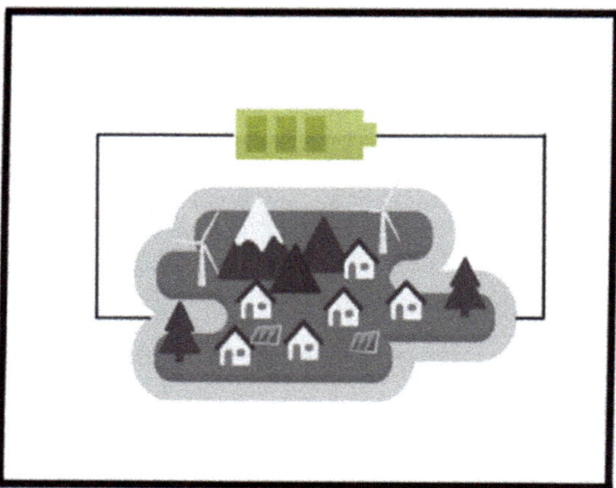

Figure 3. Local energy community.

In the local energy community mode, householders become prosumers who not only generate and consume individually, but also for and from the community's pool of energy. This allows for engaging with energy as a 'common good', or a 'common pool resource' [16]. Managing the 'common energy pool' at the community level implies new practices which include the monitoring of not only individual but also community-wide demand and generation; timing-of-demand to match local renewable energy availability (in storage); and energy sharing or peer-to-peer trading between community members.

Theoretically, local energy community storage can be organized in various ways. The local energy community may consist of a pre-existing energy cooperative that decides to add storage to its local renewable energy generation. In the pilot projects we studied, however, the batteries were owned, operated and controlled by other parties than the community itself, requiring little involvement of communities and households. Community energy storage with batteries in its present phase is still experimental, taking place in pilots and living labs. One of the reasons for the absence of 'commercial' variants of this mode are the regulatory barriers to peer-to-peer trading within a community, and to energy collectives becoming their own supplier [12]. In the Netherlands and the United Kingdom, however, regulatory sandboxes are now in place that enable the first communities to experiment with peer-to-peer supply [17]. In conclusion, community energy storage in principle offers a range of possibilities to organize energy supply and demand at decentral level. Different forms of (community) co-ownership of storage technologies (and generation units) can be imagined, as well as partnerships between energy suppliers and cooperatives; for example, energy suppliers could partner with cooperatives to supply the deficit at moments when the community's energy demand is higher than supply.

4.3. Mode 3: Smart Grid Integration

The smart grid integration (Figure 4) mode centers on the increasing problems grid management faces with the ongoing growth of renewable generation at the domestic scale. Grid assets at this scale are not necessarily suited for greater and volatile flows to and from the household. This can be accommodated by making more intelligent use of the grid assets and domestic devices in place with the help of IT, which is the 'hype' [18] called the smart grid. In the smart grid, the demand of households is no longer something that is simply predicted and accommodated by the grid; demand becomes something to be managed and steered at level of the individual household. The flexibility of domestic energy usage becomes an asset to be maximally unlocked and used towards efficient grid management. Domestic energy storage capacity is an ideal flexibility tool from the point of view of the DSO: storage can buffer peaks and troughs in domestic energy demand without requiring the involvement of householders or interfering in their energy use. The rationale of this mode is therefore to align the workings of the batteries (and other household appliances) with the needs of the grid.

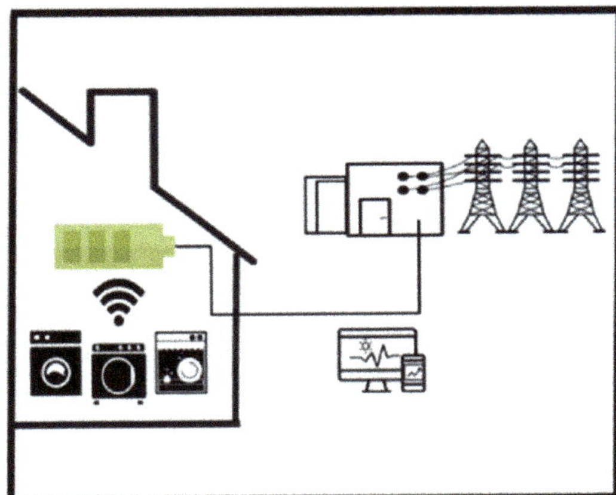

Figure 4. Smart grid integration.

Within the smart grid, householders are assigned a role as (active or passive) micro-managers with some responsibility to manage the impact they have on the grid. While they might actively shift some energy usage in reaction to more variable grid tariffs, the smart home with battery storage can also automate some of these decisions. Householders thus 'share' their batteries with the grid, allowing external control of the (dis)charging the battery.

As a result of automation and external control batteries may end up as black boxes, obfuscating the flows of renewable energy in the home and thereby creating a number of new uncertainties for householders. In smart grid pilot project Jouw Energie Moment, many participants critiqued the unintuitive information they were provided with: 'The only thing we pick up on with respect to that battery, is when it is 'humming', which means it is doing something.' The batteries would seemingly switch randomly switch between charging, discharging and neutral, never reaching full charge. Another householder stated: 'I just have no clue of what does what. And whether or not the battery is providing us any benefits.' In this respect, many householders stated that 'naturally, one would preferably want to be self-sufficient'. However, they were unclear if the batteries were contributing to this objective.

Since DSOs are barred from fulfilling "market-able" roles, the batteries are most likely controlled by an intermediate market actor like an aggregator. Domestic storage and other 'smart appliances' in the home thus become tools for grid supporting services. If householder insight into the functioning of home batteries (and other smart energy technologies) is insufficient, householders may come to see them as external or even invasive tools for solving others' problems [19]: 'At the moment it feels as if I help to solve a logistical problem for the project. I have found space in my home for someone else's experiments. But if I benefit... how can I see that? In effect I can't. I only see a big battery and hear a humming sound.'

4.4. Mode 4: Virtual Energy Community

The fourth mode –virtual energy community (Figure 5)- has parallels with the local energy community mode. The situation in which householders possess a battery system to increase their individual self-sufficiency while still relying on conventional energy suppliers to cover additional needs is seen as unsatisfactory. The rationale therefore is to link householders and optimize the use of self-produced energy within the community. While in the local energy community mode members live in the same local area, the virtual community members consist of geographically dispersed households. The members' energy devices (including solar panels, storage devices) are connected via smart meter technologies to a digital platform that allows for the monitoring and exchange of surplus energy. The first real world applications are now emerging (e.g., SonnenCommunity, Schwarmdirigent). One of these virtual energy communities, established by a battery storage provider, is presented as 'a community of [battery owners] who are committed to a cleaner and fairer energy future'. The same provider states that 'as a [member of our community], you don't need your conventional energy provider anymore—you are independent' [20]. In these framings, householders become not only prosumers in a virtual energy community, but also 'part of the energy future'. The goal of the virtual energy community is to meet the energy demand of the community with energy that is generated by the community itself.

Figure 5. Virtual energy community.

In this mode, ownership of the battery is with the individual householders, but the solar surplus that is produced when the batteries are full and/or the stored energy is 'shared' with others. It is important to note here that the sharing or exchanging of energy is virtual: The network does not consist of separate cables between members, but of a digital platform that enables virtual exchange via the existing grid. The meaning of 'sharing' therefore is complex. As one interviewed virtual community member put it: 'the idea is good. With [my friends] I spoke about it, they are part of it. Then I said, when there's sun at your place, I'm using your power. It's certainly a good idea, as the solar power that is stored, that is too much, can also be used on a place where it rains. But it's all virtual, it's not physical. The energy does not move from one place to the other, but okay, it doesn't matter'.

In the examples we studied, households were not actively engaged in energy exchange in the sense that they needed to decide on when and with whom to enter transactions; the process was managed by a third party—the aggregator—and often highly automatized. It is the responsibility of the aggregator to make sure that the demand within the virtual community matches the supply, so choices and decisions about the distribution of energy are made by this intermediary actor. The exchange of energy is not disclosed or made actionable to householders in the sense that they get insight in for example the current availability of community energy or get rewarded or sanctioned for their energy behavior. What is requested from households is to provide access to their energy data: the energy production, consumption and storage practices of members are monitored, and together with weather forecasts, used to make predictions of supply and demand in the community.

4.5. Mode 5: Electricity Market Integration

In the fifth and final mode, electricity market integration (Figure 6), the problem is defined in economic terms: due to competition on free electricity markets and growing renewable energy generation, electricity markets have become increasingly volatile. Batteries allow people to exploit this volatility, because the electricity flow can be temporarily halted, captured, and released again at a later point in time. The rationale of this mode is to align the workings of the batteries with energy market demands in order to create financial benefits for battery owners. In our research, we did not find any commercial variants of this mode yet, but there are examples of trials such as the Dutch pilot project City-zen. The households with batteries do not trade individually because the capacity an individual household can have available is too small. Instead, the participating households are aggregated to form a collective of householders. The aggregator in the Dutch project uses a Virtual Power Plant as the underlying technical infrastructure and explained that 'with all 50 participants, we want to create a large community. This community will be seen as one energy producing or consuming unit' [21]. In the project, the batteries loaded from the grid when prices were low and exported the electricity to the grid when prices were high. Energy thereby became a (tradeable) commodity and householders were ascribed a role as an economic actor who 'acts' according to market rhythms and logics. In the City-zen project, it appeared that for many householders this role as a market actor was at tension with their initial motivation to acquire solar panels for environmental reasons. As one householder explained: 'I didn't first go green with these things to now only think about money!'

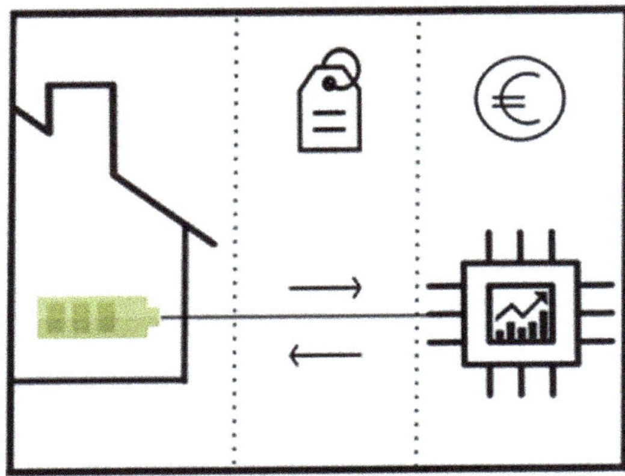

Figure 6. Electricity Market Integration.

In theory, in this mode the batteries could be owned by householders as well as third parties. The householders provide (stored) energy, their energy data, as well as the control over the charging and discharging of the battery to an intermediary party in exchange for a monetary reward. The intermediary acts as an aggregator of a group of households and trades on their behalf, by using historical and real-time energy consumption and production data from households in order to make accurate predictions of the amount of energy each household has available for trading. Householders thus engage in trading but this activity does not require specific skills or competences from them, nor does it require or stimulate them to actively adjust their energy consumption practices.

5. Discussion

5.1. Comparing the Five Storage Modes

Our identification of storage modes shows that a variety of different combinations of home battery storage technology and social organization is currently emerging. In addition to the already more established individual energy autonomy mode, providers are developing new modes that enable energy sharing and providing energy services to the energy system. The five modes we have distinguished differently engage householders in energy production and consumption through storage, in terms of the practices householders are enabled to engage in, and with regard to their relations with the conventional energy system and other householders (see Table 1).

Table 1. Five modes of energy storage, including the real-world examples in which fieldwork was conducted.

Storage Mode		Householder Engagement		Real-World Example
Modes	Energy Practices	Relation to Conventional Energy System	Engagement Level	Title
Individual energy autonomy	Self-consumption	Autonomous	Individual	Sonnenbatterie (DE)
Local energy community	Self-consumption and sharing	Autonomous	Collective	project ERIC * (UK), SWELL * (UK)
Smart grid integration	Providing grid services (and possibly self-consumption)	Integrated	Individual/collective	Jouw Energie Moment (NL)
Virtual energy community	Self-consumption and sharing	Autonomous/Integrated	Collective	SonnenCommunity (DE)
Market integration	Trading (and possibly self-consumption)	Integrated	Individual/collective	City-zen (NL)

*: no interviews with householders.

First, each mode affords particular energy practices for householders to engage in. In the individual energy autonomy mode, householders engage in self-consumption of stored energy within their household. In the other four modes, self-consumption is complemented with energy sharing, providing grid services, and trading.

Second, the modes entail particular relationships of householders to the conventional energy system. In the individual energy autonomy and local energy community modes, the aim is to increase self-sufficiency at household or community level, and in the ultimate case create a local microgrid. This idea of storage facilitating greater energy autonomy is opposite to the logic of integration that underpins the smart grid and market integration modes. In the latter modes, householders provide energy and services to actors within the energy system and thereby engage in the management of the energy system. The virtual energy community mode is less straightforward to characterize, as it fosters both autonomy and integration. While virtual energy communities may aim at autonomy from conventional energy suppliers, their geographically distributed character means that they need to rely on the public grid for sharing energy.

Third, the five storage modes also imply different types of relationships with other householders. The individual energy autonomy mode is the only mode in which householders do not engage with other householders. The two community modes (mode 2 and 4) connect householders based on shared local identity or values, in order to exchange energy among each other. The market and grid modes, on the other hand, may also aggregate individual households, but these 'collectives' engage in energy transactions with market and grid actors. For householders it may feel as if they participate on an individual basis, while in fact an aggregator treats multiple households as a pool in order to enable their participation in grid management and energy markets [22].

In the remainder of this paper, we want to draw out two important potential implications of these storage modes. Rather than discussing the implications of each mode separately, we reflect on two overarching effects that we consider to bring the most fundamental changes to how people can take part in the energy transition. First, some of the modes enable householders to engage in energy production, consumption and storage via *new collectivities* that challenge our traditional understanding of energy communities. Second, in all of the modes, a large part of the organizational 'work' around storage is performed by *intermediaries and smart technologies*, which challenges the idea of empowerment of prosumers and communities.

5.2. New Collective Material Practices

The individual energy autonomy mode is the only mode in which householders produce, store and consume energy within the bounded spaces of their own home. The other four modes comprise material practices which enable householders to form larger collectives and share their hardware and/or energy with others. Such material practices allow householders to go beyond optimizing self-consumption and exchange energy with other households or start transacting with the market or the grid. Existing local energy communities can add batteries to their renewable generation to boost local energy autonomy, but batteries can also enable the formation of new collectives of prosumers. These new collectives are a result of technical infrastructures that interconnect multiple households with batteries. Since aggregation does not require geographical proximity of the households, such new collectives can have members nation-wide as the example of the SonnenCommunity showed. The storage modes that afford collective material practices thereby bring along a range of questions about the character, aims and ideologies of these practices, and how they may and may not differ from the well-known local renewable energy generating communities.

In the literature, a common way to describe renewable energy communities is as 'those projects where communities (of place or interest) exhibit a high degree of ownership and control in renewable energy production as well as benefiting collectively from the outcomes' [23]. Such communities for example consist of local energy cooperatives that develop collective energy practices [24], such as collectively generating solar energy for local use. The aggregation of domestic batteries in particular affords new communities of interest, with new collective practices, to be formed. While the SonnenCommunity is an example of the creation of a community of like-minded users aiming at autonomy from conventional suppliers, other prosumer collectives may align their collective practices with market or grid rationalities. So just like local communities, the new collectives may be oriented towards social goals (e.g., autonomy), sustainability (green energy), and economic goals (profit seeking). An important difference is that the prosumer collectives that are now emerging are often not initiated bottom-up by citizens, but by grid operators, energy suppliers, and start-ups which have the expertise to build and manage the complex underlying technical infrastructure.

How householders can engage these new collectives may differ widely. There are prosumer collectives in which householders participate without being aware of the other 'members', for example when householders are aggregated to provide grid services. In other collectives the connections with other households are made visible in particular ways. For the SonnenCommunity, for example, the provider visualizes the location of community members on a map and shows which type of energy they generate for the community (solar, biogas). In some peer-to-peer exchange platforms consumers can even choose the peer they want to buy energy from. Emerging prosumer collectives thus shape new collectives which can take very different forms: from the aggregation of householders in collectives that remain invisible and anonymous, to a community of interest with 'members' or 'peers'. An important remaining question, however, is how inclusive these new collectives are for different types of households including lower-income households or tenants. As Ryghaug et al. [7] also argue for the case of electric vehicles and solar panels, the costs of these storage devices may mean that material participation via batteries is not equally accessible to all groups in society.

5.3. The Growing Power of Aggregators and Algorithms in New Material Energy Practices

Even though storage devices are located in households or communities, the role of householders in energy storage cannot simply be characterized as the active and aware prosumer. Most of the 'work' around energy storage is carried out by or on behalf of professionals, such as the installation and maintenance of the battery system, the monitoring and management of the battery charging strategy, and the managing of aggregated batteries. The emerging material practices surrounding storage are organized by intermediaries [25] as well as by information technologies.

Intermediary organizations, such as aggregators and green energy suppliers, play a key role in facilitating what householders can do with storage, as well as how, and with whom. Intermediaries are new players in the energy system, who act as a mediator or broker between householders and energy providers. They collectivize householders' energy consumption and production practices and enable and manage their participation in local and national energy systems. In the case of energy exchange among householders, intermediaries may arrange the balancing of supply and demand in the community. Intermediaries thus broaden the options for householders to enter into transactions with other householders and the energy system: transactions that are either too complex, or otherwise inaccessible to (individual) households. For geographical and virtual energy communities who want to become (more) self-sufficient, increased autonomy may thus go along with new forms of dependence on intermediaries who arrange the management and operation of energy exchange. There are concerns about the extent to which householders are able to access the full market potential of their batteries, as business models offered by intermediaries may distribute burdens and revenues unfavorably [26]. Material participation by householders through the purchase of storage batteries is, in other words, not synonymous with householder empowerment.

Information technologies too are a major factor in the management and control of (networked) households with batteries. Smart metering technologies monitor householders' energy consumption, production and storage practices. Hence, it is through these technologies that the householders' energy behavior becomes visible and gets embedded in battery management. Battery charging and discharging strategies often rely on algorithms that predict a household's energy behavior based on its historical energy production and consumption data. In addition, algorithms instruct the direction of energy flows (e.g., discharge to the household, or to the grid). Algorithms may also prioritize certain types of energy (green energy, cheap energy) in the way the battery systems work. In other words, they decide which energy is allowed to flow where and when. Householders choose these 'settings' when they buy a particular storage product or service, and may fine-tune them when the battery is installed. After that, the charging and discharging processes are often automated and users have little possibilities to change settings. Information technologies thus appear as a key factor in enabling connections between local or geographically distributed households and connections with wider infrastructures such as electricity markets. In shaping which transactions can take place, how, and between which entities, digital platforms [27] are becoming a new underlying structure for organizing energy production and consumption at decentral level, with as yet unknown implications for power relations in the energy system [28].

6. Conclusions

In this paper, we discussed energy storage as a 'technology of engagement' to better understand how householders and communities through their interactions with storage technologies engage in energy transitions. Drawing on Walker and Cass, we developed the concept of 'storage mode' to examine how battery technologies can be part of diverse sociotechnical configurations. We identified the emergence of five different storage modes, which demonstrates that renewable energy storage can entail a wide variety of relationships and interactions between householders and other energy system actors. To further unpack the various roles and engagements for householders, we examined the problem definitions, practices and task divisions in the modes. Our approach highlights that people can relate to renewable energy technologies not just as supporters or protestors or users, but through a diversity of roles that actively integrate them in the wider energy system (see also [15]): as co-manager or market actor, and as communities or individuals organizing energy production and consumption at decentral scale. As a technology of engagement, energy storage thus allows householders to interact with and shape the energy system in new ways. Most of the storage modes allow prosumers with battery systems to generate not only use value (by self-consumption of stored energy), but also exchange value (by sharing and trading energy and providing grid services) [29]. Energy storage thereby leads to more options for prosumers about what they want to do with their self-generated energy and with whom.

When storage affords energy practices in which self-produced energy gets exchange value, an important question is how prosumers will relate to this. Two diverging storylines now get connected to this exchange value: the first presents self-produced energy as a potential source of revenue for householders (energy as commodity), and the second emphasizes the sharing of surplus energy with other households (energy as (common pool) resource). Future social scientific research could follow up on these storylines and analyze the "moral economies" -or in other words moral and ethical questions about the production, distribution and exchange of energy- that emerge around this newly unlocked exchange value.

In examining the ways in which the new energy practices are organized in storage modes, our framework challenges the notion of active and aware citizens owning and operating their own household or community batteries. On the one hand, energy storage enables householders to become more autonomous from conventional suppliers and to enter new exchange relationships with other householders and the energy system. On the other hand, they face new dependencies on intermediaries and opaque information technologies. As long as householders believe that aggregators and algorithms act in their interest, they may not consider this a problem. Our analysis showed, however, similar to Parra et al. [12]), that the real-world applications of energy storage are still very much provider-driven. For existing community groups, it is difficult to initiate storage projects because in most countries legal limitations and complexities block communities from supplying their own energy to its members, or to organize the distribution of energy. In this context of provider-driven storage products and services, the question for householders is if they trust it is their aspirations and interests that are taken into account.

It is with regard to this potential for alternative forms of organizing energy production and consumption that we can identify policy implications. To foster storage modes that take into account a wider range of (future) interests and aspirations of householders and communities, and enable diverse forms of energy citizenship, governments could develop policies to actively support experimentation with social organization. An example of this is the Dutch 'Experimentenregeling' which provides energy cooperatives regulatory lenience to experiment with generating, supplying and distributing energy in their own local network. At the same time, studies have shown that such community-based models face difficulties due to financial, legal, social and technical restrictions and complexities surrounding energy storage and engaging with governance circles [2,17]. Beyond regulatory leniency, two other requirements for enabling experimentation include elimination of some of the financial risks and uncertainties in order to embolden communities as initiators of pilot projects, and secondly, professional facilitation of householders and communities to enable them to articulate their interests and ambitions vis-à-vis intermediaries. The emergence of prosumer platforms too offers opportunities for co-creation by citizens. Prosumer platforms could be developed or adapted together with local or virtual energy communities to ensure that energy exchange takes place based on valuations of energy and distribution of benefits and costs that the community favors. Opening up spaces for communities to initiate and develop energy storage projects may prevent that some emerging modes become marginalized too soon, and prevent lock-in situations in which existing power relations between providers and householders are reproduced. Recognizing that energy storage (as technology of engagement) offers prosumers enticing—and sometimes conflicting—perspectives on greater energy autonomy and self-sufficiency as well as on greater systems integration, it is important to provide space for experimentation with (mixed modes of) energy storage that both empower householders and communities in the pursuit of their own sustainability aspirations and serve the needs of emerging renewable energy-based energy systems. Providers and policy makers need to recognize that the 'storage revolution' should not just be seen in technical or economic terms, but also as an experiment with multiple new ways of relating to energy and new forms of social organization of energy production and consumption.

Author Contributions: Conceptualization, S.K., R.S. and N.V.; methodology, S.K., R.S. and N.V.; validation, S.K., R.S. and N.V.; investigation, S.K., R.S. and N.V.; data curation, S.K., R.S. and N.V.; writing—original draft preparation, S.K., R.S., and N.V.; writing—review and editing, S.K. and R.S.; visualization, R.S.; project administration, S.K.

Funding: This research was funded by the Netherlands Organisation for Scientific Research NWO, grant number 408-013-3.

Acknowledgments: We are grateful to the DEMAND Center at Lancaster University for co-funding and hosting all three of us to do empirical work in the United Kingdom between April and June 2016. Thanks also to Walter Fraanje for assisting with the empirical work in Germany, and to Marten Boekelo for conducting the interviews in the City-zen project.

References

1. Crabtree, G. Perspective: The energy-storage revolution. *Nature* **2015**, *526*, S92. [CrossRef] [PubMed]
2. Koirala, B.P.; van Oost, E.; van der Windt, H. Community energy storage: A responsible innovation towards a sustainable energy system? *Appl. Energy* **2018**, *231*, 570–585. [CrossRef]
3. Ruotsalainen, J.; Karjalainen, J.; Child, M.; Heinonen, S. Culture, values, lifestyles, and power in energy futures: A critical peer-to-peer vision for renewable energy. *Energy Res. Soc. Sci.* **2017**, *34*, 231–239. [CrossRef]
4. Devine-Wright, P.; Batel, S.; Aas, O.; Sovacool, B.; Labelle, M.C.; Ruud, A. A conceptual framework for understanding the social acceptance of energy infrastructure: Insights from energy storage. *Energy Policy* **2017**, *107*, 27–31. [CrossRef]
5. Parra, D.; Patel, M.K. The nature of combining energy storage applications for residential battery technology. *Appl. Energy* **2019**, *239*, 1343–1355. [CrossRef]
6. Marres, N. *Material Participation: Technology, the Environment and Everyday Publics*; Springer: New York, NY, USA, 2016.
7. Ryghaug, M.; Skjølsvold, T.M.; Heidenreich, S. Creating energy citizenship through material participation. *Soc. Stud. Sci.* **2018**, *48*, 283–303. [CrossRef] [PubMed]
8. Geels, F.W. *Technological Transitions and System Innovations: A Co-Evolutionary and Socio-Technical Analysis*; Edward Elgar Publishing: Cheltenham, UK, 2005.
9. Walker, G.; Cass, N. Carbon reduction, 'the public' and renewable energy: Engaging with socio-technical configurations. *Area* **2007**, *39*, 458–469. [CrossRef]
10. Juntunen, J.K.; Hyysalo, S. Renewable micro-generation of heat and electricity—Review on common and missing socio-technical configurations. *Renew. Sustain. Energy Rev.* **2015**, *49*, 857–870. [CrossRef]

11. Rutherford, J.; Coutard, O. *Urban Energy Transitions: Places, Processes and Politics of Socio-Technical Change*; Sage Publications Sage UK: London, UK, 2014.

12. Parra, D.; Swierczynski, M.; Stroe, D.I.; Norman, S.A.; Abdon, A.; Worlitschek, J.; O'Doherty, T.; Rodrigues, L.; Gillott, M.; Zhang, X. An interdisciplinary review of energy storage for communities: Challenges and perspectives. *Renew. Sustain. Energy Rev.* **2017**, *79*, 730–749. [CrossRef]

13. Ambrosio-Albalá, P.; Upham, P.; Bale, C.S. Purely ornamental? Public perceptions of distributed energy storage in the united kingdom. *Energy Res. Soc. Sci.* **2019**, *48*, 139–150. [CrossRef]

14. Thomas, G.; Demski, C.; Pidgeon, N. Deliberating the social acceptability of energy storage in the UK. *Energy Policy* **2019**, *133*, 110908. [CrossRef]

15. Throndsen, W.; Ryghaug, M. Material participation and the smart grid: Exploring different modes of articulation. *Energy Res. Soc. Sci.* **2015**, *9*, 157–165. [CrossRef]

16. Wolsink, M. The research agenda on social acceptance of distributed generation in smart grids: Renewable as common pool resources. *Renew. Sustain. Energy Rev.* **2012**, *16*, 822–835. [CrossRef]

17. Lammers, I.; Diestelmeier, L. Experimenting with law and governance for decentralized electricity systems: Adjusting regulation to reality? *Sustainability* **2017**, *9*, 212. [CrossRef]

18. Verbong, G.P.; Beemsterboer, S.; Sengers, F. Smart grids or smart users? Involving users in developing a low carbon electricity economy. *Energy Policy* **2013**, *52*, 117–125. [CrossRef]

19. Smale, R.; Spaargaren, G.; van Vliet, B. Householders co-managing energy systems: Space for collaboration? *Build. Res. Inf.* **2019**, *47*, 585–597. [CrossRef]

20. Sonnen. Available online: https://www.sonnenbatterie.de/en/sonnenCommunity (accessed on 24 July 2017).

21. Greenspread. Available online: http://www.cityzen-smartcity.eu/ressources/smart-grids/virtual-power-plant/ (accessed on 15 November 2019).

22. Rathnayaka, A.D.; Potdar, V.M.; Dillon, T.; Hussain, O.; Kuruppu, S. Goal-oriented prosumer community groups for the smart grid. *IEEE Technol. Soc. Mag.* **2014**, *33*, 41–48. [CrossRef]

23. Barbour, E.; Parra, D.; Awwad, Z.; González, M.C. Community energy storage: A smart choice for the smart grid? *Appl. Energy* **2018**, *212*, 489–497. [CrossRef]

24. Seyfang, G.; Park, J.J.; Smith, A. A thousand flowers blooming? An examination of community energy in the uk. *Energy Policy* **2013**, *61*, 977–989. [CrossRef]

25. Verkade, N.; Höffken, J. Collective energy practices: A practice-based approach to civic energy communities and the energy system. *Sustainability* **2019**, *11*, 3230. [CrossRef]

26. Hodson, M.; Marvin, S.; Bulkeley, H. The intermediary organisation of low carbon cities: A comparative analysis of transitions in greater london and greater manchester. *Urban Stud.* **2013**, *50*, 1403–1422. [CrossRef]

27. Burlinson, A.; Giulietti, M. Non-traditional business models for city-scale energy storage: Evidence from UK case studies. *Econ. e Politica Ind.* **2018**, *45*, 215. [CrossRef]

28. Gillespie, T. The politics of 'platforms'. *New Media Soc.* **2010**, *12*, 347–364. [CrossRef]

29. Kloppenburg, S.; Boekelo, M. Digital platforms and the future of energy provisioning: Promises and perils for the next phase of the energy transition. *Energy Res. Soc. Sci.* **2019**, *49*, 68–73. [CrossRef]

Article

Energy Justice as Part of the Acceptance of Wind Energy: An Analysis of Limburg in The Netherlands

Nikki Kluskens *, Véronique Vasseur and Rowan Benning

International Centre for Integrated Assessment and Sustainable Development (ICIS), Maastricht University, 6200 MD Maastricht, The Netherlands; veronique.vasseur@maastrichtuniversity.nl (V.V.); rowanbenning@gmail.com (R.B.)
* Correspondence: n.kluskens@student.maastrichtuniversity.nl; Tel.: +31-657584916

Received: 30 September 2019; Accepted: 13 November 2019; Published: 18 November 2019

Abstract: Policy documents in Limburg stress the importance of participation and distribution of benefits in wind energy projects, but it is not clear which modes of participation and distribution of benefits are most just, both in terms of perceived justice, and in terms of justice principles. Research shows that considering justice in renewable energy transitions increases the level of acceptance. This study aims to provide insight in what modes of participation and distribution are perceived as most just and likely to create local acceptance of wind parks. The most preferred modes are being compared to the indicators of the energy justice framework in order if they meet the criteria for a fair procedure and distribution of outcomes. Based on semi-structured interviews the analysis of the data demonstrated that different modes of participation in different phases of the process are being preferred and that a balance between modes of distribution of benefits is preferred. The results indicate that the most preferred modes of participation cannot necessarily address all indicators of procedural justice and that depending on the mode of distribution of benefits and the balance between those modes indicators of distributive justice can be addressed.

Keywords: energy transition; energy justice; acceptance of wind energy; modes of participation; modes of distribution of benefits; cooperative development

---

1. Introduction

By virtue of European agreements, The Netherlands has to produce 14% renewable energy by 2020 [1,2]. In order to reach this goal, the Province of Limburg has committed itself to produce 95.5 MW of wind energy by 2020 [3–7]. The feasibility of this goal seems realistic, since one wind park has been realized and six have been permitted [4]. However, in Limburg wind energy is a renewable energy source that shows differences in the level of local acceptance: some wind energy projects show a relatively high level of acceptance whether others cause a lot of debate [8]. Local opposition to renewable energy projects is often characterised by Not In My Back Yard (NIMBY)-ism, which implies an abstract acceptance of renewable energy by the public, but 'not in my backyard' attitudes on the local and thus the concrete level [9]. It is too simplistic to assume that local opposition of citizens and non-governmental organisations (NGOs) to renewable energy projects is only a selfish consideration. Complex institutional practices are in particular relevant for explaining NIMBY-ism [10]. Factors particularly relevant for causing opposition on the local level are perceived procedural and distributional equity [10,11], in other words, how fair the energy transition is perceived. One way to look into social sustainability of wind energy transitions and address long-term developments that are morally acceptable and talk about values and equity is through the framework of energy justice. The energy justice literature claims that both procedural and distributive justice have to be taken into account in order to call an energy transition just [12].

Procedural justice entails the elements in the process of decision-making regarding the establishment of renewable energy projects. A fair decision-making, put into practice by participatory methods, has been described as being the basis of legitimate rules and outcomes [12,13].

Distributive justice entails the outcomes of the decision-making, so the perceived balance between the costs and benefits of the renewable energy project. Important aspects are that the outcomes are equally distributed, that the allocation is just, and the mode of distribution of benefits is taken into consideration [12].

At the moment Limburg has one established wind park in the municipality of Leudal, in the village of Neer. Six other wind parks (Weert, Ospeldijk, Heibloem, Egchelse Heide, De Kookepan, Venlo) have been permitted by the appropriate body, of which two are irrevocable (Ospeldijk and Heibloem). Notwithstanding the fact that in in Limburg importance of participation and profit for the local citizens has been expressed in policy documents and many wind projects in Limburg used modes of participation and distribution of benefits, projects have shown differences in local acceptance. Elements as benefits in the distribution and participation in decision-making seem relevant in the level of acceptance in Limburg and were mentioned as reasons for opposition [14]. There seems to be a difference between the private market operator approach and the (partly) cooperative approach [4,15]. Cooperative development means that instead of big private energy companies developing wind parks, local initiators, sometimes framed as citizens, in the form of a corporation develop wind parks [16]. Also in local policy documents preference was given to a substantial percentage of cooperative development of a wind park to increase the level of local acceptance [5]. The preference for local initiated renewable energy constructions has also been concluded in scientific literature, since their method is more bottom up than the top down approach of big energy companies [16]. The private and cooperative approach differ in methods of development regarding the timing of participation, the modes of participation and the distribution of benefits [17–19].

A lot of research has been done on how participation and a fair distributional of benefits increases public acceptance. In the literature different forms and practices of what in general is identified as participation can be found, such as informing, consultation and partnerships [13,20]. Likewise, multiple forms of sharing in benefits can be noticed in the literature, such as ownership, community benefits and compensation measures [11,21]. However, understanding of what factors are important for a perception of procedural and distributive justice and what modes of participation and distribution of benefits are most likely to address these factors is understudied. This research aims to provide insight in which modes of participation and distribution of benefits (and in what phase) are perceived as most just and whether these preferred modes can tackle the indicators of energy justice. This paper is structured as follows: Section 2 gives the theoretical framework, Section 3 discusses the methods being used for this research, Section 4 provides the results and Section 5 gives the conclusion and discussion.

2. Literature Review

2.1. Energy Justice

Justice considerations regarding energy systems are being discussed in the energy justice literature. It is claimed that principles of justice are seen as a requirement in order to call renewable energy transitions sustainable and that not considering justice might erode political support for energy transition efforts [22]. The relevance for justice considerations in energy transitions becomes apparent when looking at the moral implications of our fossil based and renewable based energy systems and looking at the benefits justice considerations can bring to the social acceptance of renewable energy systems. Energy systems are understood as 'multiple interconnected processes of generation and consumption' [23]. Both our fossil and renewable energy systems are to a greater or lesser extend contributing to injustices in society.

The effects of our fossil-based energy system, such as CO_2 emissions, cause injustices both at the global level and the local level and therefore have to be tackled from a moral point of view. Even though the consequences of our fossil energy system are created by many actors all over the world, the outcomes are often disproportionally felt by the less fortunate in this world. This applies as well for the people living in the global North and the global South [24]. An example of consequences of climate change disproportionally felt by the poor was visible during and after Hurricane Katrina in the Unites States. The less fortunate were less able to protect themselves against the consequences and had more difficulties to recover from the damage, as well regarding health care and reconstruction of their houses [24]. Climate change can therefore result in further inequalities between people and thus increase injustices in society.

The injustices occurring due to the fossil energy system might not be the same as the ones occurring due to new renewable energy systems, but by moving to new renewable energy systems injustices may not be addressed by simply tackling CO_2-related injustices. With the transition to renewable energy such as wind, new moral considerations can be observed. For example, the location of the wind turbine, where the profit goes, how the costs and benefits are balanced among members of society and the possible consequences for the affordability of the energy bill [12]. A conflict is noticeable between the overall societal benefits of wind energy, such as cleaner air and profit of the companies involved, and the costs of wind energy, which are concentrated on the local level [11,25].

By looking carefully at the moral implications of energy systems, injustices present due to both the old and new energy system can be phased out [23]. Even though the literature provides different conceptualisations of the concept of energy justice [22,26], they all coincide on that the key aspects of energy justice are distributive and procedural justice. Both modes of justice ought to be present in order to call an energy system or transition towards a new energy system just [12]. The framework of energy justice developed by Mundaca Busch and Schwer makes it possible to assess both old and new energy systems by their identification of indicators for energy justice [23]. These indicators can be traced back to both procedural and distributive aspects.

2.2. Procedural Justice

Procedural justice entails the elements in the process of decision-making regarding the establishment of renewable energy projects. Justice is not only the greatest possible outcome of distribution, but also entails the way in which it is distributed [27]. The rationale of that a due process is a prerequisite for just outcomes is that the processes of institutions shape the outcomes of these institutions. This relates to the bias in outcomes of decisions when processes ignore or do not include the ones affected and thus stays in hands of relative powerful groups [28]. Procedural justice is according to the literature being realized by relevant stakeholder participation [12]. In general participation entails citizens involvement in decision-making processes [13]. The degree of participation in wind energy projects is a determining factor for the level of perceived fairness in a decision making procedure and affects the outcomes of that process [13].

Important aspects of meaningful stakeholder participation are (1) who is included and (2) the degree of involvement. Additionally, the meaningfulness of the degree of participation is dependent on the timing and the frequency of the involvement of stakeholders [13]. Participation cannot be identified as one general concept but is divided in different modes based on the degree of influence stakeholders have. The most influential starting point of different definitions of public participation and citizen empowerment is 'The Ladder of Citizen Participation' constructed by Arnstein. According to Arnstein there are different forms of citizen involvement, which vary in their ability to create inclusion and meaningful influence in decision-making processes. In total she identifies eight forms of citizen participation namely: (1) Manipulation; (2) Therapy; (3) Informing; (4) Consultation; (5) Placation; (6) Partnership; (7) Delegated Power and (8) Citizen Control. She argues that only the upper three levels (6, 7, and 8) create citizen power and a real influence in decision-making [20]. Table 1 shows the forms of participation and its definitions identified by Arnstein. These forms form the basis of identifying and categorising used and preferred modes of participation in wind park developments in Limburg.

Table 1. Forms of participation identified by Arnstein [20].

Form of Participation	Definition
Manipulation	Manipulation is about educating the people about the policy idea and/or problem. The people being educated have usually no legitimate function or power. The policy plan is being sold.
Therapy	This form of participation puts emphasis on curing the people participating from their ideas. The goal is to adjust the disagreement showed by the citizens
Informing	This form is about informing people of 'their rights, responsibilities and options' [20]. The citizens can ask questions, but there is no receptiveness for the opinion of the citizens. It is a one-way flow of, often technical, information from the decision maker to the citizen.
Consultation	In this form the opinion of citizens is asked, but not necessarily taken into account. Policy options are not available, just consultation on one policy option takes place. The scope of options is already limited by the people in power. There are no mechanisms to assure that their opinions will be taken into account.
Placation	There is an information flow and the scope of policy options is not limited beforehand. The expectation is that there is some influence of the less powerful. However, the powerholders still decide and can outvote the powerless since they judge the legitimacy of the input.
Partnership	In this form 'the power is redistributed through negotiation between citizens and powerholders' [20]. There is agreement on structures to 'share planning and decision-making responsibilities' [20]. In this form people can initiate plans, engage in joint planning and review plans.
Delegated Power	Citizens have dominant power in the decision-making and are accountable for the project. They (citizens) have the power to put things on the agenda, such as new plans, and the powerholder has to negotiate.
Citizen Control	Citizens have control over the budget, are responsible for the process and the solution. They are in charge of the policy-making and managerial aspects. If final approval is needed from the city council, it cannot be framed as citizen control.

2.3. Distributive Justice

Distributive justice entails the outcomes of the decision-making, so the perceived balance between the costs and benefits of a renewable energy project. The idea of distributive justice finds its most influential starting point in the social justice literature. In Rawls theory of justice, justice is identified as the fair distribution of primary goods. These primary goods are 'rights and liberties, powers and opportunities, and income and wealth and should be distributed in a manner a hypothetical person would choose if, at that time, they were ignorant of their own status in society' [29]. Distributive justice in energy justice recognizes that both costs and benefits, thus the outcomes of energy systems, are equally distributed among members of society, regardless their position in society [26]. It is basically a question of allocation of technologies and allocation of outputs of these technologies [23]. Wind energy in particular is known for its national or international contribution to cleaner air and reduction of greenhouse gas emissions, but the negative effects are present at a local scale [11]. This net benefit of wind energy is usually not visible on the local level and affects how citizens perceive fairness of the allocation of costs and benefits between 'society, community, local residents and the companies involved' [30]. This leads to perceived unjust distribution of the benefits and costs and thus non-acceptance on the local level [10]. Scientific literature acknowledges three most well-known forms of distribution of benefits, that have the aim to improve the perceived distributive justice and the local acceptance of wind energy: Compensation, Community Benefits and Ownership [11]. Definitions of these modes can be found in Table 2. These forms form the basis of identifying and categorising used and preferred modes of distribution of benefits of wind energy in Limburg.

Table 2. Forms of distribution of benefits identified by the literature [11].

Mode of Distribution	Definition
Compensation measures	Compensation measures cover the negative consequences for affected individuals of wind energy projects, for example regarding the value of property or houses of affected citizens [11]. Examples exists of developers directly paying compensation for the perceived costs, but also agreements where it is guaranteed that citizens can sell their property at the current market value [11]. However, this form has not been identified as the most effective form of distribution so far, since the line between bribery and compensation is thin and thus faces the risk of creating trust issues which results in doubts regarding the fairness of this mode [11,25].
Community Benefits	Community benefits are in contrast to compensation measures not specified to just a couple of individuals, but create benefits for the whole community and thus compensate in that sense for the local consequences of wind energy projects [11]. Community benefits are based on the equality principle since the aim is to give the people involved an equal share of the benefits [30]. An example is a local reduced electricity tariff for the affected people or the community [10]. Also, annual compensation payments to the community or part of the profit going to local funds can be noticed in the literature as a form of community benefits [10,11].
Ownership	Ownership measures can be seen as the most direct form of financial participation in wind energy projects. There are different forms of citizens' financial participation in which the degree of ownership differs. It ranges from citizens investment by shares to full community ownership of a wind turbine [11]. Ownership measures in the form of shares are based on the equity principle, since the financial benefits are proportional to how big someone's share or investment is [30].

With the necessity to transform the energy system in a reasonable time, emphasis has been put on the need for public support and local acceptance. International, national and regional policy documents address the social challenges of wind energy. Participation and equal distribution of benefits are expressed as important components to create public and local support.

3. Methods

In order to get insight in the current ways of operating wind parks in Limburg and draw conclusions about the most preferred modes of participation and distribution of benefits the following conceptual framework was being used to test the data.

Within the procedural justice aspect current used mode(s) of participation, in both the cooperative approach and private market approach, were analysed. In order to be able to draw conclusions on most preferred modes of participation (and in what phase), the current used modes were compared with the factors being mentioned in the data as relevant for a perception of a just procedure, in order to conclude which mode(s) are perceived as most just and if this corresponds with the modes used in both the private and cooperative approach. The modes that are perceived as most just were tested on their ability to tackle procedural justice indicators presented in Figure 1. These indicators are bias, ability to be heard, institutional representation, access to consultation, information disclosure, objectivity & adequacy & timeliness, and mobilisation of local knowledge.

Regarding the distributive justice aspect current used mode(s) of distribution of benefits, in both the cooperative approach and private market approach, were analysed. In order to be able to draw conclusions on the modes preferred modes of distribution of benefits, the current used modes were compared with the factors being mentioned in the data as relevant for the perception of a just distribution of benefits, in order to conclude which mode(s) are perceived as most just and if this this matches the modes used in both the private and cooperative approach. The mode(s) that are perceived as most just were tested on their ability to tackle distributive justice indicators presented in Figure 1. These indicators are distribution of costs and distribution of benefits.

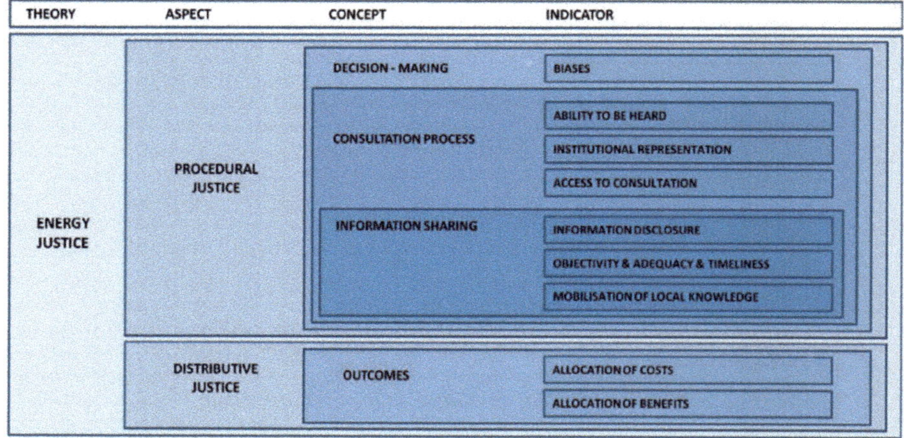

Figure 1. Overview of the conceptual and analytical framework of energy justice [12].

This study makes use of a case study approach and qualitative data collection. With the literature on the basis, a case study analysis of the already permitted wind parks in Limburg was done in order to research the procedures and the outcomes of the wind parks. Only permitted wind parks by august 2019 were included because of the necessity to have a clear boundary to compare completed processes of decision-making and distribution of benefits. Table 3 presents the seven permitted wind parks by ugust 2019. A case study approach is an appropriate tool to identify actor groups and get an elaborate understanding of which forms of participation and distribution of outcomes are used most frequently and which forms are most preferred by different stakeholder groups. In total seven wind parks in Limburg have been permitted. At present, all of them are supposed to have cooperative development elements. However, in Venlo and Neer, the starting point was a private development approach, but during the development process it started to include a cooperative development approach.

Table 3. Permitted wind parks in Limburg by August 2019.

Number	Wind Park	Established	Balance Private/Cooperation
1.	Wind park Venlo	No	Privately initiated, but promised to include cooperations.
2.	Wind park Neer	Yes	Privately initiated, but one turbine 100% of cooperation.
3.	Wind park Weert	No	Share between private company and cooperation.
4.	Wind park Ospeldijk	No	Share between private company and cooperation.
5.	Wind park Heibloem	No	100% cooperation.
6.	Wind park Egchelse Heide	No	Share between private company and cooperation.
7.	Wind park De Kookepan	No	100% cooperation.

Qualitative data in the form of semi-structured interviews was used for this research. Perceived justice, indicators for justice, and forms of participation and distribution of benefits are concepts that have subjective elements and are therefore harder to quantify. The conceptual framework of this research is more appropriate to be analysed and tested with qualitative methods to fathom different values and perceptions. Besides that, qualitative research can be useful in adding on the theories discussed in the literature review. To make sure to get valuable information out of the interview open-ended questions were the basis of getting in depth understanding of the stakeholders' perceptions, opinions, values and knowledge [31]. During the semi-structured interviews by phone, different types of open-ended questions were being asked, known in the literature as knowledge questions, feeling questions and background or demographic questions [31]. These questions (see Appendix A) aimed for getting information about the developments of the wind parks, insights in the perceived justice elements, and information of the stakeholders and their interests. The latter sort of questions were used to check whether the stakeholder list should be elaborated upon and to check the categorisation of the stakeholders based on their own input.

With the imperative approach the first stakeholders of the permitted wind parks in Limburg were identified. Actors were identified through literature and case study analysis. In order to prevent bias by the author in the first attempt of identifying stakeholders and create a fair representation of stakeholders, the imperative approach was complemented with help of the snowballing sampling technique. Snowball sampling entails that persons within the initial stakeholder categories will be interviewed and further stakeholders will be identified with help of these interviews [32]. After identifying the stakeholders and performing a double check by the author with analysis of open- ended 'background' questions they were categorised based on their role in a governance system. Based on the techniques described, stakeholders were categorised as (1) Local wind energy cooperations (2) Non-governmental organisations and (3) Citizens, which is elaborately presented in Table 4.

Table 4. Categorisation of stakeholders.

Stakeholder Categories	Stakeholders	Case Studies
Local wind energy cooperations	Energy cooperation A: Energy cooperation Newecoop	Ospeldijk
	Energy cooperation B: Energy cooperation Zuidenwind	Neer (only the Coöperwieck)
	Energy cooperation C: Energy cooperation Leudal Energie/Energy cooperation Weert Energie	De Kookepan & Weert
	Energy cooperation D: Energy cooperation Reindonk Energie	Venlo
Citizens	Citizen A: Village consultation Egchel	Egchelse Heide + Neer (including De Coöperwieck)
	Citizin B: Direct stakeholder and representative of village council Ospeldijk	Ospeldijk
	Citizen C: Village cooperation Boerderijweg	Neer (including De Coöperwieck) + Heibloem
	Citizen D: NLVOW (Dutch Association for People living near Wind Turbines)	Venlo + included in general way
	Citizen E: Village consultation Boekend	Venlo
Non-governmental organisations	NGO A: NMFL (Nature and Environment Federation Limburg)	Venlo + included in general way
	NGO B: NMFL (Nature and Environment Federation Limburg)	Weert + included in general way
	NGO C: LLTB (Limburg Agriculture and Greenery Federation	Heibloem + included in general way

In total 32 stakeholders were emailed, of which 12 positively responded. The point of saturation was determined on how much new information was yield after adding new stakeholders. Furthermore, with the aim to have a fair representation of stakeholders, the point of saturation also included taking into account that every category of stakeholders consisted of about the same amount of interviewees. To guarantee a fair balance and the unlikeliness of obtaining new information by adding one more identical or too powerful stakeholder the point of saturation was reached by 37.5% response.

For the analysis of the data collected the coding method was used. With help of the software program Atlas.ti reoccurring concepts and themes could be looked for in order to analyse the ideas expressed by the interviewed stakeholders. By constant comparison it was checked if the identified concepts fitted into identified themes. The collected themes were structurally compared with the aim to define categories. The interviews gave elaborate insights in among others perceived justice and preferred justice. With the insights gained from the coded interviews the conceptual framework, presented in Figure 2, was tested. Examples of the categories are: mode of participation, mode of distribution, and factors for a perception of justice. Codes were among others phase of involvement, policy options, individual benefits, and influence. The full list of the codes identified can be consulted in Appendix B.

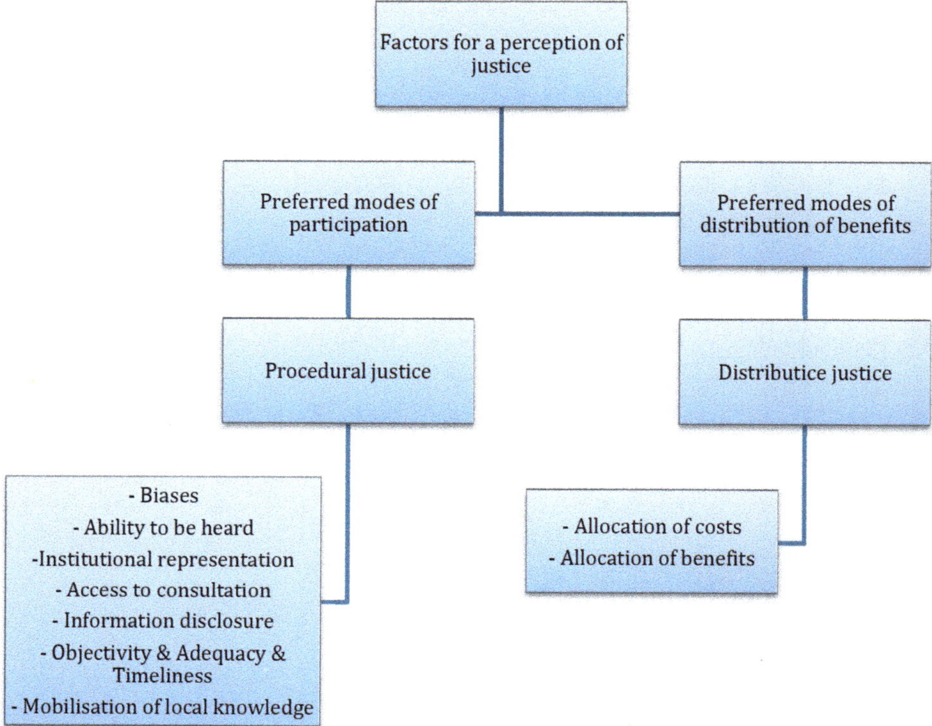

Figure 2. Conceptual framework of this research.

4. Results

Insights from the coded interviews resulted in the collection of the following elements: the currently used modes of participation and distribution of benefits and factors for a perception of fairness in both the procedure and the distribution of benefits. This section addresses the results regarding participation and distribution of benefits.

4.1. Participation

4.1.1. Used Modes of Participation in Limburg

Based on the interviews three main phases of wind energy project could be identified, namely the development of ideas phase, of which the location choice is a sub-category, the consultation process at the municipality, which can be identified as the implementation phase, and the distribution of benefits, which can be identified as the exploitation phase. It differs per development approach (private or cooperative) what modes of participation are used in what phase. On top of that, it could be notified that used modes of participation differ per stakeholder category. This section describes the three phases of a wind park development in which in every phase a distinction is made between a private and cooperative development approach and stakeholder categories.

Phase 1 (Development of Ideas phase)

Private development

Within a private approach the development of ideas phase, including the location choice, citizens predominantly experience the mode information as mode of participation. Stakeholders express the technical details about wind energy being mentioned, such as cast shadow and noise, but miss the context of the renewable energy goals of the municipality and the argumentation about the location choice. Data shows that citizens are in general not included via consultation or 'higher' modes of participation in this phase.

For NGOs a balance between cooperation and placation as mode of participation is noticeable in this phase. In most cases they are involved in the policy making phase of setting criteria for wind energy projects. In some wind energy projects the policy options, such as the location of the wind park or development criteria, are not already narrowed down by the people in power. Even though this shows elements of placation, their influence differs per case. In general they are considered as important partners also in the development of ideas phase, but the degree of their interest being taken into account differs per wind energy project.

(Partly) Cooperative development

In general, citizens are being informed in this phase, mostly materialised in the form of an information evening. Again, in the majority of the cases in Limburg, the location or the estimated project area of the wind park has at this time already been determined without citizen influence. So the decision on the location seems to be without modes of citizen involvement higher than information. Despite the cooperative development using the same mode of participation as the private approach regarding the location choice, the cooperative approach seems to involve citizens differently throughout other aspects in this phase of development. The information disclosure, transparency of the initiators, the frequency of involvement and the inclusion of citizens' interests are significantly more elaborate. Except from the location choice, citizen involvement can be characterised as consultation and sometimes even shows elements of placation, since policy options regarding the location are already narrowed down, but citizens' opinions are being asked for. The degree to which citizens' visions are being heard and create an expectation of having influence defines whether elements of placation are being present and differs per wind park development. Full placation could not be identified since the scope of options is already limited.

In the cooperative approach one element in this phase is being characterised by the partnership mode of participation, namely the division of ground compensation, which is a form of distribution of benefits. Usually the distribution of benefits takes place in the exploitation phase, but data confirms that in a lot of cooperative cases elements of distribution of benefits are already present in the first phase of the development of a wind park and make use of participatory methods. Even though citizens have no influence on the location of a wind turbine, directly affected landowners of the searching area have influence, in the form of co-decision, on the budget of the ground compensation, before the decision of the exact location of a wind turbine is been taken. Not only the landowners of the field where the wind turbine is going to be established are getting compensated, but all the landowners of the searching area, so before the location has exactly been decided upon, decide together how they are going to divide the money of the ground compensation budget. By giving a limited number of citizens, namely landowners in the searching area, considerable power to co-decide on the division of ground compensation, this element in the first phase of the development of a wind park can be characterised as partnership.

While in a private development approach, greater variations regarding modes of involvement between citizens and NGOs are visible, this difference seems to diminish in a cooperative development. The involvement of NGOs balances between consultation and placation, since in general their opinion is being asked for, even regarding the location choice, but the expectation that their interest being taken into account is context specific. This shows that for NGOs the modes of participation being used, in both the private and cooperative development approach are comparable.

Phase 2 (Implementation phase)

Private development

In general, the moment of any more influential form of participation for citizens is the moment when the location of the wind turbine is already determined but needs the approval of the city council. That is the moment for citizens to express their opinion via a public participation procedure. This mode of participation can be qualified as consultation, since the scope of options is already limited, since the location of the wind turbine has already been narrowed down to one option. Despite the fact that in this procedure citizens' opinions are being sought as there is no mechanism to ensure that citizen opinions will be taken into account or have particular power to influence the decision making process.

Interestingly NGOs seem not to be unambiguous regarding the expectation that their opinion is being considered and taken into account. Consultation and elements of placations are visible, in which the degree of elements of placation being present differs per wind energy project.

(Partly) Cooperative development

The same public participation procedure at municipal level is visible in a cooperative development approach. Even though citizens' opinions are better heard in the development-of-ideas-phase the consultation process in the implementation phase on the municipal level is being criticized. They express the absence of any mechanism to ensure that citizens' opinions are being taken into account. Besides that, the overarching interest of the municipality to reach the renewable energy goals is seen as a limitation to balance the interests correctly and influences the expectation of citizens that their opinion is being considered and taken into account. Again, for NGOs no significant difference could be noticed between a cooperative and a private development approach in this phase.

Phase 3 (Exploitation phase)

Private development

Concerning the exploitation phase, so after the permit has been given, the data seems to suggest that no public is involved anymore in any further phase of the process of a privately developed wind energy project. Neither citizens nor NGOs seem to have a role in the exploitation phase where usually the division of profit is being materialized.

(Partly) Cooperative development

Data confirms that in a cooperative developed wind park there is a role for citizens in the exploitation phase. This phase shows elements of partnership since there is an agreement to share the decision-making responsibilities to a certain extent, namely the division of money within the mode of distribution of benefits. For example, in the case of community benefits, co-decision is present on how the money within that mode is going to be spend. Citizens do not necessarily have influence on what modes of distribution there are going to be, or the budget of the mode, but they have influence on the division of the budget within these modes of distribution.

For NGOs it could not significantly be concluded that they have a role in this phase in a cooperatively developed wind park. Interestingly this is the only phase where NGOs seem to be subject to a lower mode of involvement than citizens. However, there is one example present where they are involved in the exploitation phase in the form of a working group deciding on how to divide the budget for nature repair measures.

Table 5 shows an oversight of the used modes of participation in wind development in Limburg.

Table 5. Current modes of participation in wind development in Limburg [1].

Phases of Wind Park Development	Private Development Approach	Cooperative Development Approach
Phase 1 (development of ideas)	Information	Information + Consultation + Elements of placation + (Elements of partnership *)
	Information + Balance between consultation and placation	Information + Balance between consultation and placation
Phase 2 (implementation)	Consultation	Consultation
	Consultation	Consultation
Phase 3 (exploitation)	-	Elements of partnership
	-	-

Note: Grey: used modes of participation for citizens; Blue: used modes of participation for NGOs; *: only concerns a small group of people, namely landowners of the search area.

4.1.2. Factors Important for a Perception of Procedural Justice

Regardless their position in the governance system and the development approach of the wind park, stakeholders gave similar answers to the question of what factors are important for a perception of fairness in the process. First of all, stakeholders mentioned the factors timing of the involvement and feeling taken seriously. Those two factors are interconnected. An important note is that almost all stakeholders mentioned wanting to be involved in the process of a wind park development as soon as possible and that this consultation is easily accessible for every stakeholder. With the ability to think along regarding alternative wind park locations in an early phase and the guarantee that their opinion is being asked and taken into account the factors 'as soon as possible' and 'feeling taken seriously' would be met. Feeling taken seriously is linked to that the interest and claims of the stakeholders are taken into account. This does not necessarily mean that they want to have a (co-) decision-making role, but they want at least a facilitated influence on the process by seeing their interests being represented. Being taking seriously starts with collectively informing the people and good communication. Without transparency people cannot become informed by the overall project according to the data. Also inclusivity and collectivism, which entails being involved all at the same time and with different stakeholders together, is mentioned as a factor of importance to create a fair process. On top of that, the adequacy of information could be identified as a factor being important for a correct information flow and a perception of fairness in the process.

Based on the factors being mentioned as important for a perception of procedural justice and supplemented with one factor being mentioned as important for a perception of distributive justice, concerning co-decision in the division of benefits, the stakeholder express their preference for the mode of information in the development-of-ideas-phase, but supplemented with placation in the division of the location, and partnership in the division of benefits in the development of ideas- and exploitation phase. However, a preference for the latter mode of participation could not be concluded among non-governmental organisations.

Table 6 shows an oversight of the preferred modes of participation per phase in wind park development in Limburg.

4.2. Distribution of Benefits

4.2.1. Used Modes of Distribution of Benefits in Limburg

Private development

Data shows that in the private market approach ground compensation for the landowner where the turbine is going to be located and individual compensation for residents is the most common form of distribution of benefits. Due to the small percentage of this individual compensation compared to the total profit data shows that in most cases the biggest percentage of the revenues goes to the market operator. No restrictions exist where this company has to be located, which means profit can also end up in other countries. Next to individual compensation, it is not impossible that a private market operator makes use of a community fund, which entails benefits measurements for the region. In some wind energy projects even private market operators give the option to participate financially in a wind park, but other requirements apply such as a minimum amount of money investment. In general, the perception exists that regardless the mode of distribution of benefits being used the distribution of benefits is not equally balanced in a privately developed wind park, taking into account the small percentage of the revenues ending up locally.

Table 6. Preferred modes of participation per phase in wind park development in Limburg.

Phases of Wind Park Development	Preferred Modes of Participation
Phase 1 (Development of ideas)	Information + Placation + (Partnership *) Stakeholders want to be collectively informed and included when the location of the wind park has not yet been determined, so that there is still an open discussion about the location. Besides that, they want their interests to be taken into account. They would like to think along about alternative locations. This does not mean that they necessarily want to take part in the decision-making process, which they generally still see as a task for the municipality. In conclusion they want a broader scope of options and they want their interest being taken into account.
Phase 2 (Implementation)	Decision making in hands of the municipality
Phase 3 (Exploitation)	Partnership Data shows that the only part where citizens prefer a higher form than 'placation' as a form of participation is regarding decisions on the distribution of benefits. They do not necessarily want to take part in deciding on the budget or what modes of distribution will be available, but they do want to determine what will happen to the money within a certain mode of distribution. * In the 'development of idea' phase, citizens indicate that they would also like to make a collective decision about the distribution of the ground compensation. However, this concerns a small group of landowners that is included in this division in this phase. Interestingly, the NGO did not indicate that it would prefer partnership as a form of participation at any stage.

Note: * shows that partnership is also preferred in the first phase of wind park development, but only concerns a small group of stakeholders, namely landowners.

Cooperative development

As in a private approach, different modes of distribution of benefits are present in cooperative cases in Limburg. In almost all cases collective ground compensation is present. Whereas in the private approach only the landowner of location of the wind turbine would receive ground compensation the cooperative approach includes the searching area too in the ground compensation budget, which means inclusion of the haze parcels of the estimated location too. Furthermore, all wind parks show collective compensation measures in the form of a community fund. The budget of this fund is most of the time determined by following at least the NWEA norm of 50 cent per MWh. Within this community fund members of the cooperation can bring in ideas for projects they want to realize in the region. Most of the time the projects have to fulfil required elements of liveability or sustainability. It depends per project whom is included to receive the benefits of the community fund. Data shows differences regarding individual compensation measures between cooperations. While in some cases individual compensation measures are present, others explicitly do not make use of such a mode.

Different formats of community benefits are visible in a few cooperations, such as an energy fund and a nature fund, present in Ospeldijk and De Kookepan. These modes of distribution of benefits are not visible in every cooperatively developed park but are elaborate forms of letting the profit end up locally. For example, in Ospeldijk the intention is to facilitate the energy transition of the region with the energy fund budget and is increase in nature quality economically facilitated with the nature fund in De Kookepan.

The last form which is present in a majority of the cooperatively developed cases in Limburg is that as a member of the cooperation you can participate financially, by investing money in the energy cooperation with yield as return favour for the investment. Regardless the mode of distribution, with a cooperative development approach the benefits of the wind energy projects will in general land locally.

Data shows that even though the intention is to let the profit end up locally, no requirements are set in general on investing money in an energy cooperation. This means there is no guarantee that only local citizens will invest, which might weaken the intention to let the profit end up locally.

Table 7 shows the current modes of distribution of benefits of wind parks in Limburg.

Table 7. Current modes of distribution of benefits of wind parks in Limburg.

Private Developed Wind Park	Cooperative Developed Wind Park
Individual compensation (not transparent)	Individual compensation (transparent)
Community fund	Community fund
Ownership in the form of financial participation	Ownership in the form of financial participation
	Energy fund *
	Nature fund *

* These modes are only present in two cooperatively developed wind parks in Limburg.

4.2.2. Factors Important for a Perception of Distributive Justice

Regardless the development approach of the wind park, stakeholders express that the most important factor for a perception of a fair distribution of benefits is that the profit ends up locally. Also, trust and a fair treatment in the process were mentioned as factors that influence the perception of fairness in the distribution of benefits.

Besides that, a factor specifically mentioned by citizens, is the partnership mode of participation being present in the decision-making concerning the sharing of the benefits. It is not necessarily clear if they want to co-decide on what modes there are going to be or the budget of it, but it can be concluded that they want to co-decide on how the money is going to be distributed within a certain mode. Moreover, they express that there has to be a certain balance between the budget available per mode. This entails for example that money available for ground compensation and individual compensation are in equal proportion. This balance of budgets links with the importance of transparency of the modes and its budgets, which is another factor that creates a fair perception of the distribution of outcomes.

Data shows that there are differences between the focus of the distribution of benefits between a privately developed wind park and cooperatively developed wind park. Privately developed wind parks especially differ in the way they balance the local and non-local profit, and the transparency of the money flow. An example of the lack of transparency is the non-collective approach regarding the ground compensation, where it is not clear to the haze parcels what amount of money the landowner received. With a private approach it is not impossible that some of the profit ends up locally, but the percentage of what ends up locally and what not is not in proportion according to the data. In cooperatively developed parks data shows that the starting point is to let the profits end up locally. Different modes of distribution of benefits are recognizable in order to achieve this. In Limburg it differs per cooperative case which modes of distribution of benefits are present. As well individual compensation, community benefits and investment opportunities are identifiable in the case studies.

Taking into account the factors being mentioned as important for a perception of distributive justice it can be concluded that stakeholders do not prefer one mode of distribution of benefits over the other, they preferably see a combination of modes in which they have influence regarding the division of the budget and that are transparently balanced.

5. Conclusions and Discussion

Taking into account the implication of the most preferred modes of participation per phase of development and comparing it with the procedural justice indicators the most preferred modes of participation, where citizens and non-governmental organisations are adequately and correctly informed, and have an advisory role which is being taken into account, can address among other things the indicators 'ability to be heard', 'information disclosure', 'objectivity & adequacy' and 'mobilisation of local knowledge'. When the inclusiveness and timing of the invitation to be involved are considered also the indicators 'timeliness', 'institutional representation' and 'access to consultation' can be addressed with these modes. Taking into account the preference for placation over higher modes of participation in the first phase of development of a wind park, the indicator 'biases' in decision-making is harder to address. With powerholders, in this case the municipality, still being able to judge the legitimacy and outvote the input of citizens and Ngo's in the second phase of the development of a wind park, it is not guaranteed that their interests are properly considered. The double role of the municipality to on the one hand fulfil the renewable energy goals and on the other hand weigh the interest of stakeholders makes critiques regarding their incompetence to balance the interest properly plausible. Considering the preference for partnership as mode of participation in the division of benefits does not change this. This higher mode of participation concerns only a small element of the decision- making regarding the distribution of benefits in the and does not take away the risk of biases in the overall decision-making process regarding a wind park.

With no preference for one mode of distribution of benefits over the other, no general answer on their ability to tackle the distributive justice indicators can be given. A combination of modes addressing the factors being mentioned as important for a perception of distributive justice has the preference. Regardless of the mode preference is given to let the profits land locally and the possibility of co-decision within the modes of distribution of benefits. The relevance of a combination of modes might become apparent when looking at the ability of individual modes to address all factors of importance for a perception of distributive justice. With no general conclusion on what mode is most appropriate to address these factors a critical note has to be made by the 'investment' mode of distribution. Even though data confirms that the factor that profit has to land locally is of importance, the 'investment mode' cannot guarantee this completely. With no requirements on whom can invest, also actors not living in the region are able to invest and yield profit. This makes the ability of this mode to address factors important for a perception of distributive justice questionable. Figure 3 shows what indicators of procedural justice could be tackled by the most preferred modes of participation and shows that it could not be confirmed which indicators of distributive justice could be tackled.

This research contributed to the existing scientific literature by getting (1) insight in what modes of participation are most preferred in what phase of a wind park development by making nuances regarding what modes are perceived as most just in regard to specific elements within the process. In view of different aspects of the decision making it showed for example that regarding the location choice another mode of participation is being preferred than regarding the distribution of benefits. Besides that, (2) it showed that a combination of modes of distribution of benefits is being preferred over one specific mode, but that the ability of individual modes to address all factors of importance for a perception of distributive justice is being questioned. Subsequently, this research showed that (3) there is a discrepancy between the most preferred modes of participation and their ability to address the procedural justice indicators.

In view of other research, the following similarities and differences can be identified:

Concerning the modes of participation, the research of Langer, Decker and Menrad confirms, even though the modes have been categorised differently, that information, consultation (in this research identified as placation), cooperation (in this research identified as partnership) have a positive influence on the acceptation of wind turbines on the local level. Moreover, they conclude that transparency, information as well as inclusion of citizens in the decision-making enhance the level of acceptation [13]. The latter could not be concluded out of the data of this research, since a distinction on what they want to co-decide has to be made and turns out to be only regarding the distribution of benefits. Regarding the ability of the preferred modes of participation to tackle procedural justice indicators further research has to find out if higher modes of participation are more likely to address the indicator 'biases' since that question was out of the scope of this research.

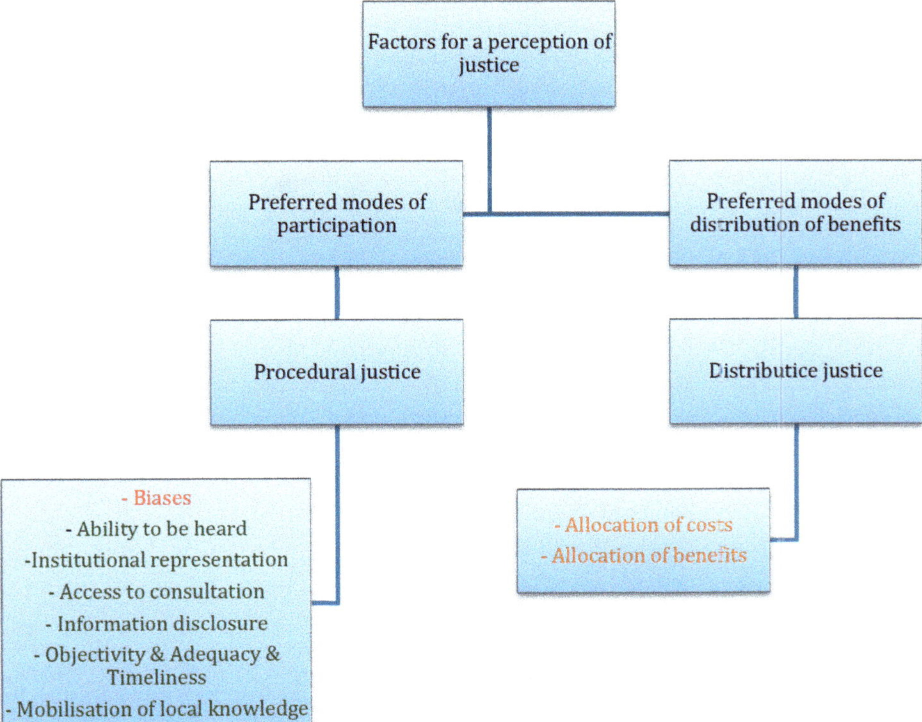

Figure 3. Tested conceptual framework. (red means this indicator is not likely to be tackled; green means this indicator can be tackled; orange means it cannot significantly be confirmed these indicators can be tackled).

Regarding modes of distribution and related factors that enhance acceptation research conducted by Lienhoop confirms that the most important factor to create acceptation is that profit ends up locally. Comparable to the data results in this research, Lienhoop's research confirms that humans do not necessarily always act to maximize their own benefits. Transparency and a combination of modes seems more desirable [10]. Lienhoop confirms that only financial investment as a mode is not desirable, since it can create the risk of not being affordable for everyone. This is comparable to the results in this research, where the possibility of this mode to tackle distributive justice indicators is being questioned. Further research has to find out whether the investment mode of distribution of benefits is an appropriate form to distribute the profits equally. As mentioned already, with the investment mode critical notes have to be taken into account whether the indicator of a just allocation of benefits can be guaranteed. Concerns are that this mode facilitates investments from other regions and is more attractive for higher incomes. This raises questions on whether this form is appropriate to tackle the indicators of distributive justice. Further research has to find out if this mode is also able to address the equality principle. Even though other studies show that the mode of community benefits leads to the highest level of acceptance, this is being questioned in this research, with concluding remarks emphasizing the importance of a variation of modes [25].

For this research a conceptual framework was being used and a comparison between perceived justice and the energy justice framework was being made. Out of the data it could be concluded that regarding the procedural justice indicators it matches with the factors being mentioned as important for procedural justice. However, the indicator 'biases' was mentioned as a barrier in the second phase, but data could not confirm stakeholders want a higher form of participation in that phase to tackle this indicator.

The data in this research did only show differences between initially privately developed wind parks and (partly) cooperative wind parks. However, the scope of this research did not allow a comparison between 100% cooperatively developed wind parks and partly cooperatively developed wind parks. Further research is suggested in order to draw conclusions whether there is a difference in perception of justice between 100% cooperative wind parks and partly cooperative wind parks.

Further testing of the conceptual framework and the energy justice framework on other renewable energy transitions and systems, for example solar fields, is recommendable in order to test if indicators of the energy justice framework will be matched in other circumstances and if the perception of justice differs per source of renewable energy development.

Author Contributions: Conceptualization, N.K.; methodology, N.K.; validation, V.V. and R.B.; formal analysis, N.K.; investigation, N.K.; resources, N.K.; data curation, N.K.; writing—original draft preparation, N.K.; review and editing, V.V. and R.B.; visualization, N.K.; supervision, V.V. and R.B.; project administration, N.K.

Funding: This research received no external funding

Acknowledgments: I want to thank the Province of Limburg for facilitating the technical materials available to conduct the research.

Conflicts of Interest: The authors declare no conflict of interest.

Appendix A

Translation of the interview questions in English:

1. Can you tell me about the development of the plans for this wind farm?
2. Can you tell what is your role concerning the wind park?
3. How were you informed about the plans of the wind project?
4. When were you informed of the plans for this wind project?

   a. (Question for clarification; Was the decision about where the project would be realized already taken?)
   b. (Question for clarification; And if so, by whom?)

5. What are the advantages of this wind energy project?

   a. (Question for clarification: How does the community benefit from the wind project?)

6. What are the disadvantages of this wind energy project?

   a. (Question for clarification; In what way does the community experience disadvantages of this wind project?)

7. What is your opinion on the distribution of benefits of the wind energy project? (profit, employment, cost of electricity)
8. What is your opinion about the distribution of the disadvantages of the wind energy project? (maintenance, environmental disadvantages)
9. What do you think is the best/most effective way to share the costs and benefits of a wind energy project?
10. Which factors are important to you in the distribution of the benefits/disadvantages of wind energy projects?

    a. (Question for clarification; Which factors are important to have the feeling that the distribution of costs and benefits is sound/fair?)

11. Who are involved/have been able to participate in the development and decision-making of this wind energy project?
12. Can you tell how you are involved/have been able to participate in the plans for the wind project?

    a.   (Question for clarification; Or in the decision-making process?)

13. What did you think of the way in which you were involved/have been able to participate in the decision-making process?
14. What did you think of the timing of your involvement in the decision-making process of the wind project?
15. What was your influence on the decision-making process?

    a.   (Question for clarification; Can you tell about your influence on the decision-making process?)
    b.   (Question for clarification; How were your interests taken into account?)
    c.   (Question for clarification; Was there room for other views?)

16. What do you think is an effective way to get involved in the decision-making process/to participate in the decision-making process?
17. How do you want to be involved in a decision-making process?

    a.   (Question for clarification; Which way of involvement/participation do you prefer?)

18. Which factors are important to you in the decision-making process to feel that a decision has been made in a sound/fair way?

Appendix B

List of Codes and Categories

Categories	Codes
Mode of participation	- Private approach - Cooperative approach - Phase of involvement - Influence (Opinion asked/Opinion taken into account) - Policy options
Modes of distribution	- Private approach - Cooperative approach - Individual compensation (Ground compensation + Individual resident compensation) - Community benefits - Investment and yield

Categories	Codes
Factors relevant for perception of procedural justice	- Phase of involvement - Collectively involved - Influence (Opinion taken into account) - Transparency - Information disclosure - Access to consultation - Inclusivity
Factors relevant for perception of distributive justice	- Locality - Balance of modes - Involvement in mode - Transparency - Trust in process

References

1. Planbureau Voor de Leefomgeving. Analyse van Het Voorstel Voor Hoofdlijnen van Het Klimaatakkoord-Hoofdstuk 14 Elektriciteit. 2018. Available online: https://www.klimaatakkoord.nl/documenten/rapporten/2018/09/28/pbl-analyse-elektriciteit (accessed on 10 April 2019).
2. Rijksoverheid. Wind op Land; Europees Beleid. Available online: https://www.windenergie.nl/beleid-en-regels/europees-beleid (accessed on 24 May 2019).
3. Provinciale Staten Limburg. Voor de Kwaliteit van Limburg; Provinciaal Omgevingsplan Limburg (POL2014). 2014. Available online: https://www.limburg.nl/onderwerpen/omgeving/omgevingsvisie/provinciaal/pol-2014-inclusief/ (accessed on 7 July 2019).
4. Provincie Limburg (n.d.). Windenergie. Available online: https://www.limburg.nl/onderwerpen/duurzame-energie/windenergie/ (accessed on 5 April 2019).
5. Regionale Energie Strategie. Noord en Midden- Limburg. 2019. Available online: https://www.regionale-energiestrategie.nl/kaart+doorklik/noord-+en+midden+limburg/default.aspx (accessed on 13 April 2019).
6. Regionale Energie Strategie. Zuid- Limburg. 2019. Available online: https://www.regionale-energiestrategie.nl/kaart+doorklik/zuid+limburg/default.aspx (accessed on 13 April 2019).
7. Zuidelijke Rekenkamer. Energie in Transitie: Een Vergelijkend Onderzoek Naar de Inzet van de Provincies in de Energietransitie. 2018. Available online: https://www.zuidelijkerekenkamer.nl/wp-content/uploads/2018/03/Bijlage-1-Eindrapport-Energie-in-Transitie.pdf (accessed on 5 April 2019).
8. Gemeentegrensoverschrijdende Ontwikkeling Windpark Greenport Venlo in Venlo en Horst aan de Maas. 2018. Available online: https://www.venlo.nl/venlo-windpark-greenport (accessed on 13 June 2019).
9. Devine-Wright, P. Beyond NIMBY-ism: Towards an integrated framework for understanding public perceptions of wind energy. *Wind Energy* **2005**, *8*, 125–139. [CrossRef]
10. Lienhoop, N. Acceptance of wind energy and the role of financial and procedural participation: An investigation with focus groups and choice experiments. *Energy Policy* **2018**, *118*, 97–105. [CrossRef]
11. Olsen, B.E. Renewable energy: Public acceptance and citizens' financial participation. In *Elgar Encyclopedia of Environmental Law*; Edward Elgar Publishing Limited: Dortmund, UK, 2016; pp. 476–486.
12. Mundaca, L.; Busch, H.; Schwer, S. 'Successful' low-carbon energy transitions at the community level? An energy justice perspective. *Appl. Energy* **2018**, *218*, 292–303. [CrossRef]
13. Langer, K.; Decker, T.; Menrad, K. Public participation in wind energy projects located in Germany: Which form of participation is the key to acceptance? *Renew. Energy* **2017**, *112*, 63–73. [CrossRef]
14. Pauw, R. Enthousiasme en Ongenoegen Over Limburgse Windmolens. NOS, 13 March 2019. Available online: https://nos.nl/artikel/2275750-enthousiasme-en-ongenoegen-over-limburgse-windmolens.html (accessed on 21 July 2019).

15. Reijn, G. 'Niemand wil zo'n Molen in Zijn Achtertuin. Maar als er Iets Tegenover Staat, is Het een Ander Verhaal'. Volkskrant, 13 August 2018. Available online: https://www.volkskrant.nl/economie/niemand-wil-zo-n-molen-in-zijn-achtertuin-maar-als-er-iets-tegenover-staat-is-het-een-ander-verhaal~{}bc2fd7ad/ (accessed on 23 June 2019).
16. Boon, F.P.; Dieperink, C. Local civil society based renewable energy organisations in the Netherlands: Exploring the factors that stimulate their emergence and development. *Energy Policy* **2014**, *69*, 297–307. [CrossRef]
17. Etriplus. Windpark Greenport Venlo. Available online: http://www.etriplus.nl/wind/ (accessed on 6 April 2019).
18. Gemeente Leudal. Windenergie. Available online: https://www.leudal.nl/windenergie (accessed on 12 April 2019).
19. Holsgens, J. Werken Aan Het Draagvlak Voor een Windpark in Heibloem. *Leudal Nieuws*. 5 March 2018. Available online: https://leudal.nieuws.nl/2018/03/05/werken-aan-draagvlak-windpark-heibloem/ (accessed on 26 July 2019).
20. Arnstein, S.R. A ladder of citizen participation. *J. Am. Inst. Plan.* **1969**, *35*, 216–224. [CrossRef]
21. Musall, F.D.; Kuik, O. Local acceptance of renewable energy—A case study from southeast Germany. *Energy Policy* **2011**, *39*, 3252–3260. [CrossRef]
22. Williams, S.; Doyon, A. Justice in energy transitions. *Environ. Innov. Soc. Trans.* **2019**, *31*, 144–153. [CrossRef]
23. McCauley, D.; Ramasar, V.; Heffron, R.J.; Sovacool, B.K.; Mebratu, D.; Mundaca, L. Energy justice in the transition to low carbon energy systems: Exploring key themes in interdisciplinary research. *Appl. Energy* **2019**, *233*, 916–921. [CrossRef]
24. Murphy, K. The social pillar of sustainable development: A literature review and framework for policy analysis. *Sustain. Sci. Pract. Policy* **2012**, *8*, 15–29. [CrossRef]
25. García, J.H.; Cherry, T.L.; Kallbekken, S.; Torvanger, A. Willingness to accept local wind energy development: Does the compensation mechanism matter? *Energy Policy* **2016**, *99*, 165–173. [CrossRef]
26. Jenkins, K.; McCauley, D.; Heffron, R.; Stephan, H.; Rehner, R. Energy justice: A conceptual review. *Energy Res. Soc. Sci.* **2016**, *11*, 174–182. [CrossRef]
27. Frankena, W.K. The concept of social justice. In *Justice in General*; Vallentyne, P., Ed.; Routledge: New York, NY, USA, 2003; pp. 63–83.
28. Rawls, J. *A Theory of Justice*, Rev. ed.; Oxford University Press: Oxford, UK, 1999.
29. Sovacool, B.K.; Dworkin, M.H. Energy justice: Conceptual insights and practical applications. *Appl. Energy* **2015**, *142*, 435–444. [CrossRef]
30. Langer, K.; Decker, T.; Roosen, J.; Menrad, K. Factors influencing citizens' acceptance and non-acceptance of wind energy in Germany. *J. Clean. Prod.* **2018**, *175*, 133–144. [CrossRef]
31. Rosenthal, M. Qualitative research methods: Why, when, and how to conduct interviews and focus groups in pharmacy research. *Curr. Pharm. Teach. Learn.* **2016**, *8*, 509–516. [CrossRef]
32. Reed, M.S.; Graves, A.; Dandy, N.; Posthumus, H.; Hubacek, K.; Morris, J.; Stringer, L.C. Who's in and why? A typology of stakeholder analysis methods for natural resource management. *J. Environ. Manag.* **2009**, *90*, 1933–1949. [CrossRef] [PubMed]

 *energies*

Article

Facilitating the Energy Transition—The Governance Role of Local Renewable Energy Cooperatives

Donné Wagemans *, Christian Scholl *  and Véronique Vasseur *

International Centre for Integrated assessment and Sustainable development (ICIS), P.O. Box 616,
6200 MD Maastricht, The Netherlands
* Correspondence: d.wagemans@alumni.maastrichtuniversity.nl (D.W.);
christian.scholl@maastrichtuniversity.nl (C.S.); veronique.vasseur@maastrichtuniversity.nl (V.V.);
Tel.: +31-(0)-43-3882659 (C.S.)

Received: 30 September 2019; Accepted: 31 October 2019; Published: 1 November 2019

Abstract: The governance role of local renewable energy cooperatives (LRECs) in facilitating the energy transition remains under-scrutinized in the scholarly literature. Such a gap is puzzling, since LRECs are a manifestation of the current decentralization movement and yield a promising governance contribution to a 'just energy transition.' This paper presents a study of the governance roles of LRECs in the province of Limburg, the Netherlands. Building on existing work on the cooperative movement and energy governance, we, first, develop a conceptual framework for our analysis. The framework is built around three key interactions shaping these governance roles, between (1) LRECs and their (potential) members, (2) LRECs and the government and (3) LRECs with other LRECs. The results of an online survey and qualitative interviews with selected cooperatives led to the identification of five key governance roles that these cooperatives take up in the facilitation of the energy transition: (1) mobilizing the public, (2) brokering between government and citizens, (3) providing context specific knowledge and expertise, (4) initiating accepted change and (5) proffering the integration of sustainability. The paper concludes by reflecting on the relevance of our findings in this Dutch case for the broader 'just transition' movement.

Keywords: energy transition; local renewable energy cooperatives; governance roles; citizen participation; mixed methods

---

1. Introduction

Community action and involvement in the transition towards a society based on sustainable renewable energy has increased significantly during the last decade, leading to changes in how energy systems are integrated into societies around the world [1]. Spurred not least by concerns about the negative effects of fossil fuels which pollute the biosphere, reinforce the greenhouse gas effect in the atmosphere and upset the balance of the hydrosphere. As opposed to traditional fossil fuel based energy, renewable energy (RE) originates from naturally replenished resources such as sunlight, wind, rain, tidal movements and geothermal heat [2]. Similar to many other European countries, the Netherlands has recently experienced the emergence of local renewable energy initiatives. These initiatives are community efforts that mean to transform the energy sector to make it more decentralized, democratic and sustainable [3–5]. A distinct type of a community effort is the Local Renewable Energy Cooperative (LREC). In recent years, cooperatives have been created to promote the use of renewable energies, most notably in Canada [6], the United Kingdom [7], Denmark [8], Belgium [9] and Germany [10]. Cooperatives are autonomous associations of citizens who collaborate voluntarily to meet their common economic, social and cultural needs and aspirations through a system of businesses that are jointly owned and democratically controlled [2].

As the energy industry is getting more diverse and decentralized these LRECs are one of the visible ongoing developments. By integrating equity and other concerns, they carry the additional potential of contributing to a 'just (energy) transition,' which according to recent scholarship transcend established concerns of a 'energy transition' [11]. Following the well-known concept of cooperatives which emerged in the United Kingdom in the 19th century [12], LRECs empower citizens outside the energy industry with the opportunity to bundle resources to implement renewable energies while also participating in cooperative energy consumption. As renewable energy is becoming increasingly relevant in many countries, LRECs are gaining ground. For instance, the European Federation of Renewable Energy Cooperatives (*REScoop*) represented 1240 LRECs in 2018 with a total of 650,000 European citizens as members. Given this remarkable success of LRECs, it is expedient to find out how the governance of and by these cooperatives plays a role in the facilitation of the energy transition [13].

Bauwens [9] researched the cooperative energy movement in Flanders, specifically investigating the role of cooperative members and the heterogeneity of their motivations and the implications this has on their level of engagement. He concluded that while cooperative members are often considered as one homogenous group, several categories of members with differing motivations and levels of participation can be distinguished. Our research did not include the perspective of the cooperative members as Bauwens did. Therefore, the identified governance roles only reflect the perspective of the cooperative board. It is suggested that future research will also include the perspective of the cooperative members to verify the identified governance roles.

Research conducted by Hoicka and MacArthur [6] investigated community energy projects within Canada and New Zealand. More specifically focusing on the participation of indigenous people. Their research investigated the role of incumbent resources, actors and the political environment to investigate the differing functions of community energy initiatives. Their research concluded that community energy initiatives play an important role in overcoming challenges of uneven economic development, inequality and fuel poverty similar to the results of our research in Limburg. These issues are especially prevalent in countries with a colonial history which differs it from the Netherlands. There are however similarities between the role of LRECs in peripheral areas such as Limburg and the uneven economic development and inequalities mentioned by Hoicka and MacArthur in rural communities within Canada and New Zealand.

While our research took a specific focus on the Dutch experience of the governance of energy transition, it can be seen within the wider movement of the democratization of energy and a just energy transition. This movement calls for more participatory forms of energy provisions, including local autonomy over energy in decentralized systems such as seen in LRECs. Energy cooperatives are expected to play a strong part in this movement as they are owned and managed by the members of their members and reflect the priorities of their communities as indicated by Stephens [14]. The identified governance roles found within this research align with these expectations for LRECs as they partially reclaim the energy infrastructure shifting toward more direct community-level economic benefits. At the same time, they contribute to the democratization movement by moving away from interests that concentrate wealth and power

In 2018 the Netherlands counted 484 energy cooperatives, an increase of 20% compared to 2017 [15]. Almost 70,000 citizens are currently a member of such a cooperative. The generally defined goal of these local renewable energy cooperatives is to involve citizens to participate in practices concerning energy saving, production and trade, with the proceeds of these activities flowing back to the local community as much as possible [15]. These efforts are vastly different from the status quo practices which often involve large energy providers with little to no connection to the local community. In many cases renewable energy projects will therefore experience resistance by local community members as they experience negative externalities of these projects and are not actively involved in the project themselves nor do they share in the benefits [16,17].

Similar to the rest of the Netherlands, the province of Limburg is in the process of working towards the energy transition in an attempt to contribute to the efforts set forth in the Paris Agreement to limit global temperature rise to well below 2 °C as compared to pre-industrial levels. In order to reach this goal, all the signatory parties including the Netherlands pursue a shift toward low carbon economies with a focus on using renewable energy sources, reducing energy demand and increasing energy efficiency levels. In specific terms the Netherlands as part of the European Union is committed to a 'binding target of an at least 40% domestic reduction in greenhouse gas emissions by 2030 as compared to 1990' [18]. As stated in the Nationally Determined Contribution of the EU and its Member States. This includes the effort to reduce energy emissions which has been described as the low carbon energy transition [19]. Additionally, the European Commission's RED II directive came into force in November 2016. This directive sets an overall target of 32% renewable energy consumption by 2030. RED II mentions activities of individual and collective self-consumption through renewable energy communities such as LRECs to collectively facilitate local participation in the energy system and attain the set targets [20].

Energy Cooperatives in Limburg have been supported by the *Natuur en Milieufederatie Limburg* (NMF), the Nature and Environmental Federation Limburg, since 2012 through the *Servicepunt Energie Lokaal Limburg* (SELL). SELL is a service centre that aims to support energy initiatives in order to accelerate the energy transition [21]. After the first cooperative initiatives were started, local projects soon began to take shape with the first cooperative wind turbine being built in Limburg in 2015 and five more in the works. Additional solar projects have been sprouting up in multiple municipalities as well. According to the national energy monitor '*HIER opgewekt*' (Dutch for "generated here"), the future of local energy cooperatives in Limburg is looking very promising as there are relatively few objections and short lead times. This resulted in an almost 35% growth of energy cooperatives in 2018 where the number of energy cooperatives in Limburg increased from 13 to 20 [22]. Therefore, Limburg provides a promising environment for researching the governance by LRECs.

As of 2017, there are multiple LRECs in Limburg with more than a hundred members generating revenues exceeding € 100,000. The cooperatives have plans to have a combined capacity of more than 70 MW within the near future together, aiming to circulate profits within the local community, support social goals and promote the liveability of small local communities [23]. These are ambitious goals. However, the feasibility of these goals will depend on whether the cooperatives can successfully facilitate the governance of the energy as well, as bad governance practices which could potentially smother the potential of LRECs [24].

In this paper, LRECs are analysed from a governance perspective. Thereby, we build on recent attempts by scholars who have started to scrutinize governance processes in community energy projects [7] and mapped legal governance issues of energy sector innovations by community energy services [5]. By studying LRECs in Limburg, we want to get to know in which ways local renewable energy cooperatives contribute to the renewable energy transition from a governance perspective. We conceive of governance as the process of steering society and the economy through collective action and in accordance with common goals [25]. More specifically certain dimensions of the 'governance paradigm' include: inclusion of institutions and actors from and beyond government; blurring roles and responsibilities; power dependence in relationships between institutions, autonomous self-organising networks of actors; and, governing with new techniques to steer and guide, rather than utilising command or authority [26]. By studying the role LRECs play in governance, we want to find out what their actual contribution is to the governance of the energy transition, how this contribution is hampered and how it could be amplified. Despite deeper insights in the governance roles taken up by LRECs, we also expect the results to contribute to give LRECs a better and more reflected place in the overall governance of the energy transition. While our research placed a specific focus on the Dutch experience of LREC governance the results will also be discussed in the wider context of the (energy) democratisation and 'just transition' movement.

For this research, governance is studied by analysing it at a more concrete level of three interactions between an LREC and other parties reflecting the polycentric environment of power in which they operate, as indicated by Meadowcroft [27]. This environment is characterized by power being decentralised and distributed to different groups who collectively determine the direction of developments. The first relationship is that of the LREC and its members (a) for example, opportunities for member input, the second of the LRECs amongst each other (b) for example, utilization of knowledge sharing opportunities, the third is the relationship between the LREC and the government (c) for example, regularity of meetings with city councils. These relationships are displayed in Figure 1.

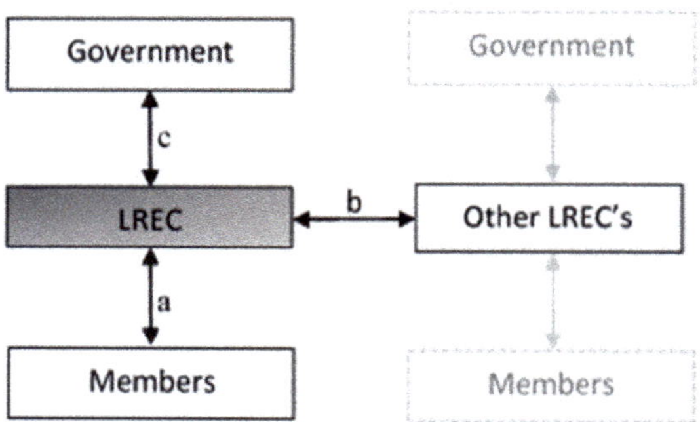

Figure 1. Key governance relations (a, b, c) that are the focus of the proposed research (authors' own).

2. Research Design and Methods

The first phase of the research consisted of a literature review. First, a search for several key words regarding energy cooperatives and local energy initiatives were submitted to online academic databases such as: EBSCO, Web of Science, JSTOR, Scholar, Sage and Springer in search of relevant, peer-reviewed and timely articles with a focus on the European context of energy cooperatives. Based on the popularity of these articles as determined by the number of references, several articles where selected for the initial preparatory study. Second, by following up on key concepts, frequent authors and referred grey literature found in the initial study, a more extensive body of literature was created. The cumulative result of this study was the identification of several key governance dimensions that are relevant for local renewable energy cooperatives.

In the second phase of the research, the key dimensions of the LRECs within Limburg were investigated through an online survey created using the Qualtrics software for collecting and analysing data online. The online survey started with several short questions inquiring about general data on the cooperative, including amongst others the focus, the years of operation and the number of members. The data collected was helpful for having a general overview of the energy cooperatives in Limburg. In addition, the survey investigated the different governance dimensions and their importance for the cooperative. Respondents were able to provide answers on a 5-point Likert scale ranging from 'not at all important' to 'very important' [28]. The option 'not applicable' was included to prevent respondents having to choose a level of agreement for statements that do not apply to their case. The score was then used for the typology development.

There are several advantages of utilizing the Likert scale. First, the responses are easily quantifiable. Second, respondents are not forced to take a definite stand on a particular topic but to respond in a degree of importance. This is supposed to make answering easier for the respondent. Third, the responses accommodate neutral or undecided feelings on importance. Because of these reasons, the Likert scale offers a quick, efficient and suitable method for data collection [29]. The survey was distributed to all 23 known cooperatives in the province of Limburg. This led to 11 distinct responses from 11 different cooperatives. The response rate was 43%. We wanted as many cooperatives as possible to answer the survey to have an extensive amount of data available for the typology development. Preferably, the cooperatives should have different focusses, sizes, locations and years of operation.

In the third phase and based on the survey results, a typology has been developed in two steps. First, the survey data was analysed to identify the greatest polarity of responses. If cooperatives provide vastly different answers regarding the importance of a particular governance dimension, this could indicate that this dimension is suitable for typology development. If respondents indicate very similar preferences for one of the governance dimensions, this could indeed be a very important dimension but was not considered suitable for differentiating amongst the cooperatives. Out of the most polarizing governance dimensions indicated by the widest response ranges, two have be selected as the main variables for the typology development.

As the answers the respondents corresponded to an interval scale, an average score could be calculated for the two determining governance dimensions. These scores will correspond to a coordinate on an x-and y-axis. Resulting in a scatterplot indicating the positions of each individual LREC and the degree to which they fit into a certain type. This allowed for a clear and simple overview of the distribution of the research population as well as a straightforward indication of the most deviating cases, which qualified for further inquiry through interviews.

The fourth phase of the research consisted of two separate (group) interviews with five senior controlling cooperative members of two distinct cooperatives. The interviews were performed in a semi-structured format and inquired about the cooperatives and their different governance interactions at an in-depth level. The semi-structured format allowed the interviews to unfold in a conversational manner, offering the interviewees the chance to pursue issues they feel are important [30]. Interviews lasted ca. 90 min and, were audio-recorded and summarized afterwards. Four distinct LRECs were selected for follow up interviews based on the results of the survey and additional information they provided. As selection criteria, the cooperatives either had the most extreme scores in their respective category, were surprising outliers or they provided information indicating interesting governance perspectives which required further investigation. Out of the four selected cooperatives two agreed to an interview. These were *Duurzaam Roerdalen* and *Duurzaam Maasgouw*, established in 2017 and 2018, respectively. Both cooperatives are relatively young each having less than 100 members.

3. Governance by Local Energy Collectives

In order to identify the key dimensions that are relevant for analysing governance by local renewable energy cooperatives, we reviewed the literature of the two fields of energy governance and cooperative movements. During this search, we identified key dimensions for the analysis of the role of cooperatives in energy governance. These are: participatory practices, democratic decision making, mobilizing capacity, professionalization, legitimacy, collaboration with governmental institutions, support networks and the policy context. Each of them is briefly outlined below. For our analytical framework presented at the end of this section, these dimensions are then grouped together under the three key interactions of cooperative governance.

3.1. Participatory Practices

Participation is one of the key elements of environmental governance that contributes to better decision making [27,31]. It is recognized that issues regarding sustainability require the involvement of the public. Participation provides this link between the public and the governance of the energy transition in this instance. A strong public participation in environmental governance increases the commitment among stakeholders by providing a stronger sense of ownership. When stakeholders are allowed to voice their opinions and insert them into a project, this strengthens their belief in the cooperative project as well as fostering increased acceptance of any measures taken. An example is the increased acceptance of wind turbines if they are managed through participatory citizen initiates such as cooperatives [7]. In addition, some argue that the right to participate in matters concerning the protection of the environment such as the phasing out of fossil fuels for renewable energy, is a procedural right that should be considered as incorporated in the fundamental right to environmental protection [31]. From this perspective, governance of the energy transition is expected to operate by a framework of fairness, inclusivity and equality, which calls for the engagement of the public.

Yet, participation can take many forms, as already indicated by Arnstein [32] when she devised the participation ladder for the level of citizen involvement in government decision making. This ladder ranges from full scale citizen power (or member power when translated to cooperatives) trough forms of tokenism to nonparticipation. While the top rungs of the ladder indicating citizen power would be more in line with the cooperative movement philosophy, empirical testing might reveal disparities amongst the different energy cooperatives in Limburg.

3.2. Democratic Decision-Making

Participatory practices are closely linked to the way in which energy cooperatives organize internal democracy. Democratic control is one of the seven principles of the cooperative movement [33]. It is defined as the governance of an organization by its members through majority decision-making. The cooperative movement as a general rule employs the rule of '1 member = 1 vote' this eliminates the possibility that members with higher investments trump the decisions, leaving members with a smaller investment without decision making power [33].

In cooperatives, internal democracy includes consideration of rights and corresponding responsibilities. It also encourages the fostering of a "spirit of democracy" [33] within the cooperative. This spirit of democracy has proven to be a challenging task but it is considered to be socially valuable and essential. The major benefit is that it contributes to deepening democratic roots within civil society.

It is important that it is recognized that a democratic process, in itself, is no guarantee for competence. A fundamental characteristic of sustainable democratic systems is that democracy requires the protection of sound governance codes, democratic laws procedures and processes, similar to formalized models of organization management. Organizing internal democracy can be considered a key dimension of cooperatives. Cooperatives tend towards deliberative and participatory forms of democracy with constant engagement of members in day-to-day decision making according to the cooperative principles. Members are involved in proposing and approving fundamental strategic policy decisions and able to hold elected representatives on boards or committees and senior executives to account. One of the biggest challenges facing cooperatives is creating a culture that accepts and encourages debate rather than stifling it. Debate should be seen as a sign of a healthy democracy that encourages members to become an active part of the cooperative [33,34].

Cooperatives can take advantage of technological developments. Especially advances in modern mobile as well as internet communication make it easier to actively engage members in the democratic process of the cooperative [33].

3.3. Mobilizing Capacity

According to the European Commission [35] citizens are at the core of the energy transition. Citizens should take ownership of the transition, benefit from new technologies to reduce their bills, participate actively in the market and be protected when vulnerable. Since the European market is transforming from a centralized market dominated by large utilities to a decentralized market with millions of citizens that are active or prosumers, citizen involvement in the energy transition becomes more likely [35].

Cooperatives allow in different ways for the mobilization of citizens for investment in sustainable energy and for projects where energy is provided by citizens. Renewable energy cooperatives have transformed the energy landscape in many European countries while also consequentially contributing to revitalizing local economies and creating local jobs [36]. These mobilized energy communities deliver a significant share of renewable energy investments, promote local development and increase public support of renewable energy. For example Germany, where renewables deliver 40.4% of the country's electricity [37]. Nearly every second kWh of this renewable electricity is generated by a broad range of citizen initiatives. Therefore, it revitalizes the local economy while also generating jobs within the local domain [36].

3.4. Professionalization

The increasing scope and scale of cooperative projects has led to a trend of growth and a desire for professionalization in the Dutch cooperative energy section. Elzenga and Schwencke [38] indicate a clear wish among community initiatives to professionalize. There is a shift from providing energy saving services to more ambitious projects of energy production. According to Hermans and Fens [39], there is an increase in projects that include the actual supply of electricity. Thereby, cooperatives become electricity producers and take a role in service provision.

Almost two thirds of the cooperatives situated in the Netherlands deliver electricity to their members and customers through a resale construction. However, two recently founded cooperatives have taken this supply function to a new level. The cooperatives *NLD* and *DE Unie*, established respectively in 2013 and 2014, did acquire a supplier license. This enables them to act as an electricity utility and purchase electricity on the wholesale market to provide to its customers. The members of these two cooperatives are existing local wind and energy cooperatives who now no longer require mediation of a conventional commercial energy company to supply electricity to their members [40].

In order to obtain a supplier license, the cooperative has to comply with a list of stringent rules and regulations. This includes conditions set by the Consumer & Market Authority [41]. In order to comply with these regulations, a high level of organizational, financial and technical expertise are required for the cooperatives to meet their legal obligations for supplying electricity. Therefore, the acquirement of a supplier license by these two cooperatives forms an adequate illustration of the ongoing professionalization across energy cooperatives [40].

However, this trend towards professionalization does not apply to every cooperative as not all cooperatives have the ambition to increase the scope or scale of their projects. Seyfang, Park and Smith [42] state that 'although some groups do have ambitions to expand and grow, others are simply providing local solutions to local needs as an end in itself and have no desire to expand' (p. 988). This appears to be the case for the Netherlands, where scholars identified a tension between the small-scale idealists who prefer local small-scale solutions and the more commercially oriented cooperatives who would like to scale up local renewable energy projects [43].

The degree of professionalization is a relevant governance dimension as it is an indicator as to how far citizen initiatives take over utility services formally managed by either public governments or private businesses, indicating a blurring roles and responsibilities which are a key aspect of governance.

3.5. Legitimacy

The International Co-operative Alliance's (ICA) Guidance Notes on the Cooperative Principles state that openness, transparency and accountability are important for good democratic governance [33]. These three concepts are grouped together according to the ICA's approach, as they are firmly related to each other with effects on one having immediate impacts on the others. Together these three concepts reflect a sense of trust and legitimacy of the LREC. According to the ICA these three concepts are essential for any cooperative to be legitimate and thus effective [33]. Cooperatives should make agendas and write down minutes of meetings of their elected committees and boards. These should become available to members. However, there are types of information that cannot easily be shared openly. This could be because of commercial sensitivity, regulatory requirements or respect for employee privacy. However, within these limitations cooperatives should ensure that members have the opportunity to debate and hold the board accountable for decisions. Elected representatives should present regular statements of account, financial reports and performance reports to their members. These should be presented in such a way that it is understandable for laypeople [33].

Since democratic member control is a key differentiating characteristic of cooperatives in comparison to conventional investor or shareholder-owned businesses, cooperatives should aspire to be open, transparent and accountable. This increases trust and legitimacy which is key for the success of the cooperative [44]. The democratic practices of an energy cooperative should be subject to critical assessments which can be achieved through cooperative-specific audits [33].

3.6. Collaboration with Governmental Institutions

Local renewable energy cooperatives collaborate with multiple governance levels and in multiple ways. Common amongst these is a collaboration through knowledge sharing at the local to regional level [45]. Cooperatives could be key in assisting municipalities to switch towards renewable energies. Yet, municipalities are often reluctant to work together with cooperative solutions initiated by citizens. In turn, many cooperatives are unsure about what to expect from municipalities. Even so, there seems to be a strong desire amongst both parties to find a way to work together while as of now they are still unsure about how this should happen [45].

Collaboration with cooperatives yields advantages for regional and local governments. Energy cooperatives have proven to be effective at mobilizing citizens in energy production and saving solutions. This indicates an opportunity for taking large steps towards the energy transition. One example of such a cooperation is the municipality of Haarlem. In this city there are five jointly managed roofs with solar PV installations, resulting in a combined amount of ca. 2000 solar PV panels [45]. According to the municipal policy makers in Haarlem, people who have been actively involved with citizen led energy initiatives will experience a lasting sustainability effect. This mindset could be key for meeting the national climate targets [45]. Haarlem is therefore a good example for collaboration between local governments and energy cooperatives.

Cooperatives, on the other hand, can also benefit from collaboration with local and regional governments. An example is the *Leudal Energie* cooperative, which initiated a project to change thousands of traditional lightbulbs within their community to more energy efficient LED-lights. In addition, they operate two solar PV stations on the roofs of local schools and are working on a local wind turbine. The *Leudal Energie* board indicated that the municipality is of great help when it comes to realizing these projects [45].

Collaboration with governmental institutions is not only about securing additional funding. The network that becomes available through collaboration can be as valuable as financial assistance. The *Leudal Energie* cooperative, for example, started working together with local housing cooperatives through mediation by the municipality. The municipality also assisted in the search of suitable fields for solar installations and offered the roof of the city hall for solar developments. Finally, the municipality contributes to the outreach of the cooperative by communicating successes on their website and to local newspapers [45].

The support from the community and other actors is an important dimension for cooperatives. On the local level, support from local residents and other local organizations such as schools, sports associations and community centres, create vital opportunities for cooperatives to find members as well as a strong basis for developing projects. Local businesses also form important partners. Shops, local installation businesses, restaurants or farms all provide valuable additions to a cooperative network [46]. Additionally, cooperatives also work together with larger commercial parties such as energy providers, for example, by reselling the electricity from an energy company to the cooperative members through resale construction [40].

On the other hand, a lack of community support or a limited network of other actors can cause great challenges for a cooperative. The absence of local resident support could result in public apathy, the NIMBY-effect ("Not in My Back Yard") and other forms of community resistance [47]. Even if cooperatives have the intention to generate benefits for the local community, this does not convince all residents and could even be regarded by some as bribery [17,38]. This can be illustrated by the case of the *Energie-U* cooperative in Utrecht. The cooperative worked in commission of the municipality on the development of a wind farm near the city for almost two years before the project was cancelled. The city council decided against the wind farm due to strong local resistance, which indicates the impact of the absence of local support for the success of cooperatives [38,48].

Strong support networks are thus of key importance for energy cooperatives. Not only on the local but also on the regional, national and international level, cooperatives provide and receive help from a range of organizations. Many energy cooperatives work together with other cooperatives to form a supportive network. An example of this is the *REScoopNL* network, which aims to support renewable energy cooperatives to make them successful [49]. These networks provide a knowledge sharing and mutual learning environment by providing 'distinctive expertise that is not readily available elsewhere' [50] (p. 4403). Networks also have the added benefit of allowing for a joint lobby force in cooperation with other initiatives [43].

The *REScoop* organization is also active at the European level. With a network of 1500 European renewable energy cooperatives, representing a combined 1,000,000 citizens *REScoop* wishes to empower citizens to achieve energy democracy by representing their voice, supporting start-up cooperatives, providing services and promoting the LREC business model. *REScoop* promotes collaboration amongst European cooperatives [51]. Cooperative networks are thus active from the regional to the international level.

3.7. Policy Context

The context in which energy cooperatives operate is shaped by government regulations and therefore forms an important dimension for their analysis. There is a wide body of laws, policies and regulations that together form the regulatory and policy context or as titled by Bakker [40] the 'rules of the game.' These are the conditions under which the interaction between LRECs, their members, society and the wider governance system takes place.

Navigating along these rules can be tough for cooperatives as the current electricity law in the Netherlands dates back to 1998 and therefore is often unsuited for these changes in the energy system. The main structure of the law has remained unchanged and although several amendments were made throughout the years current rules and regulations do not always seem to be equipped to handle the rapidly changing role of civil actors in the energy market. This is especially true for community energy actors such as cooperatives who produce their own electricity, as this means that the consumer will operate within the regulated domain [40,52]. Dutch law dictates that each consumer needs to have an energy company that is, a party with a supplier license, to cover their electricity demand. However, as the term "prosumers" [53] suggests, a growing number of previous consumers in the electricity supply chain are now also producers, by taking part in a small energy company next to buying energy. In the Netherlands, it is still challenging for consumers to acquire a license needed to legally fulfil the supplier role. As indicated before, some energy cooperatives manage to obtain such a license. This requires a highly professionalized organization of the cooperative, which is often challenging for citizen's initiatives and may come with adverse effects concerning internal democracy.

The policy context also is critical where it comes to the amount of taxes consumers pay for electricity. In the Netherlands, individual producers generating their own electricity are exempted from this tax through feed-in tariffs. Where producers of electricity receive around six to seven eurocents for each kWh produced, the consumers pay roughly 20 eurocents for each kWh [40]. The difference is caused by distribution costs and taxes. When a consumer has a solar panel that produces electricity, they can deduct the produced energy from their total energy bill [54]. This saves roughly 20 cents per kWh of electricity generated by the solar panel, as this electricity is for direct use and not distributed through grid. On the other hand, if solar panels produce more than the users demand, the excess electricity is compensated with six or seven cents per kWh [54]. Up until 2016, it was legally impossible to generate energy anywhere outside of your personal, privately owned property. This was challenging for energy cooperative as due to these regulations they had to base a profitable business plan on the low six to seven cent rate. This has changed since 2016 with the introduction of the 'postcoderoos' (Dutch for 'zip code area arrangement') postal code regulation, providing more room for LRECs [40]. The 'postcoderoos' regulation provides a tax rebate which gives members of an energy cooperative a discount on their energy bills. If consumers invest in the generation of renewable energy within their area, they have a right to this rebate. The condition is that participants have to live within certain nearby postal zones of the project. Participants may only use a maximum of 10,000 kWh/year [55].

Multiple cooperatives in the Netherlands have also been pushing for self-delivery (e.g., *NLD* and *DE Unie*). This is the direct delivery of power to their members without having an energy company working as an intermediary. Thereby, cooperative members can avoid VAT and energy taxes. The institutional framework forms a barrier in this case as self-delivery is against the rules of the game and is not allowed under the current legal framework [54].

These rules of the game are of vital importance to the local renewable energy cooperatives as decisions made at higher government levels-over which they have little to no say- could create opportunities for energy cooperatives or, conversely, limit their ability to operate. Therefore, we consider the "rules of the game" a critical dimension for the analysis of the governance by energy cooperatives. In recent developments, however, cooperatives have been increasingly involved in the development of policy, indicating a blurring line between the roles of citizens and governments in shaping the energy transition [56].

4. Analytical Framework for Studying Governance by LRECs

In order to analyse the governance by LRECs and their influence of the facilitation of the energy transition key interactions between LRECs and respectively, their members, government and other LRECs were identified. Figure 2 works out the three key interactions of LRECs—with their members, other LRECs and governments—in a more detailed way integrating the dimensions identified in the literature review above. This figure served as analytical framework for this article and guided the analysis of the research data. Whereas we do not want to claim that this conceptual framework is an exhaustive and perfect representation of all the complex governance interactions that LRECs engage in, we believe that it sufficiently specifies the dimensions that are relevant for an analysis of their governance roles. The analytical framework also guided our development of a typology of LRECs which can serve to structure future comparative research into these governance roles.

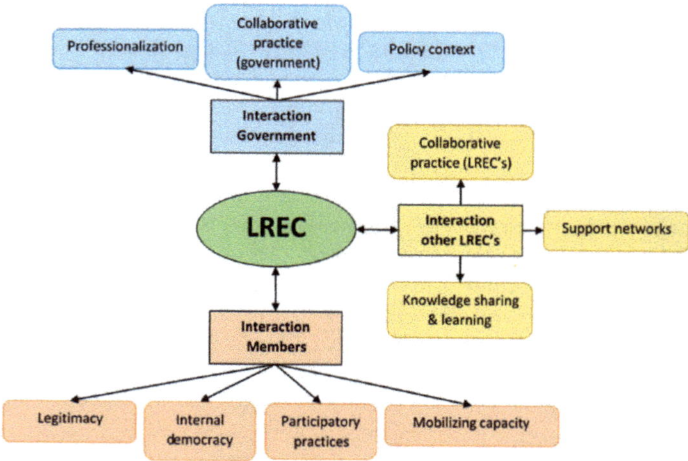

Figure 2. Conceptual framework of local renewable energy cooperative (LREC) interactions and governance dimensions.

The framework can be explained as follows. There are three main interactions between local renewable energy cooperatives and other governance actors, similar to the upwards, downwards and sideways interactions as indicated by Meadowcroft [27]. First, there is the upward interaction between members and the LREC. Here the members of the cooperative steer the cooperative through collective action. The LREC itself is however still in control. Similar to how citizens steer the state through a representative democracy while the state is still the official source of legitimate power, LRECs are controlled by their members who are represented in the board. Thus, this reflects a form of sub-level governance. Here, key dimensions are—democracy, participation, openness, transparency and accountability and the mobilization of communities. Secondly, the LRECs also cooperate with each other through sideways interactions such as collaboration, knowledge sharing practices and networks through a set of different governance dimensions that operate at the meso-level. These reflect the interaction amongst LRECs. This interaction is very similar to cooperation between cities in city networks such as found in conventional governance literature [57,58]. Third, there is the downward interaction at the macro-level between LRECs and the state. These interactions are characterized again by collaboration-which now represents collaborative efforts between cooperatives and the state, professionalization and the rules of the game reflecting the entire law and policy environment in which the LRECs operate.

The conceptual framework served as a visual tool initially with the purpose to assist in the design of the interview grid, which covered all these different dimensions of interactions. During the interview phase, these interactions where the main thread that guided the semi-structured interviews, thereby ensuring that the cooperatives could discuss all the three interactions in more depth with a focus on their governance roles.

5. Towards a Typology of LREC Governance

Based on the data collected from the surveys, we attempted to develop a typology of differentiation. In order to determine the two key characteristics to be used for typology development, the variables with the biggest range amongst respondent answers are of importance. If the range is rather large, this means that respondents have given vastly different answers within this group. If the cooperatives provide different answers, this is interesting for typology development as it indicates that the cooperatives differ on this area. Hence, this approach employs diversity as the guiding principle for selecting the two analytical dimensions.

The two groups with the widest difference in responses were the categories "collaboration" and "ambition" and therefore served as the central axes for the typology development. Any other combination of two variables would have resulted in a less clear delineation between the cooperatives of this dataset. "Collaboration" reflects the degree to which cooperatives wish to collaborate with other cooperatives and government institutes. "Ambition" reflects the difference between cooperatives preferring small scale projects with a local impact and those which would like to have a larger scale impact. When these two variables are plotted against each other (see Figure 3), the following observations can be made. Almost all cooperatives are located in the *important/important* quadrant of this figure. Only one cooperative is located in the *not important/important* quadrant. However, this cooperative is still located rather close to the median of the collaboration axis, indicating that cooperation is still somewhat important. Therefore, based on this figure no clear typologies can be developed as that would require the cooperatives being spread out more across the four quadrants and ideally more towards the extremes of these quadrants.

Therefore, based on the available data no sufficient evidence for the creation of different governance typologies can be identified. Fortunately, the survey data revealed several other interesting issues, which helped to guide the further research on LRECs governance roles. It is however recognized that a sample of 11 cooperatives is rather small and therefore there is a large chance of making a Type II Error where the cooperatives were actually different but it is concluded that they are not [59]. Additionally, there could be sub-forms of clustering that could provide relevant outcomes for devising typologies. For example, the ambitions variable could be divided in a material and idealistic aspect and the collaboration in institutionalised and ad hoc aspect. However, based on the data provided by the conducted survey no such conclusions can be drawn and the usefulness of this typology will have to be determined by future studies.

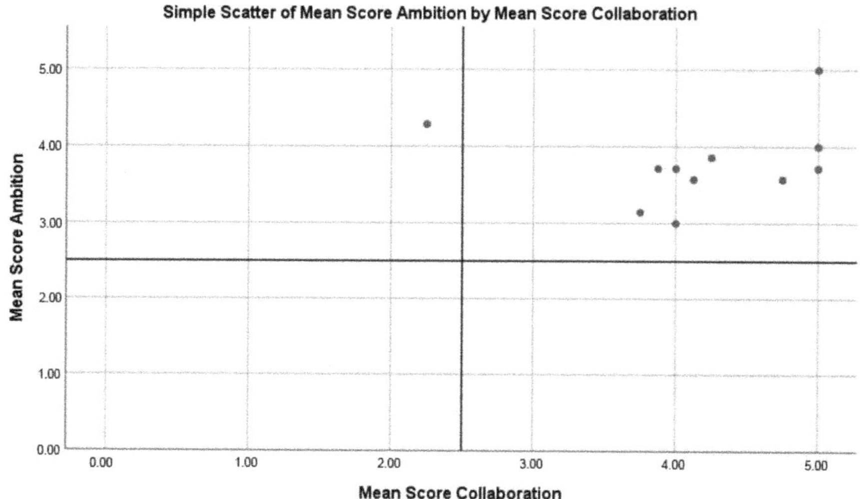

Figure 3. Scatterplot of cooperatives by collaboration and ambition characteristics.

6. The Five Governance Roles of LRECs

In this section the results of the empirical research are synthesized. First, we discuss the five key governance roles of the researched cooperatives: they facilitate the energy transition by (1) mobilizing the public, (2) brokering between government and citizens, (3) providing context-specific knowledge and expertise, (4) initiating change and (5) proffering the integration of sustainability. Second, comparisons will be drawn based on differences in approaches amongst the cooperatives. Third, the results of the study will be used to identify several good practices for the governance of the energy transition through local renewable energy cooperatives.

The role of local renewable energy cooperatives for the facilitation of the energy transition can be distilled to five different key roles that these cooperatives provide based on their interactions with the public and members, the government, other cooperatives and the established energy sector. These activities are as follows: mobilizing the public, brokering between government and citizens, providing context-specific knowledge and expertise, initiating change and proffering the integration of sustainability. These roles reflect several key interactions of the governance dimensions represented in the conceptual framework. In the following section the specific roles will be explained in more detail.

6.1. Mobilizing the Public

The first governance role of cooperatives aligns with the interaction between the cooperative and its members which here reflects a wider scope based on the cooperatives' experiences extending further to the general public and prospective members as well. This role is the mobilizing of the public.

LRECs play a critical role in mobilizing citizens for the energy transition. Through various activities these cooperatives raise awareness and build support for the energy transition. Where a government tries to steer for a more renewable energy system through policies which often do not directly speak to citizens on an individual level, cooperatives attempt to mobilize people directly. Through personal advice, information days and public events at the local level, cooperatives promote the energy transition at a level that speaks directly to citizens.

A major asset for LRECs is that they provide the opportunity to participate on a voluntary basis. Whereas regulations will often require people to adopt sustainable practices because they have to, cooperatives offer this opportunity for people that want to become more sustainable. The key here is that people can choose for themselves if they want to participate and in how far they wish to participate.

Both interviewed cooperatives believe that this voluntary approach is more effective at having people actively engage with sustainability than a commanding approach which often goes paired with a lot of citizen push-back.

Despite the potential for citizen mobilization, many cooperatives admit that it is only a small group of citizens that are actively involved within the cooperative while other members are only interested and do not actively participate. As a result, many of these cooperatives struggle with mobilizing these citizens and having them play an active role. On the other hand, there are plenty of examples where even a small number of active participants in a cooperative managed to mobilize vast amounts of resources and funding through a large group of members, according to the interviewed cooperatives. One such example is the success of *Leudal Energie* who gathered large amounts of funding for a wind turbine initiative [60]. While not all these members might be active participants, the cooperatives report that even association with renewable energy projects positively effects the member's attitude towards sustainability.

Therefore, it can be concluded that LRECs play a role in mobilizing citizens for the facilitation of the energy transition. While this role might seem trivial, it is not to be underestimated. Governments and private sector operators have attempted to mobilize citizens in an effort to promote renewable energy specifically and sustainability more generally. Successful mobilization is key for ensuring the energy transition, conventional efforts so far have only had limited success according to the cooperatives and scholars [61,62].

6.2. Brokering between Government and Citizens

The second governance role for cooperatives reflects both the interactions that the cooperative has with the government as well as with its members. This role is that of broker between the government and citizen.

LRECs build bridges between citizens and the local government. The cooperatives indicated that they have a direct connection to their respective municipalities. There are regularly planned meetings between the members of the cooperative board and dedicated civil servants, the municipal council or the aldermen. Through these meetings the cooperative board can voice their opinion, plans and current activities but most importantly, represent the voice of their members who in turn represent the local community. This is especially effective if the cooperative is situated in small municipalities in rural areas where the cooperative members form a relatively larger portion of the population and the electorate and are therefore considered a serious voice and potential partner by the local government.

The cooperatives also function as broker by providing support for navigating government regulations and bureaucracies. While many citizens might want to conduct their own renewable energy projects (e.g., the instalment of a solar tracker or a small-scale solar farm on their property), they are often daunted by the associated regulatory requirements and the navigation off bureaucracies. This could hinder them in doing the project or even make them lose heart completely. In these specific cases, the cooperative can provide advice and help these initiators to realize their projects through its experiences as well as closer collaboration with municipal or regional government officials.

6.3. Providing Context-Specific Knowledge and Expertise

The third governance role of the cooperative spans all three interactions and constitutes the provision of context specific knowledge and expertise.

The LRECs play a key role in adapting the overall energy transition plan to a tangible and on local level. They do this not by stamping one blueprinted idea on every situation they find but by looking at the specific context of the situation, providing a targeted advice. As an example given by one of the interviewees: 'Some installers of solar panels attempt to convince people to invest in a set solar PV panels for their roof without first analysing the roof itself. This has resulted into new panels being installed on a roof that required extensive maintenance within the next two year period. This maintenance required the solar panels to be removed from the roof again, costing the homeowner a lot of money. This homeowner now has a negative view of solar panels and therefore sceptical of the renewable energy transition.' The cooperatives attempt to prevent these situations by looking at these specific circumstances such as the condition of the roof, before the advice of installing solar panels.

By providing this context specific knowledge and expertise, the cooperatives are able to provide better customized solutions. In general, this results in better attitudes of citizens towards sustainable technologies, a better strategy for achieving the sustainability goals and a more cost-effective method of facilitating the energy transition.

6.4. Initiating Socially Accepted Change

The fourth governance role for LRECs also works across all three interactions. This role is the initiation of accepted changes which cooperatives do through working with governments, other cooperatives as well as members and the public. Initiating socially accepted change differentiate LRECs from conventional command-and-control approaches towards the energy transition as it has a higher degree of social acceptance and perceived legitimacy. These conventional approaches are often only accepted on a limited basis by the public, however, LRECs often initiate projects for the energy transition that are accepted by the local population. As a cooperative member of *Duurzaam Maasgouw* stated 'if you tell people they have to do something they will often resist, however if you involve them and allow them to do things voluntarily, you get much more support and cooperation.' This involvement allows the cooperatives to initiate change that is socially accepted by local communities.

First, LRECs initiate change by kick-starting projects. The interviewed cooperatives noted that they create momentum for the energy transition by starting projects. The cooperatives look for and create opportunities within their local area for renewable energy projects. For example, by securing public rooftops for collective solar installations or by looking for suitable land for other renewable energy projects such as solar or wind farms. Even though some cooperatives indicated that they do not wish to exploit the projects themselves, they do provide concrete plans for local entrepreneurs, governments and project developers. The cooperatives have noted that if they provide plans that have been fully developed and researched in detail there is a big change these plans will become realized, 'if we go to the municipality with a fully developed plan they don't have to do much work themselves, thus, they will often continue with these sort of plans.' (Interviewee cooperative *Duurzaam Roerdalen*).

Second, LRECs work together in larger regional and national networks. Within these networks the cooperatives share knowledge, projects, successes and obstacles that they might be facing. This is not only relevant for the cooperatives themselves but it also allows the larger scale network organizations to represent a collective lobby of these energy cooperatives to push for change. For example, several cooperatives noted that they often run into bureaucratic rules which do not seem to fulfil any sort of function but hinder their ability to operate. Together with other cooperatives this issue was discussed during network meetings which resulted in the network pushing for changes in regulation. If this is successful the process for conducting the energy transition will be streamlined. Therefore, LRECs besides kick-starting energy projects also initiate policy change.

This initiating role where cooperatives provide fully developed plans for renewable energy projects are especially valuable if there is a lack of knowledge or incentive within the local area. In such a case the cooperative could present their plans to for example the municipality which may not have had the resources or dedicated civil servants to go through the development of such a plan. The cooperatives indicate that as long as the plans they provide are sound and developed in detail, municipalities are much more likely to follow-through and realize renewable energy projects that otherwise would have never been developed.

Third, LRECs foster local acceptance. This through offering people a voice in the development and running of renewable energy projects–as well as potential financial opportunities. LRECs are much more likely to generate local support than other parties. For example, large scale energy businesses might face a lot of resistance when attempting to create a new renewable energy project due to a lack of involvement and mutual mistrust [44]. The cooperatives believe that this resistance is mainly occurring because people only experience the negative externalities of these projects and cash flowing away from the region to large corporations. LRECs on the other hand strive to keep cash flows within the local community. When conducted in this way, the local population does not only suffer the negative effects of these projects but also get to share it is benefits. This inclusion of local citizens in energy projects leads to vastly different attitudes towards energy projects within one's vicinity, as noted by the cooperatives. They state that people could protest against the construction of a wind turbine while at the same time being extremely positive of the exact same turbine built in the exact same spot but (partially) funded by the citizens themselves. This suggests that cooperative initiatives are an effective tool against the dreaded 'Not in My Back Yard' NIMBY effects as discussed in Olsen [17].

6.5. Proffering the Integration of Sustainability

The fifth governance role of the energy cooperatives is the integration of broad perspective on sustainability going beyond the installation of more renewable energy capacity. This role reflects the interaction that the cooperatives have with both the government and their members.

Many of the cooperatives prefer to categorize themselves as a sustainability cooperative rather than just an energy cooperative. The idea of a sustainability cooperative is preferred as it represents a much broader view of what the cooperative considers the challenges ahead are and the potential solutions that it can employ to address those challenges. An interviewee stated that 'We prefer to think of ourselves as sustainability cooperatives as we want to do much more than just energy. It is true that energy is currently the most popular topic which attracts people but in the future we want to expand our focus.' (Interviewee cooperative *Duurzaam Maasgouw*). Where the main focus of the energy transition approach currently lies on a shift to renewables and direct energy saving measures, the cooperatives attempt to reflect a more integrated view on sustainability.

The cooperatives state that taking this more integrated view is a much better approach for ensuring sustainability and reaching the Sustainable Development Goals. According to the interviewed cooperatives, the idea that making the energy system sustainable is only about installing more renewable electricity generation capacity is a mistaken one. The cooperatives believe that it is not a realistic goal to build large amounts of renewable energy projects such as wind turbines and solar farms to provide electricity to only one municipality. Therefore, they pursue a more integrated form of sustainability in the hopes off achieving better results than those of the approach focused on by the established energy transition approach. They state that just because you can claim a certain amount of carbon credits for an energy initiative, does not mean that the project was sustainable and an effective contribution towards the 2050 goals.

The energy cooperative therefore certainly play a role in providing a more integrated view on sustainability for the facilitation of the energy transition. They go beyond what is required by regulation and subsidy requirements in the hope of working towards a more effective energy transition approach. In total the five different cooperative roles are: mobilizing the public, brokering between government and citizens, providing context specific knowledge and expertise, initiating accepted change and proffering the integration of sustainability. These roles are displayed in Figure 4.

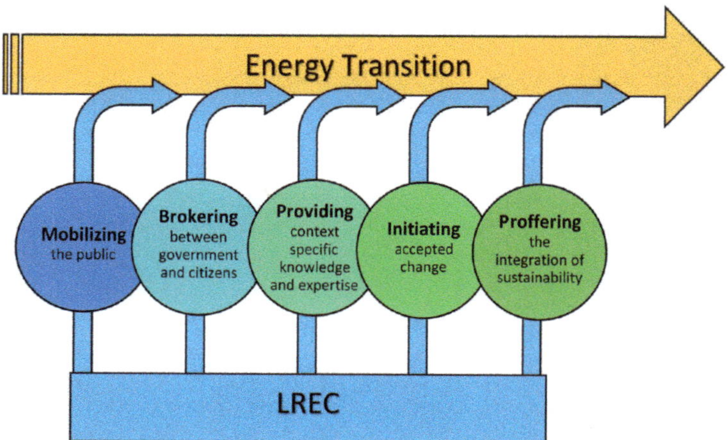

Figure 4. The five governance roles of LRECs for facilitating the energy transition (authors' own).

7. Conclusions and Discussion

This paper set out to inquire in which ways LRECs contribute to the renewable energy transition from a governance perspective. A conceptual framework is developed for analysing the governance roles that cooperatives play in the renewable energy transition. The framework is built around three key interactions shaping these governance roles, between (1) LRECs and their (potential) members, (2) LRECs and the government and (3) LRECs with other LRECs. Based on the survey and in-depth, semi-structured interviews in the province of Limburg, the Netherlands, five different roles for LRECs for facilitating the energy transition could be distilled. These indicated roles are: (1) mobilizing the public; (2) brokering between government and citizens; (3) providing context specific knowledge and expertise; (4) initiating accepted change; and (5) proffering the integration of sustainability. These five roles are a new addition to academic literature as the literature review did not reveal any peer-reviewed articles that attempted to identify the governance roles that LRECs fulfil in the energy transition.

Mobilizing the public is the first role, where cooperatives play a role in actively involving citizens for the energy transition. The second role is brokering between government and citizens, here cooperatives play a role in translating government policy to the citizen level for implementation. Simultaneously the cooperative acts as a representative of its members, voicing the citizens' opinions to the government. Third, providing context specific knowledge and expertise, here local energy cooperatives leverage their local embeddedness and personal approach towards facilitating the energy transition to provide context specific solutions. Fourth, initiating accepted change, here the energy cooperatives fulfil a role initiating projects within their communities. As the projects are initiated from the community and community members can have a say in how the project develops, there is a bigger chance that the project is accepted. This prevents any protest from within the community and builds support for a sustainable energy transition. And the last (fifth) role is proffering the integration of sustainability, here cooperatives take an active role is advocating for a more integrated sustainability approach towards the energy transition by focusing on factors beyond just energy generation.

Based on our empirical analysis, we distilled key factors for success for local energy cooperatives. First, they need to be locally embedded and try to be part of the community they are trying to serve. This provides the cooperative with unique knowledge, connections, as well as the goodwill of local citizens preventing potential NIMBY-effects [47]. A second success factor is that the cooperatives often have regular and direct communications with municipal or regional governments, this allows them to provide their insights and push for change at controlling government levels, this is especially important if the cooperatives manage to collaborate with the government to conduct sustainability projects. While regular collaboration with a municipality enables local small scale energy projects such as improvements of home insulations or small arrays of solar panels, collaboration with regional governments have the potential for larger projects such as wind-turbines or solar farms. The third success factor is honesty and transparency where the openness and honest trustworthy advice that the cooperatives try to provide earns them respect and trust from local citizens. The final success factor is related to this as it is non-commercial interests, whereas many other energy initiatives have a direct incentive to sell a certain product, the cooperatives try to steer away from these biases, leading to more citizen acceptance. While many cooperatives are focusing on small-scale projects, a small number of cooperatives such as *Leudal Energie* are working on major projects such as a wind farm, indicating the potential of cooperatives for contributing significantly to the energy transition. This success will be largely dependent on the cooperative's ability to mobilize citizens, government actors and resources resulting from these success factors and handling the following barriers.

Next to success factors the cooperatives also face certain barriers. The first one is mobilizing people, here the cooperatives indicated that is difficult to interest people in taking an active role in the cooperative and working towards the energy transition. Many people are interested but cannot find the time to actively participate. Resulting in cooperatives which mostly consist off and are run by pensioners. Thereby, excluding large sections of the population. A second barrier specifically limiting cooperative community energy projects is the lack of a dense grid network throughout most of the province of Limburg. This means that if a cooperative for example would like to start a solar farm, they will have to pay for the connection of that farm to the grid. As the network is not very dense in Limburg, such a connection might have to be very long and thus expensive. The final barrier for energy cooperatives is certain inhibiting regulations. The cooperatives indicated that there are certain regulations which do not seem to have a clear purpose but limit their ability to operate. An example of this is that cooperatives have to jump through several regulatory hoops in order to provide volunteers with a small compensation for their time or costs. Fortunately, the cooperatives work together in networks such as *REScoop* in an effort to change these regulations. However, these regulations inhibit their ability to facilitate the energy transition by diverting their attention and resources.

These success factors and barriers are a direct result of the empirical research conducted in this study. Therefore, these success factors and barriers reflect the specific context of the cooperative movement in Limburg. Studies within other regions might find additional success factors or barriers that could either complement in contrast those found in this thesis. A study conducted in the communities of Zschadraß and Nossen, Germany, by Musall & Kuik [63] also concluded that the cooperative model indeed increases acceptance of renewable energy measures. This study however did not identify potential governance related barriers that could mitigate the success of cooperative energy initiatives.

Research conducted by Elzenga and Schwencke [38] did discuss the challenging relation between LRECs and local municipalities as both parties are still looking for their roles. However, their research did not consider this relation as part of the brokering process where the cooperatives represent the citizens in collective steering with municipalities. Our conclusion that this brokering roles takes place to the benefit of both parties aligns with the conclusions of Jonker et al. [45] who concluded that this collaboration affords municipalities and LRECs to take larger steps towards the energy transition.

Research conducted by Olsen [17] investigated a novel community energy typology, through analysis of several technical and social dimensions in Scotland. The results show that whilst the Scottish community energy sector contains a diverse range of motivations, technologies and social practices, the sector is dominated by groups who utilize local energy generation to achieve local socio-economic development, aligning with our conclusions for LRECs. Olsen also attempted to devise a typology of community energy initiatives. Her research had a broader perspective however with a more general focus on laws and regulatory forms. It did not focus specifically on a governance which this study does.

We conclude that based on our conducted research LRECs fulfil the following five governance roles regarding the facilitation of the energy transition: mobilizing the public, brokering between government and citizens, providing context specific knowledge and expertise, initiating accepted change and proffering the integration of sustainability. These roles were distilled from mixed method research containing of literature research, a survey and in-depth interviews in the province of Limburg, the Netherlands. The identified roles are a new addition to the academic literature.

We recommend that future research expands the research scope to include more cooperatives beyond the borders of Limburg. Additional participants will be necessary to develop a robust typology which distinguishes the LRECs based on certain governance criteria. Furthermore, as this study only investigated the perspective of the LRECs themselves, future research should investigate whether the identified governance dimensions in this paper are also recognized by other parties such as government institutions and the cooperative members. Finally, the discovered governance interactions and governance roles should be tested in other studies and fields to investigate their robustness regarding these interactions.

Author Contributions: Conceptualization, D.W. and C.S.; methodology, D.W., C.S. and V.V.; validation, D.W.; formal analysis, D.W.; investigation, D.W.; resources, D.W.; data curation, D.W.; writing—original draft preparation, D.W.; writing—review and editing, C.S. and V.V.; visualization, D.W.; supervision, C.S. and V.V.

Funding: The research presented in this paper received funding from the European Union's H2020 Research and Innovation program under grant agreement number 727642 ("ENERGISE"). The sole responsibility for the content of this paper lies with the authors.

Conflicts of Interest: The authors declare no conflict of interest.

References

1. Mey, F.; Hicks, J. Community Owned Renewable Energy: Enabling the Transition Towards Renewable Energy? In *Decarbonising the Built Environment: Charting the Transition*; Newton, P., Prasad, D., Sproul, A., White, S., Eds.; Springer: Singapore, 2019; pp. 65–82. [CrossRef]
2. Viardot, E. The role of cooperatives in overcoming the barriers to adoption of renewable energy. *Energy Policy* **2013**, *63*, 756–764. [CrossRef]
3. Proka, A.; Hisschemöller, M.; Loorbach, D. Transition without Conflict? Renewable Energy Initiatives in the Dutch Energy Transition. *Sustainability* **2018**, *10*, 1721. [CrossRef]
4. Van Veelen, B. Negotiating energy democracy in practice: Governance processes in community energy projects. *Environ. Politics* **2018**, *27*, 644–665. [CrossRef]
5. Heldeweg, M. Normative Alignment, Institutional Resilience and Shifts in Legal Governance of the Energy Transition. *Sustainability* **2017**, *9*, 1273. [CrossRef]
6. Hoicka, C.E.; MacArthur, J.L. From tip to toes: Mapping community energy models in Canada and New Zealand. *Energy Policy* **2018**, *121*, 162–174. [CrossRef]
7. Van Veelen, B. Making Sense of the Scottish Community Energy Sector—An Organising Typology. *Scott. Geogr. J.* **2017**, *133*, 1–20. [CrossRef]
8. Bauwens, T.; Gotchev, B.; Holstenkamp, L. What drives the development of community energy in Europe? The case of wind power cooperatives. *Energy Res. Soc. Sci.* **2016**, *13*, 136–147. [CrossRef]
9. Bauwens, T. Explaining the diversity of motivations behind community renewable energy. *Energy Policy* **2016**, *93*, 278–290. [CrossRef]
10. Holstenkamp, L.; Kahla, F. What are community energy companies trying to accomplish? An empirical investigation of investment motives in the German case. *Energy Policy* **2016**, *97*, 112–122. [CrossRef]
11. McCauley, D.; Heffron, R. Just transition: Integrating climate, energy and environmental justice. *Energy Policy* **2018**, *119*, 1–7. [CrossRef]
12. Walton, J.K. Revisiting the Rochdale Pioneers. *Labour Hist. Rev.* **2015**, *80*, 215–248. [CrossRef]
13. Hentschel, M.; Ketter, W.; Collins, J. Renewable energy cooperatives: Facilitating the energy transition at the Port of Rotterdam. *Energy Policy* **2018**, *121*, 61–69. [CrossRef]
14. Stephens, J.C. Energy Democracy: Redistributing Power to the People through Renewable Transformation. *Environ. Sci. Policy Sustain. Dev.* **2019**, *61*, 4–13. [CrossRef]
15. Schwencke, A.M. *Lokale Energie Monitor 2018*; Hier Opgewekt: Utrecht, The Netherlands, 2018.

16. NOS. Drentse ondernemer windmolenpark krijgt bedreigingen en stopt. Available online: https://nos.nl/artikel/2277775-drentse-ondernemer-windmolenpark-krijgt-bedreigingen-en-stopt.html#targetText=Drentse%20ondernemer%20windmolenpark%20krijgt%20bedreigingen%20en%20stopt,de%20Drentse%20Monden%20en%20Oostermoer (accessed on 7 March 2019).

17. Olsen, B.E. Renewable energy: Public acceptance and citizens' financial participation. In *Elgar Encyclopedia of Environmental Law*; Edward Elgar Publishing Limited: Cheltenham, UK, 2016; pp. 476–486.

18. European Union. *Intended Nationally Determined Contribution of the EU and its Member States*; NDC Registry: Berlin, Germany, 2015.

19. Hoppe, T.; de Vries, G. *Social Innovation and the Energy Transition*; Multidisciplinary Digital Publishing Institute: Basel, Switzerland, 2019.

20. Frieden, D.; Tuerk, A.; Roberts, J.; D'Hebermont, S.; Gubina, A. Collective Self-Consumption and Energy Communities: Overview of Emerging Regulatory Approaches in Europe. Available online: https://www.compile-project.eu/wp-content/uploads/COMPILE_Collective_self-consumption_EU_review_june_2019_FINAL-1.pdf (accessed on 31 October 2019).

21. Servicepunt Energie Lokaal Limburg (SELL). Servicepunt Energie Lokaal Limburg. Available online: https://www.nmflimburg.nl/sell (accessed on 31 October 2019).

22. HIER Opgewekt. Ondersteuning Energiecoöperaties Per Provincie. Available online: https://www.hieropgewekt.nl/kennisdossiers/ondersteuning-energiecooperaties-per-provincie (accessed on 31 October 2019).

23. Geenen, H. Brief lokale energie coöperaties inzake ondersteuning SELL names lokale duurzame energie coöperaties in Limburg. Available online: https://www.nmflimburg.nl/brief-lokale-energie-coperaties-inzake-ondersteuning-sell (accessed on 31 October 2019).

24. Upham, P.; Shackley, S. The case of a proposed 21.5 MWe biomass gasifier in Winkleigh, Devon: Implications for governance of renewable energy planning. *Energy Policy* **2006**, *34*, 2161–2172. [CrossRef]

25. Torfing, J.; Peters, B.G.; Pierre, J.; Sørensen, E. *Interactive Governance: Advancing the Paradigm*; Oxford University Press on Demand: New York, NY, USA, 2012.

26. Stoker, G. Governance as theory: Five propositions. *Int. Soc. Sci. J.* **2018**, *68*, 15–24. [CrossRef]

27. Meadowcroft, J. Who is in Charge here? Governance for Sustainable Development in a Complex World*. *J. Environ. Policy Plan.* **2007**, *9*, 299–314. [CrossRef]

28. Gray, D.E. *Doing Research in the Real World*; Sage: Thousand Oaks, CA, USA, 2013.

29. LaMarca, N. The Likert Scale: Advantages and Disadvantages. *Field Res. Organ. Psychol.* **2011**, 1–3.

30. Longhurst, R. Interviews: In-Depth, Semi-Structured. In *International Encyclopedia of Human Geography*; Elsevier: Amsterdam, The Netherlands, 2009; pp. 580–584. [CrossRef]

31. Du Plessis, A. Public participation, good environmental governance and fulfilment of environmental rights. *Potchefstroom Electron. Law J. Potchefstroomse Elektron. Regsblad* **2008**, *11*, 1–34. [CrossRef]

32. Arnstein, S.R. A ladder of citizen participation. *J. Am. Inst. Plan.* **1969**, *35*, 216–224. [CrossRef]

33. International Co-operative Alliance [ICA]. *Guidance Notes to the Co-operative Principles*; International Co-operative Alliance: Brussels, Belgium, 2013.

34. Fairchild, D.; Weinrub, A. Energy Democracy. In *The Community Resilience Reader*; Springer: Heidelberg, Germany, 2017; pp. 195–206.

35. European Commission. *A Framework Strategy for a Resilient Energy Union with a Forward-Looking Climate Change Policy*; European Commission: Ispra, Italy, 2015.

36. REScoop MECISE. Mobilising European Citizens to Invest in Sustainable Energy—Clean Energy for All Europeans. Available online: https://www.rescoop.eu/blog/mobilising-european-citizens-to-invest-in-sustainable-energy (accessed on 31 October 2019).

37. Burger, B. *Net Public Electricity Generation in Germany in 2018*; Fraunhofer Institute for Solar Energy Systems ISE: Freiburg im Breisgau, Germany, 2019.

38. Elzenga, H.; Schwencke, A.M. Lokale energiecoöperaties: Nieuwe spelers in de energie. *Bestuurskunde* **2015**, *24*, 17–26. [CrossRef]

39. Hermans, P.; Fens, T. Self-organisation in the residential electricity domain. 2013; 1–28, Unpublished work.

40. Bakker, M.D. New Energy Alliances; Exploring the Partnerships between Local Energy Cooperatives and Energy Companies in the Netherlands. Master' Thesis, Radboud Universiteit, Nijmegen, The Netherlands, 2016.

41. Autoriteit Consument & Markt. Vergunning Aanvragen Voor Het Leveren van Energie aan Kleinverbruikers. Available online: https://www.acm.nl/nl/onderwerpen/energie/energiebedrijven/vergunning-aanvragen-bij-acm/ (accessed on 31 October 2019).

42. Seyfang, G.; Park, J.J.; Smith, A. A thousand flowers blooming? An examination of community energy in the UK. *Energy Policy* **2013**, *61*, 977–989. [CrossRef]

43. Schwencke, A.M. Energieke bottomup in lage landen. Available online: http://asisearch.nl/wp-content/uploads/2016/11/2012-ESSAY-Energieke-BottomUp-in-Lage-Landen-Schwencke-21082012-FINAL.pdf (accessed on 31 October 2019).

44. Walker, G.; Devine-Wright, P.; Hunter, S.; High, H.; Evans, B. Trust and community: Exploring the meanings, contexts and dynamics of community renewable energy. *Energy Policy* **2010**, *38*, 2655–2663. [CrossRef]

45. Jonker, F.; van Beek, J.; de Jong, E.; Brasser, A. De Energiecoöperatie als Samenwerkingspartner in de Gemeentelijke Energietransitie. Available online: https://www.hieropgewekt.nl/kennisdossiers/energiecooperatie-als-samenwerkingspartner-in-gemeentelijke-energietransitie (accessed on 31 October 2019).

46. Van der Schoor, T.; Scholtens, B. Power to the people: Local community initiatives and the transition to sustainable energy. *Renew. Sustain. Energy Rev.* **2015**, *43*, 666–675. [CrossRef]

47. Devine-Wright, P. Beyond NIMBYism: Towards an integrated framework for understanding public perceptions of wind energy. *Wind Energy* **2005**, *8*, 125–139. [CrossRef]

48. Elzenga, H.; Schwencke, A.M. *Energy Cooperatives: Aims, Operational Perspective and Interaction with Municipalities—The Energetic Society in Action*; PBL Netherlands Environmental Assessment Agency: The Hague, The Netherlands, 2014.

49. REScoopNL. Over REScoopNL. Available online: http://www.rescoop.nl/over-rescoopnl/ (accessed on 24 May 2019).

50. Walker, G. What are the barriers and incentives for community-owned means of energy production and use? *Energy Policy* **2008**, *36*, 4401–4405. [CrossRef]

51. REScoop. *The Energy Transition to Energy Democracy: Power to the People—Final Results Oriented Report of the Rescoop 20-20-20 Intelligent Energy Europe project*; REScoop: Antwerp, Belgium, 2015.

52. Tanrisever, F.; Derinkuyu, K.; Jongen, G. Organization and functioning of liberalized electricity markets: An overview of the Dutch market. *Renew. Sustain. Energy Rev.* **2015**, *51*, 1363–1374. [CrossRef]

53. Parag, Y.; Sovacool, B.K. Electricity market design for the prosumer era. *Nat. Energy* **2016**, *1*, 16032. [CrossRef]

54. Boon, F.P. Local is Beautiful: The Emergence and Development of Local Renewable Energy Organisations. Master's Thesis, Utrecht University, Utrecht, The Netherlands, 2012.

55. HIER Opgewekt. Postcoderoosregeling-de Regeling in Het Kort. Available online: https://www.hieropgewekt.nl/kennisdossiers/postcoderoosregeling-regeling-in-het-kort(accessedon24May2019). (accessed on 24 May 2019).

56. Janssen, M. The contribution of Local Energy Cooperatives to the Energy Transition in The Netherlands. Master's Thesis, Supervisor Radboud University, Nijmegen, The Netherlands, 2018.

57. Parnell, S. Defining a global urban development agenda. *World Dev.* **2016**, *78*, 529–540. [CrossRef]

58. Satterthwaite, D. Successful, safe and sustainable cities: Towards a New Urban Agenda. *Commonw. J. Local Gov.* **2016**, 3–18. [CrossRef]

59. Hinton, P.R. *Statistics Explained*; Routledge: London, UK, 2014.

60. Leudal Energie. Windpark De Kookepan. Available online: https://leudalenergie.nl/windpark-de-kookepan (accessed on 31 October 2019).

61. Hoppe, T.; Graf, A.; Warbroek, B.; Lammers, I.; Lepping, I. Local Governments Supporting Local Energy Initiatives: Lessons from the Best Practices of Saerbeck (Germany) and Lochem (The Netherlands). *Sustainability* **2015**, *7*, 1900–1931. [CrossRef]

62. Hendriks, C.M. On Inclusion and Network Governance: The Democratic Disconnect of Dutch Energy Transitions. *Public Adm.* **2008**, *86*, 1009–1031. [CrossRef]

63. Musall, F.D.; Kuik, O. Local acceptance of renewable energy—A case study from southeast Germany. *Energy Policy* **2011**, *39*, 3252–3260. [CrossRef]

Concept Paper

An Exploratory Agent-Based Modeling Analysis Approach to Test Business Models for Electricity Storage

Seyed Ahmad Reza Mir Mohammadi Kooshknow [1,*] **, Rob den Exter** [2] **and Franco Ruzzenenti** [1]

[1] Energy and Sustainability Research Institute Groningen, University of Groningen, 9747 AG Groningen, The Netherlands
[2] Stored Energy, 3125 BN Schiedam, The Netherlands
* Correspondence: s.a.r.mir.mohammadi@rug.nl

Received: 14 February 2020; Accepted: 25 March 2020; Published: 2 April 2020

Abstract: Electricity storage systems (ESSs) are potential solutions to facilitate renewable energy transition. Lack of viable business models, as well as high levels of uncertainty in technology, economic, and institutional factors, form main barriers for wide implementation of ESSs worldwide and in the Netherlands. Therefore, the design of business models for an ESS is necessary for the development of ESSs. We elaborated on this problem before, and developed a design space for business models of ESSs in the context of the Netherlands. This conceptual paper provides a further view on barriers and uncertainties of ESS development in the Netherlands through the involvement of a business practitioner, elaboration of goals, objectives, and testing of ESS business model designs, suggests and provides a theoretical foundation for combining agent-based modeling and exploratory modeling analysis as a method to test and explore ESS business models, and provides an abstract conceptual agent-based model design thereof. This work can be used as a foundation of detailed design and implementation of models for testing ESS business models in the Netherlands and worldwide.

Keywords: electricity storage system; business model test; agent-based modeling; exploratory modeling analysis

---

1. Introduction

Electricity storage systems (ESSs) are among the suggested solutions to manage the variability of renewable energy productions and stability of grids. Despite the potential of ESS, the implementation of ESSs worldwide (except pumped-hydro storage) is still small in size due to technical, institutional, and economical challenges [1,2]. Power system flexibility can be defined as "the extent to which a power system can adapt electricity generation and consumption as needed to maintain system stability in a cost-effective manner" [3]. The stability of the system can be maintained if we guarantee that the volume of supply and demand is equal at all locations and at every moment in time. An ESS is capable of solving the problem of mismatches in time of generation and consumption section of the power system as it is capable of keeping already-generated electricity and re-generating it at better times. Therefore, on the generation side, an ESS can help to manage problems of variable generation such as wind generation by providing a firm output. Across the grids, an ESS enables peak shaving by discharging electricity near heavily loaded points. In addition, it enables the arbitrage of electricity among various markets for electricity and its services. Moreover, at the consumer side (behind the meters), an ESS helps to manage the time-of-use of electricity for cost reduction. We can find three general sets of global challenges for the development of ESSs. The first set of challenges are technical challenges. The development of technologies suitable for desired applications is a challenge as no ES technologies are currently suitable for all applications. In addition, most current ES technologies are still under development and they are not matured yet. The second challenge is the low penetration of variable renewables in the electricity systems. In [1], we highlighted that with a low share of variable renewables, variations can be solved by the grid or cheaper flexibility solutions, and we explained that the current energy portfolio in the Netherlands, which consists of a high share of natural gas and coal, does not motivate solutions such as an ESS to offer flexibility to the market. The third set of challenges includes economic and business challenges. Here, the first challenge is the high costs of ESS, and in turn, the high levelized cost of energy (LCOE). The fact that most ES technologies are not matured yet partially justifies the high cost of ESS. The ability of an ESS for competing with other solutions of power system flexibility, such as cheap demand response, puts ESSs under question. In addition, a lack of viable business models, as well as uncertainty and unsupportiveness of regulatory frameworks, are barriers of development for ESSs [4–6]. Thus, developing business models for ESSs could improve the market penetration of incumbent ES technologies and increase their economic scaling up, with a positive feedback on both production costs and the marketability of renewable energy. Arguably, good business models (third challenge) could lead to a positive feedback loop on two fronts: increasing the economy of scale for ESSs (first challenge) and expanding market share of renewables (second challenge). This is why we deem it to be of paramount importance to effectively approach the lack of business models as a possible, viable solution to unlock the potential of ESSs and exit the current stalemate.

In previous work, institutional challenges and business model design alternatives for ESSs in the Netherlands have been examined [1]. A business model describes "the rationale of how an organization creates, delivers, and captures value" [7]. To design business models, it is necessary to define business goals, identify business model alternatives, develop tests, and select among the alternatives using tests. While in the previous work we developed a design space as a set of alternatives, the objective of this paper is to elaborate on the goals, objectives, and constraints and the involved uncertainties, as well as analysis and selection of testing methods for an ESS business model design, by means of a combination of agent-based modeling (ABM) and exploratory modeling analysis (EMA). In doing so, we benefited from the collaboration with a practitioner of the ESS business in the Netherlands who is also the co-author of this paper.

In Section 2, we will elaborate on ESS business challenges and uncertainties in the Netherlands. Identifying uncertainties is critical for the design and development of business models. In Section 3, we will have a generic view on a design process, and its meaning for designing business models. We will explain the goals, objectives, and constraints for designing business models, as well as considerations for testing business models. Then, in Section 4, we will outline how to combine ABM and EMA as a suitable approach for testing ESS business models. In addition, we will provide an abstract conceptual design for such a test. Finally, in Section 5, some conclusions will be drawn.

2. ESSs in the Netherlands: Status Quo Analysis

Organizations with an interest in energy storage technology tend to approach their investment decision from an operational point of view. Mainly, they are interested in how energy storage can lower their energy bill or solve operational constraints.

2.1. Products, Services, and Value Propositions

ESS offer organizations the opportunity to lower their energy bills. To some extent this is possible due to increased self-consumption of generated solar energy. Instead of feeding excess energy back into the grid for a low fee, the energy can be stored and used when necessary. However, due to the relatively low energy prices per kWh in the Netherlands compared to the high cost for the ESS, this is not yet a viable business model.

A second and more promising application is peak shaving. A behind-the-meter ESSs can act as a buffer and reduce peak demand at the connection point. This reduces charges from the grid operator. This application becomes more relevant with increased electrification of facilities, e.g., due to increased demand for EV-charging. Already, there are some viable business cases in certain demand ranges [8,9].

Third, an ESS can be used to benefit from the ancillary services and the associated balancing markets. Especially, the revenue that can be unlocked in the market for primary reserve (FCR) has been the most significant driver of the business case for most large scale (>1 MW) energy storage projects in the Netherlands. However, there is only a limited demand for these services from the TenneT transmission system operator (TSO), which has resulted in quite a decline in average weekly revenues over the last few years [10].

Fourth, ESSs are used as a source of emergency or mobile power. Increasingly, ESSs are suitable as a realistic replacement for conventional diesel-powered generators. More stringent norms and legislation for CO_2 and NOx emissions are driving companies to search for alternatives. Additional benefits are low noise and no fumes that are generated compared to diesel generators. Still, there are challenges from an operational point of view, since the energy capacity and power output per Euro invested in an ESS is quite high compared to diesel gen-sets.

Combining a few of the above-mentioned applications of ESSs is key for financial viability. Only then, an ESS is used to its full potential. Large projects at the Amsterdam Arena or Cars Jeans Stadion have both succeeded because of this so-called revenue stacking. Though, both projects have also been subsidized to some extent, to close the financial gap.

2.2. State of the Art

From a technology point of view, lithium-ion ($LiNiMnCoO_2$ and $LiFePO_4$)- based systems have been dominant. The value chain is dominated by a relatively small amount of cell manufacturers, power conversion system manufacturers and system integrators. Lithium-ion based systems tend to offer the lowest LCOE in most use cases. Also, the technology offers attractive technical characteristics, such as energy density and rate of (dis)charge [11,12].

In the Netherlands, the market for energy storage is still in the introduction phase. No official data of ESS-installations are tracked by authorities. Estimates, however, based on the EnergystorageNL database [13], reveal that there have been less than 10 large-scale (>1 MW) ESS installations, and about 15 medium-scale (100 kW–1 MW) projects. Smaller-scale projects, amongst home batteries, are estimated at about 500 installations. In Germany, the market is more mature with every second residential PV-installation being complemented with a home battery. Estimates are that there are now more than 120,000 (home) batteries in Germany [14].

Most large- and medium-scale projects in the Netherlands have received some kind of subsidy (local, national or even European) [15]. Without these subsidies, most projects are not viable from a financial point of view.

2.3. Technology Potentials

Downsides of lithium-ion based ESSs are the dependency on (rare) earth materials, lack of a recycling industry, limited lifespan and limited potential for large scale seasonal energy storage. Other technologies are expected to complement lithium-ion based systems, resulting in various energy carriers that are co-existing.

Most promising as an energy carrier is hydrogen. The variety of applications, high energy density (when compressed or liquefied) and scalability makes it a key storage technology. Combined with a fuel cell and hydrogen storage tank, a hydrogen-based ESS has the potential to be a very relevant asset in future energy storage markets. As of now (early 2020), these systems tend to be far too expensive for commercial applications. In particular, the cost of electrolysers that are needed to convert electricity to hydrogen, as well as the cost of fuel cells, need to drop significantly [10].

Vanadium redox flow batteries are slowly entering the markets. They offer the opportunity of storing energy in non-toxic, non-hazardous fluids, and consist of widely available elements. A long lifespan is promised, as well as the opportunity to fully discharge the ESS (li-ion based systems can discharge up to 90%). As of early 2020, about 5–10 companies were ready to offer such systems in the Netherlands. The costs are still quite high compared to lithium-ion based systems, and performance characteristics such as c-rate and energy density cannot yet compete with lithium-ion ESSs.

Finally, quite a few other technologies are becoming commercially available. Amongst others, there are salt-water batteries and nickel-iron batteries that are showing promising value for money. In all cases, further scale-up and mass production will be necessary to fully benefit from these technologies.

2.4. Barriers for ESS Development

The main constraint in the Dutch market for energy storage is that there is no urgency for the implementation of the technology. The grid is very reliable, with power outages of less than 0.1% of the time per year. So far, the penetration of renewables in the energy mix is quite low, which makes it still relatively easy for grid operators to include renewable energy in the grid. Therefore, the value of the flexibility that can be provided by ESSs is not yet acknowledged by market mechanisms (for more details, see [1]).

Despite significant price drops for lithium-ion cells in the past 10 years, batteries are still quite expensive, especially when compared to the cost of solar panels or wind turbines that they are supposed to complement. Prices range from €500–€1.000 per kWh installed, with significant economies of scale being present. For example, a 500 kWh system requires a CAPEX of about €375,000. If an investor would put such an amount in a solar array, there is a predictable cash flow and payback period. Investing in ESSs is, from an investor's perspective, much riskier since the cash flow is very uncertain [6].

Subsidies helped to kick-start the market for ESSs. In order to receive some kind of subsidy, the innovative application of ESSs are required by most authorities. Thus, there is no consistent subsidy in place that can always be applied by investors in ESS equipment. This makes it harder for new projects to obtain subsidy since most demonstrations of ESS-applications have been subsidized already.

Finally, there are constraints from a legal point of view. Norms and standards are still under construction [16], resulting in unclear procedures on how to assess safety issues by, for instance, fire departments or insurance companies.

2.5. Uncertainties

The uncertainties concerning the ESS business can be classified into four groups. Legal uncertainties are the first group of uncertainties. The rules and regulations of ownership of ESSs, and the safety of some electricity storage technologies provide challenges for ESS business. The second group of uncertainties are the fiscal framework, as there is no clear view on the development of subsidies or tax schemes. Technical uncertainties form the third group of uncertainties. The development of new ES technologies and their technical characteristics such as lifetime and maintenance, as well as the integration of the ES technologies to the grid can influence the ESS business. Last but not least, market uncertainties are inherent uncertainties in most businesses as the supply, demand, and prices in ES market, electricity market and balancing market are uncertain.

The status quo analysis for ESS business in the Netherlands, indicates that not all possible applications of ESSs received attention due to economics, technical, or legal challenges. Business model innovation could be a key to overcome the current barriers and unlock the potentials of ESS. In the following sections, we will outline an approach for designing a business model under deep uncertainties.

3. Design Process for Business Models of ESS

Designing business models is required in order to overcome some ESS business challenges. Osterwalder & Pigneur identified four general goals for business model innovation [7]: (1) satisfying existing but unanswered market needs; (2) bringing new technologies, products, or services to market; (3) improving, disrupting, or transforming an existing market with a better business model; (4) creating an entirely new market. For ESS, the main requirement can be considered as to enter and sustain in the market (goal 2) because an ESS is not part of the market yet. Transforming the current electricity market to a more flexible one (goal 3) could be the next goal of ESS business model innovation.

Similar to any other design project, certain steps should be followed for designing business models. Figure 1 illustrates a generic conceptual framework for design. The framework includes five main steps: (1) determine goals, (2) determine objectives, (3) determine constraints, (4) develop design space, and (5) tests for goals. In this framework, objectives are goals that need to be optimized, constraints are the binary goals to be met, and the design space illustrates a set of variables and components. In this framework, test means to determine to what extent the objectives and constraints are met by a design [17].

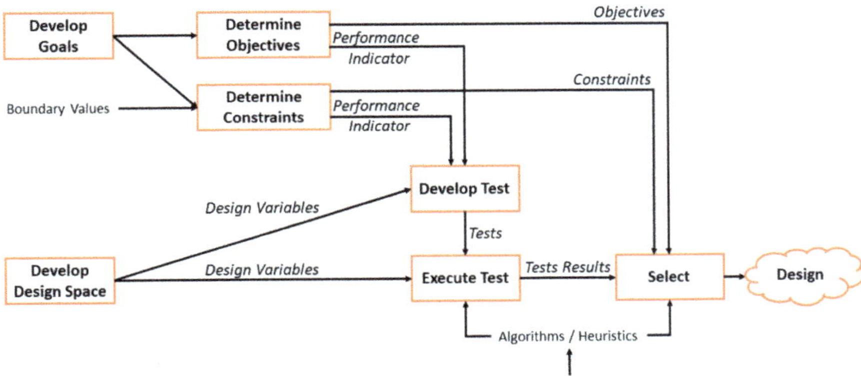

Figure 1. Generic design process reproduced from [17].

We previously covered some steps for designing business models for ESSs in the context of the Netherlands in [1] where we developed a design space (including a map of single-application business models for ESS) and we identified some constraints such as institutional, technical, and location constraints. In this paper, we follow up the design process by elaborating on goals, objectives, constraints, and the development of tests.

3.1. Goals, Objectives, and Constraints

Most successful designs, in various domains, share certain goals. A framework originally developed at IDEO, a global design company, indicates that successful designs provide balance among three generic goals of (1) feasibility, (2) viability, and (3) desirability [18]. For the process of business model design, feasibility indicates whether it is possible to provide a product or service, desirability indicates whether customers will value (or pay for) the products or services, and viability indicates whether a business is financially sound. In addition to these three criteria, it is necessary for business models to conform to external conditions in the environment such as legal conditions, macro-economic situations, etc. Therefore, another goal for a successful business model can be framed as "adaptability" [19].

The design goals need to be translated into measurable objectives and constraints to be optimized and met, respectively. In economic terms, the primary objective of a firm is maximizing its profit and other objectives finally serve this primary objective [20]. Profitability is a measure of the economic viability of business models. In addition, feasibility of business models for ESSs mainly depends on the well-functioning of the ESS devices and technology for target products. In the business or policy modeling, ESSs are considered as black boxes with aggregated parameters such as capacity, power rating that influence the transformations of inputs to outputs. Such parameters directly or indirectly influence or limit the operation, and in turn, the profitability of ESS. In analyzing the business model of ESSs, technical feasibility and its relevant considerations can be considered as assumptions or hypotheses of the business model. Furthermore, the desirability of business models depends on the number of customers who are willing to pay for a product or a service. Higher desirability in a business model design influences the profitability of the model. Therefore, similar to feasibility, desirability can be considered as an assumption that influences economic viability. Finally, the adaptability of business models influences all other goals. It may limit the desirability, feasibility, and viability of the business models. Figure 2 illustrates the causal/limiting relationships among the four aforementioned goals.

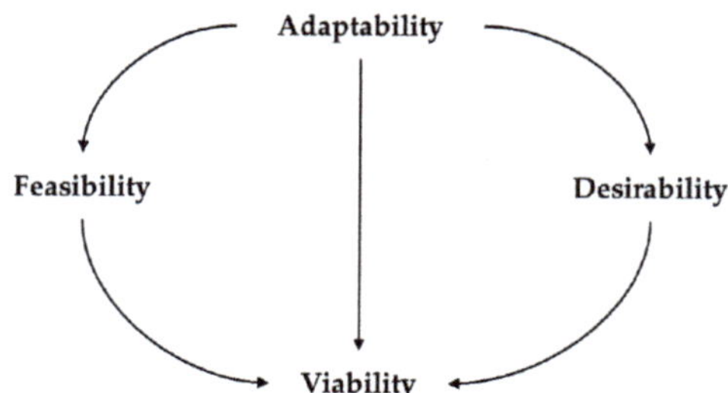

Figure 2. Relationship among the goals of business model design.

3.2. Developing Tests for Business Models

3.2.1. Meanings of Business Model Test

Tests of business models have some differences with engineering tests. In a generic design process framework (see Figure 1), testing a design involves measuring to what extent the design meets the objectives and constraints of the design and it needs to be done before selecting the final design. The test is done either on a prototype (computer simulation or material artifact) or on a complete product, depending on the test expenses. The meaning of a test in engineering designs seems to be very straight forward, however in the business model innovation literature, testing is not only about checking whether a business idea works and meet the goals, but it is also about checking if the business hypotheses (things need to be true for an idea to work but have not been validated yet) are valid [21,22]. In engineering designs, assumptions about many physical forces and characteristics are already validated in the natural sciences, but many assumptions in the business model designs still need to be validated. Therefore, testing business models are more or less about validating their assumptions.

3.2.2. Directions of Business Model Test

Four design goals for business models provide directions for testing them as well. It is possible to relate elements of business models to each direction. In our previous work, we elaborated on some design variables for ESS business models in the context of the Netherlands in which those elements were mostly selected from Business Model Canvas (BMC) [1]. BMC is a popular framework for analysis, design, and communication of business models. BMC contains nine blocks to illustrate the elements of the business models and their relationships [7]. Figure 3 illustrates the relationship between business model elements and directions of business model tests. The goal, "desirability", can be explored through BMC blocks of "customer segments", "customer relationships", "channels" and "value proposition"; the goal "feasibility" can be explored through BMC blocks of "key resources", "key activities", and "key partners"; and the goal "viability" can be explored through BMC blocks "cost structure" and "revenue streams" [19,23]. Finally, the goal adaptability can be explored through the analysis of the business model environment [19].

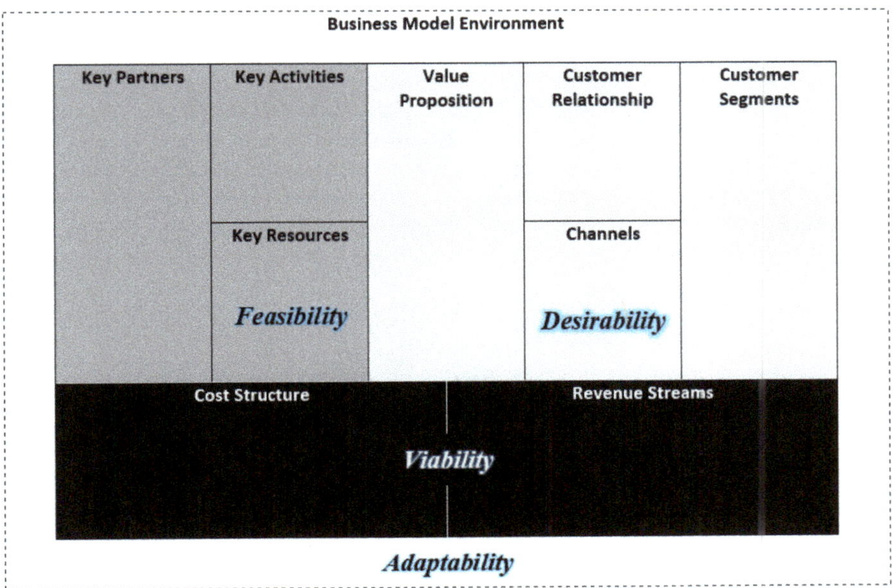

Figure 3. Relationship between business model elements and directions of business model test [19,23].

3.2.3. Yellow Hat before Black Hat

The directions of a test come with a question on their weight and priority. Engineers and product developers may tend to think about a business by looking first at its technical feasibility, whereas the most important direction for investors is viability and the economic calculations. On the other hand, businesspeople prioritize desirability and market considerations arguing that "for every one of our failures, we had spreadsheets that looked awesome" [22], and policymakers may focus only on the adaptability of a business as they usually have no direct interest in the business.

It is wise to investigate the advantages of new ideas before analyzing their disadvantages. Edward de Bono in his book "six thinking hats", in which directions of effective thinking are represented by six colored hats, maintains that:

"In an assessment situation, it makes sense to put the yellow hat (hat of benefits and value) before the black hat (hat of caution and critics). If, under the yellow hat, you cannot find much value to the idea, there is no point in proceeding further. On the other hand, if you find much value under the yellow hat and then proceed to the black hat and find many obstacles and difficulties, you will be motivated to overcome the difficulties because you have seen the benefits. But if you start off by seeing all the difficulties, then your motivation is totally different [24]."

The interdependence of business model design goals (see Figure 2) reveals the fact that the viability of a business model depends on the other goals. If we seek to identify the advantages of business models before disadvantages, we can explore the potential benefits on the viability side. If there are not attractive or no potential benefits, it will not make sense to continue testing the assumptions of the other test directions.

4. Modeling for Testing ESS Business Models

Models may help business designers to test several assumptions of viability for business models. If models illustrate that there will be "hopes" for sustainable profit, further tests on other directions can be conducted to reduce risks to the business. In addition, lessons of modeling activities may provide insights on weights, importance, and priority of tests on the other directions.

4.1. Modeling Approach

4.1.1. Model Requirements

A model for exploring and testing ESS business models must meet several requirements. First of all, the model should represent the electricity and services markets in terms of complex socio-technical systems. Therefore, it needs to entail several and diverse social and technical components. Second, the model should be capable of demonstrating co-evolution of the markets and ESSs as they are expected to influence each other. Third, the model should enable the analyst for ex-ante analysis of the effects of scenarios and policies under deep uncertainty as there is little information on various aspects of ESSs as a new solution in the market.

4.1.2. Exploratory Modeling Analysis (EMA)

Developed by Bankes in the RAND Corporation in 1990s, Exploratory Modeling and Analysis (EMA) is a research methodology that employs computational experiments to study systems by systematic exploration of the consequences of uncertainties, including uncertainties in parameters, structures, or methods [25–27]. EMA contrasts with "consolidative modeling" in which a model is built by consolidating known facts into a single best-estimate set and is used as a surrogate for the real system [25].

An initial point and a driver in the development of EMA was admitting that using models as surrogates of the real-world systems was not always possible [25,28,29]. In the presence of barriers to experimental validation, significant uncertainties, or strong non-linearity, outcomes of consolidative modeling are rather poor and unreliable [25,29,30].

EMA is a solution for coping with the significant uncertainties, which are presented in the literature under various names such as deep uncertainty or sever uncertainty [27]. Deep uncertainty can be described as a situation in which analysts or decision-makers do not know or cannot agree on the appropriate conceptual models, the probability distribution of uncertain variables and parameters, and valuation of alternative outcomes [31]. In another view, it can be described as a situation in which the researcher is "being able to enumerate multiple alternatives without being able to rank order the alternatives in terms of how likely or plausible they are judged to be" [32]. This definition of deep uncertainty applies to several aforementioned uncertainties in Section 2. EMA copes with such uncertainties by conducting extensive computational experiments and calculating the outcomes of a set of plausible models, which is formed by varying assumptions, parameters, and methods [25,28].

Contrary to consolidative modeling, EMA not only can be used to answer, "what if" questions, but it can also answer questions such as "under what conditions a behavior may occur?", and "what are the plausible future dynamics in a phenomenon?" [27].

The EMA process consists of the following steps [33]: (1) conceptualization of the decision problem and the associated uncertainties, (2) development of a set of simple models, (3) specification of the targeted uncertainties, (4) analysis of the behaviors and model outcomes, (5) identification of the combinations of uncertainties which results in interesting behaviors, (6) assessment of model quality under the combinations of uncertainties, and (7) qualitative or quantitative communication of the typical futures from the combinations of interest.

EMA can be applied to several modeling paradigms, such as systems dynamics, agent-based modeling, etc., [27]. In this work, we suggest agent-based modeling as an appropriate paradigm to explore and test business models.

4.1.3. Agent-Based Modeling (ABM)

Agent-based modeling (ABM) is a computational modeling paradigm that enables us to describe how an agent will behave in a controlled environment [34]. In this paradigm, an agent is a "thing that interacts with other things" [35,36]. An agent-based model creates "an artificial world of heterogeneous agents and enables investigation into how interactions between these agents, and between agents and other factors such as time and space, add up to form the patterns seen in the real world" [37].

An agent-based model consists of agents, an environment, and interactions among agents or agents and the environment [34]. Agents are autonomous and encapsulated entities situated in a particular environment, they have goals and are capable of flexible actions. They have states (also called properties, attributes, etc.), internal rules for change of the states, and actions and behaviors. Agents can take actions that influence either themselves, other agents, or the environment. An environment is a place in which agents live, and it contains information, structure, and time (the latter can be either considered as part of the environment or an independent element of the whole model) [36].

Agent-based modeling is one of the best modeling approaches to model complex adaptive systems [36]. Among several modeling paradigms to study complex systems, ABM makes a better predictive approach as (1) it enables the capture of more complex structures and dynamics, (2) it makes the study possible even in absence of global interdependence data, and (3) it is easier to maintain and evolve [38]. Generally, ABM is applicable for studying systems if (1) the problem has a distributed character, (2) the sub-systems (can be elements, components, agents, etc.) in the study operate in a highly dynamic environment, and (3) the sub-systems have to interact in a flexible way [36,39].

Therefore, ABM is a powerful paradigm for the studying and the evaluation of electricity storage business models considering the distributed control across the electricity systems and markets, the inherent dynamics (in demand, supply, prices, weather, rules, and regulations, etc.), and the increasing importance of flexibility in current and future electricity systems.

4.1.4. Exploratory Agent-Based Modeling Analysis

In this research, we suggest a combination of agent-based modeling and exploratory modeling analysis to address business model design for ESSs for a number of reasons. First of all, this combination provides the best fit to the model requirements that we elaborated on earlier in this section. Most current studies on the economics of an ESS talk about business cases and they adopt static models with several assumptions about external factors of a business (e.g., [40]). Using exploratory modeling analysis, we incorporate more external factors and their underlying (deep) uncertainties in our analysis.

Moreover, most current models cannot capture the dynamics, randomness, or even chaos in a variable, such as electricity market price, whereas agent-based models can easily capture such phenomena. In addition, the electricity market price is not only influenced by the existence and performance of ESSs on the market but also influences the economics and existence of ESSs. Many static models for evaluating the economics of projects cannot capture such a feedback loop, whereas this will be easily possible in the ABM.

Last but not least, despite the potential of EMA to provide researchers with insights on business model problems, to our knowledge this approach has not been adopted for such a purpose yet. Adoption of such an approach can not only lead to interesting results, but it can also come with surprises that pave a path for future research.

4.2. Agent-Based Model Design to Test ESS Business Models

To explore and test business model designs for ESSs, we consider the following organization for our research model.

4.2.1. Agents

Agents can be classified into two groups of institutional agents and physical components. Institutional agents may make commercial decisions whereas physical components make no decision and they follow rules of physical operations. The decision-making agents in the model are energy companies, end consumers, markets and the transmission system operator (TSO). Physical components include power plants, loads, ESSs, transmission grid (T-Grid), and distribution grids (D-Grid).

4.2.2. Interactions

Interactions within the model (in the form of agent–agent and agent–environment) include:

- Control (right of utilization or ownership of physical components by institutional agents)
- Supply contracts (contract between institutional agents and end consumers to supply of end consumers)
- Wires (physical delivery of electricity)
- Information (message transfer among institutional agents)
- Payment links (cash flow among institutional agents)

4.2.3. Model Environment

The environment of the model has a network structure in which agents are nodes and interactions form links among the nodes. Modeler initializes the model by defining the number and type of agents and the interactions among them.

TSO is connected to T-Grids through links. Energy companies may be connected to power plants or ESSs through control links if they enact a power producer role, or they can control no physical component if they enact only a retailer role. Large consumers are connected to large loads via control links, while they may control ESSs as well. Power plants, ESSs, large loads, and D-Grids are connected to a T-Grid via wire links. In addition, small loads are connected to D-Grids via wire links. The market agent is connected to TSO, energy companies, and large consumers via information links. There are also payment links from the market to the energy companies and large consumers.

The spatial location of agents is not relevant to the present modeling approach (a further modeling requirement could include the physical sustainability of a grid's balance according to the geographical location of loads, distributed/centralized generations and transmission network, leading to a spatially embedded ABM). The model can run for years, where each year is represented by 288 time-steps to capture the dynamics over 24 h along 12 days (representing 12 months) and enable the modeler for inter-temporal analysis.

4.2.4. Agent's Actions and Behaviors

The model environment publishes the weather forecasts and fuel price data.

Energy companies and large consumers make decisions and a plan of electricity generation or consumption (also charge and discharge of ESSs) and make offers or bids to the wholesale and balancing markets. The market agent collects all offers and bids through information links, it clears the wholesale market and specifies the market price and accepted volumes, and it communicates the market results to the market participants. The market also publishes CO_2 certificates and prices based on the collected bids. The agent market collects electricity payments from buyers and transfers electricity revenue to the seller via payment links. The agent market also sends a balancing bid ladder to TSO for later use.

Energy companies and large-scale consumers send orders of execution of generation or consumption through the control links towards physical components including power plants, ESSs, and large loads. At the same time, small loads start to consume or generate electricity (if micro-RES is attached). Physical components deliver electricity to or receive electricity from the connected grids.

If an imbalance occurs in the T-Grid, the T-Grid sends a signal to the TSO about the location and volume of imbalances through control links. TSO selects some energy companies or large consumers according to the bid ladder to provide balancing energy and notify them through the market agents. Then, the notified balancing parties adjust their generation or consumption to balance supply and demand in the grids. TSO asks the market to fine the parties with balancing deviations and pay parties who provided balancing energy. The market collects the fines and sends the balancing revenues through payment links.

The model environment calculates carbon emissions. The market also collects CO_2 payments and fines the parties with extra CO_2 emission and collects the fines through money links.

The model environment calculates and updates the statistics of electricity generation and consumption, and financial status of institutional agents.

The experiments will explore the effects of different business models designs on the profitability of economic agents, CO_2 emissions in the whole model, and reliability of the whole system under uncertainties in technologies and energy resources, fuel prices, demand, weather, and regulations.

4.2.5. Relationship between the Agent-Based Model and Business Models

As mentioned earlier, the business model is "the rationale of how an organization creates, delivers, and captures value" [7]. Therefore, business models can be encoded in an agent-based model by defining a set of states, rules, behaviors, and interactions for agents that represent commercial organizations. For example, for an agent that represents an energy company, which owns an ESS for wholesale arbitrage:

- A value proposition such as electricity is a property of the agent;
- Customer segments may be composed of other agents that represent other energy companies, energy consumers, or an energy market operator (if it buys electricity on behalf of others);
- Channels to interact with customers may be composed of a set of links such as information exchange with a market operator or end consumer as well as a set of wire links for delivery of electricity to the customer;
- Key resources may be defined as physical entities (agents) connected to an agent via control links. They can also be defined as states/ properties of agents;
- Key activities may be translated to a set of internal behaviors of an agent as well as a set of interactions with other physical or institutional agents. Activities such as production, and ESS (dis-)charge are among key activities of the business models which are encoded as internal behaviors of the agent and influence its interactions with other agents;
- Revenue streams may be represented by incoming payment links from other agents (for compensation of the delivery of a specific product or service);
- Cost structure included costs calculated from internal behaviors of the agent and outgoing payment links to other agents or the environment.

4.2.6. Experimentation Plan

One advantage of agent-based modeling is that it enables modelers to look at a problem from various perspectives. Using the aforementioned model, one can see the problem from the perspective of an energy company by exploring and discovering the effects of the business model designs (independent variable of the research) on patterns of profitability of the energy companies (more specifically, the value of the company and NPV of ESS projects), as well as the green image of the company (e.g., its contribution to CO_2 emission reduction). Similarly, the experiments from the perspective of policymakers will explore and discover the effects of various business model designs on the patterns of dynamics in electricity prices, CO_2 emissions in the whole model, and reliability of the whole system (e.g., in terms of the number of blackouts) under uncertainties in technologies and energy resources, fuel prices, demand, weather, and regulations (which are the parameters of the model).

5. Conclusions

In this conceptual article, we provided a view on the testing of business models for electricity storage systems (ESSs). First, the current business perspectives on ESSs were explained and the necessity for designing new business models for ESSs was highlighted. Next, the design process for business models was elaborated on, where the meaning of a business model test was clarified, and directions of tests and their priority were introduced and discussed. Then, a combination of agent-based modeling (ABM) and exploratory modeling analysis (EMA) was suggested as a way of testing and exploring the economic viability of business models under various uncertainties of the market, technologies, and business environment.

The provision of a business perspective on ESS business uncertainties, analysis of the capability of ABM and EMA to test business models under deep uncertainty, and provision of an abstract agent-based model design were the main contributions to this paper. Implementation of the model and conducting business model test experiments will be the next step of our research.

Unless we expect that ESSs will be forcibly introduced or copiously funded by states, in a market economy, only viable business models can channel investors' money to ESSs and create the condition for their development. In the same vein, it is our convincement that deep uncertainties concerning the business space for ESSs represent a major bottleneck to the development of the full potential for ESS capacity, in the Netherlands and elsewhere. In a vicious circle, the lack of sufficient buffering capacity in the network is hindering the penetration of renewable energy as much as the insufficiency of renewable sources in the energy mix hinders the demand for ESSs. In this paper, we made a joint effort between academics and business to dissect the problem and envisaged a possible approach to theoretically tackle such uncertainties and facilitate the creation of new business models. Technology is under development and in some cases economical, business and institutional changes are the next steps to be taken, and we hope to contribute to this effort, with this and future work.

Author Contributions: Conceptualization, S.A.R.M.M.K. and F.R.; Methodology, S.A.R.M.M.K.; Formal analysis, S.A.R.M.M.K. and R.d.E.; Investigation S.A.R.M.M.K. and R.d.E.; Writing—original draft preparation, S.A.R.M.M.K., R.d.E. and F.R.; writing—review and editing, S.A.R.M.M.K. and F.R.; visualization, S.A.R.M.M.K.; supervision, F.R.; project administration, F.R.; funding acquisition, S.A.R.M.M.K. and F.R. All authors have read and agreed to the published version of the manuscript.

Funding: This research and the APC were funded by University of Groningen project code 190152210.

Conflicts of Interest: The authors declare no conflict of interest. The funders had no role in the design of the study; in the collection, analyses, or interpretation of data; in the writing of the manuscript, or in the decision to publish the results.

References

1. Mir Mohammadi Kooshknow, S.A.R.; Davis, C.B. Business models design space for electricity storage systems: Case study of the Netherlands. *J. Energy Storage* **2018**, *20*, 590–604. [CrossRef]
2. Aneke, M.; Wang, M. Energy storage technologies and real life applications—A state of the art review. *Appl. Energy* **2016**, *179*, 350–377. [CrossRef]
3. Papaefthymiou, G.; Grave, K.; Dragoon, K. *Flexibility Options in Electricity Systems*; Ecofys: Berlin, Germany, 2014.
4. Gallo, A.B.; Simões-Moreira, J.R.; Costa, H.K.M.; Santos, M.M.; Moutinho dos Santos, E. Energy storage in the energy transition context: A technology review. *Renew. Sustain. Energy Rev.* **2016**, *65*, 800–822. [CrossRef]
5. Anuta, O.H.; Taylor, P.; Jones, D.; McEntee, T.; Wade, N. An International Review of the Implications of Regulatory and Electricity Market Structures on the Emergence of Grid Scale Electricity Storage. *Renew. Sustain. Energy Rev.* **2014**, *38*, 489–508. Available online: https://www.sciencedirect.com/science/article/abs/pii/S1364032114004432?via%3Dihub (accessed on 13 February 2020). [CrossRef]
6. DNV GL. *Safety, Operation and Performance of Grid-Connected Energy Storage Systems.* 2017. Available online: https://rules.dnvgl.com/docs/pdf/DNVGL/RP/2017-09/DNVGL-RP-0043.pdf (accessed on 13 February 2020).
7. Osterwalder, A.; Pigneur, Y. *Business Model Generation: A Handbook for Visionaries, Game Changers, and Challengers*; John Wiley & Sons: Hoboken, NJ, USA, 2010; ISBN 9780470876411.
8. McKinsey & Company How Battery Storage Can Help Charge the Electric-Vehicle Market. Available online: https://www.mckinsey.com/business-functions/sustainability/our-insights/how-battery-storage-can-help-charge-the-electric-vehicle-market# (accessed on 13 February 2020).
9. ALFEN Shell Ultrafast EV Charging Service to Incorporate Alfen Energy Storage. Available online: https://alfen.com/en/news/shell-ultrafast-ev-charging-service-incorporate-alfen-energy-storage (accessed on 14 February 2020).

10. van der Veen, A.; van den Noort, A.; Kranenburg-Bruinsma, K. *Marktpotentie Groene Waterstof Productiefaciliteiten*; DNV.GL, TNO: Groningen, The Netherlands, 2018.
11. Baker, D.R. Why Lithium-ion Technology is Poised to Dominate the Energy Storage Future. Available online: https://www.renewableenergyworld.com/2019/04/03/why-lithiumion-technology-is-poised-to-dominate-the-energy-storage-future/#gref (accessed on 14 February 2020).
12. Temple, J. Why Lithium-ion May Rule Batteries for a Long Time to Come. Available online: https://www.technologyreview.com/s/611982/why-lithium-ion-may-rule-storage-technology-for-a-long-time-to-come/ (accessed on 13 February 2020).
13. Energy Storage, N.L. Energieopslagprojecten in Nederland. Available online: https://www.energystoragenl.nl/projects (accessed on 7 February 2020).
14. Hockenos, P. In Germany, Consumers Embrace a Shift to Home Batteries—Yale E360. Available online: https://e360.yale.edu/features/in-germany-consumers-embrace-a-shift-to-home-batteries (accessed on 7 February 2020).
15. GIGA STORAGE GIGA STORAGE VERSNELT DE ENERGIETRANSITIE MET DE KRACHTIGSTE BATTERIJ VAN NEDERLAND. Available online: https://giga-storage.com/2019/12/04/persbericht-giga-rhino/ (accessed on 13 February 2020).
16. PGS Veilige Opslag van Lithium-ion Batterijen. Available online: https://publicatiereeksgevaarlijkestoffen.nl/nieuws/veilige-opslag-lithiumion-batterijen.html (accessed on 7 February 2020).
17. Herder, P.M.; Stikkelman, R.M. Methanol-Based Industrial Cluster Design: A Study of Design Options and the Design Process. *Ind. Eng. Chem. Res.* **2004**, *43*, 3879–3885. [CrossRef]
18. Brown, T. *Change by Design: How Design Thinking Can Transform Organizations and Inspires Innovation*; HarperCollins Publishers: New York, NY, USA, 2009; ISBN 978-0-06-176608-4.
19. Strategyzer Strategyzer Webinar: The Basics of Testing Business Ideas. Available online: https://www.youtube.com/watch?reload=9&v=IFjZkRbXhiA (accessed on 18 June 2018).
20. Parkin, M. *Microeconomics*, 10th ed.; Pearson Series in Economics; Addison-Wesley: Boston, MA, USA, 2012; ISBN 9780131394254.
21. Osterwalder, A.; Pigneur, Y.; Bernarda, G.; Smith, A.; Papadakos, T. *Value Proposition Design: How to Create Products and Services Customers Want*; John Wiley & Sons, Inc.: Hoboken, NJ, USA, 2014; ISBN 9781118968055.
22. Johnson, M.W. *Seizing the White Space: Business Model Innovation for Growth and Renewal*; Harvard Business Press: Boston, MA, USA, 2010; ISBN 1422124819.
23. Jeffries, I. Free eBook: Building A Strong Business Model. Available online: https://isaacjeffries.com/blog/2017/10/14/free-ebook-building-a-strong-business-model (accessed on 1 July 2019).
24. de Bono, E. *Six Thinking Hats*; Penguin Books Limited: London, UK, 2017; ISBN 9780241336878.
25. Bankes, S. Exploratory Modeling for Policy Analysis. *Oper. Res.* **1993**, *41*, 435–449. [CrossRef]
26. Bankes, S.; Walker, W.E.; Kwakkel, J.H. Exploratory Modeling and Analysis. In *Encyclopedia of Operations Research and Management Science*; Gass, S.I., Fu, M.C., Eds.; Springer US: Boston, MA, USA, 2013; pp. 532–537, ISBN 978-1-4419-1153-7.
27. Kwakkel, J.H.; Pruyt, E. Exploratory Modeling and Analysis, an approach for model-based foresight under deep uncertainty. *Technol. Forecast. Soc. Chang.* **2013**, *80*, 419–431. [CrossRef]
28. Agusdinata, B. *Exploratory Modeling and Analysis: A Promising Method to Deal with Deep Uncertainty*; Delft University of Technology: Delft, The Netherlands, 2008.
29. Walker, W.E.; Haasnoot, M.; Kwakkel, J.H. Adapt or perish: A review of planning approaches for adaptation under deep uncertainty. *Sustainability* **2013**, *5*, 955–979. [CrossRef]
30. Van Der Pas, J.W.G.M.; Walker, W.E.; Marchau, V.A.W.J.; Van Wee, G.P.; Agusdinata, D.B. Exploratory MCDA for handling deep uncertainties: The case of intelligent speed adaptation implementation. *J. Multi-Criteria Decis. Anal.* **2010**, *17*, 1–23. [CrossRef]
31. Lempert, R.J.; Popper, S.W.; Bankes, S.C. *Shaping the Next One Hundred Years*, 1st ed.; RAND Corporation: Pittsburgh, PA, USA, 2003; ISBN 9780833032751.
32. Kwakkel, J.H.; Walker, W.E.; Marchau, V.A.W.J. Classifying and communicating uncertainties in model-based policy analysis. *Int. J. Technol. Policy Manag.* **2010**, *10*, 299–315. [CrossRef]
33. Kwakkel, J.H.; Auping, W.L.; Pruyt, E. Dynamic scenario discovery under deep uncertainty: The future of copper. *Technol. Forecast. Soc. Chang.* **2013**, *80*, 789–800. [CrossRef]
34. Wilensky, U.; Rand, W. *An Introduction to Agent-based Modeling: Modeling Natural, Social, and Engineered Complex Systems with NetLogo*; MIT Press: Cambridge, MA, USA, 2015; ISBN 9780262731898.

35. Shalizi, C.R. Methods and Techniques of Complex Systems Science: An Overview. In *Complex Systems Science in Biomedicine*; Deisboeck, T.S., Kresh, J.Y., Eds.; Springer US: Boston, MA, USA, 2006; pp. 33–114, ISBN 978-0-387-33532-2.

36. van Dam, K.H.; Nikolic, I.; Lukszo, Z. *Agent-Based Modelling of Socio-Technical Systems*; Springer Publishing Company, Incorporated: Dordrecht, The Netherlands, 2012; ISBN 10: 9400749325; ISBN 13: 9789400749320.

37. Hamill, L.; Gilbert, N. *Agent-Based Modelling in Economics*; John Wiley & Sons: Chichester, West Sussex, UK, 2016; ISBN 9781118456071.

38. Borshchev, A.; Filippov, A. From System Dynamics and Discrete Event to Practical Agent Based Modeling: Reasons, Techniques, Tools. In Proceedings of the 22nd International Conference of the System Dynamics Society, Oxford, UK, 25–29 July 2004.

39. Van Dam, K.H. *Capturing Socio-Technical Systems with Agent-Based Modelling*; Delft University of Technology: Delft, The Netherlands, 2009.

40. Schroeder, A.; Traber, T. The economics of fast charging infrastructure for electric vehicles. *Energy Policy* **2012**, *43*, 136–144. [CrossRef]

MDPI

St. Alban-Anlage 66

4052 Basel

Switzerland

Tel. +41 61 683 77 34

Fax +41 61 302 89 18

www.mdpi.com

Energies Editorial Office

E-mail: energies@mdpi.com

www.mdpi.com/journal/energies